International Conference on Resonances in Condensed Matter: Altshuler100

Journal of Physics: Conference Series Volumes 324

Kazan, Russia
21-25 June 2011

ISBN: 978-1-61839-294-7
ISSN: 1742-6588

Printed from e-media with permission by:

Curran Associates, Inc.
57 Morehouse Lane
Red Hook, NY 12571

Some format issues inherent in the e-media version may also appear in this print version.

Printed by Curran Associates, Inc. (2011)

For permission requests, please contact the Institute of Physics
at the address below.

Institute of Physics
Dirac House, Temple Back
Bristol BS1 6BE UK

Phone: 44 1 17 929 7481
Fax: 44 1 17 920 0979

techtracking@iop.org

Additional copies of this publication are available from:

Curran Associates, Inc.
57 Morehouse Lane
Red Hook, NY 12571 USA
Phone: 845-758-0400
Fax: 845-758-2634
Email: curran@proceedings.com
Web: www.proceedings.com

TABLE OF CONTENTS

Preface

The International Conference Alt100 - "Resonances in Condensed Matter" was held in Kazan, Russian Federation, from 21-25 June, at Kazan (Volga Region) Federal University. The conference was devoted to the centenary of Professor, Corresponding member of Academy of Sciences of the USSR S A Altshuler (24.09.1911 - 24.01.1983), well known for his pioneer works together with E K Zavoiskii on EPR, prediction of neutron magnetic moment (with I E Tamm) and acoustic magnetic resonance.

The program of the conference covered experimental and theoretical studies of

(i) electron paramagnetic resonance (EPR), nuclear magnetic resonance (NMR), ferromagnetic and antiferromagnetic resonances (FMR, AFMR) in condensed matter, enhanced nuclear magnetic resonance in Van-Vleck paramagnets (ENMR), optical spectroscopy of paramagnetic crystals;
(ii) magnetic relaxation;
(iii) bulk and local magnetic properties of transition metal oxides and intermetallic compounds;
(iv) modern applications of magnetic resonance techniques and methods.

The conference attracted 120 participants from 14 countries. There were 26 invited talks and 36 regular oral talks in the morning and afternoon sessions, and 70 poster presentations. A book of Abstracts is available on the conference website (http://alt100.ksu.ru). The Memorial Session involved talks from B I Kochelaev (Chairman of the Organizing Committee) and A A Manenkov and reminiscences about S A Altshuler given by A A Kaplyanskii, V A Kutuzov, Yu R Pol'skii and I N Pen'kov. The memorial talks and Kaplyanskii' speech are presented in this volume.

The papers submitted for this Proceedings were carefully considered by at least two referees, and we are very grateful to referees for their quick and very thoughtful responses. A few papers were eliminated because of referee reports. Finally 41 papers have been included. We thank the international advisory committee members very much for their recommendations of speakers for Alt100 and especially members of the Program Committee (M V Eremin, S I Nikitin, Yu N Proshin, L R Tagirov, G B Teitelbaum) and the Local Committee Chairman M S Tagirov for their valuable help. We also appreciate the behind-the-scenes work of the Local Committee members in preparing the conference site, web page, mail correspondence, arrangements for travel and accommodation, excursions and conference dinner, and handling the financial support for participants.

The conference was supported by the Government and the President of Tatarstan Republic, the Russian Academy of Sciences, the Russian Fund for Basic Research and Bruker Ltd (Moscow, Russia).

A V Dooglav
B Z Malkin
D A Tayurskii

Semen A. Altshuler: scientist, mentor, teacher

Boris I. Kochelaev

Kazan Federal University, Kremlevskaya, 18, Kazan 420008, Russian Federation

E-mail: bkochelaev@gmail.com

Abstract. International Conference "Resonances in Condensed Matter" is devoted to 100 years of the birthday of the Corresponding member of Academy of Sciences of the USSR, Professor of the Kazan University Semen Alexandrovich Altshuler (1911-1983). He is well known by pioneer works on EPR, the prediction and grounds for an existence of the neutron magnetic moment, the prediction and the theory of the acoustic paramagnetic resonance, and as a founder of the Kazan scientific school "Magnetic radiospectroscopy of condensed matter" (with E.K. Zavoiskii and B.M. Kozyrev)

1. Childhood, youth, education

Semen Altshuler was born 24 September in Vitebsk (Belorussia). It is remarkable that it was a native town for many outstanding persons: Field-Marshal Alexander Menshikov (the closest retainer to the Emperor Peter I), a painter Marc Shagal, academicians Zhores Alferov, Vitalii Goldanskii, Boris Zaharchenya. Later the Altshuler's family moved to Nizhnii Novgorod, where Semen Altshuler graduated the school (1927) being 15 years old.

In his school testimonial was written: "... showed abilities for critical thinking and a strongly pronounced attraction to studies mathematical subjects in abstract aspects." Semen Altshuler decided to become a physicist, however he was too young to enter the university. One year Altshuler studied in the Industrial school.

In summer of 1928 Altshuler came to Kazan to join the university. He was lucky since at that time the regular session of the USSR-Association of physicists took place at the Kazan University. Many famous physicists were present: P. Dirac, M. Born, P. Debye, L. Brillouin, A. Ioffe, L. Mandelstam, Ya. Frenkel... Altshuler was impressed by the greetings of guests given by prof. Ulyanin in German, English and French.

Altshuler studied with enthusiasm. Kazan University had a brilliant mathematical school which was founded by the great mathematician N.I. Lobachevsky, the originator of a non-Euclidean geometry. So, Altshuler got here a good mathematical education. He took an active part in a work of the scientific student's society and became its chairman. He graduated the university in 1932 and became a post-graduate student.

2. The first steps in science and teaching

Semen Altshuler dreamed of theoretical physics. However, the only theorist at the university professor A.D. Goldhammer left at that time Kazan for Leningrad. Altshuler decided to find Igor Tamm in Moscow, whom he knew only due to his excellent text-book on the theory of electricity. As a matter of fact Tamm was an outstanding theorist; he became later a Nobel Prize winner. After some short tests Tamm suggested possible topics for research: one in physics of metals, another in quantum chemistry and the third in physics of atomic nucleus.

It was a romantic time in physics: only few years left after creation of quantum mechanics, neutron and positron have been just discovered (1932); nuclear physics started to develop. Altshuler had chosen the third problem from suggested. Tamm proposed a problem to analyze optical spectra of atoms with nucleus having odd number of neutrons on the basic of an idea that the neutrons have magnetic moment. Altshuler started with enthusiasm the development of a quantum theory of magnetic properties of elementary particles and atomic nuclei. As a result the magnetic moment of the neutron was predicted and its value was correctly estimated [1]. The proposal of magnetic moment for an electrically neutral particle was so unusual that even Niels Bohr, who visited Moscow in 1934, could not accept this idea. Nevertheless, the magnetic moment of the neutrons was discovered later and magnetically polarized beams of neutrons are used now as a powerful method to study the condensed matter.

Contacts with Tamm were important for Altshuler not only in the field of science. He was a teacher for him in the full sense of this word. Tamm's rich individuality, his enthusiasm for science, his views of principle, and his tact in relation with people had a profound influence on Altshuler. Their meetings, correspondence continued till the Tamm's decease.

In 1934 Altshuler was recalled back to Kazan University for lecturing on theoretical physics. He had ten lectures a week; nevertheless, he finished his thesis in 1936. Shortly after that the collaboration with E.K.Zavoiskii and B.M.Kozyrev was established, which developed soon into the tight friendship. Zavoiskii was investigating at that time the energy absorption of radiowaves in liquids and solids by his very sensitive electronic method.

In 1939 S.A.Altshuler reported at the seminar about experiments of Rabi on measuring of nuclear magnetic moments of atomic beams using the resonant method. Zavoiskii became very interested and the team decided to look for nuclear magnetic resonance in condensed matter using protons of water. At the beginning 1941 the first NMR signal were obtained; however, they were not stable… The team needed several months more, but the war started…

3. The Second World War stops the scientific work

Soon after the beginning of the war most of the institutions of the Academy of Sciences had to move from Moscow and Leningrad to Kazan University. The Zavoiskii's laboratory was closed down. S.A. Altshuler as a volunteer joined the army.

Zavoiskii was able to resume his research work only in 1943. However, he did not continue the search for NMR, but began studies of the electron paramagnetic relaxation in orthogonal magnetic fields. These measurements led him to the discovery of the electron paramagnetic resonance (EPR) in 1944. The EPR rather soon became a powerful method to study the condensed matter on a microscopic level in physics, chemistry, geology, biology, medicine. The EPR discovery was one of the most important achievements in physics of the 20th century. At least five the following discoveries related to magnetic resonance in condensed matter were crowned by the Nobel Prizes. In particular, the first one among them was NMR, discovered by E.M.Purcell. H.C.Torrey and R.V.Pound in 1946 using the installations similar to that of Zavoiskii.

Let us return to our story about S.A. Altshuler. In July 1941 the Hitler's troops took his native town Vitebsk, at the beginning of October they were approaching Moscow. Altshuler deemed it is his duty to defend own country, though he had an official release from army as a university teacher, the family (the wife Eugeniya and the daughter Tanya who was only nine months old), an interesting work and scientific community. After some training S.A. Altshuler got an appointment to the First Separate Fighter Antitank Artillery Brigade of the Reserve of General Head-quarters. This unit was used to close breaches of our defense or for a break-through of the enemy front. One of officers of a similar unit recollected P.M. Borodin (later a professor of physics): "Our autos with guns were moving to meet an advanced enemy. Often they had to enter into a battle immediately without any preparation. However it was not usual battle with an artillery cannonade and crackling of machine-guns. The defenseless anti-tank gun is waiting in silence allowing the armored tank to come closer. When the

Journal of Physics: Conference Series **324** (2011) 011002 doi:10.1088/1742-6596/324/1/011002

duel between the gun and tank started, the gun is able to fire only 2-3 shots. If they are beside the tank, then the gun and the crew will be annihilated."

Since December 1942 till the end of the war S.A. Altshuler participated in heavy battles from Kursk till Easten Prussia. Some fragments from Altshuler's letters to his wife Eugeniya give some impression of these events.

April 1943. "...It seems to happen that I will participate in decisive battles, which shall lead Hitler's villains to the catastrophe... "

From the decree concerning the first Altshuler's order (Red Star): "... Near village Samodurovka the antitank regiment on 11-12 July (1943) had a severe battle with new German tanks T-6. Comrade Altshuler showed fearlessness by personal example. The regiment destroyed 10 enemy's tanks..."

During the month (24.06.44-24.07.44) Altshuler participated in the raid about 900 km till Brest mostly in the rear of enemy. Before it he wrote:

24.06.44. "What is going on around! ...I could write a lot, but it is not allowed and I have no time. The only what I want to tell to you. In the next several weeks don't wait letters from me... All my thoughts about you and Tanyusha! So long! Be happy!

S.A. Altshuler finished the war as a Major and was awarded by three Orders of the Patriotic War, one Order of the Red Star and many medals.

4. Back to the university, new fields of research and teaching

S.A. Altshuler left the army only in 1946 according to the Moscow Physics Institute inquires. However, he decided to return to Kazan University. Altshuler joined to Zavoiskii and Kozyrev in their work on EPR. This common work lasted not very long, since Zavoiskii left Kazan in 1947 to participate in the atomic project.

Altshuler started to develop a theory of the spin-lattice interaction and relaxation in paramagnetic crystals. In 1952 the studies of spin-lattice relaxation led Altshuler to the prediction of a new physical phenomenon, namely, the acoustic paramagnetic resonance (APR) [2]. The main idea of the APR concerned the selective absorption of the energy of the acoustical wave in paramagnet which occurs under the coincidence of the sound frequency with the interval between the quantum energy levels of the paramagnetic ion or the nucleus with nonzero spin.

The nuclear acoustic resonance was observed in 1955 by Proctor and Tantilla. The electron acoustic resonance was detected in 1959 in the USA by Jackobson, Shiren and Tucker. The APR discovery was the beginning of a new field in physics: *quantum acoustics*. The Altshuler's Doctorate was submitted for approval in 1955 at Physical Institute of Academy of Sciences in Moscow. In 1956 he became a professor.

Experimental studies in the field of the EPR were resumed at the Kazan University due to Altshuler's efforts. He involved to this research work several gifted postgraduate students from other fields. Some of them participated in the interpretation of experimental results.

An important step was made in 1956 when M.M. Zaripov and Yu.Ya. Shamonin detected and interpreted the EPR spectrum of trivalent chromium in a ruby. The analogous research was performed independently by A.M. Prokhorov and A.A. Manenkov in Moscow. Later the ruby became "the touchstone" of quantum electronics. A.M. Prokhorov and N.G. Basov together with C. Towns got the Nobel Prize in this field.

In 1954-1961 S.A. Altshuler headed the chair of the experimental and theoretical physics. He started to deliver a new course of lectures "Quantum theory of paramagnetism". In 1956 was created a group of students in theoretical physics, to which I joined too. To improve their mathematical education Altshuler invited the well known algebraist professor V.V. Morozov to deliver lectures on the group theory. The first graduation of the theorists was in 1957.

5. A new turn: the head of experimental laboratories

In 1950s important engineering applications of crystals with paramagnetic impurities, as the working substance of quantum amplifiers of microwaves (maser), were advanced. Masers produced a revolution in cosmic communications and radio-astronomy due to the extremely low level of background noises.

The Magnetic Radio-Spectroscopy (MRS) laboratory (1957) and the chair of magnetic radio-spectroscopy and quantum electronics (1961) were established at Kazan University for the research and for the training of highly qualified specialists in this field (the head of the chair– M.M.Zaripov). All these institutions (involving the chair of Theoretical Physics) formed at that time a joint collective body under the leadership of S.A. Altshuler.

At first, before the laboratory succeeded in growing the synthetic crystals, minerals from the geological museum were investigated with help of professor V.M. Vinokurov, a specialist on mineralogy. On occasions the magnificent jewels from women's decorations belonging to relatives and friends got into the cavities of spectrometers. A number of crystals were recommended and applied later as working substances in quantum paramagnetic amplifiers.

However, the most important results of these investigations consisted in forming a large team of highly qualified specialists in the scope of magnetic radio-spectroscopy and optics of solids. This group included M.M. Zaripov, L.Ya. Shekun, Yu.Ye. Polsky, A.L. Stolov, V.G. Stepanov, I.N. Kurkin A.A. Antipin, R.M. Valishev, A.Kh. Khasanov, M.A. Teplov, L.D. Livanova and many others.

In 1973 the head of the MRS chair M.M. Zaripov was appointed as a director of the Kazan Physico-Technical Institute. S.A. Altshuler decided to take care of the MRS chair; to be the head of the theoretical physics chair he suggested to me.

6. Development of the scientific school at the Kazan University

It is evident that the Kazan School of Magnetic Radiospectroscopy started by the EPR discovery of E.K. Zavoiskii. Its following development was going under the leadership of S.A. Altshuler at the Kazan University and B.M. Kozyrev at the Kazan Physical-Technical Institute.

After discovery of the APR the most impressive results of S.A. Altshuler were related to so-called Van Vleck's paramagnets (in collaboration with M. Zaripov, M. Teplov, L. Aminov, B. Malkin, Yastrebov and others) and non-linear spin-phonon phenomena (with R. Valishev, A. Khasanov, B. Kochelaev). His idea to use the Van Vleck's paramagnets to achieve super low temperatures [3] was very successive (up to tenths of a millidegree K). Altshuler's idea to use the Mandelshtam-Brillouin scattering of the laser light to observe the resonant phonons in solids resulted in the discovery of an avalanche of phonons radiated by the saturation of the wing of the EPR line by powerful microwave impulses. The effective temperature of these phonons raised to several million degrees, whereas the temperature of the sample was retained at a level of 2 K [4].

Altshuler initiated and supported all trends in magnetic spectroscopy: NMR, EPR, optical and gamma resonance spectroscopy. "Magnetic Seminar", headed by Altshuler, was playing an important role in developing of his school.

The books of Altshuler and Kozyrev related to the electron paramagnetic resonance [5] helped to spread the resonant methods in our and other countries. It was translated into German, English and Polish.

Altshuler had a rare intuition in physics. He did not believe that fundamental results in physics could be obtained only by mathematical calculations without understanding of phenomena in a simple way. An important feature of his school is a tight collaboration experimentalists and theorists.

Altshuler took care of his post-graduates without patronizing them, trying to explain to them the nature of physical phenomena rather than training them in calculation techniques. He did it so tactfully that a student began to think that he came to the solution of a problem himself. Altshuler always related to his colleagues and students with kind attention which favored the conversation. He was using only polite form in conversation (in German "Sie"); Russian equivalent to "Du" was used with close friends only.

Altshuler never hurried and was never late, whether it was an official session or a meeting with a post-graduate, or a friendly visit. He usually was among the first guests, and he never changed his habits.

Now the scientific school "The Resonance Spectroscopy of Condensed Matter" is recognized as one of the leading school of Russia. It includes researchers of the five chairs: Theoretical physics, Magnetic radio-spectroscopy and quantum electronics, General physics, Physics of molecular systems, Solid state physics.

7. National and international contacts

Altshuler was well known among the national and international magnetic resonance and other scientific communities. However the international contacts had difficulties during the Soviet time. For example, in 1965 N. Bloembergen being in Moscow by the invitation of the rector of the Moscow University R.V. Khokhlov could not come to Kazan to meet Altshuler and myself (I was Bloembergen's postdoc at Harvard university during 1963-64 academic year), since Kazan was a closed city at that time. Altshuler and I had to go to Moscow for this meeting.

For the first time Altshuler participated in the international conference in 1966 (the Congress AMPERE in Ljubljana) with a plenary lecture. He met there many well known researcher in the field of magnetic resonance: Anatole Abragam, Alfred Kastler, Alex Müller, Dan Bolef and others.

In 1968 the Congress AMPERE was organized in Grenoble. The Ministry of education of the Soviet Union included to the soviet delegation Altshuler, Zaripov and me from the Kazan University. Nevertheless Altshuler could not go because the Communist Party committee of the Tatar Republic did not give permission. When we met Anatole Abragam, he related this incident to the anti-Semite campaign at that time.

In 1969 Altshuler managed to organize in Kazan a conference dedicated to 25 years of the EPR discovery. For the first time (after 1928) to Kazan came well known scientists from other countries: K. Gorter, C. Jeffries, M. Strandberg, B. Judd, A. Abragam, A. Kastler, K.A. Müller, D. Bolef, and many others. Later Altshuler participated also at the Congresses AMPERE in Bucharest, Turku, Heidelberg, and Tallinn.

S.A. Altshuler was a chairman of the Council "Radiospectroscopy of condensed matter" and a member of the Council "Physics of low temperatures" (Academy of Sciences of the USSR); a member of the editorial board of "Journal of Experimental and Theoretical Physics"; a chairman and member of different Councils at the Kazan University.

8. Not only the science…

Altshuler started a tradition to organize seminars somewhere outside the city (say, on Volga river). After the report the participants took part in inconceivable play with a ball, the play being something between football and basketball. Altshuler often happened to be the goalkeeper. He seemed to enjoy himself in the young company. He joked, told stories, and sang old student songs (having a strong baritone) which he knew a lot.

It was known, also, that if a musical event took place in Kazan, one would certainly meet him there together with his friend a mathematician B.L. Laptev. His favorite hobby was also reading novels, especially dedicated to historical and biblical events.

Altshuler's jubilees were celebrated cheerfully, with enthusiasm: a lot of songs, humor, cartoons… According to this tradition we celebrated his 100th birthday. Besides the International Conference "Resonances in the condensed matter" the premiere of a humorous opera "Eugeny Zavoisky or the Epochs Paradoxical Resonance (EPR-2)" took place at the Cultural Center of the Kazan University. This opera was created by professor of the magnetic radio-spectroscopy chair N.I. Silkin. It is based on the student's songs of the Kazan, Moscow and Petersburg universities and contains a lot of humor and irony. A performance of the opera was realized by the producer of the Kazan Academic Theater of opera and ballet E. Treskin, who graduated the Physics Department of the Kazan University in 1969 and later became a professional opera-soloist. The premiere had a great success.

Many years are over since that time when Semen Alexandrovich Altshuler left us, but many meetings with him, conversations, and events recall to me so clear like they took place yesterday. I will be always grateful that fate brought us together with my Teacher.

9. Instead of Conclusion

Here I site some quotations from letters of well known persons addressed to S.A. Altshuler on different occasions.

"*...We have always found your work a source of great stimulation. There can be no doubt that you have not only done the first work in the field of acoustic paramagnetic resonance, but your work is of the most basic type...*"

Prof. M.W.P. Strandberg
MIT, Boston, USA, 1961

"*Our personal contacts with you always give rise to sensation of a resonant sympathy and delicacy; it is symbolic that you discovered an acoustical paramagnetic resonance, every conversation with you retains a sharp resonant trace.*"

A.S. Borovik-Romanov, I.M. Lifshits, A.F. Andreev, et al.
Institute of Physical Problems, Moscow, 1971

"*I want cordially congratulate you with your election as a Corresponding member of the Academy. It makes me glad for many reasons besides, of course, recognition of your scientific deserts. It is not the last of them that our Section enriched itself by the person of really high moral properties, in which it so needs.*"

E.M. Lifshits
Institute of Physical Problems, Moscow, 1976

"*You are highly intellectual and harmonic person... It is very pleasant and joyful for us to note that you among your wide interests always gave a place for music, an intercourse with it became for you a vital requirement...*"

Professor N.G. Zhiganov,
Rector of Conservatory, Kazan, 1981

"*...I remember S.A. Altshuler very well... I have the greatest regard for him as an outstanding physicist, a pioneer in the field of ESR, and as a man highly respected by all those who had the privilege of interacting with him, which includes myself.*"

Anatole Abragam
France, 2001

References

[1] Tamm I and Altshuler S 1934 *Doklady AN SSSR* **1** 455–460 (Russian and German)
[2] Altshuler S A 1952 *Doklady AN SSSR* **85** 1235–1238 (Russian)
[3] Altshuler S A 1966 *Pis'ma v ZhETF* **5** 177–180 (Russian)
[4] Altshuler S A, Valishev R M, Kochelaev B I and Khasanov A Kh 1971 *Pis'ma v ZhETF* **13** 535–538 (Russian)
[5] Altshuler S A and Kozyrev B M 1961 *Electron Paramagnetic Resonance* (Moscow: Fizmatgiz) (Russian); 1963 (Leipzig: B.G. Teubner Verlagsgeselschaft) (German); 1964 (New York, London: Acad. Press) (English)

Semen A. Altshuler (1979)

The dissertation is finished (1936)

Major S.A. Altshuler (1945)

With N. Bloembergen and B. Kochelaev at the Moscow University (1965)

With C.J. Gorter and C.D. Jeffries at the Conference in Kazan (1969)

With A. Kastler on Volga-river during the Conference in Kazan (1969)

Journal of Physics: Conference Series **324** (2011) 011003 doi:10.1088/1742-6596/324/1/011003

Reminiscences about S. A. Altshuler

A A Kaplyanskii

A. F. Ioffe Physical-Technical Institute, Saint-Petersburg, 194021 Russia

E-mail: a.kaplyanskii@mail.ioffe.ru

Abstract. This is the authorized translation of Prof. A Kaplyanskii's speech at the opening ceremony of Alt100 conference.

I am grateful to Organizers of the Conference devoted to the centenary of the outstanding scientist and beautiful person, Semen Alexandrovich Altshuler, for the invitation. Conference opens just at the day of the 70th anniversary from the beginning of the Great Patriotic War. In this War, young physicist S.A.Altshuler of 30 years old participated as a volunteer. He showed great fortitude in battles and was marked by the most prestigious military awards. Nowadays, on the Victory Day, workers of the A.F.Ioffe Physical-Technical Institute in Saint-Petersburg recollect about evacuation of the Institute to Kazan, to the building of Kazan university, recollect about help of Kazan colleagues, including the lecturer S.A.Altshuler, who have taken care about evacuated scientists from Leningrad. Thus, collaboration of S.A.Altshuler with Saint-Petersburg originated since that hard time. As it is known, links between the researchers appear, first of all, due to common scientific interests. Spectroscopy of crystals containing paramagnetic ions with unfilled electronic shells was such a common field for us. Semen Alexandrovich had been already engaged in studies of these crystals long ago, from 1940-1950 years, after the discovery of electron paramagnetic resonance by E.K.Zavoiskii. In 1952 S.A.Altshuler discovered acoustic paramagnetic resonance. My work with these materials began much later, after I had defended in 1957 the PhD thesis connected with the spectroscopy of excitons in semiconductors and fulfilled in A.F.Ioffe Physical-Technical Institute under the tutorship of Euvgenii Fedorovich Gross. For becoming acquainted with physics of insulating crystals containing impurity ions with unfilled electronic shells, I am obliged to Petr Petrovich Feofilov who has been carrying out optical investigations of rare earth doped fluorite crystals at the State Optical Institute. This branch of science began to develop quickly from 1960 after the laser based on ruby crystal had been discovered. The First All-Union Symposium on Spectroscopy of Crystals activated by rare earth and iron group ions was held in 1965. Symposium was organized by Sofiya Vladimirovna Grum-Grzhimailo at the Institute of Crystallography, and P.P.Feofilov was a chair of the Program Committee. S.A.Altshuler also attended this Symposium. In co-authorship with Maksud Zaripov, they presented the introductory talk devoted to applications of EPR in investigations of activated crystals. My talk was also devoted to possible applications of the methods of optical spectroscopy in external magnetic and electric fields and under uniaxial deformations in the studies of crystals. This talk was prepared in common with Boris Petrovich Zakharchenya, and I have to mention here about collaboration of B.P.Zakharchenya, my friend and partner during all my work in Physical-Technical Institute since our post-graduate studies under E.F.Gross tutorship, with Kazan physicists. B.P.Zakharchenya warmly describes his

Journal of Physics: Conference Series **324** (2011) 011003 doi:10.1088/1742-6596/324/1/011003

meetings with S.A.Altshuler and his pupils - B.Kochelaev, K.Salikhov, B.Malkin, A.Khasanov, I.Silkin- in the book published on the 90-th anniversary of S.A. Altshuler birthday. From 1965, fourteen Symposiums on Spectroscopy of Crystals were held, and always Kazan physicists were active participants, including the last years one which was arranged in Saint-Petersburg. Twice these Symposiums were held in Kazan.

When speaking about our collaboration with S.A.Altshuler and his pupils, it is pleasant to remember about so called shekunneaki - by this word we named meetings with Lev Yakovlevich Shekun (the pupil of S.A.Altshuler who died being only 36 years old) at the apartment of P.P.Feofilov in Saint-Petersburg. I have to mention one more active participant of our common works with Kazan colleagues, our collaborator Valentin Semenovich Vikhnin, who also prematurely passed away. Contacts with S.A.Altshuler post-graduated student Boris Zalmanovich Malkin, the theorist, were especially useful for us in A.F.Ioffe Physical-Technical Institute. He helped us to interpret experimental results obtained not only in the studies of impurity insulating crystals but also in some other investigations.

Life presented me not too many meetings with Semen Alexandrovich, but every time I was deeply impressed by his beautiful human qualities, his wisdom, based on rich life experience, benevolent attitude and interest to interlocutor, including young interlocutors - we had the age difference of 20 years. In 70ths S.A.Altshuler sent his younger daughter, Nina, to go through period of training in our laboratory at Physical-Technical Institute. I think this fact gives the best evidence for his kind attitude to our scientific activity. Works of S.A.Altshuler in the field of acoustic paramagnetic resonance that had been defined by him as the energy transfer from sound waves to a system of magnetic particles when quantum of energy of a lattice vibration equals the difference between energies of magnetic levels, were the most important for us. Just this energy quantum was named as phonon by academician I.E.Tamm, who was a scientific tutor and a mentor of S.A.Altshuler. Some kind of peculiar resonance in time happened in 1974: when the discovery of acoustic paramagnetic resonance was officially registered, we just began to study resonance transfer of energy from acoustic vibrations to electronic subsystem of impurity centers, however, not at resonance conditions for magnetic sublevels and phonons, but for resonance transitions with much higher energies between energy levels of impurity ions split in the crystal field. Thus we shifted frequencies of relevant acoustic phonons to the terahertz band. Just in this our first work, following the method of K.F.Renk and future Nobel prize winner J.Deisenhofer, who studied quantum transitions of impurity ions in ruby induced by phonons from heat pulses, we found high anisotropy of interaction between acoustic phonons and chromium ions in the excited states. This work gave rise to start of long standing investigations of non equilibrium acoustic phonons with terahertz frequencies in impurity insulators and some later in semiconductors as well. S.A.Altshuler took a keen interest to our results which in many cases were mutually supplementary with his investigations of acoustic paramagnetic resonance where phonons with gigahertz frequencies played the dominant role. I have to mention here also about some bright results obtained by S.A.Altshuler in 70-ths in paramagnetic resonance studies when these studies were supplemented by methods of optical spectroscopy. In 1971, the avalanche-like generation of resonant phonons was detected in Mandelshtam-Brillouin scattering of light at the saturation of the wings of EPR lines, and 10 years later the light scattering in paramagnets induced by spin-flip transitions had been observed. It is very pleasant for me to remember about these studies because the review article by S.A.Altshuler, B.I.Kochelaev and A.Kh.Khasanov "Studies of spin systems by means of light scattering in paramagnetic crystals" was published in the volume "Spectroscopy of solids containing rare earth ions" (North-Holland, Amsterdam 1987) where I was one of the Editors. This review was published when S.A.Altshuler had passed away, and his last work was very important and useful for us at that time. The volume mentioned above belongs to the series Modern problems in condensed matter sciences initiated by V.M.Agranovich and A.A.Maradudin, both Western and Soviet scholars contributed to the series, and each volume had accordingly two editors. At that time, publication of this series was of great importance for establishing of international collaboration between scientists from different countries.

Peer review statement

All papers published in this volume of *Journal of Physics: Conference Series* have been peer reviewed through processes administered by the proceedings Editors. Reviews were conducted by expert referees to the professional and scientific standards expected of a proceedings journal published by IOP Publishing.

EPR and development of quantum electronics

A A Manenkov

Prokhorov General Physics Institute RAS, Moscow, Russia

E-mail: manenkov@kapella.gpi.ru

Abstract. A role of electron paramagnetic resonance in development of quantum electronics is discussed. Basic principles and history of masers are briefly described. Spin-levels of paramagnetic ions in crystals as a very suitable object for active media of solid-state masers (called as EPR-masers) and physical processes in EPR-masers (population inversion of energy states) are analyzed. This analysis demonstrates a significant role of relaxation processes in multi-level spin-systems for efficient maser action. In this context peculiarities of spin-lattice and spin-spin cross relaxation processes in multi-level systems are analyzed. Development of EPR-masers and their application in radioastronomy and far-space communication systems are briefly described.

1. Introduction

Studies of electron paramagnetic resonance phenomenon (EPR) discovered by E. K. Zavoiskiy in 1944[1] made a very strong impact on a development of modern physics and other sciences (chemistry, biology, etc), and very wide applications. These studies significantly affected also the development of quantum electronics. By now a great number of review papers and books was published in which different aspects of EPR physics and various applications including in quantum electronics were described (see, for instance [1, 3, 4]). In one's turn the development of quantum electronic (QE) had a positive feedback – it significantly influenced further development of researches in EPR field.

In this paper the author aims to analyze briefly both aspects of EPR-QE connection. The plan of the paper includes:

- description of basic principles of QE and a brief history of masers,
- analysis of spin levels of paramagnetic ions in crystals as an object of active media of solid-state masers (we will call this type of masers as EPR-masers)
- analysis of physical processes in EPR-masers: inversion of spin level populations
- a brief description of EPR-masers development and application
- analysis of relaxation processes in multi-level spin-systems.

2. Basic principles and brief history of quantum electronics

Quantum electronics as a new field of physics has originated in a middle of 50-th of XX century from the research works in microwave spectroscopy of gases. Two research groups, Basov and Prokhorov in the USSR and Townes and coworkers in the USA, have led independently to an idea of using

[1] The history of the discovery is described in details by Zavoiskiy collaborators S. A. Altshuler and B. M. Kozyrev[1], and his daughter N. E. Zavoiskaya [2]

stimulated emission of radiation of molecules in molecular beams to generate an electromagnetic radiation [5,6].

The first molecular oscillator (called later a maser), based on this principle, has been realized on the ammonia molecular beam [6]. An inversion of energy level populations relative to a thermodynamically equilibrium state, required for exceeding stimulated emission over stimulated absorption, was obtained by passing the molecular beam through an inhomogeneous electric field. A positive feedback in the oscillator was realized in a microwave resonator through which molecules in an upper energy state were passed.

In 1955 a new method for population inversion in molecular beams was proposed by Basov and Prokhorov [7] – the method based on excitation of molecules by auxiliary radiation (pumping). An essence of the method is explained in figure 1 for a 3-level molecule system in which auxiliary radiation with frequency $\nu_{aux.}$ induces a resonance transition between low and upper levels and at high enough power produces the inversion of a populations of one of these levels and an intermediate level. Thus, the oscillation conditions at 1→2 transition frequency $\nu_{osc.}$ are produced.

Figure 1. Schematic illustration of population inversion in two configurations of energy levels using auxiliary radiation at frequency $\nu_{aux.}$; $\nu_{osc.}$ is signal frequency at which oscillation can be obtained. After Ref. [7].

Though this method has been proposed by Basov and Prokhorov for molecular beams, it is, obviously, universal one -applicable to any multi-level atomic systems. Basing on this idea the author of these words analyzed a possibility of population inversion in the ground-state spin-level system of Cr^{3+} ions in ruby which by that time (1955) have been well studied: energy levels and corresponding wave functions have been determined [8,9]. An analysis showed that the spin-level system of Cr^{3+} in ruby is very suitable for realization of e-m pumping method. The results of this analysis became a basis for our subsequent works on EPR-masers (see below for more details).

In 1956 Bloembergen analyzed the conditions for population inversion in a three-level spin-system by e-m pump in continuous-wave regime [10]. His analysis has shown that e-m pump can be very effective method for creation of solid-state masers.[2]

Later on the author of these words has generalized Bloembergen analysis considering besides stationary condition for inversion also a transient process of inversion during a pump pulse [14]. Since such a consideration is of interest for more complete understanding of pumping process in both stationary and non-stationary regimes we will present below its results in some details.

The three-level system with energies $E_3>E_2>E_1$ was considered [14] in which pump radiation at frequency $\nu_{31}=(E_3-E_1)/h$ induced transitions between levels E_1 and E_3 producing, at definite conditions, the population inversion for 3→2 or 2→1 transitions. A change in time of level populations n_i (i=1,2,3) under the pump radiation and spin-phonon interaction is described by the following rate equations:

[2] Note that N.Bloembergen came to the idea of population inversion by auxiliary radiation proceeding from other considerations: he refers in [10] to Pound's and Overhauser's works [11,12] on nuclear magnetic resonance and nuclear polarization (see also Bloembergen's paper [13] in which he discusses especially his approach).

$$\dot{n}_1 = -(w_{13} + w_{12} + W_{13})n_1 + w_{21}n_2 + (w_{31} + W_{31})n_3$$
$$\dot{n}_2 = -(w_{21} + w_{23})n_2 + w_{12}n_1 + w_{32}n_3 \qquad (1)$$
$$\dot{n}_3 = -(w_{31} + w_{32} + W_{31})n_3 + (w_{13} + W_{13})n_1 + w_{23}n_2$$

where W_{13} , W_{31} and w_{ik} are transition probabilities induced by pump radiation and spin-phonon interactions, respectively. According to a quantum theory of thermodynamic equilibrium in atomic systems, the probabilities of radiation-induced and phonon-induced transitions obey the conditions: $W_{13}=W_{31}=W$ and $w_{ik} = w_{ki}\exp\left(-\frac{E_k - E_i}{kT}\right)$.

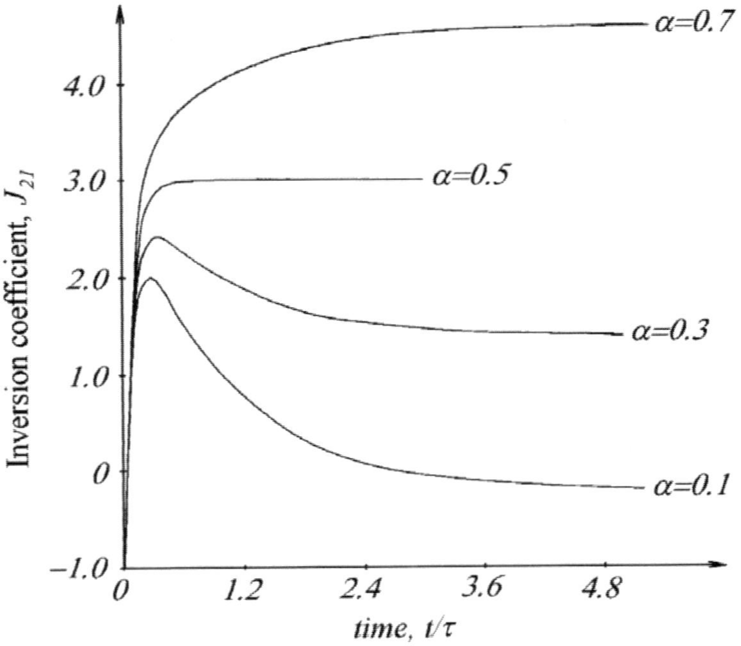

Figure 2. Temporal evolution of inversion coefficient in active medium during pumping, switched on at t = 0, at various values of relaxation parameter $\alpha = \omega_{23}(\omega_{23} + \omega_{21})^{-1}$. The parameter $\tau = (2/3)(w_{23} + w_{21})^{-1}$ is relaxation time. After Ref. [14].

From a solution of rate equations the following expression for the inversion coefficient $J_{21}(t)$ = $-x(t)/x_0$ (where $x(t)$ = n_2-n_1 is a current population difference of levels 2 and 1 and $x_0=(n_2-n_1)_0$ is an initial population difference of these levels at t = 0, when the pump pulse is just switched on) has been obtained:

$$J_{21}(t) = -\frac{v_{31}}{2v_{21}}\exp\left[-\left\{2W + \frac{3}{2}(\omega_{23} + \omega_{21})\right\}t\right] + \frac{v_{31}}{2v_{21}}(1-2\alpha)\exp\left[-\frac{3}{2}(\omega_{23} + \omega_{21})t\right] + \left(J_{21}\right)_s \qquad (2)$$

where $\alpha = \frac{\omega_{23}}{\omega_{23} + \omega_{21}}$, $(J_{21})_s$ is the stationary inversion coefficient:

$$(J_{21})_s = \frac{\nu_{31}}{\nu_{21}}\alpha - 1. \qquad (3)$$

The inversion coefficient is the important parameter characterizing amplification properties of an active medium: at $J_{21}>0$ it amplifies an incident radiation, whereas at $J_{21}<0$ it absorbs the radiation. A value and temporal dependence of the inversion coefficient significantly vary with the relaxation parameter α. Figure 2 demonstrates these variations of $J_{21}(t)$ computed from (2).

The presented analysis shows that relaxation processes in active media are very important for determining conditions for most efficient maser action. Above theoretical results have been confirmed in experimental studies of amplification characteristics of ruby and Cr^{3+}:TiO_2 masers [14]. For example, figure 3 shows an output signal shape observed in the Cr^{3+}:TiO_2 maser (λ_{signal} = 10 cm) at pulsed pumping.

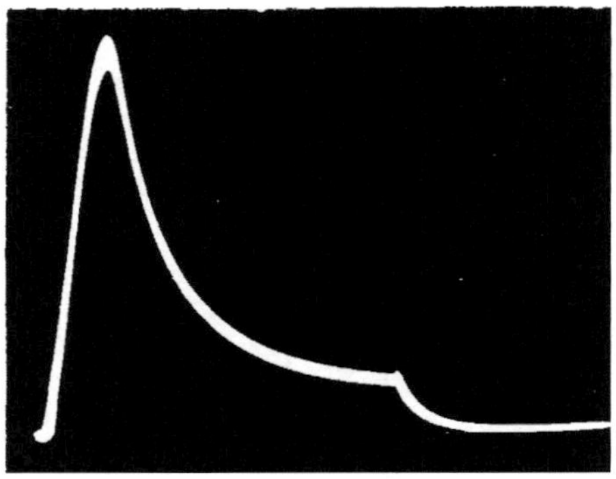

Figure 3. Oscillogram of the output signal observed in Cr^{3+}:TiO_2 maser amplifier at pulsed pumping. A sharp increase of the signal is seen at the beginning of pump followed by its transient decrease to the end of pump and relaxation decay after its switching off. Duration of the pump pulse is 8 msec. After Ref. [14].

3. Development and application of EPR-masers

Several research groups in USA and USSR have realized the first EPR-masers in 1957-1958. As active media they used different paramagnetic ions in crystals: Gd^{3+} in ethylsulfate [15], Cr^{3+} in cyanide[16, 17] , and Cr^{3+} in corundum (ruby) [18-20].

Later on the other crystals as maser materials have been studied: Cr^{3+} and Fe^{3+} in TiO_2 and tungstates, Gd^{3+} in CaF_2 and others (see [4] for a review). The studies have shown that ruby is the most effective maser material possessing a unique combination of properties: a large crystalline field splitting of the ground-state (Kramers doublets), low dielectric losses, a long spin-lattice relaxation time, a high thermal conductivity, high chemical and mechanical stability. Using this material, masers in a wide spectral range, from dm to mm wavelengths, have been created. One of the first laboratory ruby masers created at Lebedev Physical Institute is shown in figure 4.

Masers as amplifiers of electromagnetic radiation possess an ultimately low intrinsic noise (which is due to spontaneous emission in an active medium) [4]. This important property makes them very attractive for application as high-sensitivity receivers in the microwave spectral range. In 1960-1970th at some academic and industrial laboratories of the USSR ruby masers have been worked out and applied in radioastronomy and space communication systems [21]. Owing to their use a number of new important results in space research has been obtained.

In particular, detail studies of space hydrogen radiation at 21 cm wavelength and discovery of new radiation lines of highly-excited hydrogen atoms at 8 mm spectral range (Sorochenko et al. [22]) gave valuable data on a distribution and characteristics (temperature, density, dynamics) of hydrogen in Galaxy.

Studies of maser emission from space at 1.35 cm (Matvienko et al. [23-25]) discovered earlier by Townes and coworkers [26], have presented very interesting data on a water content in some space sources. The use of EPR-maser amplifiers in planet radar systems allowed to obtain new data on characteristics of Mercury, Venus, Mars, Jupiter (Kotelnikov et al. [27]).

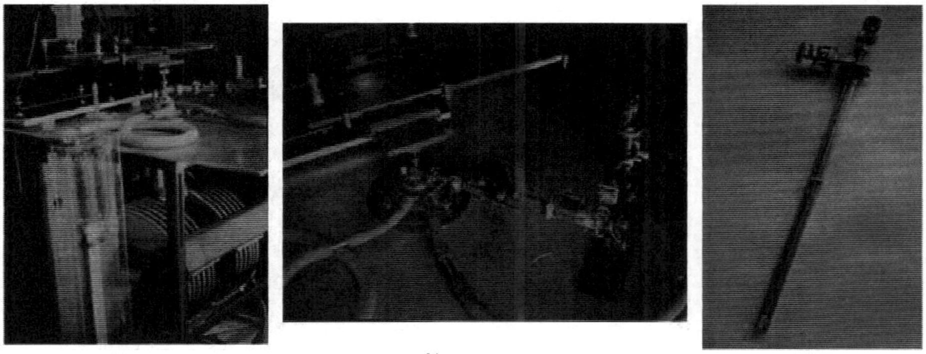

a) b) c)

Figure 4. Ruby maser created at Lebedev Physical Institute in 1958: a -general view; b, c – fragments; b - signal and pump waveguides, pump source (klystron), c –two-frequency resonator. Signal and pump wavelengths are 15 cm and 2.21 cm respectively. After [19].

4. From masers to lasers

A successful development of microwave EPR-masers stimulated an extension of quantum electronics principles to shorter wavelength region. The fact that the first optical maser (laser) has been realized [28] on ruby as the active material was not accidental. Indeed, to that time (1960) physics of microwave ruby masers was well developed and optical spectra of this crystal were studied in details.

Creation of the first laser led to a rapid development of quantum electronics. Intensive studies were started of lasers using different type materials as active media: crystals and glasses activated with ions of transient group elements, atomic and molecular gases, semiconductors, etc.

Investigations in EPR and EPR-maser fields determined many directions in solid-state laser studies. In particular, EPR spectroscopy became an important tool for studies of crystals doped with Fe and RE group ions as active laser materials. Similarity of many processes in masers and lasers was very useful in studies of laser dynamics.

In one’s turn, investigations in maser-laser field stimulated further development in EPR physics. In particular, studies of physical processes in EPR-masers and solid-state lasers (population inversion, transient processes, saturation effects, etc) stimulated a development of studies of relaxation processes in crystals. Some results obtained in these studies are briefly described in the next section. They relate to features of relaxation processes in multi-level spin-systems.

5. Relaxation processes in multi-level systems

Make several general comments on a nature and a role of relaxation processes in multi-level spin-systems.

Two types of interactions determine a thermodynamic equilibrium in such the systems: spin-lattice and spin-spin cross relaxation. The first type interaction depends naturally on a crystal lattice temperature, and a character of this dependence is determined significantly by a mechanism of a spin-photon interaction. For elucidating the fundamental mechanisms of such interaction, a temperature dependence of a spin-lattice relaxation time in paramagnetic crystals was a subject of many theoretical and experimental studies still in pre-maser period, and especially in the maser-time in a connection with practical problems of quantum electronics. It relates, specifically, to the low (liquid helium) temperature range where, as a rule, EPR-masers work.

As far as the second type interactions, spin-spin cross-relaxation, interest to their studies appeared in connection with quantum electronics. Bloembergen and co-authors were the first who paid an attention to a role of cross-relaxation in spin-systems. They presented a theory of resonance cross-relaxation [29]. This work stimulated studies of cross - relaxation phenomena in more complex multi-level systems. Both types of interaction, spin-lattice and spin-spin cross-relaxation, play the fundamental role in physics of EPR-masers determining a possibility of population inversion in spin-energy levels, transient processes and saturation effects in maser amplifiers and oscillators. Cross-relaxation processes are characteristic ones also for multi-level systems used in solid-state lasers. A comprehensive analysis of the role of relaxation processes in masers was presented in many review papers and books (see, for example, [3, 4]).

After these general comments we will present below two results obtained in studies of spin-lattice and spin-spin cross - relaxation in multi-level systems.

1. In studies of spin-lattice relaxation the peculiarities of temperature dependence of relaxation times in liquid helium temperature range have been revealed: strong (exponential) dependence for lower doublets [30, 31] and, in contrast, very weak dependence for upper doublets [31]. The first of these results, the exponential temperature dependence of relaxation times, have been interpreted on the basis of two different models: as a resonance two-phonon Raman process [32, 33] (called in some papers as Orbach process) and as a specific case of one-phonon resonance process in multi-level systems [34]. A comprehensive analysis of these models based on quantum field theory has shown [35] that the latter is the most adequate model. Essence of this model is illustrated in figure 5 on an example of relaxation in a three – level system.

Relaxation transition probabilities w_{ik} between corresponding levels (i, k =1,2,3) in this system are supposed to be due to the one-phonon resonance processes. An analysis has shown [34] that relaxation kinetics of populations of energy levels in this system has, in general, a complex (non-exponential) character depending on all transition probabilities involved. However, in some specific cases of ratio between probabilities it can be one-exponential.

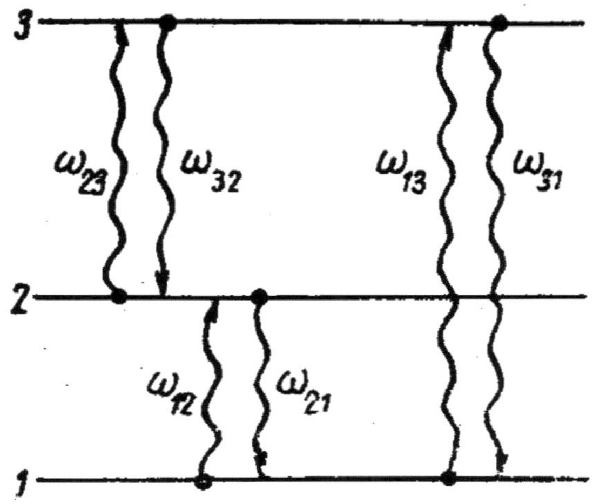

Figure 5. Relaxation transitions in tree-level system (wavy lines with corresponding transition probabilities w_{ik})

For example, at w_{12}, $w_{21} \ll w_{23}$, w_{32}, w_{13}, w_{31} the levels 1 and 2 relax via the upper level 3. In the other case, if w_{23}, $w_{32} \ll w_{12}$, w_{21}, w_{13}, w_{31}, the levels 2 and 3 relax via the lower level 1. Temperature dependence of relaxation rates predicted for these two cases, respectively for lower and upper doublets, are as follows:

$$T_1^{-1} \propto [\exp(\Delta/kT) - 1]^{-1} \quad (4)$$
$$T_1^{-1} \propto [1 - \exp(-\Delta/kT)]^{-1} \quad (5)$$

where Δ is an energy gap between levels 1 and 3.

Experimental data on the temperature dependencies of relaxation times observed in some crystals doped with ions of iron group and rare earth ions [30, 31, 36] are in a good agreement with these predictions confirming the adequacy of the model suggested in [34].

2. In studies of relaxation processes in crystals, containing multi-level paramagnetic ions, strongly pronounced cross-relaxation effects were observed. Thus, in ruby (four-level system) besides resonance (two-spin) energy exchange between spin levels more complex multi-spin processes-harmonic and combinative- have been revealed [37]. Figure 6 illustrates such the processes.

If the energy difference between a, b and a', b' levels satisfy the condition $\Delta E_{ab} \cong \Delta E_{a'b'}$ the spin-spin interaction leads to the resonance energy exchange in this system. At other ratios of the level energies: $m\Delta E_{ab} \cong n\Delta E_{a'b'}$ or $m\Delta E_{bd} \cong n\Delta E_{ab} + l\Delta E_{cd}$, where m,n,l are numbers of spins participating in the processes, harmonic or combinative cross-relaxation takes place.

Theoretical analysis based on a kinetic model of cross-relaxation showed a good agreement with experimentally observed features [37] demonstrating a high efficiency of such processes in multi-level systems. Cross-relaxation processes revealed at rather high concentrations of paramagnetic ions, influence significantly on inversion of populations of energy levels in maser active media, and, hence, on the efficiency of the maser.

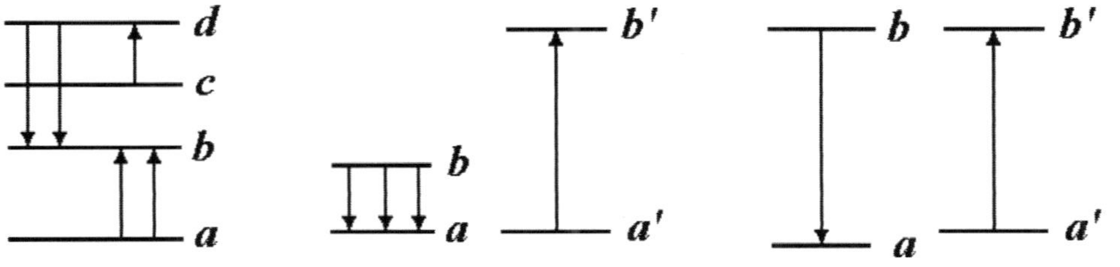

Figure 6. Schematic illustration of cross-relaxation processes revealed in ruby [37]: (from right to left) resonance (two-spin), harmonic (four-spin), and combinative (five-spin).

A significance of cross-relaxation processes refers in equal degree to both microwave and optical masers using ion-doped crystals as active materials. In this context note that multi-particle cross-relaxation processes similar to those revealed in spin-systems [37] were observed also in optical levels of rare earth ions in crystals [38, 39].

6. Conclusions

Summarizing the brief analysis presented in this paper of the role of electron paramagnetic resonance in development of quantum electronics emphasizes several the most important, to our view, points.

The idea to use multi-level spin systems of paramagnetic ions in crystals as the active media of quantum amplifiers turned out very fruitful. It has led to creation of microwave quantum amplifiers possessing ultimately low intrinsic noise. Such amplifiers found important practical applications in space research.

Application of auxiliary radiation for population inversion in spin-system levels was the first realization of this principle of obtaining atomic states with negative temperature. This has led not only to creation of microwave masers but also stimulated further progress in quantum electronics – creation of laser.

EPR-spectroscopy is one of the most effective tools in studies of active media for solid-state lasers.

In one's turn, development of quantum electronics significantly influenced further development of researches in EPR field, in particular for studies of relaxation processes in multi-level systems.

References

[1] Altshuler S A and B.M.Kozyrev B M 1972 *Electron paramagnetic resonance in compounds of transient group elements* Second edition (Moscow: Nauka)
[2] Zavoiskaya N E 2007 *History of one discovery* (Moscow: ITD Group)
[3] Siegman A E 1964 *Microwave solid-state masers* (New-York-San Francisco-Toronto-London: McGraw-Hill Book Company)
[4] Karlov N V and.Manenkov A A 1966 *Quantum amplifiers*, edited by A.M.Prokhorov (Moscow: Institute of Scientific Information)
[5] Basov N G and Prokhorov A M 1954 *JETP* **27** 431
[6] Gordon J P, Zeiger H J and Townes C H 1954 *Phys. Rev.* **95** 282
[7] Basov N G and Prokhorov A M 1955 *JETP* **28** 249
[8] Manenkov A A and Prokhorov A M 1955 *JETP* **28** 762
[9] Manenkov A A 1955 *Paramagnetic resonance in some compounds of iron-group and rare earth elements,* Ph.D. dissertation (Moscow: Lebedev Physical Institute Acad. Sci. USSR)
[10] Bloembergen N 1956 *Phys. Rev.* **104** 324
[11] Pound R V 1950 *Phys. Rev.* **79** 685
[12] Overhauser A W 1953 *Phys. Rev.* **92** 411
[13] Bloembergen N 2005 *J. Mod. Optics* **52** 1553
[14] Manenkov A A 1965 *Paramagnetic resonance in crystals and problems of quantum electronics,* doctor of science dissertation (Moscow: Lebedev Physical Institute Acad. Sci. USSR)
[15] ScovilH E, Feher G and Seidel H 1957 *Phys. Rev.* **105** 762
[16] Culver W H 1957 *Science* **125** 810
[17] McWhorter A L, Meyer J W and Strum P D 1957 *Phys. Rev.* **108** 1642
[18] Zverev G M, Kornienko L S, Manenkov A A and Prokhorov A M 1958 *JETP* **34** 1660
[19] Prokhorov A M, Manenkov A A, Zverev G M and Kornienko L S 1958 *Report on creation of paramagnetic amplifiers at* λ *=15 cm and* λ *= 21 cm* (Moscow: Lebedev Physics Institute Acad. Sci. USSR)
[20] Makhov G, Kikuchi C H, Lambe J and Terhune R W 1958 *Phys. Rev.* **109** 1999
[21] Manenkov A A and Shteinshleiger V B 1977 *Quantum amplifiers and their use in radio receiver systems of far-space communication and radioastronomy*, in Annual of Soviet Encyclopedia, N 21, p. 566, S.M.Kovalev editor (Moscow: Sov. Encyclopedia)
[22] Gordon M A and Sorochenko R L 2002 *Radio recombination lines: their physics and astronomical applications*, Astrophys. and Space Laboratory, v. 282 (Dordreht: Kluwer Acad. Publ)
[23] Matvienko L I et al. 1980 *Letters Astronom. J.* **6** 662
[24] Gufrida T S et al. 1981 *Letters Astronom. J.* **7** 358
[25] Matvienko L I et al. 1988 *Letters Astronom. J.* **14** 468
[26] Cheung A C et al. 1969 *Nature* **221** 626
[27] Kotel'nikov V A et al. 1980 *Problems of Modern Radio Engineering and Electronics,* editor V.A.Kotelnikov (Moscow: Nauka)
[28] Meiman T H 1960 *Nature* **187** 493
[29] Bloembergen N, Shapiro S, Pershan P S and Artman J A 1959 *Phys. Rev.* **114** 445
[30] Finn C B P, Orbach R and Wolf W P 1961 *Proc. Phys. Soc.* **77** 261
[31] Manenkov A A, Milyaev V A and Prokhorov A M 1962 *Fiz. Tverd. Tela* **4** 388
[32] Orbach R 1961 *Proc. Roy. Soc.* A **264** 458
[33] Aminov L K 1962 *JETP* **42** 783 (1962 *Sov. Phys. JETP* **15** 547).
[34] Manenkov A A and Prokhorov A M 1962 *JETP* **42** 1371 (1962 *Sov.Phys. JETP* **15** 951)

[35] Hernandez R and Walker M B 1972 *Canad J. Phys.* **50** 440
[36] Manenkov A A and Orbach R (editors) 1966 *Spin-lattice relaxation in ionic solids* (New York: Harper and Row Publishers)
[37] Manenkov A A and Prokhorov A M 1962 *JETP* **42** 75
[38] Basiev T T, Doroshenko M E and Osiko V V 2000 *Pis'ma JETP* **71** 14
[39] Basiev T T, Doroshenko M E, Osiko V V and Prokhorov A M 2001 *JETP* **120** 1362

Spin Noise Spectroscopy

E. B. Aleksandrov[1] and V. S. Zapasskii [2]

[1] Ioffe Physico-Technical Institute, Saint-Petersburg,194021 Russia

[2] Fock Institute of Physics, Saint-Petersburg State University, Saint-Petersburg,198504 Russia

E-mail: zap@vz4943.spb.edu

Abstract. Development of Spin Noise Spectroscopy (SNS) beginning from the first experiment on detecting magnetic resonance in the Faraday-rotation noise spectrum in sodium vapor up to its recent achievements is briefly reviewed.

1. Introduction

Spin noise spectroscopy (SNS) is now considered as a new branch of magnetic resonance spectroscopy based on detecting spontaneous fluctuations of magnetization by means of Faraday-rotation technique. The first experiment that demonstrated feasibility of such an approach was performed in 1981 [1]. For us, the story of what is now called 'spin noise spectroscopy' has begun in the middle of seventies when we started working on achieving the limiting (shot-noise-limited) polarimetric sensitivity. In the experiment [1], we detected spontaneous noise of transverse magnetization of sodium atoms in a magnetic field at Larmor frequency. In this communication, we briefly consider the reasoning underlying this first experiment, review the conceptual and methodological aspects of this experimental approach, and discuss some possible ways of its further development.

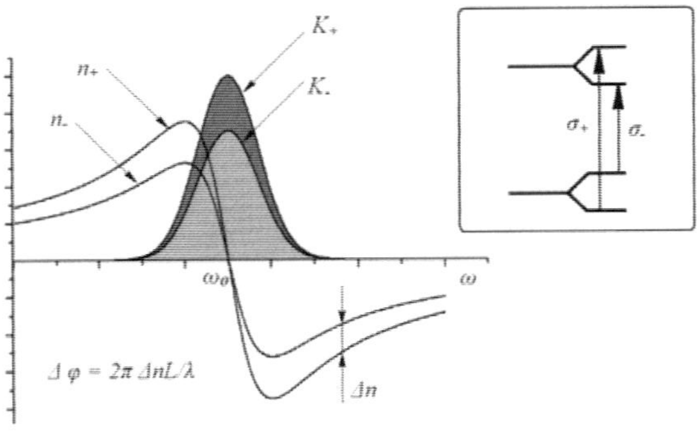

Fig. 1. Formation of "paramagnetic" part of the Faraday rotation. A difference in populations of the ground-state sublevels creates difference in optical absorption for two circular polarizations, which (in conformity with Kramers-Kronig relations) makes different refractive indices for two circular polarizations in the transparency region near the absorption line. This 'magnetic circular birefringence' is revealed as rotation of the polarization plane (Faraday rotation). Magnetic splitting of the levels (giving rise to the 'diamagnetic' magnetooptical effects) is ignored.

2. Conceptual background

In 70s, when the experiment [1] was planned, all the fundamental information needed to predict its feasibility and to evaluate the magnitude of the effect was available. Information capacity of the noise spectroscopy seemed to be, in principle, clear. Correlation characteristics of the 'noise signal' and its spectrum were known to contain information about dynamics of the process underlying the observed signal, and, in conformity with the fluctuation-dissipation theorem, to be closely related to the appropriate susceptibility spectrum of the system. Some experiments of this kind, demonstrating potentialities of the noise technique, have been already fulfilled [2]. As for the magnetization noise, the fundamental possibility of its observation in the nuclear spin systems was mentioned by Bloch in 1946 in its classical monograph "Nuclear Induction". Of course, it was not essential that Bloch spoke about nuclear, rather than electronic, spins. The more so, that possibility of observation of regular signals of coherent spin precession in atomic system by means of magneto-optics was also well justified in experiments on optical pumping [3]. So, the question was of more practical, rather than fundamental, nature: it was not clear whether it is possible experimentally to detect the spin noise of essentially macroscopic systems for reasonable times with sufficiently high signal-to-noise ratio and, after all, whether it has any sense.

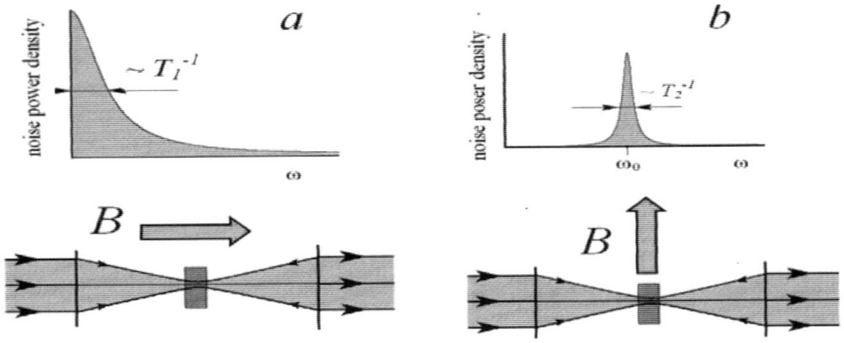

Fig. 2. Measuring spin noise in two configurations: (a) in a longitudinal magnetic field (Faraday configuration) and (b) in a transverse magnetic field (Voigt configuration).

At that time, we were well familiar with magneto-optics of atomic and solid-state paramagnets and knew that the Faraday rotation (or, more exactly, its "paramagnetic" part) is proportional to magnetization of the spin-system and can be used for magnetic measurements. So, by measuring fluctuations of the Faraday rotation, we could, in principle, measure fluctuations of magnetization of the system or what is called 'spin noise'.

3. Experimental geometries

The first idea was to measure fluctuations of longitudinal magnetization (Fig. 2a), which may, in principle, give important information about longitudinal relaxation time T_1 and about its dependence on temperature, magnetic field and other external perturbations. Information of this kind was usually obtained from the EPR data. This experimental approach was, in fact, close to the paramagnetic relaxation technique employed by the Dutch physicist Gorter, before the advent of magnetic resonance, for measuring magnetic susceptibility of paramagnets in parallel fields [4]. The essential difference between these two approaches was that in our case the system was not supposed to be perturbed by the external oscillating field.

After a short discussion, however, we came to conclusion that this idea was not good, at least, from the methodical viewpoint, because it implied measuring the "noise signal" at low frequencies where the measurements are usually hampered by many other sources of noises, and the peak of the noise power density at zero frequency should be observed anyway. So, more reasonable was to measure *transverse* fluctuations of magnetization. In this geometry (the so-called Voigt configuration, Fig. 2b), a random transverse fluctuation of the magnetization should precess around the magnetic field during a time interval of the order of T_2 (dephasing time or transverse relaxation time) and then be replaced by another random realization of the transverse magnetization. The linearly polarized probe light, in this geometry, should experience preferential oscillations of the polarization plane azimuth at the frequency of Larmor precession of the spin. In other words, in this configuration, we expected to observe a peak, in the Faraday rotation noise spectrum, at the frequency of magnetic resonance and, thus, to obtain the magnetic resonance spectrum of the unperturbed system.

4. Problems of sensitivity

The main requirements to experimental arrangement for such measurements, dictated by the necessity to achieve the greatest polarimetric sensitivity, implied

1) maximizing the probe beam intensity (to reduce relative level of the shot noise, proportional to $N^{-1/2}$, where N is the number of photons);
2) minimizing the level of excess light-intensity noise;
3) minimizing the total number of spins contributing to the Faraday rotation (to increase relative value of the fluctuation). This requirement strongly differs from what is usually needed in measurements of susceptibility, when signal increases with the size of the system;
4) maximizing specific Faraday rotation (per spin);
5) maximizing total Faraday rotation for the completely aligned spin system.

The last item follows from the fact that the "noise signal" proportional to $n^{-1/2}$ (n is the number of spins) is calculated as a relevant fraction of *total rotation*, i.e., of the rotation of completely oriented spin-system. One can easily see that the total rotation may substantially differ for samples with the same number of spins (n), but with different shape (Fig.3). Therefore, to maximize the "noise signal" for a fixed number of spins, the sample volume probed by the light should be maximized in length and minimized in cross section or, in other words, the light beam passing through the sample should be tightly focused.

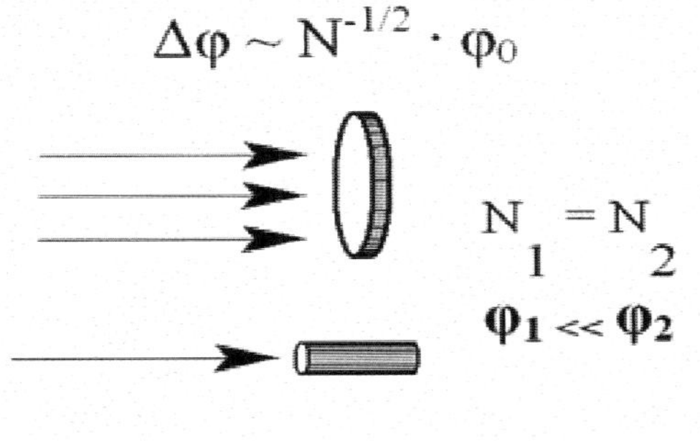

Fig. 3. Shape-controlled spin noise signal. Magnitude of the signal ($\Delta\phi$) depends not only on the factor $N^{-1/2}$, but also on the value of total Faraday rotation provided by completely aligned spin system ϕ_0. Therefore, the probing light beam should be preferentially tightly focused inside the sample.

It is noteworthy that, at that stage, we seriously discussed the possibility of using thermal source (incandescent lamp) for detecting these small polarization fluctuations. The point was that the intensity fluctuations of laser sources, at that time, exceeded the shot noise level by 4 - 6 orders of magnitude, and, as a result, the laser sources, with all their unquestionable merits, seemed unsuitable for ultimately sensitive polarimetric measurements. Still, these two types of the light sources were so incomparable in their brightness, that we stopped at lasers.

The problem of excess noise suppression for laser sources was solved using several expedients [5], the most substantial of them being balanced detector.

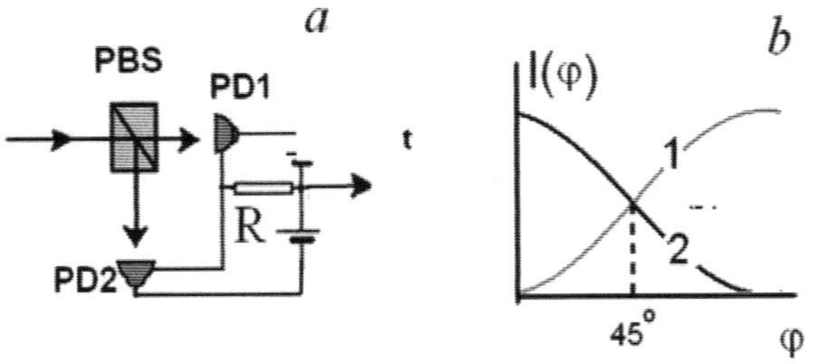

Fig. 4. Schematic of the balanced detector (a) and dependence of photocurrents of the two photodetectors on azimuth of the polarization plane of the incident light.

5. Balanced detector

The detection unit of the polarimetric system consisted of a polarization beamsplitter with its two outputs coupled to two photodetectors included into a differential circuit so that their photocurrents were subtracted at the load resistor (Fig. 4). When the plane of polarization of the incident light made an angle of 45^0 with polarizing directions of the beamsplitter, photocurrents of the two photodetecctors were cancelled at the load resistor, while the signals resulted from oscillations of the incident beam polarization plane rotation, with opposite phases on the detectors, were summed up. In this way, we have managed to suppress the excess noise by three orders of magnitude and to achieve the shot-noise limit of polarimetric sensitivity lying, for the laser power of several hundreds mW, in the range of 10^{-8} rad (see [5] and references therein). In those experiments, the balanced detector, which is now produced commercially, was applied to polarimetric measurements, to the best of our knowledge, for the first time.

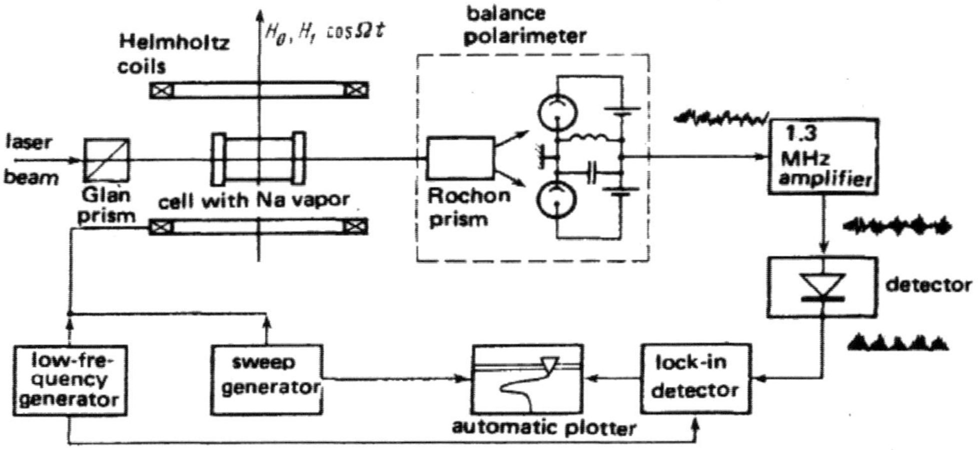

Fig. 5. Schematic of the experimental setup (from [1]) used for detecting magnetic resonance in the Faraday rotation noise spectrum of sodium atoms.

6. The first experiment

The experiment on detection of magnetic resonance in the Faraday rotation noise spectrum [1] was performed on sodium atoms at a pressure of ~ 10 Torr. Schematic of the experimental setup is shown in Fig. 5. As a light source, we used a cw dye laser tuned to close vicinity of the D1 or D2 line of Na. The cell with sodium vapor was placed into a transverse magnetic field created by a system of Helmholtz coils. The field was slightly modulated at a frequency of ~ 10^2 Hz. The output signal of the balanced detector was selectively amplified at a frequency of 1.3 MHz, then quadratically detected, lock-in amplified at the frequency of field modulation, and recorded as a function of the applied magnetic field. Figure 6 shows two typical plots of the spin-noise resonance recorded with a time constant of 2 s. The probe beam was tuned to the edge of D1 line or to the midpoint between the two lines (for more details, see Ref. [1]).

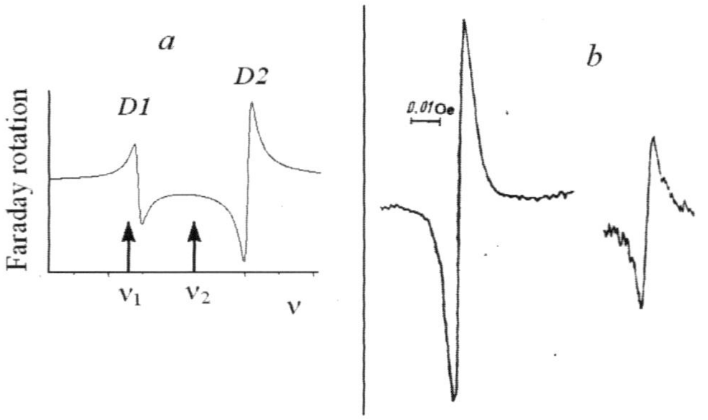

Fig. 6. Schematic of the Faraday rotation spectrum of sodium atoms and spectral positions of the probe beam (a) for the two experimental plots of the EPR signal in the Faraday rotation noise (b).

7. Perturbative or not?

One of the questions that arose after successful realization of the experiment was whether this technique could be considered nonperturbative or not. At first glance, it looked absolutely passive and nonperturbative because it was based on detecting Faraday rotation in the transparency region, and the probe beam, as it seemed, did not touch, in any way, the atoms under study. But, on the other hand, it could not be nonperturbative for evident reasons: The initially monochromatic light, after passing through the spin system, becomes modulated. Therefore, it acquires sidebands with shifted frequencies. It means that some photons of the beam experience inelastic scattering, which implies energy exchange between the light and spin-system. To resolve this contradiction, we addressed our famous theorist V. Perel, who immediately accepted this problem for consideration and very soon, in coauthorship with Gorbovitskii, has published the paper [11], where it has been shown that the noise of the Faraday rotation at the frequency of Larmor precession can be considered as a result of coherent forward Raman scattering of the incident laser beam. So, this process can be regarded, to a considerable extent, nonperturbative, because the probe light interacts with macroscopic fluctuations of the system, and does not select any particular spins to flip. The technique becomes more and more perturbative as we pass from macroscopic ensembles to microscopic ones.

8. Nuclear spin noise

Soon after our work on detection of spin-noise resonance in atomic vapor, an ideologically the same experiment has been performed on a nuclear system [7]. As a spin system, in this experiment, was used an ensemble of nuclei ^{35}Cl in a $NaClO_3$ crystal. The experiment was performed at 4 K, and the magnetization noise was detected by means of SQUID. This was the first experiment that demonstrated resonant spin noise predicted by F. Bloch on a nuclear system. Here, the number of spins contributing to the signal was much larger than in our experiment with electron spins, and the sensitivity of the detection system was much lower. As a result, the peak of spin noise at the frequency of nuclear quadrupole resonance of ^{35}Cl was detected with the signal-to-noise ratio of a few units for the accumulation time of 7 hrs. Still, further technical advancement in this area made it possible to observe the nuclear spin noise at room temperature [8, 9] and even to develop, on its basis, the method of nuclear spin noise imaging [10].

9. Primary development of SNS

For a decade or so, our publication remained practically unnoticed. The Faraday-rotation noise spectroscopy as a new technique of ESR detection was appreciated at the end of last century, first of all, by a group of researchers from Los Alamos National Laboratory [11], who reproduced the experiments with alkali atoms and announced the spin noise spectroscopy as an efficient probe of spin dynamics and magnetic resonance. In [12], it was shown, in particular, that magnetic resonance in the Faraday rotation noise spectrum can be also detected in the longitudinal magnetic field. This may occur for the ESR transitions allowed in the ac magnetic field oscillating along the dc magnetic field. This result can be easily comprehended in the context of the fluctuation-dissipation theorem, which establishes direct relationship between the spectra of susceptibility and fluctuations (in this case, both with respect to longitudinal magnetization).

A highly important step in development of SNS was made by Oestreich et al. [13] who have managed to successfully apply this technique to semiconductors. At this stage, these experiments demonstrated more a fundamental applicability of this experimental approach to semiconductors rather than its practical importance, because the signal became visible after its accumulation for 20 hrs.

10. A breakthrough in SNS – FFT spectrum analyzer

A real breakthrough in this field of research has happened when, in the system of data acquisition, the sweeping spectrum analyzer was replaced by the one with the real-time fast Fourier transform processing system [14]. The conventional sweeping spectrum analyzer, as is known, measures the noise only in a very narrow (resolved) frequency interval at a time, thus disregarding most of the information coming to the photodetectors. The FFT spectrum analyzer, on the contrary, digitizes the signal in the whole operation bandwidth of the system (with a sampling rate of around 10^9), performs FFT of the signal in real time, and averages the spectrum thus obtained. As a result, the whole bulk of information contained in the signal is being processed and stored.

This technical advance allowed one to increase sensitivity of SNS by a few orders of magnitude and made it suitable, in particular, for studying spin dynamics in low-dimensional semiconductor systems (quantum wells, quantum dots, etc.) highly important for the present-day technological applications.

11. GHZ and THZ spin noise spectroscopy

An interesting idea was proposed in [15] to overcome limitations associated with the bandwidth of optical detectors. For that purpose, it was proposed to use a pulsed laser with a high repetition rate instead of a cw laser, as a source of probe light. In this case, we can observe, in the spectrum of the detected signal, not only the peak at the frequency of magnetic resonance f_R, but also the peaks at the frequencies shifted from f_R towards lower frequency by multiples of the repetition rate f_0. As a result, the frequency of the spin noise resonance under study can be transferred to the frequency range $[0, f_0]$. This experimental approach allowed the authors of [15] to detect spin noise at Larmor frequencies up to 16 GHz. Under these conditions, the total bandwidth of the spectrum analyzer and of the photodetectors may not exceed the repetition rate f_0. At the same time, the measured spin dephasing times are limited, in this technique, by approximately half of the repetition rate. It is important that application of this technique does not reduce sensitivity of SNS as compared to its conventional version.

Another remarkable approach to solution of the problem of widening of the bandwidth of the Faraday-rotation-based SNS was proposed in [16]. The authors of this idea proposed to use, for this purpose, an extremely broad bandwidth of optical femtosecond pulses and to measure correlation characteristics of the magnetization of the medium with a very high time resolution. For that purpose, the probe beam was proposed to form from pairs of ultrashort femtosecond pulses (not resolved by the detection system) and to measure the Faraday rotation noise power as a function of time delay between the pulses. This technique allows one to directly detect decay of the transverse magnetization in time domain and to expand the accessible frequency range up to several terahertz.

12. Cavity enhancement of spin noise

In [17], it was shown that the Fabry-Perot cavity may be highly sensitive to modulation of its intracavity anisotropy at frequencies multiple of its intermode spacing. This effect can be easily understood in terms of light pulse traveling over the cavity back and forth and has much in common with the mode-locking effect in lasers. Figure 8 shows frequency dependence of the polarization modulation gain factor for different positions of the anisotropic element inside the cavity (for more detail, see [17]). This effect can evidently be used in SNS for selective amplification of the polarization modulation signal. Amplification, in this case, is performed in optical (rather than electronic) channel, with the gain factor being approximately equal to Q-value of the cavity ($\sim 10^2 - 10^3$).

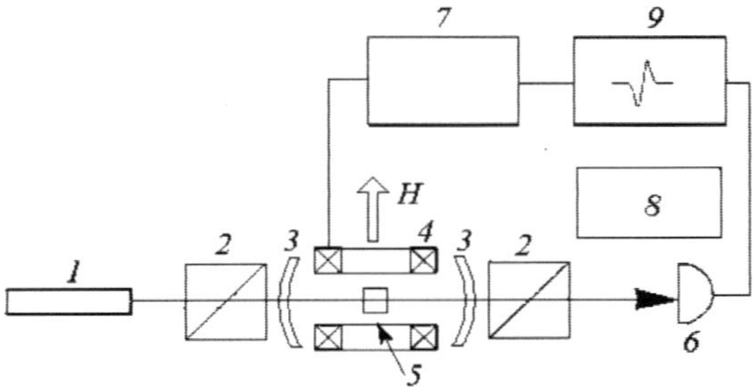

Fig. 7. Schematic of the all-optical spin noise spectrometer: (*1*) laser, (*2*) polarizers, (*3*) mirrors of the optical cavity, (*4*) magnetic coils, (*5*) sample, (*6*) photodetector, (*7*) magnetic field sweeping unit, (*8*) magnetic field modulation unit, and (*9*) detection system.

More interesting, however, is to use this effect for creating an all-optical spin-noise spectrometer [17]. The idea of this proposal is as follows (Fig. 8). Let the sample under study be placed inside an optical cavity. A linearly polarized laser beam incident on the cavity hits one of its longitudinal modes and thus passes through the cavity with no loss. At the exit of the cavity, we place a linear polarizer crossed with the polarizer at the entrance, so that no light can passes through the system. Now, if we start sweeping the transverse magnetic field applied to the sample, then, at the moment when Larmor precession frequency of the spins becomes equal to (or multiple of) the intermode frequency of the cavity, oscillations of the polarization plane at the exit of the cavity appear to be strongly enhanced, and the light starts to pass through the output polarizer. In this case, it does not matter how high is the frequency of these oscillations. We detect only dc component of the transmitted light (or modulated light intensity if the transverse field is modulated). A unique property of this arrangement is that it allows one to detect magnetic resonance in the Faraday rotation noise spectrum at any frequency without any broadband electronics and broadband photodetectors. What we observe is just a result of mutual resonance of the spin system and Fabry-Perot cavity. A serious drawback of this proposal is that the transmitted light intensity, at small rotation angles, is quadratic in angle and, for small angles of rotation, becomes negligibly small.

13. Conclusions

At present, due to a number of technical advances, spin noise spectroscopy turns into a standard and fairly unique tool of magneto-resonance investigations. With the advent of FFT spectrum analyzers into arsenal of technical means of SNS, a considerable portion of labor has been reload on the shoulders of electronics, and the information of a rather sophisticated nature became fairly chip. The absence of any HF oscillators allows one not only to avoid perturbation of the system, but also to substantially simplify technical structure of the setup and to make it much more convenient in handling. High sensitivity of this technique in combination with high spatial resolution makes it unique for studying spin systems in micro- and nanostructures. A sort of 3D spin-noise tomography has been demonstrated in [18], where the medium was probed by the waist of a tightly focused laser beam, which provided the main amount of the noise signal. In [19], it was shown that the time-of-flight broadening may be highly important in SNS of semiconductor materials. Spin noise spectroscopy is now successfully applied for studying electron statistics and evaluating the degree of electron localization [20, 21].

The main research on SNS is now concentrated around semiconductor systems of different morphology, whose spin-related properties are considered to be highly promising for applications in photonics and information science. Still, SNS also finds application in studying, along with diluted paramagnets, spin glasses and ferromagnetic fluids. As a nonperturbative technique, it is considered highly efficient in studies of ultracold atoms. In spite of close connection between the fluctuation and susceptibility spectra, formulated by the fluctuation-dissipation theorem, their information capacities are not identical. In many cases it may appear important that the measurements of susceptibility imply coherent perturbation of the system, while fluctuations are caused by its random incoherent thermal motion.

We did not mean, by any means, to give here a comprehensive review of this topic. We just tried to show that the idea of detecting magnetic resonance in the Faraday-rotation noise spectrum, that looked, at the beginning, as an academic trick, useful primarily for tutorial purposes, turned into a highly efficient practical tool of magneto-spectroscopic research.

References

[1] Aleksandrov E B and Zapasskii V S 1981 *Sov. Phys. JETP* **54** 64
[2] Aleksandrov E B, Golunev Yu M, Lomakin A V, and Noskin V A 1983 *Sov. Phys. – Usp.* **26** (8) 643
[3] Happer W *Rev. Mod. Phys.* 1972 **44** 169
[4] Gorter C J 1946 *Paramagnetic Relaxation* (Leyden)
[5] Zapasskii V S *J. Appl. Spectrosc.* **37** (2) 857
[6] Gorbovitskii B M and Perel V I 1983 *Opt. Spectrosc,* **54** (3) 229
[7] Sleator T, Hahn E L Hilbert C and Clarke J 1985 *Phys. Rev. Lett.* **55** (17) 1742; 1987 *Phys. Rev.* B **36** (4) 1969
[8] McCoy M A and Ernst R R 1989 *Chem. Phys. Lett.* **159** 587
[9] Guéron M and Leroy J L 1989 *J. Magn. Reson.* **85** 209
[10] Müller N and Jerschow A 2006 *Proc Natl Acad Sci U S* A. **103** (18) 6790
[11] Crooker S A, Rickel D G, Balatsky A V and Smith D 2004 *Nature* **431** 49
[12] Mihaila B, Crooker S A, Rickel D G, Blagoev K B, Littlewood P B and Smith D L 2006 *Phys. Rev.* A **74** 043819
[13] Oestreich M, Romer M, Haug R G and Hagele D 2005 *Pys. Rev. Lett.* **95** 216603
[14] Römer M, Hubner J H, Oestreich M 2007 *Rev. Sci. Instrum.* **78** 103903
[15] Müller J M, Römer M, Hübner J and Oestreich M 2010 *Phys. Rev.* B **81** 121202 (R)
[16] Starosielec S and Hägele D 2008 *Appl. Phys. Lett.* **93** 051116
[17] Zapasskii V S and Prshibelskii S G 2011 *Opt. Spectrosc.* **110** (6) 917-924
[18] Römer M, Höbner J H, Oestreich M 2009 *Appl. Phys. Lett.* **94** 112105
[19] Möller J M, Römer M, Schuh D, Wegscheider W, Höbner J H abd Oestreich M 2008 *Phys. Rev. Lett.* **101** 206601
[20] Crooker S A, Cheng L, Smith D L 2009 *Phys. Rev.* **B 79,** 035208
[21] Römer M, Bernien H, Möller J M, Schuh D L, Höbner J and Oestreich M, 2010 *Phys. Rev.* **B 81** 075216

Journal of Physics: Conference Series **324** (2011) 012003 doi:10.1088/1742-6596/324/1/012003

Dynamic nuclear polarization: Yesterday, today, and tomorrow

V A Atsarkin

Kotel'nikov Institute of Radio Engineering and Electronics RAS, 11 Mokhovaya Str., 125009 Moscow, Russia

e-mail: atsarkin@mail.cplire.ru

Abstract. A review is given of the history and basic principles of dynamic nuclear polarization (DNP) in condensed matter, starting from the Overhauser pioneering work and up to the modern applications in chemistry and medicine. Different physical mechanisms of DNP, such as the Overhauser effect, solid effect, electron-nuclear cross relaxation, and thermal mixing are considered. Further prospects of the DNP techniques for future are discussed.

Dynamic nuclear polarization (DNP) consists in transferring high spin polarization from the electrons of paramagnetic impurities to the host nuclei by means of microwave irradiation close to EPR frequency. This technique, firstly suggested for condensed matter by A. Overhauser in 1953 [1], became a powerful method of modern science, being applicable to a wide variety of urgent problems such as polarized targets for particle physics, giant increase in NMR sensitivity, spin thermodynamics in solids, *etc*. Keeping in mind the memorial type of this Conference dedicated to the Centinary of Professor S. A. Altshuler, this report is focused primarily on historical aspects, beginning from the Overhauser's discovery and up to contemporary "new Renaissance". Main attention will be paid to various physical mechanisms of DNP, and especially to the role playing by spin-spin interactions. Some traditional and novel applications will be touched as well.

The restricted volume of this article dictates laconic style. In fact, we do not pretend to any exhaustive review with a vast list of references. Instead, a conceptual sketch is proposed, including the main milestones only, with citing preferably to books, review articles, and the most principal papers.

Let us start with definitions. Consider a solid sample containing host nuclei with nuclear spins I and paramagnetic impurities with electron spins S. Spin polarization $P_{I,S} = <\hat{I}^z, \hat{S}^z> /(I, S)$ is defined as a fraction of spins (nuclear or electronic) directed along the external magnetic field $\boldsymbol{H} \| \mathbf{z}$. For spins ½, the polarization values at the equilibrium temperature T_0 are

$$P_{I,S}^0 = \tanh\left(\frac{\hbar\gamma_{I,S} H}{2k_B T_0}\right) \qquad (1)$$

where $\gamma_{I,S}$ are the nuclear and electronic magnetogyric ratios, respectively, and common notations for Planck and Boltzmann constants are used. One can see that, even at high magnetic field of 5 T and very low temperature T_0 = 0.5 K, the equilibrium polarization of protons is merely about 0.01, whereas $P_S^0 \approx -1$.

As mentioned above, the DNP techniques enables one to transfer high polarization from electrons to nuclei. To do this, Albert Overhauser suggested [1] to saturate the EPR line with resonant microwaves and employ specific relaxation processes with simultaneous flips of electronic and nuclear spins in opposite directions. Such relaxation can be governed by the scalar (contact) hyperfine interaction of the $A(\boldsymbol{I}\cdot\boldsymbol{S})$ type. To compensate the energy mismatch, the correlation function should have considerable spectral density $J(\omega_S + \omega_I) \sim J(\omega_S)$, where $\omega_S=-\gamma_S H$ and $\omega_I=\gamma_I H$ are, respectively, the EPR and NMR frequencies. Since the ω_S value is commonly high enough, the Overhauser Effect (OE) is characteristic for systems with fast motion of electron spins, such as metals and materials with strong S-S exchange. The enhancement of nuclear polarization, $E = P_I/P_I^0$, was predicted to amount ω_S/ω_I (at maximum), that is about 660 for protons.

The Overhauser's idea was soon confirmed by T. P. Carver and C. P. Slichter who performed DNP experiments on Li metal [2]. A bit later, A. Abragam suggested another version of this effect employing dipole-dipole electron-nuclear interactions [3]. This kind of OE works in liquids and provides the maximum enhancement of $-\omega_S/2\omega_I$. The Overhauser mechanism of DNP was also applied successfully to the nucleus-nucleus spin-spin interaction; this version, known as Nuclear Overhauser Effect (NOE), is widely used in the NMR spectroscopy [4].

The next step in the DNP history is related to the work by Abragam and Proctor published in 1958 [5] (see also the famous Abragam's book [6]). Instead of employing specific electron-nuclear relaxation, they proposed to use the electron-nuclear transitions induced by microwave pumping at the frequencies $\omega_S \pm \omega_I$. Formally, these "satellite" transitions are forbidden, but become slightly allowed due to the dipole-dipole interactions between paramagnetic center and neighboring nuclei, owing to the operators like $\hat{I}^{\pm}\hat{S}^{z}$. The authors [5] named this mechanism "Solid Effect" (SE) because of its applicability mostly to non-conducting solids. The maximum enhancement of nuclear polarization due to SE amounts to $\pm\,\omega_S/\omega_I$, the sign being dependent on the hyperfine satellite chosen for saturation. An essential condition for successful SE experiment is nuclear spin diffusion [7] which provides the spreading of enhanced polarization from the nearest neighbors of paramagnetic centers throughout the sample volume. Besides, the electron spin-lattice relaxation time T_{1S} should be short enough (much less than the nuclear one, T_{1I}), in order to provide numerous electron-nuclear flips *via* every paramagnetic center.

The SE is most effective when the EPR line-width δ_S is much less than ω_I, so the "forbidden" satellites are well resolved. A sketch of typical $E(\omega_p)$ dependence for such a case is shown in figure 1.

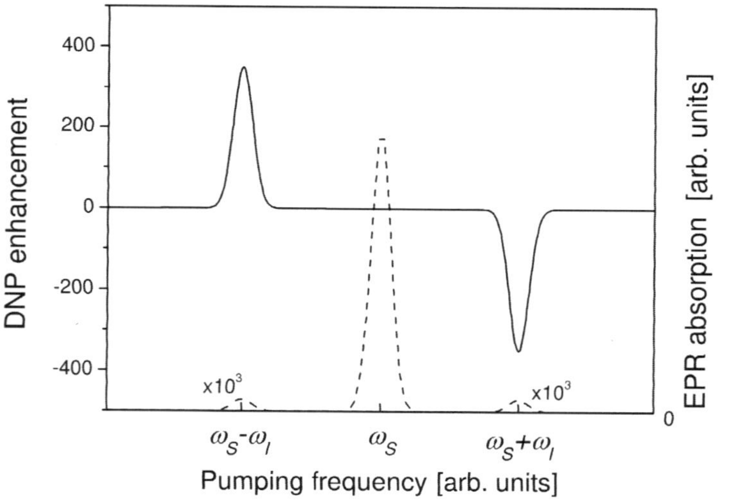

Figure 1. Schematic of the well-resolved Solid Effect. The DNP enhancement (E) as a function on pumping frequency (solid curve, left scale) and the EPR absorption spectrum with the central and satellite lines (dashed curve, right scale).

Numerous results of this type were obtained on various single crystals, such as $La_2Mg_3(NO_3)_{12}\cdot 24H_2O$ (LMN) doped with Nd^{3+} [8] and others, where the EPR line was suffisiently narrow. The SE was thoroughly studied worldwide and successfully applied in building the first high-polarized proton targets working at liquid helium temperatures in magnetic fields up to 2.5 T. An excellent description of the DNP physics and applications as of the early 60-s was presented by C. D. Jeffries; his monograph [9] became a handbook for the DNP community.

At this stage, however, some difficulties arose. To provide high enough concentration of protons in a polarized target, it was necessary to turn from single crystals, such as LMN, to some frozen alcohols, like glycerol, ethylene glycole, *etc*. The EPR lines of stable radicals or paramagnetic ions in such systems are inhomogeneously broadened due to the anisotropy of g-factor. The electron dipole-dipole interactions between paramagnetic centers also contributes to the EPR line width and, in any case, leads to effective spectral diffusion (electron spin-spin cross relaxation) throughout the line. As a result, the "forbidden" satellites overlap with each other and with the central (allowed) EPR transition, and cannot be saturated separately. At the first glance, this should strongly diminish or even cancel the DNP effect.

As is often the case, the solution was found owing to the same factor which created the problem, namely, the spin-spin interaction. In 1963, A. V. Kessenikh and A. A. Manenkov suggested the 3-spin Cross Effect (CE) [10], involving simultaneous flips of two electronic and one nuclear spins, see figure 2. Such a process is caused by the dipole-dipole operators like $\hat{S}_a^{\pm}\hat{S}_b^{\mp}I^{\pm}$ and requires the energy matching,

$$\omega_S^a - \omega_S^b = \omega_I \tag{2}$$

so the paramagnetic species *a* and *b* should have different g-factors. It should be emphasized that, in contrast to the Solid Effect, now it is sufficient to saturate an *allowed* (not *forbidden*) transition corresponding to one or another electron spin species, the resulting nuclear polarization being equal to P_S^a-P_S^b. In particular, equation (2) can be applied to two spin packets separated by the ω_I distance inside the inhomogeneously broadened EPR line with the width $\delta_S^{\mathrm{inh}} > \omega_I$. In this case, the 3-spin Cross Effect becomes much more effective than SE.

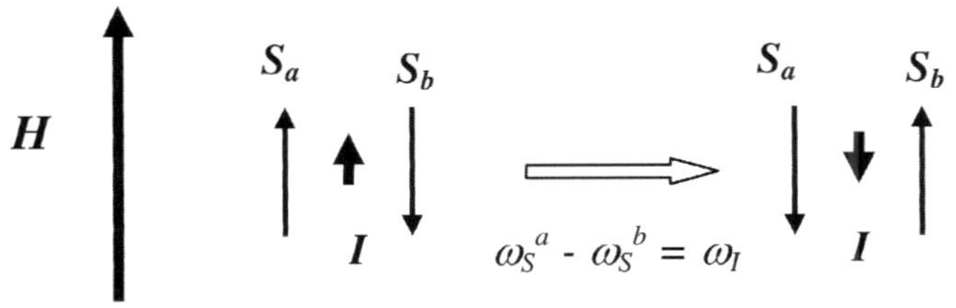

Figure 2. An elementary act of the 3-spin electron-nuclear cross relaxation.

Some experimental evidences for the 3-spin CE, as well as more detailed theoretical discussion were published by various authors [11]. However, this was the first step only. Evidently, a lot of spin packets in the inhomogeneous EPR line do not satisfy the condition (2). To take into account all spin-spin processes, including the ordinary electron spin-spin cross relaxation, a strict many-particle theory was needed, based on quantum statistics and spin thermodynamics.

Such theory was advanced by B. N. Provotorov in 1961 [12]. According to the Provotorov's equations, saturation of magnetic resonance with some detuning from the central frequency results in strong cooling of the quasi-equilibrium subsystem formed by the secular part of the spin-spin (mostly, dipole-dipole) interactions. This so-called spin-spin reservoir has a quasi-continuous spectrum in the range of the mean local (dipolar) frequency and can be characterized by its own temperature which, generally, differs from the ordinary Zeeman spin temperature and can be either positive or negative, depending of the saturation detuning. In real samples, the NMR frequency ω_I often falls into the range of the characteristic local frequency ω_{SS} of the electron spin-spin reservoir (SSR). This coincidence, together with the electron-nuclear dipole-dipole interactions, leads to the thermal mixing (TM), that is, to equalizing the SSR temperature T_{SS} with nuclear Zeeman temperature T_{IZ} and thus to the corresponding increase in nuclear polarization $P_I = \tanh(\hbar\omega_I / 2k_B T_{IZ})$. This process is illustrated in figure 3, where the energy spectrum of the SSR is represented as quasi-continuous bands which broaden the Zeeman sublevels of a paramagnetic center.

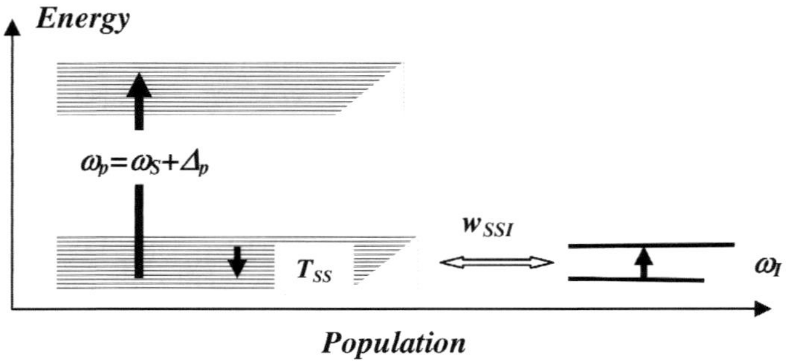

Figure 3. Diagram of the TM DNP mechanism (for the case of $T_{SS} < 0$). Horizontal segments reflect populations of corresponding energy levels in the electronic (left) and nuclear (right) systems. The arrows show transitions induced by the microwave pumping and I-S dipolar interactions.

As was shown by Provotorov [13], a strong decrease in $|T_{SS}|$ can be also achieved by saturation one of the EPR lines (a) coupled with the other one (b) with dipole-dipole cross-relaxation, the energy difference $\hbar(\omega^a - \omega^b)$ being transrerred to (from) the SSR. This approach is fully applicable to the inhomogeneously broadened EPR line, when the cross-relaxation between spin packets is faster than spin-lattice relaxation and so leads to effective spectral diffusion throughout the line. Thus, this case is somewhat similar to homogeneous dipolar broadening.

The TM mechanism of DNP was firstly suggested by Kozhushner and Provotorov in 1964 [14]. Further development of the theory was performed by Buishvili [15], Borghini [16], Kozhushner [17] and others. Kozhushner [17] named this mechanism "Dynamical Cooling", whereas in the West the term TM is commonly used. Systematic experiments dedicated to clearing up the role playing by the electron spin-spin temperature in DNP (and, generally, in EPR in solids) were carried out in the Moscow, Leiden and Saclay groups, see the review articles [18-22] and monographs [23, 24].

To get a deeper insight into the physical nature of the TM mechanism, let us consider the elementary act of thermal mixing, see figure 4. Here the flip-flop of two electron spins (S_2 and S_3) is accompanied by simultaneous flip of a nuclear spin I. Such a process is governed by the dipole-dipole operators like $\hat{S}_2^{\pm}\hat{S}_3^{\mp}\hat{I}^{\pm}$, the energy matching being provided by the equality $\omega_I=\Delta_{23}$, where Δ_{23}/γ_S is the difference of the local fields created on the S_2 and S_3 spins by one more electron spin S_1. Extending to the many-spin system, one can say that a portion of the SSR energy is transferred to the nuclear Zeeman subsystem, and *vice versa*. Thus, the thermal mixing can be considered as generalization of the 3-spin CE to real many-particle spin-spin interactions.

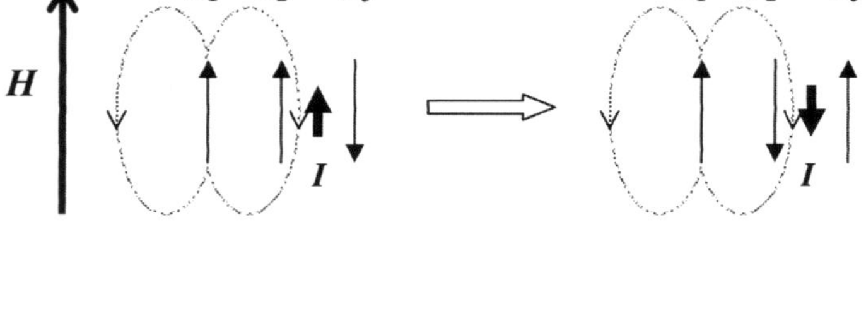

Figure 4. Elementary act of thermal mixing: simultaneous flips of the nuclear spin I and electron spins S_2, S_3 in the local field of spin S_1 (the flux produced by S_1 is shown with dotted lines).

Journal of Physics: Conference Series **324** (2011) 012003 doi:10.1088/1742-6596/324/1/012003

The spin temperature approach enables one to calculate the DNP enhancement by the method of thermal balance. The most convenient way is to consider the problem in the rotating reference frame (RF) which rotates about z axis at the frequency of microwave pumping, $\omega_p = \omega_S + \Delta_p$, in the sense of the electronic Larmor precession [25]. The conversion to the RF is realized by means of the unitary operator $\hat{U} = \exp(i\omega_p \hat{S}^z t)$ which acts on the electron Zeeman subsystem (SZ) only, the other ones being unaffected. In the RF, the frequency and, consequently, the spin temperature of the electron Zeeman system are multiplied by the factor of $-\Delta_p/\omega_S$. As a result, the corresponding equilibrium temperature in the RF is as low as

$$T_0^{RF} = -\frac{\Delta_p}{\omega_S} T_0 \tag{3}$$

The corresponding block diagram is shown in figure 5.

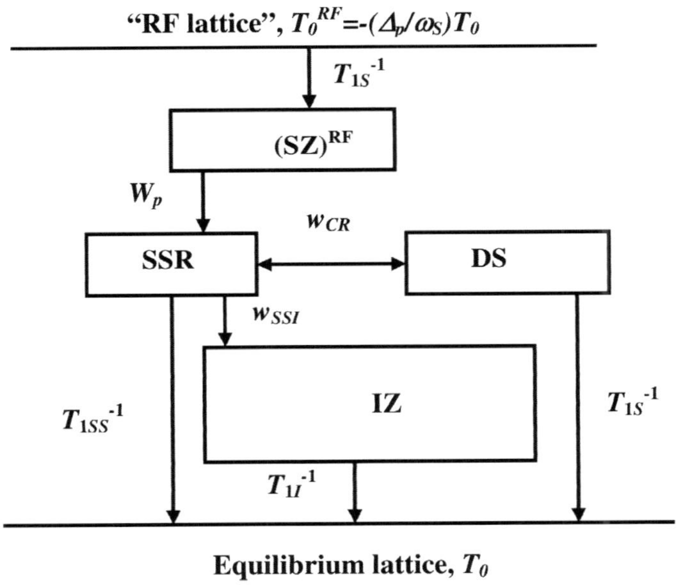

Figure 5. Thermal balance diagram of the Thermal Mixing DNP mechanism. The rotating reference frame for electron spins is used. The difference system (DS) is added in the case of inhomogeneous broadening of the EPR spectrum, see the text.

In the RF, there are two "lattice thermostates", one of them (strongly cooled) being coupled to SZ^{RF} with the relaxation rate T_{1S}^{-1}, and another (at the equilibrium temperature T_0) connected to the nuclear Zeeman system (IZ) and SSR with corresponding spin-lattice relaxation rates, T_{1I}^{-1} and T_{1SS}^{-1}. The coupling between SZ and SSR is provided by the microwave pumping with the rate W_p of the resonant microwave absorption, and the thermal mixing between SSR and IZ occurs as discussed above with the rate w_{SSI}. In the case of inhomogeneous EPR broadening with fast spectral diffusion, one more subsystem, the "difference" one (DS) should be added and coupled to the SSR with the rate w_{CR} of the SS cross-relaxation. Using these coupling rates and corresponding thermal capacities determined from characteristic frequencies and spin concentrations of the subsystems involved [23-25], one can calculate the steady state enhancement of nuclear polarization. In the high temperature approximation ($P_{S,I}$<<1) one gets:

$$E = \frac{-\omega_s \Delta_p}{\Delta_p^2 + M_2^{inh} + a\omega_{SS} + f\omega_I^2} \tag{4}$$

where M_2^{inh} is the second moment of the inhomogeneous EPR line; $a = T_{1S}/T_{1SS}$; and f is the "leakage factor" accounting for extrinsic nuclear spin-lattice relaxation. According to equation (4), the maximum positive and negative enhancements are attainable at

$$\Delta_p^{\max} = \mp\left(M_2^{inh} + a\omega_{SS} + f\omega_I^2\right)^{1/2} \tag{5}$$

and amount to

$$E_{\max} = \frac{\omega_S}{2\Delta_p^{\max}} \tag{6}$$

When the SS cross-relaxation (spectral diffusion) inside the EPR spectrum is less effective, the microwave pumping produces a "burned hole" in the line contour. Corresponding calculations show [26] that, in the limiting case of very narrow "hole", the $E(\Delta_p)$ dependence resembles the derivative of the EPR absorption and thus can be asymmetric in some cases. Similar results were obtained for the 3-spin CE as well [21, 23].

To check the validity of the TM mechanism experimentally, two methods can be used. First, one can compare directly T_{SS} with T_{IZ}. This approach was used in [27], where the *SS* temperature of Cr^{3+} paramagnetic ions in the Al_2O_3 crystal was determined from the asymmetric shape of the saturated EPR line and found to be equal to T_{IZ} of the ^{27}Al nuclei polarized at T_0=1.8 K and $\omega_p/2\pi$ =9.5 GHz. To our knowledge, this was the first experimental confirmation of the TM DNP mechanism. Later, similar evidences were demonstrated on various materials [28-30, 21, 22].

Another method consists in comparison of spin temperatures of different nuclei present in the same sample upon DNP. Such experiments were performed, for example, on ^{7}Li and ^{19}F nuclei in LiF crystals doped with paramagnetic defects (T_0=0.74 K; H = 2.5 T) [31]. The Zeeman spin temperatures of different nuclei were found to be equal at the steady state, evidencing for the thermal mixing with SSR. Other examples can be found in references [18, 21, 23, 24].

Thus, to the end of the 20th Century, the thermal mixing was recognized as an efficient DNP mechanism working at low (helium) temperatures in materials used in polarized nuclear targets, as well as in nuclear ferro- and antiferromagnets elaborated by A. Abragam, M. Goldman and their collaborators [24]. By that time, the DNP studies seemed to be a rather completed field, not very attractive for further development. However, a sudden burst happened just at the start of the new Millennium.

Now, the main goal is a giant increase in NMR sensitivity combined with the most modern NMR techniques, such as magic-angle spinning (MAS), coherent multi-pulse methods, cross-polarization, 2D spectra, and so on. The success was achieved owing to new technical options, such as extremely high magnetic fields provided by superconducting magnets and, especially, powerful microwave sources in the short millimeter range, named gyrotrons, which supply up to 5 W c. w. in the range of 100-600 GHz. As a result, the DNP experiments became compatible with a rotor for magic-angle spinning and other modern NMR equipment. Of particular value is also the possibility of the EPR saturation not only at liquid helium, but at liquid nitrogen and even room temperatures as well.

Though the first attempts to combine the MAS NMR with DNP were realized as early as 1985 [32], the most successful DNP experiments in high magnetic fields were performed only recently by the group of R. G. Griffin, one of the leaders of the "new wave". Since the author of this presentation did not participate in the recent DNP research (except for minor consulting and discussion), the consideration will be restricted to several imressive examples. More detailed information can be found in the review articles [33, 34] and references therein, as well as in the special issues of the APMR and PCCP journals where these articles were published.

The most striking idea realized in recent years is the technique of temperature jump, named also the DNP-dissolution or hyperpolarization method. This technique, proposed firstly by Ardenkjaer-Larsen *et al.* [35], consists of three steps. First, high nuclear polarization is achieved by DNP at very low temperatures and high magnetic field. Then the polarized sample is extracted momentary from the cryostate and melted with superheated water (or, alternatively, by *in situ* laser irradiation [36]). And, finally, the liquid sample can be transported quickly, by pressured gas, to the MRI system, where the polarized solution is injected into the patient body. Surprisingly, the enhanced nuclear polarization remains nearly the same during the whole procedure, with the loss of 20-25% only. As a result, the effective enhancement relative to the equilibrium polarization at room temperature amounts to 50000, with the relaxation rate of the order of 1 minute. This allows to monitor the metabolism processes *in vivo*, in particular, in cancer treatment.

Another line of inquiry relates to the combination of DNP with MAS NMR. A typical result obtained at 90 K and 5 T was presented by van der Wel et al. [37]. The object of study was a nano-crystalline peptide fragment; the sample was immersed in frozen alcohol doped with free radicals. The enhancement of proton polarization, $E\approx150$, acquired in the polarizing medium, penetrated the peptide crystal due to spin diffusion and then was transferred to the ^{13}C nuclei *via* the cross-polarization technique. As a result, the well-resolved 2D spectra of ^{13}C were obtained. Another example relates to the ^{15}N MAS NMR spectrum in bacterial rhodopsin taken at the same temperature and field [38]. Without DNP, a poor noisy trace was obtained after 3 days of accumulation, whereas an excellent spectra, including the 2D ones, were achieved with 30 minutes of accumulation under the DNP conditions. A giant increase of sensitivity is evident. Worthy of mention are also the DNP-enhanced ^2H-^{13}C solid-state NMR correlation spectroscopy [39], the DNP experiments in very high magnetic fields up to 14 T [40], and a lot of applications of DNP to biochemical and medical problems, see the above-cited Journals and references.

Now let us discuss on the dominant physical mechanisms of DNP as it looks today. Strong arguments in favour of the 3-spin CE were advanced recently by the Griffin group [33, 41]. They showed that the highest polarization enhancement for protons can be achieved with the use of biradicals, such as TOTAPOL biradical and others, which provide a splitting of the EPR line due to the intramolecular SS interaction between two unpaired electrons. The splitting was found to be of the order of the proton NMR frequency in the field of 5 T, thus favouring to the CE. A slight asymmetry in the $E(\Delta_p)$ dependence was observed for protons, thus evidencing against the effective spectral diffusion in the inhomogeneous EPR line, see equations (4) – (6). At the same time, the DNP experiments performed at liquid He temperatures on the samples containing ^1H, ^{13}C, ^6Li and ^7Li nuclei showed the equalizing of all nuclear spin temperatures, unambiguously pointing to the TM mechanism [42].

In fact, the dominance of one or another mechanism is determined by competition between spectral diffusion (SS cross relaxation) inside the inhomogeneous EPR line and electron spin-lattice relaxation. The TM mechanism dominates when the rate of SS cross relaxation between the spin packets separated by ω_I exceeds the spin-lattice relaxation rate, $w_{CR}(\omega_I) > T_{1S}^{-1}$. This condition seems to be fulfilled at low enough temperatures (slow spin-lattice relaxation) and/or for low NMR frequencies (low enough H or γ_I). Under the opposite condition, the 3-spin CE can be the most effective DNP mechanism, what is perhaps the case for the proton DNP at 90 K and 5 T [43]. Note, however, that the $E(\Delta_p)$ dependence observed on the ^{13}C nuclei at the same experimental conditions [43], as well as for ^2H DNP with the use of a radical with narrower EPR line [39], was found to be nearly symmetric and, according to our estimations, agrees qualitatively with the TM model. Anyway, an accurate comparison of the relative efficiency of the 3-spin CR and TM DNP mechanisms under realistic conditions seems to be of interest for further study. The quantum mechanical treatment presented by Hu *et al.* [41] may be considered as the first step which should be proceeded with strict analysis of the relaxation and many-particle spin dynamics.

Journal of Physics: Conference Series **324** (2011) 012003 doi:10.1088/1742-6596/324/1/012003

And, finally, a few words about future. It seems that the steep rising of DNP development is far from its finish. A lot of new methods and applications are knocking at the door, including the multi-pulse and rotating-frame techniques adopted from modern NMR [33] and revival of the classical Overhauser effect in liquids, now in extremely high fields [44]. The last but not least are the optical methods. In particular, the use of the optically excited paramagnetic triplet states in molecular crystals with DNP performed through the electron spin locking [45] is of considerable interest.

In conclusion, the DNP methods are on the steep rise, and new successes are prospective in near future.

References

[1] Overhauser A W 1953 *Phys. Rev.* **92** 411

[2] Carver T P and Slichter C P 1956 *Phys. Rev.* **102** 975

[3] Abragam A 1955 *Phys. Rev.* **98** 1729

[4] Neuhaus D and Williamson M P 2000 *The Nuclear Overhauser Effect in Structural and Conformational Analysis* 2nd ed (New York: Wiley)

[5] Abragam A and Proctor W G 1959 *C. R. Acad. Sci.* **246** 2253

[6] Abragam A 1961 *The Principles of Nuclear Magnetism* (Oxford: Clarendon)

[7] Khutsishvili G R 1969 *Sov. Phys.Uspekhi* **11** 802

[8] Schmugge T J and Jeffries C D 1965 *Phys. Rev.* A **138** 1785

[9] Jeffries C D 1965 *Dynamic Nuclear Orientation* (New York: Wiley)

[10] Kessenikh A V, Manenkov A A 1963 *Sov. Phys. Solid State* **5** 835

[11] Kessenikh A V, Lushchikov V I, Manenkov A A and Taran Y V 1963 *Sov. Phys. Solid State* **5** 321
Kessenikh A V, Manenkov and A A Pyatnitskii 1964 *Sov.Phys. Solid State* **6** 641
Hwang C F and Hill D A 1967 *Phys. Rev. Lett* **18** 110
Wollan D S 1976 *Phys Rev* B **13** 3671

[12] Provotorov B N 1961 *Sov. Phys. JETP* **14** 1126

[13] Provotorov B N 1962 *Sov. Phys. JETP* **15** 611

[14] Kozhushner M A and Provotorov B N 1967 On the theory of forced dynamic nuclear polarization (In Russian) *Radiospektroskopiya Tverdogo Tela: Proc. All-Union Conf. on Magnetic Resonance (Krasnoyarsk, 1964)* ed S A Altshuler *et al* (Moscow: Atomizdat) p 5

[15] Buishvili L L 1965 *Sov Phys JETP* **22** 1277

[16] Borghini M 1968 *Phys. Lett.* A **26** 242
Borghini M 1968 *Phys. Rev. Lett.* **20** 419

[17] Kozhushner M A 1969 *Sov Phys JETP* **29** 136

[18] Atsarkin V A and Rodak M I 1972 *Sov. Phys. Uspekhi* **15** 251

[19] Abragam A and Borghini M 1964 *Progress in Low-Temperature Physics* vol 4 ed C Gorter (Amsterdam: North Holland) p 3484

[20] Abragam A and Goldman M 1978 *Reports on Progress in Physics* **41** 395

[21] Atsarkin V A 1978 *Sov. Phys. Uspekhi* **21** 725

[22] Wenckebach W Th, Swanenburg T J B and Poulis N J 1974 *Physics Reports* **14** 181

[23] Atsarkin V 1980 *Dinamicheskaya Polyarizatsiya Yader v Tverdyh Dielektrikah* (Moscow: Nauka)

[24] Abragam A and Goldman M 1982 *Nuclear magnetism: Order and disorder* (Oxford: Clarendon)

[25] Goldman M 1970 *Spin Temperature and Nuclear Magnetic Resonance in Solids* (Oxford: Clarendon)

[26] Atsarkin V A and Demidov V V 1979 *Sov. Phys. JETP* **49** 1104

[27] Atsarkin V A, Mefed A.E. and Rodak M I 1967 *JETP Letters* **6** 359
Atsarkin V A, Mefed A.E. and Rodak M I 1968 *Phys. Lett.* A **27** 57
Atsarkin V A, Mefed A.E. and Rodak M I 1969 *Sov. Phys. JETP* **28** 877

[28] Wenckebach W Th, van den Heuvel G M, Hoogstraate H, Swanenburg T J B and Poulis N J 1969 *Phys. Rev. Lett.* **22** 581
[29] Atsarkin V A 1970 *Sov. Phys. JETP* **31** 1012
[30] Atsarkin V A and Ryabushkin O A 1973 *JETP Letters* **17** 71
[31] Cox S, Bouffard V and Goldman M 1973 *J. Phys. C: Solid State Phys* **6** L100
[32] Wind R A, Duijvestijn M J, Lugt C V D, Manenschijn A, Vriend J 1985 *Prog. Nucl. Magn. Reson. Spectrosc.* **17** 33
[33] Barnes A B, Paëpe G De, van der Wel P C A, Hu K-N, Joo C-G, Bajaj V S, Mak-Jurauskas M L, Sirigiri J R, Herzfeld J, Temkin R J and Griffin R G 2008 *Appl. Magn. Reson.* **34** 237
[34] Griffin R G and Prisner T F 2010 *Phys. Chem. Chem. Phys.* **12** 5737
[35] Ardenkjaer-Larsen J H, Fridlund B, Gram A, Hansson G, Hansson L, Lerche M H, Servin R, Thaning M and Golman K 2003 *Proc. Natl. Acad. Sci. U.S.A.* **100** 10158
[36] Joo C-G, Hu K-N, Bryant J A and Griffin R G 2006 *J. Am. Chem. Soc.* **128** 9428
[37] van der Wel P C A, Hu K-N, Lewandowski J and Griffin R G 2006 *J. Am. Chem. Soc.* **128** 10840
[38] Bajaj V S, Hornstein M K, Kreischer K E, Sirigiri J R, Woskov P P, Mak M, Herzfeld J, Temkin R J and Griffin R G 2007 *J. Magn. Reson.* **190** 86
[39] Maly T, Andreas L B, Smith A A and Griffin R G 2010 *Phys. Chem. Chem. Phys.* **12** 5872
[40] Matsuki Y, Takahashi H, Ueda K, Idehara T, Ogawa I, Toda M, Akutsu H and Fujiwara T 2010 *Phys. Chem. Chem. Phys.* **12** 5799
[41] Hu K-N, Debelouchina G T, Smith A A and Griffin R G 2011 *J. Chem. Phys.* **134** 125105
[42] Kurdzesau F 2009 Some methods of dynamic nuclear polarization for use in metabolic imaging *Ph. D. Thesis* 4330 (Lausanne: EPFL)
[43] Maly T, Miller A-F and Griffin R G 2010 *Chem. Phys. Chem.* **11** 999
[44] Villanueva-Garibay J A, Annino G, van Bentum P J M and Kentgens A P M 2010 *Phys. Chem. Chem. Phys.* **12** 5846
[45] Henstra A and Wenckebach W T 2008 *Molecular Physics* **106** 859

Journal of Physics: Conference Series **324** (2011) 012004 doi:10.1088/1742-6596/324/1/012004

Spin relaxation in disordered media

F S Dzheparov

Institute for Theoretical and Experimental Physics, B. Cheremushkinskaya 25, Moscow 117218, Russia

E-mail: dzheparov@itep.ru

Abstract. A review is given on theoretical grounds and typical experimental appearances of spin dynamics and relaxation in solids containing randomly distributed nuclear and/or electronic spins. Brief content is as follows.
Disordered and magnetically diluted systems. General outlines of the spin transport theory. Random walks in disordered systems (RWDS). Observable values in phase spin relaxation, free induction decay (FID). Interrelation of longitudinal and transversal relaxation related to dynamics of occupancies and phases.
Occupation number representation for equations of motion.
Continuum media approximation and inapplicability of moment expansions. Long-range transitions vs percolation theory.
Concentration expansion as a general constructive basis for analytical methods. Scaling properties of propagators. Singular point. Dynamical and kinematical memory in RWDS.
Ways of regrouping of concentration expansions. CTRW and semi-phenomenology.
Coherent medium approximation for nuclear relaxation via paramagnetic impurities.
Combining of memory functions and cumulant expansions for calculation of FID.
Path integral representations for RWDS.
Numerical simulations of RWDS.
Spin dynamics in magnetically diluted systems with low Zeeman and medium low dipole temperatures.
Cluster expansions, regularization of dipole interactions and spectral dynamics.

1. Introduction

We will discuss statically disordered media, this means that constituent particles (atoms, ions and so on) do not participate in significant translational motions during the relaxation time under discussion, and therefore their positions are fixed (frozen) in the main approximation. The relaxation is assumed as forced by magnetic dipole interactions typical for nuclear magnetic (NMR) and electron paramagnetic (EPR) resonance studies. Considered systems are suggested as translational invariant in average.

In the simplest situation, when considered system consists of spins, placed in static fluctuating fields, we have inhomogeneous broadening of resonance line – very old and simple one-particle problem that nevertheless produces new results up to now [1]. Other systems of interest are much more complex for analytical studies and we will try to clarify the nature of this complexity here. Our consideration will be concentrated on peculiarities of theoretical methods which are necessary for adequate description of observable phenomena. More extended description of theoretical and experimental results can be found in [2, 3].

We will distinguish below disordered and magnetically diluted spin systems.

Typical nuclear disordered spin system is formed by impurity spins randomly placed in sites of diamagnetic (in electronic degrees of freedom) matrix crystal when host nuclei have faster phase relaxation and flip-flop transitions than impurities. Important example of such systems is formed by nuclei ^{6}Li in the single crystal ^{7}Li^{19}F. The system is accessible for direct experimental β-NMR study due to unique coincidence of g-factors of stable nuclei ^{6}Li and β-active nuclei ^{8}Li (β-nuclei) [4, 5]. Therefore, we will consider below a model system (consisting of ^{6}Li nuclei in a LiF single crystal with addition of one β-nucleus ^{8}Li) where nuclear polarization transfers from initially polarized ^{8}Li nucleus to the nearest nonpolarized ^{6}Li nuclei and then migrates over other ^{6}Li nuclei and might return back to the ^{8}Li. Difference in speeds of processes in disordered subsystem ^{8}Li-^{6}Li and matrix spins ^{7}Li^{19}F produce a possibility to describe the polarization transfer within the subsystem by master equations [6]

$$\frac{\partial p_{i0}}{\partial t} = -\sum_j \left(\nu_{ji} p_{i0} - \nu_{ij} p_{j0} \right) = -\left(Ap\right)_{i0}, \quad p_{i0}\left(t=0\right) = \delta_{i0}, \qquad (1)$$

where $p_{i0} = \left\langle I_i^z \right\rangle$ is the quantum statistical average value of the z-component (polarization) of the i-th nucleus of the system ^{8}Li-^{6}Li, placed at $\mathbf{r}_i$ (i=0 corresponds to ^{8}Li and i≠0 to ^{6}Li). Polarization of β-nucleus $\left\langle p_{00}(t) \right\rangle_c$ is a measurable value, and $t = 0$ corresponds to creation of the nucleus in the sample via (n,γ)-reaction. Here $\left\langle ... \right\rangle_c$ means averaging over random positions of β-nuclei in the sample, that can by replaced by ensemble averaging over random positions of all impurity nuclei ^{8}Li-^{6}Li. For qualitative discussion we use simplified polarization transfer rates of the form $\nu_{i\neq j} = \nu_0 r_0^s / r_{ij}^s$, where ν_0 is transfer rate at nearest distance r_0, $\mathbf{r}_{ij} = \mathbf{r}_i - \mathbf{r}_j$, and $\nu_{jj} = 0$. Dipole transport corresponds to s=6. In this case discussed problem has a name "random walks in disordered system (RWDS) with dipole transitions". Other details, necessary for description of real experiments can be found in cited references. It should be noted, that from more general point of view our problem presents "site problem" in RWDS in difference from "bond problem", where positions $\mathbf{r}_j$ fill full crystal lattice, but transition rates ν_{ij} are independent for different i and j.

Typical magnetically diluted nuclear systems are formed by spins ^{29}Si and ^{13}C in insulator silicon and diamond crystals correspondingly. Electron paramagnet systems formed by impurity spins in diamagnetic media have the same nature of spin evolution, but corresponding Hamiltonians are more complicated due to anisotropy of g-factors and influence of fine and hyperfine interactions. Therefore, for the sake of brevity, we will discuss below homonuclear magnetically disordered systems mainly. Examples of corresponding experimental studies can be found in [7-10]. Observable values for these systems are free induction decay (FID)

$$F(t) = \left\langle I_+(t) I_- \right\rangle_0 / \left\langle I_+ I_- \right\rangle_0 \qquad (2)$$

or/and magnetic resonance form function

$$g(\omega) = \int \frac{dt}{2\pi} \exp\left(i\omega t\right) F(t), \qquad (3)$$

where $I_+ = \sum_{j=1}^N I_j^+$, $I_- = (I_+)^+$, $I_+(t) = \exp(iH_d t) I_+ \exp(-iH_d t)$, $\left\langle A \right\rangle_0 = Tr\left(A\rho\right) = Tr\left(A\right)/Tr(1)$ for any operator A, and high temperature approximation for density matrix ρ was applied. We use traditional notations for other parameters [11], for example, $I_j^+ = I_j^x + iI_j^y$ and I_j^α is α–component of spin operator for j-th nucleus placed at $\mathbf{r}_j$, N is total number of spins in the sample and the Hamiltonian is of the form

$$H_d = \frac{1}{2} \sum_{ij} b_{ij} \left(I_i^z I_j^z - \frac{1}{3} \mathbf{I}_i \mathbf{I}_j \right), \; b_{i\neq j} = \frac{3\gamma^2 \hbar^2}{2 r_{ij}^3} \left(1 - 3\cos^2 \vartheta_{ij}\right) = \frac{b_0 r_0^3}{r_{ij}^3} \left(1 - 3\cos^2 \vartheta_{ij}\right), \; b_{ii} = 0, \qquad (4)$$

where ϑ_{ij} is an angle between $\mathbf{r}_{ij}$ and external static magnetic field $\mathbf{H}_0 = (0,0,H_0)$. Here and below we use $\hbar = 1$, but sometimes it is written explicitly.

Eqs. (1) and (2) present examples of longitudinal and transversal (phase) relaxation. Processes of these types are not independent. For example, transfer rate ν_{ij} is connected with phase relaxation of impurity nuclei, because it is proportional to $\int dt \left\langle I_i^+(t) I_j^-(t) I_j^+ I_i^- \right\rangle_0 \exp(\Delta_{ij} t)$, and in main approximation $\left\langle I_i^+(t) I_j^-(t) I_j^+ I_i^- \right\rangle_0 = \left\langle I_i^+(t) I_i^- \right\rangle_0 \left\langle I_j^-(t) I_j^+ \right\rangle_0$. Here Δ_{ij} is corresponding difference of Larmor frequencies, and time dependence is produced by interaction of impurities with z-components of host spins, which should be presented by longitudinal correlation functions and so on. Important simplification consists in neglecting of influence of impurities on host spins evolution that is correct, if impurity concentration is small. Similar interrelation of longitudinal and phase relaxation exist in magnetically diluted systems as well, but both processes are realized in disordered system here.

2. Occupation numbers

Effective formalism for work with random spin systems is occupation number representation [12]. Occupation number $n_{\mathbf{x}}$=1(0) if the crystal site $\mathbf{x}$ is (is not) filled by a spin of disordered system. When all $n_{\mathbf{x}}$=1 the spins fill out all crystal lattice. In disordered spin systems $n_{\mathbf{x}}$ is random variable with $0 \le \langle n_{\mathbf{x}} \rangle_c = c \le 1$, and with independent occupation of different sites, i.e.

$$\langle n_{\mathbf{x}} n_{\mathbf{y}} \rangle_c = c\delta_{\mathbf{xy}} + c^2(1-\delta_{\mathbf{xy}}), \quad \left\langle \prod_{j=1}^{m} n_{\mathbf{x}_j} \right\rangle_c = c^m. \tag{5}$$

Last relation implies that all $\mathbf{x}_j$ are different, and coinciding positions can be excluded applying the identity $(n_{\mathbf{x}})^2 = n_{\mathbf{x}}$. Other important relation: $\varphi(n_{\mathbf{x}}) = \varphi(0) + n_{\mathbf{x}}(\varphi(1) - \varphi(0))$. It is valid, if $\varphi(0)$ and $\varphi(1)$ exist.

Evolution (1) for disordered system can be rewritten now as

$$\frac{d}{dt}\tilde{P}_{\mathbf{xy}} = -\sum_{\mathbf{z}} \left(n_{\mathbf{z}} \nu_{\mathbf{zx}} \tilde{P}_{\mathbf{xy}} - n_{\mathbf{x}} \nu_{\mathbf{xz}} \tilde{P}_{\mathbf{zy}} \right) = -\left(\tilde{A}\tilde{P} \right)_{\mathbf{xy}}, \quad \tilde{P}_{\mathbf{xy}}(t=0) = \delta_{\mathbf{xy}} \frac{n_{\mathbf{y}}}{c}. \tag{6}$$

Here propagator $\tilde{P}_{\mathbf{xy}}(t) = n_{\mathbf{x}} \tilde{P}_{\mathbf{xy}}(t) = \tilde{P}_{\mathbf{xy}}(t) n_{\mathbf{y}}$ gives polarization of the lattice site $\mathbf{x}$ at time point t when initially the site $\mathbf{y}$ was polarized, and $\nu_{\mathbf{zx}} = \nu_{ij}(\mathbf{r}_i = \mathbf{z}, \mathbf{r}_j = \mathbf{x})$. Equation (6) is a direct consequence of Eq. (1), that is evident, if we omit all empty sites for which $n_{\mathbf{x}}$=0. Average propagator $P_{\mathbf{xy}}(t) = \langle \tilde{P}_{\mathbf{xy}}(t) \rangle_c$ produces information about observable values. For example, $\langle p_{00}(t) \rangle_c = P_{\mathbf{yy}}(t) = P_{00}(t)$.

The Hamiltonian (4) of magnetically diluted system became a form

$$H_d = \frac{1}{2} \sum_{\mathbf{x},\mathbf{y}} n_{\mathbf{x}} n_{\mathbf{y}} b_{\mathbf{xy}} \left(I_{\mathbf{x}}^z I_{\mathbf{y}}^z - \frac{1}{3} \mathbf{I}_{\mathbf{x}} \mathbf{I}_{\mathbf{y}} \right), \tag{7}$$

(where $I_{\mathbf{r}}^{\alpha}$ is α-component of spin, placed at lattice site $\mathbf{r}$, and $b_{\mathbf{zx}} = b_{ij}(\mathbf{r}_i = \mathbf{z}, \mathbf{r}_j = \mathbf{x})$) and FID (2) can be rewritten as

$$F(t) = \left\langle \langle I_+(t) I_- \rangle_0 \right\rangle_c / \left\langle \langle I_+ I_- \rangle_0 \right\rangle_c, \quad I_+ = \sum_{\mathbf{x}} n_{\mathbf{x}} I_{\mathbf{x}}^+. \tag{8}$$

Here self-averaging of $F(t)$ is applied, that is correct for macroscopic samples.

3. Continuum media approximation and inapplicability of moments expansions

Both types of spin systems have simple and very instructive exactly solvable examples with experimental realization.

If nondiagonal part of operator $\tilde{A}$ in (6) is neglected, then the equation has evident solution, and exact ensemble averaging can be fulfilled:

$$P_{\mathbf{xy}}(t)=\delta_{\mathbf{xy}}Q(t),\ Q(t)=\left\langle \exp\left(-\sum\nolimits_{\mathbf{r}} n_{\mathbf{r}}v_{\mathbf{r}0}t\right)\right\rangle_c=\exp\left(\sum\nolimits_{\mathbf{r}}\ln\left(1-c\left(1-\exp\left(v_{\mathbf{r}0}t\right)\right)\right)\right). \quad (9)$$

If we calculate FID in magnetically diluted system applying Anderson's Hamiltonian

$$H_A=\frac{1}{2}\sum\nolimits_{\mathbf{x},\mathbf{y}} n_{\mathbf{x}}n_{\mathbf{y}}b_{\mathbf{xy}}I_{\mathbf{x}}^zI_{\mathbf{y}}^z,$$

then $n_{\mathbf{x}}I_{\mathbf{x}}^+(t)=\exp(iH_At)n_{\mathbf{x}}I_{\mathbf{x}}^+\exp(-iH_At)=n_{\mathbf{x}}I_{\mathbf{x}}^+\exp\left(i\sum\nolimits_{\mathbf{r}} n_{\mathbf{r}}I_{\mathbf{r}}^z b_{\mathbf{rx}}t\right)$, and for spins $I_j=\frac{1}{2}$ we receive

$$F(t)=F_A(t)=\prod\nolimits_{\mathbf{r}}\left[1-c\left(1-\cos\left(\tfrac{1}{2}b_{0\mathbf{r}}t\right)\right)\right]=\exp\left\{\sum\nolimits_{\mathbf{r}}\ln\left[1-c\left(1-\cos\left(\tfrac{1}{2}b_{0\mathbf{r}}t\right)\right)\right]\right\}. \quad (10)$$

Simple derivation of relations (9) and (10) can be found in [2, 3].

Very important continuum media approximation (CMA) corresponds to limit $c\to 0$ and prime cell volume $\Omega\to 0$ at a fixed value of impurity density $n=c/\Omega$. Calculation of $Q(t)$ and $F_A(t)$ in CMA corresponds to retaining of leading terms in expansion of logarithms in powers of c together with substitution $\sum\nolimits_{\mathbf{r}}\to\int\frac{d^dr}{\Omega}$, where d is a space dimension of the lattice. As a result, in this limit

$$Q^{CMA}(t)=\exp\left(-(\beta t)^{d/s}\right),\qquad \beta=v_0r_0^s\left(n\frac{\sigma_d}{d}\Gamma\left(1-\frac{d}{s}\right)\right)^{s/d}, \quad (11)$$

$$F_A^{CMA}(t)=\exp\left(-\left|D_At\right|^{d/3}\right),\qquad D_A(d=3)=\frac{2\pi^2}{3\sqrt{3}}\gamma^2\hbar n. \quad (12)$$

For other dimension $d\leq 2$ Anderson's constant $D_A\sim c^{3/d}$ and it depends on orientation of the sample relative external field $\mathbf{H}_0$.
Relations (11) were first derived for $d=3$ and $s=6$ by Foerster [13] in studies of luminescence quenching, and relations (12) were obtained by Anderson and Abragam for $d=3$ [11, 14].

Four main consequences should be extracted from these results.

1) Dependence on time is nonanalytical at $t=0$ contrary to evident analytical dependence, realized before configurational averaging.

2) All dependence of effective rates of the processes on concentration consists in proportionality of these rates to transfer rates at average distance $r_{av}=(\Omega/c)^{1/d}$.

3) If $c\ll 1$, but $v_0t\gg 1$ (or $b_0t\gg 1$), then deviation $1-Q(t)$ (or $1-F_A(t)$) is not small, and $Q(t)\approx Q^{CMA}(t)$ (or $F_A(t)\approx F_A^{CMA}(t)$). More precisely: $\left(\ln Q(t)-\ln Q^{CMA}(t)\right)/\ln Q(t)\sim c\ll 1$ and $\left(\ln F_A(t)-\ln F_A^{CMA}(t)\right)/\ln F_A(t)\sim c\ll 1$.

4) Decay of $Q(t)$ and $F_A(t)$ is fast enough to indicate, that application of a percolation theory is hopeless for such problems. For comparison we can indicate, that the percolation theory works [15] for other type of transition rates of the form $v_{\mathbf{xy}}=v_0\exp\left(-\left|\mathbf{x}-\mathbf{y}\right|/\rho\right)$, when for large v_0t [16]

$$\ln\left\langle \exp\left(-\sum\nolimits_{\mathbf{r}} n_{\mathbf{r}}v_{\mathbf{r}0}t\right)\right\rangle_c\sim -n\left(\rho\ln\left(v_0t\right)\right)^d$$

that is much more slow decay, than $Q(t)$ and $F_A(t)$. Main achievement of percolation theory consists in determining of a constant κ_{perc} in exponential dependence of conductivity (which is proportional to diffusion coefficient D) on concentration: $D \sim \exp\left(-\kappa_{\text{perc}} n^{1/d} / \rho\right)$. This accuracy is insufficient for problems of dipole transport.

First property reflects existence of infinite transfer rates at small distances i.e. $v_{ij}\left(r_{ij} \to 0\right) \to \infty$ and $b_{ij}\left(r_{ij} \to 0\right) \to \infty$. Second statement means that $\beta \sim v_{ij}\left(r_{ij} = r_{av}\right)$ and $D_d \sim b_{ij}\left(r_{ij} = r_{av}\right)$. Third consequence indicates, that if $c \ll 1$, then CMA produces satisfactory approximation to exact results in most important time region, where deviations of observables from their initial values are not very small.

If $c = 1$ (ordered system), then spin dynamics is based on calculation of first terms of expansion of observables in time series and in consequent rearrangement of these series in physically adopted forms. Corresponding coefficients for FID have name "moments" [11], because they can be connected with moments of form function $M_n = \int d\omega \omega^n g(\omega)$. Examples of effective realization of this program can be found in [17, 18]. Consequences, listed above, indicate that this strategy is impossible for systems with strong disorder, where $c \ll 1$, because main constructive approximation for this case (CMA) produces infinite moments. Other equivalent point of view states, that finite number of moments can describe only small part (~c) of the total variation of observables during the evolution that is clearly seen in Eqs. (9) and (10).

4. Concentration expansions

Constructive way to treat disordered spin systems with small impurity concentration c consists in application of expansions of observables in terms of powers of concentration instead of time Taylor's series [19, 20]. Simple derivation of the expansion was invented in [19]. It is based on linearity in occupation numbers of the operator $\widetilde{A}$ in (6)

$$\widetilde{A}_{\mathbf{xy}} = \sum\nolimits_{\mathbf{z}} A^{\mathbf{z}}_{\mathbf{xy}} n_{\mathbf{z}}, \quad A^{\mathbf{z}}_{\mathbf{xy}} = \delta_{\mathbf{xy}} v_{\mathbf{zy}} - \delta_{\mathbf{zx}} v_{\mathbf{xy}}, \tag{13}$$

and on application of linear symmetrization superoperator $S^{\times}$ [21] which acts on operators $A^{\mathbf{z}}$ and gives a possibility to remember the order of noncommuting operators $A^{\mathbf{z}}$ in solution of master equation (6)

$$\widetilde{P}_{\mathbf{xy}}(t) = \left(\exp\left(-\sum\nolimits_{\mathbf{z}} n_{\mathbf{z}} A^{\mathbf{z}} t\right)\right)_{\mathbf{xy}} \frac{n_{\mathbf{y}}}{c} \equiv \left(S^{\times} \exp\left(-\sum\nolimits_{\mathbf{z}} n_{\mathbf{z}} A^{\mathbf{z}} t\right)\right)_{\mathbf{xy}} \frac{n_{\mathbf{y}}}{c}. \tag{14}$$

Now operators $A^{\mathbf{z}}$ commute under the symbol $S^{\times}$ and averaging on occupation numbers is possible with the result

$$P_{\mathbf{xy}}(t) = \left(S^{\times} e^{-A^{\mathbf{y}} t} \prod\nolimits_{\mathbf{z} \neq \mathbf{y}} \left(1 + c\left(e^{-A^{\mathbf{z}} t} - 1\right)\right)\right)_{\mathbf{xy}}.$$

Expansion of last relation in powers of c is simple, and it contains terms of the form

$$\left(S^{\times} e^{-A^{\mathbf{y}} t} \prod\nolimits_{j=1}^{m} e^{-A^{\mathbf{z}_j} t}\right)_{\mathbf{xy}} \equiv \left(\exp\left(-A^{\mathbf{y}} t - \sum\nolimits_{j=1}^{m} A^{\mathbf{z}_j} t\right)\right)_{\mathbf{xy}},$$

which are easily connected with solution of initial Eq. (6) for the case, when only $m+1$ sites with positions $\mathbf{y}, \mathbf{z}_1, \ldots, \mathbf{z}_m$ are filled by impurity spins. As a result

$$P_{00}(t) = \sum\nolimits_{m=0}^{\infty} \frac{c^m}{m!} \sum\nolimits_{\mathbf{z}_1, \ldots, \mathbf{z}_m}^{/} \sum\nolimits_{k=0}^{m} (-1)^{m-k} C_m^k p_{00}^{(k+1)}\left(0, \mathbf{z}_1, \ldots, \mathbf{z}_k; t\right), \tag{15}$$

$$P_{\mathbf{x}\neq 0}(t) = c \sum\nolimits_{m=0}^{\infty} \frac{c^m}{m!} \sum\nolimits_{\mathbf{z}_1, \ldots, \mathbf{z}_m}^{/} \sum\nolimits_{k=0}^{m} (-1)^{m-k} C_m^k p_{10}^{(k+2)}\left(0, \mathbf{x}, \mathbf{z}_1, \ldots, \mathbf{z}_k; t\right). \tag{16}$$

Here $p_{00}^{(k+1)}(0,\mathbf{z}_1,...,\mathbf{z}_k;t)$ is solution p_{00} of Eq. (1) for the problem of k+1 spins, placed at sites $\mathbf{r}_0=0,\mathbf{z}_1,...,\mathbf{z}_k$, $p_{10}^{(k+2)}(0,\mathbf{x},\mathbf{z}_1,...,\mathbf{z}_k;t)$ is solution p_{10} for the problem of k+2 spins, placed at $\mathbf{r}_0$=0, $\mathbf{r}_1$=**x**, $\mathbf{z}_1,...,\mathbf{z}_k$, C_m^k is binomial coefficient. It is supposed here that for k=0 variables $\mathbf{z}_1,...,\mathbf{z}_k$ are absent, and for m=0 the variables and corresponding summations are absent. The prime in coordinate summation indicates that coincidences between spin positions are forbidden and summations are limited by common big cubic volume V. Therefore symmetrization with respect to all $\mathbf{z}_1,...,\mathbf{z}_m$ should be fulfilled before calculation of the limit $V\to\infty$.

It can be shown, that relations (15) and (16) are very general and they can be derived [19, 20, 22] from the fact, that any reasonable symmetric function of occupation numbers $Z(\{n_\mathbf{r}\})$ can be expanded into series of productions of occupation numbers using the identity $\varphi(n_\mathbf{x})=\varphi(0)+n_\mathbf{x}(\varphi(1)-\varphi(0))$. Therefore

$$Z(\{n_\mathbf{r}\})=Z^{(0)}+\sum_{m=1}^{\infty}\frac{1}{m!}\sum_{\mathbf{r}_1,...,\mathbf{r}_m}^{\prime}\left(\prod_{j=1}^{m}n_{\mathbf{r}_j}\right)\sum_{k=0}^{m}C_m^k(-1)^{m-k}Z^{(k)}(\mathbf{r}_1,...,\mathbf{r}_k) \quad (17)$$

Averaging of this relation consists in substitution $\prod_{j=1}^{m}n_{\mathbf{r}_j}\to\left\langle\prod_{j=1}^{m}n_{\mathbf{r}_j}\right\rangle_c=c^m$ that (in application to the propagator $P_\mathbf{xy}(t)$) produces relations (15) and (16).

Calculation of the FID can be carried out in similar way. Indeed, if we introduce a correlation function

$$F_\mathbf{xy}(t)=\left\langle n_\mathbf{x}n_\mathbf{y}\left\langle I_\mathbf{x}^+(t)I_\mathbf{y}^-\right\rangle_0\right\rangle_c\cdot\left(c\left\langle I_\mathbf{y}^+I_\mathbf{y}^-\right\rangle_0\right)^{-1}, \quad (18)$$

then we can receive for it relations, similar to (15) and (16), and than calculate

$$F(t)=\sum_\mathbf{x}F_\mathbf{xy}(t). \quad (19)$$

The correlator $F_\mathbf{xy}(t)$ presents orthogonal circular polarization of the site **x**, if initially the polarization existed at site **y** only.

Free induction decay $F(t)$ can be calculated as well directly using Eq. (17).

If $\nu_0t\gg1$, then $p_{00}^{(k+1)}(0,\mathbf{z}_1,...,\mathbf{z}_k;t)$ and $p_{10}^{(k+2)}(0,\mathbf{x},\mathbf{z}_1,...,\mathbf{z}_k;t)$ in concentration expansions (15) and (16) became smooth functions of summation variables $\mathbf{z}_j$ and sums can be replaced by integrals in exact analogy with derivation of Eq. (11). This calculation produces correct result for $P_\mathbf{xy}(t)$ if $c\ll1$ only, because we omit forbidding on coincidence of integration variables, which is not negligible if concentration c is not small. Now, replacing the variables $\mathbf{z}_j\to(c/\Omega)^{-1/d}\,\mathbf{z}_j$ we arrive to relations

$$P_{00}(t)=f_0(\beta t),\quad P_{\mathbf{x}0}(t)=cf_1(\beta t,x/r_{av}), \quad (20)$$

which were derived in [12] and named as theorem on singularity point.

All dependence on time t, distance x, and concentration c is written here explicitly, while unknown functions $f_0(\tau)$ and $f_1(\tau,\xi)$ do not contain dependence on c. It will be recalled that $\beta\sim c^{s/d}$ and $r_{av}\sim c^{-1/d}$. For example, in application to Foerster's problem (9), (11) we have $f_0(\tau)=\exp\left(-\tau^{d/s}\right)$.

Similar relations for magnetically diluted systems are of the form

$$F_{00}(t)=\phi_0(D_At),\quad F_{\mathbf{x}0}(t)=c\phi_1(D_At,x/r_{av}). \quad (21)$$

Here D_A~$c^{3/d}$.

It is clear from (20) and (21) that expansions (15) and (16) in terms of c^m is in reality expansion in powers $(\beta t)^{md/s}$ and therefore c is not a small parameter if $\beta t \geq 1$. Similarly, concentration expansion of FID is in reality expansion in powers $(D_A t)^{md/3}$.

Physical meaning of Eqs. (20) and (21) is simple. Natural scales for time and distance are presented by β (or D_A) and average distance r_{av}. The singularity is direct consequence of the fact, that $P_{\mathbf{xy}}$ and $F_{\mathbf{xy}}$ are the polarization of lattice site and the processes take place if initial site $\mathbf{y}$ is occupied by the impurity spin certainly, while any other point $\mathbf{x} \neq \mathbf{y}$ is occupied with probability c only. Polarization on nearly placed spins become equal with increasing of the time, therefore [12]

$$\lim_{x\to 0} P_{\mathbf{x}\neq 0}(t) \sim cP_{00}(t), \quad \lim_{x\to 0} F_{\mathbf{x}\neq 0}(t) \sim cF_{00}(t). \tag{22}$$

It means, that both $P_{\mathbf{x}0}(t)$ and $F_{\mathbf{x}0}(t)$ are not smooth functions of $\mathbf{x}$ in the point $\mathbf{x}$=0. To recognize a peculiarity of this fact we should compare it with property of $P_{\mathbf{x}0}(t \to \infty)$ for $c = 1$, when migration of polarization takes place on regular lattice. Then operator $\widetilde{A}$ is invariant relative lattice translations and solution of Eq. 6 can be received by applying lattice Fourier transformation. At large t the solution go to propagator of diffusive equation and is of the form

$$P_{\mathbf{x}0}(t \to \infty, c = 1) = \frac{\Omega}{(4\pi Dt)^{d/2}} \exp\left(-\frac{x^2}{4\pi Dt}\right),$$

having no singularity similar to (22). Here D is diffusion coefficient. As a result we should expect, that if long time behavior of $P_{\mathbf{x}0}(t \to \infty)$ is diffusion like, then

$$P_{\mathbf{x}0}(t \to \infty) = \left(\frac{1}{c}\delta_{\mathbf{x}0} + (1-\delta_{\mathbf{x}0})\right) \frac{\Omega}{(4\pi Dt)^{d/2}} \exp\left(-\frac{x^2}{4\pi Dt}\right), \tag{23}$$

and $c(4\pi Dt)^{d/2} / \Omega \sim (\beta t)^{d/2}$, it means, that

$$D = \frac{\kappa}{2d} \beta r_{av}^2, \tag{24}$$

where κ is numerical coefficient, its modern values are obtained via numerical simulations, discussed below.

Calculation of terms of concentration expansions is more complex than calculation of moments, therefore very limited number of these terms is calculated up to now. Terms of order c and c^2 for $P_{00}(t)$ in real dipole RWDS problem were calculated in [6, 19] (with taking into account possible influence of external static field). Term of first order for FID $F(t)$ for correlated distribution of spin positions was considered in [20]. Next term for uncorrelated positions was determined in [23]. Possible variation of coefficient before $\mathbf{S}_i\mathbf{S}_j$ in the Hamiltonian (4) was considered there.

5. Ways of rearrangement of concentration expansions

Observables, averaged over impurity distribution, contain less information then they contain before averaging. Standard formalism for excluding of unnecessary details of description of physical evolution is projection technique of Nakajima-Zwanzig (see for example [17] and [24]). As a rule it produces convolution master equation, if we start from local in time differential equation of motion.

5.1. Projection operator technique and memory kernels for RWDS

Simplest form of the resulting master equation can be received, if projection (super)operator $\widetilde{E}$ (main function of which is calculation of average value (mathematical expectation) of observable ($\widetilde{P}_{\mathbf{xy}}(t)$ for example)) and initial condition $\widetilde{P}_{\mathbf{xy}}(t=0)$ obey the relation $\widetilde{E}\widetilde{P}_{\mathbf{xy}}(t=0)=\widetilde{P}_{\mathbf{xy}}(t=0)$. Therefore the projector can be chosen as

$$\left(\widetilde{E}\chi\right)_{\mathbf{x}}=\frac{n_{\mathbf{x}}}{c}\langle\chi_{\mathbf{x}}\rangle. \tag{25}$$

It is supposed here, that $n_{\mathbf{x}}\chi_{\mathbf{x}}=\chi_{\mathbf{x}}$. In other way, using Dirac's notations,

$$\langle\mathbf{x}|\widetilde{E}|\chi\rangle=\frac{n_{\mathbf{x}}}{c}\langle\langle\mathbf{x}|\chi\rangle\rangle_c=\frac{n_{\mathbf{x}}}{c}\langle\chi_{\mathbf{x}}\rangle,\quad \widetilde{E}=\sum\nolimits_{\mathbf{x}}|\mathbf{x}\rangle\frac{n_{\mathbf{x}}}{c}En_{\mathbf{x}}\langle\mathbf{x}|. \tag{26}$$

Here operator E produces simple averaging on occupation numbers: $Ef=\langle f\rangle_c$. It is evident, that $\widetilde{E}$ and $\overline{E}=1-\widetilde{E}$ are projectors, that means $\left(\widetilde{E}\right)^2=\widetilde{E}$, and $\left(\overline{E}\right)^2=\overline{E}$. Therefore, starting from Eqs. (6) and treating them in standard manner using introduced projection operators we arrive to convolution master equation

$$\frac{\partial}{\partial t}P=-\int_0^t d\tau M(\tau)P(t-\tau),\qquad P_{\mathbf{xy}}(t=0)=\delta_{\mathbf{xy}}, \tag{27}$$

where

$$M_{\mathbf{xz}}(t)=\left\langle\widetilde{A}_{\mathbf{xz}}\frac{n_{\mathbf{z}}}{c}\right\rangle\delta(t-\eta)-\left\langle\left(\widetilde{A}\overline{E}e^{-\overline{E}\widetilde{A}\overline{E}t}\overline{E}\widetilde{A}\right)_{\mathbf{xz}}\frac{n_{\mathbf{z}}}{c}\right\rangle_c. \tag{28}$$

Here $\eta\to+0$ in order to receive $\int_0^t d\tau\delta(t-\eta)=1$ for $t>0$. Remembering that the system has a conservation low $\sum_{\mathbf{x}}P_{\mathbf{xy}}=1$, we can separate diagonal and nondiagonal parts:

$$M_{\mathbf{xz}}(t)=D(t)\delta_{\mathbf{xz}}-N_{\mathbf{xz}}(t),\quad D(t)=\sum\nolimits_{\mathbf{x}}N_{\mathbf{xz}}(t). \tag{29}$$

The memory matrix $M_{\mathbf{xz}}(t)$ is invariant to lattice translations, therefore it can be diagonalized by lattice Fourier transformation that gives important simplification in solution of the equation.

Main strategy of application of master equations consists in approximation of the memory kernel with consequent calculation of propagator, solving obtained master equation. For example, if $c=1$, then first step was fulfilled during derivation of Eq. (6), and at second step we arrive to diffusion process, while the memory at first step had no direct indication on this diffusion. Situation in strong disorder is opposite [6].

Indeed, relations (20) will be fulfilled for $c\ll 1$ if equation on diagonal part of the propagator is separated and has a form

$$\frac{\partial}{\partial t}P_{00}=-\int_0^t d\tau D(\tau)P_{00}(t-\tau). \tag{30}$$

It means that long time asymptotics $P_{00}(t\to\infty)\sim t^{-d/2}$ is one to one related with $D(t\to\infty)$. For example if d=3, then $D(t\to\infty)\sim t^{-3/2}$ as well. Similar relations can be derived for other dimensions using Laplace transformation to solve the Eq. (30) and Tauberian theorems. Therefore, if we want to receive the long time asymptotics of $P_{00}(t)$ solving the Eqs. (27) and (28), we should know long time asymptotics of $D(t)$ that is impossible, basing on first terms of concentration expansion. Nevertheless we will obtain reasonable diffusion asymptotics for $P_{\mathbf{x}\neq 0}(t)$ applying approximation for $N_{\mathbf{xy}}(t)$, which is correct in initial region of time and if $\sum_{\mathbf{x}}x^2w_{\mathbf{x}0}<\infty$, where $w_{\mathbf{xy}}=\int_0^\infty dtN_{\mathbf{xy}}(t)$. By other words we have no possibility to calculate directly measurable $P_{00}(t)$ basing on master equation (27), (28) and concentration expansions.

We can introduce other kind of memory (geometrical, contrary to dynamical one in (27), (28)) if we will take into account, that

$\widetilde{P}_{xy}(t)=\widetilde{P}_{xy}(t)n_y=\widetilde{P}_{xy}(t;n_y=1)n_y$.

Therefore if we introduce auxiliary propagator $\widetilde{K}_{xy}(t)=c\widetilde{P}_{xy}(t;n_y=1)$, then $\left\langle\widetilde{K}_{xy}(t)\right\rangle_c=P_{xy}(t)$, and $\widetilde{K}_{xy}(t)$ obey the equation

$$\frac{d}{dt}\widetilde{K}_{xy}=-\left(\widetilde{A}^{y}\widetilde{K}\right)_{xy},\quad \widetilde{K}_{xy}(t=0)=\delta_{xy} \tag{31}$$

with much more simple initial condition, then in (6). Here $\widetilde{A}^{y}=\widetilde{A}(n_y=1)$. It is sufficient now to chose the projectors of Nakajima-Zwanzig as simple averaging E and $\overline{E}=1-E$. As a result the master equation for $P_{xy}(t)$ receive a form

$$\frac{\partial}{\partial t}P_{xy}=-\int_0^t d\tau\sum\nolimits_z M^{y}_{xz}(\tau)P_{zy}(t-\tau), \tag{32}$$

where

$$M^{y}_{xz}(t)=\left\langle\widetilde{A}^{y}_{xz}\right\rangle\delta(t-\eta)-\left\langle\left(\widetilde{A}^{y}_{xz}\overline{E}e^{-\overline{E}\widetilde{A}^{y}_{xz}\overline{E}t}\overline{E}\widetilde{A}^{y}_{xz}\right)_{xz}\right\rangle_c. \tag{33}$$

Memory of this type gives [6] reasonable diffusive asymptotics (23), but the matrix $M^{y}_{xz}(t)$ can not be diagonalized by Fourier transformation, and Eq. (32) has no solution in quadratures.

5.2. Similarities with the theory of continuum time random walks. Semi-phenomenology

The theory of Scher and Lax [16] belongs to so called continuum time random walks (CRTW) theories. It was among the first, directed on analytical treatment of RWDS problem. The theory was formulated in the form, which is equivalent to master equation (27) with memory kernel of dynamical type (29) for transition rates of the form $\nu_{xy}=\nu_0\exp\left(-|\mathbf{x}-\mathbf{y}|/\rho\right)$, and without exact correspondence to concentration expansions. The property (30) was not recognized at that time, and asymptotics $P_{00}(t\to\infty)$ was not of diffusion type, while increasing of $\left\langle x^2\right\rangle=\sum_x x^2P_{x0}(t\to\infty)\sim t$ was diffusion like. In Ref. [6] the Scher-Lax theory was updated to produce correct first term of concentration expansion and qualitatively correct diffusion asymptotics (23) for dipole transport on the basis of application of geometrical memory (32), (33). This variant received a name "semi-phenomenological theory". The memory kernel was presented as

$M^{y}_{xz}(t)=\delta_{xy}D^{y}_{z}(t)-N^{y}_{xz}(t),\quad D^{y}_{z}(t)=\sum_x N^{y}_{xz}(t)$,

and explicit form was received in Laplace representation:

$$N^{y}_{xz}(\lambda)=\left(1-\delta_{xy}\right)w_{xz}(\lambda)+\frac{\delta_{xy}}{c}w_{yz}(\lambda),\quad w_{xz}(\lambda)=c\nu_{xz}Q(\lambda+\nu_{xz})/Q(\lambda).$$

Here $f(\lambda)=\int_0^\infty dt\exp(-\lambda t)f(t)$ for any function $f(t)$, and $Q(t)=Q^{CMA}(t)$ (see (9) and (11)). Such definition of $w_{xz}(\lambda)$ is specific for Sher-Lax theory, but application of geometrical memory produce correct terms of the propagator in the first order in concentration in difference from primary version [16].

Important consequence of this analysis consists in formulation of statement, that in calculation of long time asymptotics of $P_{x\neq y}(t)$ the condition $n_y=1$ can be omitted, and then the result of Eq. (32) is of the form

$$P_{\mathbf{x}\neq 0}(t\to\infty)=\frac{\Omega}{(2\pi)^3}\int d^3k e^{i\mathbf{kx}}\exp(-\overline{M}(\mathbf{k})t),\quad \overline{M}(\mathbf{k})=\sum\nolimits_{\mathbf{x}} e^{-i\mathbf{kx}}\int_0^\infty dt M_{\mathbf{x}0}(t). \tag{34}$$

For small k we have

$$\overline{M}(\mathbf{k})=Dk^2-\sigma k^3+O(k^4). \tag{35}$$

Here D is diffusion coefficient, which depends on details of approximation, used for construction of M, while second term is produced by dipole long range nature of primary transition rates $\nu_{\mathbf{zx}}\sim|\mathbf{z}-\mathbf{x}|^{-6}$, and it should not depend on the model treatment of M. As a result, prognosis for dipole transport in impurity system was written as

$$P_{00}(t)\approx P(t)=Q(t)+\frac{\xi(1-Q(t))}{(\mu\beta(t+\tau))^{3/2}}\left(1+\frac{\varphi}{(\mu\beta(t+\tau))^{1/2}}\right). \tag{36}$$

Here coefficient ξ reflects the asymmetry of transition rates existing in real ^8Li-^6Li system ($\nu_{0j}=\xi\nu_{j0}$, $\xi=3$, that was omitted above), μ is determined by the relation $\mu\beta=4\pi Dc^{2/3}$, coefficient φ is produced by term $\sigma k^3>0$ from (35) and it is model independent (as well as σ), and τ produces coincidence of first terms of expansion of $P(t)$ in powers of $(\beta t)^{1/2}$ with exact

$$P_{00}(t)=1-\left(\frac{\beta t}{\xi+1}\right)^{1/2}+O(\beta t).$$

Relation (36) interpolates exact asymptotical behaviour of short and long times. It should be valid for other values of diffusion coefficient as well with new μ and τ, and with the same ξ and φ.

Primary experimental studies [4, 25, 26] confirmed this prediction at $\beta t\leq 5$, with taking into account 1) necessary orientation anisotropy of transition rates, 2) new values of diffusion tensor, obtained in numerical simulation [27-29, 31], and 3) influence of correlation of local fields on transition rates [25, 30]. Consequent studies [5, 31-33] revealed, that for c<10% and $\beta t\leq 30$ more correct relation is of the form $P_{00}(t)=P(t)(1+\chi(t))$, where (for static external fields $H_0\approx 200$ G) variation $\chi(t)$ is relatively small ($-0.1<\chi(t)<0.7$) and it increases the "reoscillation", presented by term in (36), which is proportional to φ.

5.3. Connections with effective (or coherent) medium theories. Nuclear relaxation via paramagnetic impurities

This class of theories is considered as mean field ones. Application of these theories to our problems with random site positions meet both all difficulties, which were listed for theories similar to CTRW, and new obstacles, connected with absence of ways to find analytical solutions of equations, which are typical for equations of main approximation. Therefore we will consider here only relatively simple problem of nuclear relaxation via paramagnetic impurities, where nuclear spin polarization can diffuse over crystal lattice and it can decay due to interaction with electron paramagnetic impurities (traps), randomly distributed on crystal sites. Examples of experimental realization of the process can be found in Refs. [11, 17, 34-38]. Basic equations of evolution are of the form

$$\frac{\partial}{\partial t}p(\mathbf{x},t)=D\Delta p(\mathbf{x},t)-\sum\nolimits_{\mathbf{z}}n_{\mathbf{z}}\nu_{\mathbf{zx}}p(\mathbf{x},t),\quad \nu_{\mathbf{zx}}=\frac{\nu_0 r_0^6}{|\mathbf{z}-\mathbf{x}|^6}=\frac{C}{|\mathbf{z}-\mathbf{x}|^6},\quad p(\mathbf{x},t=0)=1. \tag{37}$$

Here $p(\mathbf{x},t)$ is polarization of the nucleus, placed at the crystal site $\mathbf{x}$, Δ is Laplacian, D is spin-diffusion coefficient, and the angular dependence of $\nu_{\mathbf{zx}}$ is neglected together with the difference in eigenvalues of the diffusion tensor. The observable nuclear polarization (normalized to $p(t=0)=1$) is

$$\overline{p}(t)=\frac{1}{V}\int d^d x p(\mathbf{x},t)=\langle 0|G(t)|0\rangle=\langle 0|\langle G(t)\rangle_c|0\rangle. \tag{38}$$

Here d is the space dimensionality, V is the crystal volume, the symbol $|0\rangle$ presents a vector having components $\langle \mathbf{x}|0\rangle=1$ and the propagator $G_{\mathbf{xy}}(t)=\langle \mathbf{x}|G(t)|\mathbf{y}\rangle$ obeys the Eq. (37), but for initial condition $G_{\mathbf{xy}}(t=0)=\delta(\mathbf{x}-\mathbf{y})$.

Eq. (37) presents example of random walks in a system with disordered traps.

Expansion (17) of the observable $\overline{p}(t)$ in concentration powers gives in first terms

$$\overline{p}(t)=\langle 0|G^{(0)}(t)|0\rangle+n\int d^d x\left(\langle 0|G^{(1)}(\mathbf{x},t)|0\rangle-G^{(0)}(t)\right)+O(n^2)=1-M_0(t)+O(n^2), \tag{39}$$

$$M_0(t)=n\int d^d x\left(\langle 0|G^{(1)}(t)|0\rangle-1\right). \tag{40}$$

Preceding studies [11, 17, 34-36] used (for d=3) the substitution $\overline{p}(t)=1-M_0(t)+O(n^2)\to$ $\to\exp(-M_0(t))$ without justification. Analysis of two- and one-dimensional problems was absent whereas experiments were started to study fractal objects [38]. Therefore the work [39] was directed on founding of natural way for rearrangement of the expansion (39) and improving and extension of the results of Ref. 34 on systems with spatial dimension $d\le 3$. Our attention will be directed on the rearrangement of concentration expansion. Other important results can be found in [39].

To construct a theory of mean field class we represent the propagator $G(t)$ in the form

$$G(t)=\langle\exp(-(A+U)t)\rangle_c=\exp(-B(t)), \tag{41}$$

$$U(t)=\sum\nolimits_{\mathbf{z}} n_{\mathbf{z}}U^{\mathbf{z}},\quad U^{\mathbf{z}}_{\mathbf{xy}}=\delta_{\mathbf{xy}}v_{\mathbf{xz}},\qquad B(t)=At+M(t), \tag{42}$$

where $A=-D\Delta$ and the operator $M(t)$ is as-yet undefined. We can write it in the form

$$M(t)=\sum\nolimits_{\mathbf{z}} M^{\mathbf{z}}(t) \tag{43}$$

similar to the form of "potential" U in (42). It can be said that the operators $M^{\mathbf{z}}(t)$ must adequately describe the effect of traps in the so-called effective medium that appears upon averaging over the configurations of traps. It is therefore natural to assume that, on average, the propagator $G(t)$ undergoes no changes if one of the sites of the effective medium is replaced by an actual one and if the result is thereupon averaged over the distribution of traps; that is,

$$G(t)=\left\langle\exp\left(-\left(At+M(t)-M^{\mathbf{z}}(t)+n_{\mathbf{z}}U^{\mathbf{z}}t\right)\right)\right\rangle_c. \tag{44}$$

Relations (41)–(44) form a closed set of nonlinear operator equations. The solution coincides with (39) in main order in c. According to [39], it practically coincides with $\overline{p}(t)=\exp\left(-M_0(t)\right)$, if $M_0(t)$ ~ 1 and $c\ll 1$, and corrections are important for longer time.

The derivation of Eq. (44) illustrates a way of application of "coherent medium approximation" to our problems in time representation. Other way can consists in application of similar ideas to Laplace representation of the propagator $G(\lambda)$. We should expect, that

$$G(\lambda)=\left\langle(\lambda+A+U)^{-1}\right\rangle_c=\left\langle(\lambda+A+M_L(\lambda))^{-1}\right\rangle_c,\quad M_L(\lambda)=\sum\nolimits_{\mathbf{z}} M^{\mathbf{z}}_L(\lambda), \tag{45}$$

and then equation of self-consistency is

$$G(\lambda)=\left\langle\left(\lambda+A+M_L(\lambda)-M^{\mathbf{z}}_L(\lambda)+n_{\mathbf{z}}U^{\mathbf{z}}\right)^{-1}\right\rangle_c. \tag{46}$$

Probably, solution of Eqs. (41)–(44) and time version of solutions of Eqs. (45) and (46) are close in the same region ($M_0(t)\sim 1$ and $c\ll 1$) for three dimensional systems. But they are very different for d<3, and Eqs. (41) – (44) look as more preferable.

In order to complete the picture we should notice, that (1) to clarify the influence of corrections to this solution in more detail we can calculate the next ($\sim c^2$) term of the concentration expansion of $\overline{p}(t)$ or $G(t)$, and (2) there exist physical [40] and mathematical [41, 42] studies producing the law $\ln\left(1/\overline{p}(t\to\infty)\right)\sim t^{d/(d+2)}$, which is expected [44] to be valid at $\overline{p}(t)\sim 10^{-12}$. The asymptotics can be received by methods of superfield theory [45], but up to a preexponential multiplier only, as well as in Refs. 40–42. Rather artificial mathematical constructions [41, 42] receive natural simple clarifying via steepest descent method in superfield theory [45]. No way to calculate the pre-exponential factor was found. As a result, these methods were insufficient up to now for study of RWDS without traps.

5.4. Combining of memory functions and cumulant expansions for calculation of FID

Initial equations of motion (1), (6) and (37) in RWDS problem can be considered mathematically as one particle Schrödinger equation for imagine time. Therefore conceptually the problem of calculation of FID in magnetically diluted systems is more complex, because it starts (according to Eqs. (7), (18) and (19)) from many-particle Heisenberg equation of motion

$$\frac{\partial}{\partial t}n_{\mathbf{x}}I_{\mathbf{x}}^{+}(t)=i\left[H_d,n_{\mathbf{x}}I_{\mathbf{x}}^{+}(t)\right]=iH_d^{\times}n_{\mathbf{x}}I_{\mathbf{x}}^{+}(t). \tag{47}$$

Here last equality defines Liouville superoperator $H_d^{\times}$ acting in a Hilbert space, spanned on usual operators of quantum mechanics [17, 24] with scalar product $\langle A|B\rangle=\langle A^{+}B\rangle_0$, where A and B are usual operators. Projection superoperator P can be chosen as

$$PQ=\sum_{\mathbf{r}}n_{\mathbf{r}}I_{\mathbf{r}}^{+}\left\langle n_{\mathbf{r}}\left\langle I_{\mathbf{r}}^{-}Q\right\rangle_0\right\rangle_c\left(\left\langle n_{\mathbf{r}}\left\langle I_{\mathbf{r}}^{-}I_{\mathbf{r}}^{+}\right\rangle_0\right\rangle_c\right)^{-1}, \tag{48}$$

where Q is arbitrary quantum mechanical operator. Introducing $\overline{P}=1-P$ and treating Eq. (47) in standard manner we arrive to convolution master equation

$$\frac{\partial}{\partial t}F_{\mathbf{rq}}(t)=-\int_0^t d\tau\sum_{\mathbf{x}}M_{\mathbf{rx}}(\tau)F_{\mathbf{xq}}(t-\tau), \tag{49}$$

$$M_{\mathbf{rx}}(t)=\frac{1}{\left\langle n_{\mathbf{x}}\left\langle I_{\mathbf{x}}^{-}I_{\mathbf{x}}^{+}\right\rangle_0\right\rangle_c}\left\langle\left\langle\left[n_{\mathbf{r}}I_{\mathbf{r}}^{-},H_d\right]\exp\left(i\overline{P}H_d^{\times}\overline{P}t\right)\left[H_d,n_{\mathbf{x}}I_{\mathbf{x}}^{+}\right]\right\rangle_0\right\rangle_c. \tag{50}$$

We will discuss later three dimensional systems only.

Let us start from a dense system when all $n_{\mathbf{q}}=1$. Main property of the FID here consists in existence of oscillation, approximately presented by a relation [46]

$$F(t)=\frac{b\sin at}{a\sinh bt},$$

where constants a and b are of order $\sqrt{M_2}$.

We will formulate a simple approximation, which reproduce this oscillating behavior of the FID basing on known moments M_2 and M_4, and than we will generalize it to magnetically diluted systems, using first two terms of the concentration expansion, calculated in Ref. 23. Main approximation of the analysis is founded on findings of Refs. 45 and 25, were NMR form function $g_I(\omega)$ for impurity β-nuclei ^{8}Li in LiF was measured in frequency range, where $g_I(\omega)\geq 10^{-4}g_I(0)$, and the results were satisfactory described [45, 25, 30] without adjusting parameters by an extension of Anderson-Weiss-Kubo (AWK) theory, developed primary for explaining of "narrowing by motion".

5.4.1. *Dense systems.* Memory kernel $M_{\mathbf{rx}}(t)$ is most complex object in the theory. To calculate it for dense system we use next simple and natural approximation:

$$M_{\mathbf{rx}}(t) = m_{\mathbf{rx}}\chi(t), \tag{51}$$

where $m_{\mathbf{rx}} = M_{\mathbf{rx}}(t=0)$ is easily calculable, and $\chi(t)$ is yet unknown function at that $\chi(t=0)=1$.

Free induction decay of impurity nucleus is defined by a Hamiltonian having no flip-flop terms in dipole interaction of the nucleus with surrounding host nuclei. Therefore, in order to calculate $\chi(t)$ using experience of Refs. 45 and 25, we introduce auxiliary system which has modified interaction for spin I_0 only. Its Hamiltonian is

$$H_d^a = \frac{1}{2}\sum\nolimits_{\mathbf{r}\neq 0,\mathbf{q}\neq 0} b_{\mathbf{rq}}\left(I_{\mathbf{r}}^z I_{\mathbf{q}}^z - \frac{1}{3}\mathbf{I}_{\mathbf{r}}\mathbf{I}_{\mathbf{q}}\right) + \frac{2}{3}\sum\nolimits_{\mathbf{r}} b_{\mathbf{r}0} I_{\mathbf{r}}^z I_0^z. \tag{52}$$

Repeating the derivation (47)-(51) with Hamiltonian H_d^a, we receive, that master equation for $F_I(t) = F_{00}(t) = \left\langle I_0^- I_0^+(t)\right\rangle$ is decoupled from other $F_{\mathbf{rq}}(t)$, and we can write it as

$$\frac{\partial}{\partial t}F_I(t) = -\int_0^t d\tau M_I(\tau)F_I(t-\tau), \tag{53a}$$

$$M_I(t) = m_I\chi_I(t), \tag{53b}$$

where unknown $\chi_I(t=0)=1$, and m_I is easily calculable coefficient (second moment for $F_I(t)$).

Difference between the functions $\chi_I(t)$ and $\chi(t)$ should be small, of order $1/z_e$, where z_e is effective number of neighbors, therefore next main approximation is

$$\chi(t)=\chi_I(t). \tag{54}$$

Methods for calculation of FID for impurity nuclei [45, 25, 30] can be applied for independent calculation of $F_I(t) = F_{00}(t)$ here directly. They use two main approximations. The first one consists in hypothesis that we can neglect of influence of impurity spin on evolution of the local field that is similar to approximation (54). The second approximation considers the local field on impurity spin

$$\omega_{I0}(t) = \frac{2}{3}\sum\nolimits_{\mathbf{q}} b_{\mathbf{q}0} I_{\mathbf{q}}^z(t),$$

(where time dependence is defined by Heisenberg evolution) as smooth classical normal stochastic process with correlator

$$\left\langle \omega_{I0}\omega_{I0}(t)\right\rangle = \frac{4}{9}\sum\nolimits_{\mathbf{qr}} b_{\mathbf{q}0}b_{\mathbf{r}0}\left\langle I_{\mathbf{q}}^z I_{\mathbf{r}}^z(t)\right\rangle_0 = m_I\kappa(t), \tag{55}$$

$$m_I = \frac{4I(I+1)}{9\cdot 3}\sum\nolimits_{\mathbf{q}} b_{\mathbf{q}0}^2 = \left\langle \omega_{I0}^2\right\rangle_0,$$

that produces

$$F_I(t) = \exp\left(-m_I\int_0^t d\tau\,(t-\tau)\,\kappa(\tau)\right). \tag{56}$$

Last relation is main in the AWK-theory.

The time dependence of operator $I_{\mathbf{r}}^z(t)$ in (55) is Hamiltonian again, and corresponding ways to take in into account are described in [45, 25, 30]. The relation (56) can be considered as well as first term of cumulant expansion of the correlation function $F_I(t)$ in powers of local frequency $\omega_{I0}(t)$ [21]:

$$F_I(t) = \left\langle T\exp\left(i\int_0^t d\tau\omega_{I0}(\tau)\right)\right\rangle_0 \to \exp\left(-\int_0^t d\tau\,(t-\tau)\left\langle \omega_{I0}(\tau)\omega_{I0}\right\rangle_0\right).$$

As a result, if correlation function (56) is calculated, then we can

a) define the memory kernel $M_I(\tau)$ solving the Eq. (53) with known (from (56)) $F_I(t)$;

b) define from (53) the functions $\chi_I(\tau)$ and $\chi(\tau)=\chi_I(\tau)$, and

c) calculate $F_{\mathbf{rq}}(t)$ using Eqs. (49)-(51).

If the aim consists in calculation of FID $F(t)=\sum_{\mathbf{q}} F_{\mathbf{rq}}(t)$, then at the step "c" we can solve more simple master equation

$$\frac{\partial}{\partial t}F(t)=-\int_0^t d\tau M(\tau)F(t-\tau), \tag{57a}$$

$$M(t)=\sum_{\mathbf{r}} M_{\mathbf{rx}}(t)=\sum_{\mathbf{r}} m_{\mathbf{rx}}\chi(t)=\frac{9}{4}M_I(t), \tag{57b}$$

which follows directly from Eq. (49), (50) and (51). The translation invariance of the crystal was taken into account as well.

Realization of this program was fulfilled for $M_4/M_2^2=19/9$, that can be considered as reasonable universal approximation for any simple lattice. Three different forms for function $\kappa(t)$ were applied:

1) $\kappa(t)=\exp(-\tfrac{1}{4}M_2t^2)$,
2) $\kappa(t)=1/\cosh^2\left(-\tfrac{1}{2}(M_2)^{1/2}t\right)$,
3) $\kappa(t)=1/\left(1+\tfrac{1}{3}M_2t^2\right)^{3/4}$.

Existence of oscillation of the FID was stable, and relative variations of FID due to application of these $\kappa(t)$ were reasonably small. Agreement of the results with famous measurement [46] of FID in CaF_2 is better, than solution for simple Gaussian memory function represented by Fig. 1.1 in [17] for NMR form function.

In conclusion of the section we see, that existence of oscillations of $F(t)$ and absence of oscillation in $F_I(t)$ is connected directly with coefficient 9/4 in (57), which reflects transfer of transversal polarization in Eqs. (49) and (50). From mathematical point of view we have the same effect, as in equation of motion for oscillator coordinate $x(t)$ with frequency Ω_0 and friction coefficient μ, written in integral form

$$\frac{\partial}{\partial t}x(t)=-\Omega_0^2\int_0^t d\tau \exp(-\mu\tau)x(t-\tau).$$

It is evident, that with increasing of Ω_0 we have here transition from simple decay to oscillation decay at $\Omega_0=\mu/2$.

5.4.2. *Magnetically diluted systems.* In order to generalize the construction to magnetically diluted systems we should invent substitutions for relations (51), (53) and (55), because all second moments became infinite, as it was discussed in section 3. For the sake of brevity we will use Laplace representation, and according to section 4 we will apply concentration expansion. The relations (51) and (53) for memory kernels can be substituted by

$$M_{\mathbf{rx}}(\lambda)=m_{\mathbf{rx}}(\lambda)\sigma(\lambda),\quad M_I(\lambda)=m_0(\lambda)\sigma_0(\lambda), \tag{58}$$

where functions $m_{\mathbf{rx}}(\lambda)$ and $m_0(\lambda)$ contain terms of order c^1 only and they produce the terms of the same order in concentration expansions of $F_{\mathbf{rq}}(t)$ and $F_I(t)$. Following (54) we will apply the approximation

$$\sigma(\lambda)=\sigma_0(\lambda). \tag{59}$$

For independent determining of auxiliary FID $F_I(t)$ we use the fact [47], that if instead of exact calculation (10) of Anderson FID we suppose, that in any random configuration the FID is Gaussian with exact second moment, then

$$F_A(t) = \left\langle \left\langle \exp\left(i\sum\nolimits_{\mathbf{q}} n_{\mathbf{q}} I_{\mathbf{q}}^z b_{\mathbf{q}0} t\right)\right\rangle_0 \right\rangle_c \to$$

$$\to F_G(t) = \left\langle \exp\left(-\frac{1}{2}\sum\nolimits_{\mathbf{q}} \left\langle \left(I_{\mathbf{q}}^z\right)^2 \right\rangle_0 n_{\mathbf{q}} b_{\mathbf{q}0}^2 t^2 \right)\right\rangle_c = \exp\left(-0.80 D_A t\right).$$

It means that this approximation produces surprisingly exact result. Therefore, if local fields are directed along the z-axis and depend on time, we obtain qualitatively correct representation of the process applying AWK-theory in any fixed configuration of impurities with result of the form

$$F_I(t) = \exp\left(-\left(2B^2\int_0^t d\tau\left(t-\tau\right)\exp\left(-\alpha B\tau\right)\right)^{1/2}\right), \tag{60}$$

where coefficients B and α can be received from concentration expansions and the function $\kappa(\tau) = \exp(-\alpha B\tau)$ here has the same sense as $\kappa(\tau)$ in Eqs. (55) and (56). It is simplest function, which represents influence of flip-flop transitions in surrounding spins and is compatible with analytical structure of concentration expansion for d=3. Therefore choice of this simple exponential form corresponds to Gaussian variant of $\kappa(\tau)$ for dense systems. The radical in (60) represents static fluctuations of interactions due to random distribution of positions of spins. If $\alpha = 0$, then (60) coincides with exact Anderson's solution for impurity nucleus in absence of flip-flop transitions for surrounding spins, that defines $B = 2D_A/3$.

For calculation of the FID it is sufficient to solve the Eq. (57a) with memory kernel

$$M(t) = \frac{3}{2}M_I(t), \tag{61}$$

where $M_I(t)$ is solution of Eqs. (53a) and (60), in particular

$$M_I(\lambda) = \left(F_I(\lambda)\right)^{-1} - \lambda, \quad F(\lambda) = \left(\lambda + \tfrac{3}{2}M_I(\lambda)\right)^{-1}, \tag{62}$$

and the coefficient 3/2 in (61) again reflects the relation between exact first terms of the concentration expansions for $F(t)$ and $F_I(t)$.

First two terms of concentration expansion for $F(t)$, calculated in Ref. 23 are of the form

$$F(t) = 1 - D_A t + \frac{5}{9}\left(D_A t\right)^2 + O\left(\left(D_A t\right)^3\right), \tag{62}$$

that produces $\alpha = 1/2$ and, instead of (60), we have

$$F_I(t) = \exp\left(-2\left(Bt + 2\exp\left(-Bt/2\right) - 2\right)^{1/2}\right). \tag{60a}$$

Numerical analysis indicates, that in the time region where $Bt=2D_At/3>3.2$ and $F_I(t)>1/30$ the solution $F(t)$ can be approximated by a relation

$$F_{appr}(t) = \frac{2}{5}F_I(t)\left(1 + \frac{3}{2}\left(F_I(t)\right)^{5/6}\right) - \frac{1}{12}F_I(t)\left(1 - F_I(t)\right)^2, \tag{63}$$

and relative error does not exceed 1%. The error monotonically increases later with increasing of the time and it comes up to 10% for Bt=5, and 25% for Bt=10.

Calculation of the form function $g(\omega)$ can be fulfilled in terms of Fourier-Laplace transformation of $F_I(t)$. Indeed, if we introduce the functions

$$g_c(\omega) = \int_0^\infty dt \cos\omega t F_I(t), \quad g_s(\omega) = \int_0^\infty dt \sin\omega t F_I(t),$$

then

$$g(\omega)=\frac{1}{2\pi}\left(F\left(\lambda=\varepsilon-i\omega\right)+F\left(\lambda=\varepsilon+i\omega\right)\right)=\frac{6}{\pi}\cdot\frac{g_c(\omega)}{\left(\omega g_s(\omega)-3\right)^2+\left(\omega g_c(\omega)\right)^2}. \tag{64}$$

Obtained FID (63) has no oscillations on the scale $T_2=1/D_A$ and it is monotonically decreasing function of time, because it is a solution in continuum media approximation. But oscillations on the scale $t_0=1/b_0$ can be significant in real crystals even for concentration c of order several percents. Here b_0 is a constant of dipole interaction between nearest spins in the sample. To recover them it is sufficient to replace D_At in the solution by initial sum, wherefrom D_At was produced:

$$D_At\to c\sum\nolimits_{\mathbf{z}}\left(1-\cos\frac{1}{2}b_{\mathbf{z}0}t\right). \tag{65}$$

As a consequence of this substitution monotonic $g(\omega>0)$ can receive satellite maxima at $\omega\sim b_0$. Both this properties are similar to relations between FIDs (10) and (12) and between corresponding NMR form-functions.

There exist no experimental results which can be compared directly with relations (63), (64) or with their modification for real crystals. Probably it is connected with difficulties in obtaining of homogeneous polarization for magnetically diluted systems. Best results were received in measuring of spin echo for nuclei ^{29}Si in silicon in Ref. 10 and in studies of NMR form functions in Refs. 7-9. But these results are contradictory.

The FID presented on Fig.1 from [10] has decay rate $D_{[10]}=2D_A/3$ for $t<1/D_A$ instead of expected D_A. The result indicates that the sample did not correspond to the model under consideration, because theoretical predictions for this time range are reliable. Probably the value $D_{[10]}$ is a consequence of existence of inhomogeneous Knight shift. Estimations indicate that for conditions, corresponding to this measurement, scale of the shift is of order $2\pi\cdot 3\cdot 10^3$ Hz, while $D_A=2\pi\cdot 42$ Hz, and typical variation of the shift at interspin distance exceeds D_A. In these conditions flip-flop term of the dipole Hamiltonian is not secular, and it should be omitted, producing decay rate $2D_A/3$.

NMR form functions, received in Refs. 7-9, have line widths substantially larger than the theoretical expectation (contrary to [10]), that indicates on additional source of broadening and on impossibility of direct application of the Hamiltonian (4) to the samples. But authors of Ref. 9 found a way to separate additional line broadening. After this correction line width, corresponding to FID (63), is in better agreement with results [9] than line width for simple exponential Anderson FID (12).

6. Path integral representations for RWDS

Path integrals serve as common general language of modern theoretical physics. Path integral representations of propagators of RWDM did not produced up to now the results, which we can not receive by operator methods. Nevertheless this way is very useful for understanding of the place of the problem in physical theory as a hole, and to recognize its complexity. Discussion of the problem requires special technique, therefore we will make only some general remarks here.

Many different path integrals representations for RWDS are possible. We expect, that most interesting are those, where configurational averaging over random impurities positions is fulfilled analytically. Equations (6) and (13) produce general basis for this action in the way, similar to derivation of relation (9). Path integral representations of this type were introduced in Ref. 48. For the sake of brevity we will consider only simplified version of the theory, where Eq. 6 has reduced initial condition $\widetilde{P}_{\mathbf{xy}}(t=0)=\delta_{\mathbf{xy}}$ that is sufficient for studies of long time asymptotics.

"Quantum mechanical" type of corresponding reduced propagator in continuum media approximation is of the form

$$P^r_{\mathbf{xy}}(t) = \left\langle \left(\exp\left(-\sum\nolimits_{\mathbf{z}} n_{\mathbf{z}} A^{\mathbf{z}} t\right) \right)_{\mathbf{xy}} \right\rangle_c = \int D\mathbf{p}(\tau) D\mathbf{q}(\tau) \exp\left(i\int_{\mathbf{y}}^{\mathbf{x}} \mathbf{p} d\mathbf{q} - L[\mathbf{p},\mathbf{q},t] \right),$$
$$L[\mathbf{p},\mathbf{q},t] = n\int d^d z \left(1 - \exp\left(-\int_0^t d\tau A^{\mathbf{z}}(\mathbf{p}(\tau),\mathbf{q}(\tau))\right)\right), \quad A^{\mathbf{z}}(\mathbf{p},\mathbf{q}) = \nu_{\mathbf{zq}}\left(1 - \exp(-\mathbf{p}(\mathbf{z}-\mathbf{q}))\right). \tag{66}$$

Exact definition of the integral, which is very important for real calculations, can be found in [48].

The representation (66) is similar to, but more complex than path integrals in the famous polaron problems (PP) [49]. The main difference consists of multi-time action L in (66) instead of two-time action in PP, in strong singularity of $\nu_{\mathbf{zq}}$ instead of the less singular kernel $|\mathbf{z}-\mathbf{q}|^{-1}$ in PP, in additional path integral over all $\mathbf{p}(\tau)$ and in strong dependence of the infinite-fold integral (66) on the exact form of approximations by the integrals with finite multiplicity.

Representation of superfield type is of the form [48]

$$P^r_{\mathbf{xy}}(\lambda) = \left\langle \left(\left(\lambda + \sum\nolimits_{\mathbf{z}} n_{\mathbf{z}} A^{\mathbf{z}}\right)^{-1} \right)_{\mathbf{xy}} \right\rangle_c = \int D\Phi D\Phi^+ a_{\mathbf{x}} a^+_{\mathbf{y}} \exp\left(-\lambda\Phi^+\Phi - L\left[\Phi^+,\Phi\right]\right),$$
$$L\left[\Phi^+,\Phi\right] = c\sum\nolimits_{\mathbf{z}}\left(1 - \exp\left(-\Phi^+ A^{\mathbf{z}}\Phi\right)\right). \tag{67}$$

Here superfield $\Phi_{\mathbf{q}}$ has two components: $\Phi_{\mathbf{q}}=(a_{\mathbf{q}},\alpha_{\mathbf{q}})$, where $a_{\mathbf{q}}$ is usual Bose field, while $\alpha_{\mathbf{q}}$ is anticommuting Fermi field, and the product $\Phi^+ Q\Phi = \sum_{\mathbf{xq}}\left(a^+_{\mathbf{x}} Q_{\mathbf{xq}} a_{\mathbf{q}} + \alpha^+_{\mathbf{x}} Q_{\mathbf{xq}}\alpha_{\mathbf{q}}\right)$ for any operator Q. It is evident, that relation (67) defines strongly nonlinear superfield theory, which has strong divergences in continuum media approximation.

7. Numerical methods in RWDS

Main problem for numerical simulation of RWDS, and especially in determining of long time asymptotics, consists in description of infinite system in a finite computer. Effective solution was constructed in Refs. 27-29 and 31.

The method consists in substitution of infinite disordered media by a crystal with a large disordered elementary cell (supercell), containing $N_d \sim 1000$ impurity spins, and on checking of stability of the results with regard to variation of N_d. It starts from Eq. (1) and N_d impurities randomly placed in sites of a supercell, having $N = N_d / c = N_g^3$ lattice sites, and edges $\mathbf{R}_\alpha = N_g \mathbf{b}_\alpha$, where $\mathbf{b}_\alpha$ form the basis of the matrix crystal (LiF for the system ^{8}Li-^{6}Li). Then the supercell is continued periodically to cover all space. If $N_d \to \infty$, then we go back to random media. It is very important that the initial condition remains of the correct form and that it is not continued periodically. The eigenvalues problem

$$\sum\nolimits_j A_{ij}\phi_j(m) = \varepsilon_m \phi_i(m) \tag{68}$$

has Bloch's solution

$$\phi_j(m) = \exp\left(i\mathbf{k}\mathbf{r}_j\right)\chi_j(\mathbf{k},\mu), \quad m = (\mathbf{k},\mu), \tag{69}$$

where $\chi_j(\mathbf{k},\mu)$ has periodical dependence on impurity position $\mathbf{r}_j$ and $\mathbf{k}$ belongs to Brillouin zone V_B formed by all $\mathbf{k}$ satisfying the condition $\mathbf{k}\mathbf{B}_\alpha \le \pi$. Therefore, we have a finite eigenproblem for $\chi_j(\mathbf{k},\mu)$:

$$\sum\nolimits_{j=0}^{N_d-1} B_{ij}(\mathbf{k})\chi_j(\mathbf{k},\mu) = \varepsilon_m(\mathbf{k})\chi_i(\mathbf{k},\mu), \tag{70}$$
$$B_{ij}(\mathbf{k}) = \sum\nolimits_{n_\alpha} \exp\left(-i\mathbf{k}\left(\mathbf{r}_i - \mathbf{r}_j - \mathbf{R}(\mathbf{n})\right)\right) A_{ij}\left(\mathbf{r}_j + \mathbf{R}(\mathbf{n})\right), \quad \mathbf{R}(\mathbf{n}) = \sum\nolimits_{\alpha=1}^{3} n_\alpha \mathbf{R}_\alpha .$$

Here $A_{ij}\left(\mathbf{r}_j + \mathbf{R}(\mathbf{n})\right)$ is A_{ij} with substitution $\mathbf{r}_j \to \mathbf{r}_j + \mathbf{R}(\mathbf{n})$.

As a result the propagator in fixed configuration of impurities

$$p_{j0}(t) = V_B^{-1} \int_{V_B} d^3k \exp\left(i\mathbf{k}\mathbf{r}_j\right)\left(\exp\left(-B(\mathbf{k})t\right)\right)_{j0},$$

and averaged propagator

$$P_{00}(t) = V_B^{-1} \int_{V_B} d^3k \left\langle \left(\exp\left(-B(\mathbf{k})t\right)\right)_{00} \right\rangle_c, \quad (71)$$

$$P_{\mathbf{x}0}(t) = V_B^{-1} \int_{V_B} d^3k \exp\left(i\mathbf{k}\mathbf{x}\right)\left\langle \left(\exp\left(-B(\mathbf{k})t\right)\right)_{10} \right\rangle_c^{(\mathbf{r}_1=\mathbf{x})}, \quad (72)$$

where subscript ($\mathbf{r}_1$=**x**) indicates that averaging is fulfilled under condition that site **x** is occupied by an impurity spin having number 1 (the site 0 is always occupied by spin "0").

Primarily [27-29] calculations were directed on determination of the diffusion tensor via calculation of smallest eigenvalue of the operator $B_{ij}(\mathbf{k})$ solving the equation of motion with special initial condition and using modification of Adams algorithm for ordinary differential equations. After checking of stability of the results to values of $N_d \geq 500$ full diagonalization of $B_{ij}(\mathbf{k})$ and relation (71) were applied in [31]. The results are in satisfactory agreement with experimental studies of the model system ^{8}Li-^{6}Li in LiF [5, 33].

8. Spin dynamics in magnetically diluted systems with low Zeeman and medium low dipole temperatures

Development of EPR repeated in many problems findings of NMR. Particularly two temperature NMR theory of Provotorov was successfully generalized to EPR that produced a possibility to study many new effects [50]. Standard two temperature theory was developed in high temperature approximation (HTA), when $\omega_0 / T_Z \ll 1$ and $\omega_{loc} / T_d \ll 1$. Here ω_0 and ω_{loc} are Zeeman and local frequencies, and T_Z and T_d are temperatures of Zeeman and dipole reservoirs. Conditions with $\omega_0 / T_Z \geq 1$ are not seldom in EPR, therefore the theory received corresponding extension in Refs. [51-54]. Local frequency in NMR has two different definitions – from heat capacity and from resonance line width [17]. In HTA both definitions produce close results for NMR and are related with second moment $M_2 \sim \omega_{loc}^2$. In EPR the line width is scaled on D_A at any temperatures, while heat capacity in HTA is scaled on M_2 again, and $M_2 \sim D_A^2 / c$. Therefore we come to natural problem of extension of the theory to temperature region where dipole temperature is medium low, that is $D_A / T_d \ll 1$, but $M_2^{1/2} / T_d \geq 1$. This step was fulfilled in Refs. 55-57, where two temperature fluctuation-dissipation theorem was derived, Provotorov's equations were generalized to arbitrary Zeeman and medium low dipole temperatures, dipole and Zeeman heat capacities and resonance form functions were constructed on the basis of concentration expansions and fluctuation shift was revealed. More detailed information can be found in original articles and the review [58]. Applications of these results to studies of EPR in magnetic fluids [1, 59] can be indicated among more modern investigations. These references contain generalization of the theory to samples where local fields are produced by elongated aggregates of nanoparticles instead of point dipoles and corresponding experimental studies.

9. Cluster expansions and spectral dynamics

Preceding review of magnetically diluted systems was directed in main on studies of spin evolution with simple well defined initial condition (FID) or with small external perturbation (resonance form function), when resonance line can be considered as homogeneous. If resonance perturbation is not small the dynamics became more complex [58]. Important example was investigated in [60], where dipole broadened EPR line, which as a norm should be homogeneous, revealed inhomogeneous properties. In particular, in reaction on strong resonance field burning of the hole in resonance line was observed. It is important that wing of the hole was nearly exponential contrary to standard expectation of Lorentzian line wings for homogeneous dipole lines. One of important conclusion of the work consisted in understanding of importance of close spin pairs with interspin distances much smaller than average distance between spins r_{av}. Special influence of similar spin pairs was studied as well in spin disordered systems [61, 4] among investigations [62, 63, 4] of other processes beyond the Eqs. 1. Later a theory of two-spin resonance in magnetically diluted systems was constructed, where influence of close pairs determined unusual dependence of the process rate on magnitude of RF field ω_1 [64]. Importance of influence of close pairs on lack of homogeneity of dipole line was supported as well by Ref. 47.

Regular construction for consideration of similar effects is cluster analysis (or cluster expansions) developed in Ref. 65 and 66.

First of all we should underline, that concentration expansions and cluster expansions are wery different. Both methods use occupation numbers but have different aims. Destination of concentration expansions consists in expansion of observable in powers of impurity concentration c, while cluster analysis is directed on decomposition of the system itself in such constituents as spins of mass, and two-spin, three-spin and so on clusters. The decomposition admits then to separate processes with different time scale and spectra of different nature. Very important guiding line consists in elimination of divergences, existing in naive high temperature spin dynamics in continuum media approximation. First of them is dipole heat capacity and (proportional to it) second moment of usual resonance line form function.

By definition k-cluster is a group of k spins in which every spin has more strong intragroup interaction than all its out-of-group interactions. As a measure of interaction between spins i and j we use the value of b_{ij} from dipole Hamiltonian (4). With this definition a spin can belong to several clusters simultaneously, for example, to a 2-cluster, and to 3-cluster. In applications it is convenient to have a division in which each spin belongs to no more than one cluster. This is obtained if we fix the rank k_{max} of the largest cluster, and then exclude from the k-clusters with $k< k_{max}$ all those clusters that also belong to clusters of higher rank q ($k < q \leq k_{max}$). It is natural to designate the clusters thus obtained as orthogonal.

It was found [65] that if from the entire system we separate out 2- and 3-particle clusters as carriers of the discrete spectrum, the remaining spins (the totality of which we call the mass) have finite dipole heat capacity and second moment, which are determined by interactions at average distance. Therefore, in analogy with standard NMR of dense spin systems, we can expect, that spectrum of the mass of spins can be considered as continuous. As a result spins of mass produce homogeneous broadening of cluster spectra. Transitions between states of the cluster take place due to interactions with other clusters and with mass spins, and they are slow. Numerical analysis [65] shows, that orthogonal 2- and 3-clusters contain 51(1)% and 11(1)% of all spins correspondingly if k_{max}=3, but 2-clusters define all Lorentz wings of the EPR line.

The regularization of the spin-spin interactions in the mass, consisting in finiteness of the second moment, makes it possible to hope that for the mass and for the system as a whole the fastest process will be the establishment of quasiequilibrium in the Zeeman and dipole reservoirs of the mass, for which the two-temperature description is thereby preserved, while the kinetics of the clusters should be constructed with allowance for the discreteness of their spectra. Spins of the mass are still subject to inhomogeneous broadening from the interaction with the spins of the clusters, and the two-temperature description for them will be valid only over times greater than the time of establishment of equilibrium in this inhomogeneously broadened spectrum.

As a whole, the cluster expansion gives a constructive alternative to spin packets theory of Portis [67], used in [60] for interpretation of the data. The spin packet is a global concept, characterizing the entire spin system as a whole, and its shape does not depend on the intensity of the field ω_1. But the absorption of the external field occurs locally, and has an essential dependence on the real configuration of the spins that participate in each elementary act. Unlike the shape of a spin packet, the line-shape function describing the homogeneous broadening of these spins is not a self-averaging quantity, and only directly observable quantities are subject to configurational averaging. In accordance with this, we shall examine the results of Ref. 60 starting from the ideas developed above.

Transitions between states of the cluster take place due to interactions with other clusters and with mass spins, and they are slow. As a result saturation at the EPR line wing induces transitions between states of the 2-clusters, and the line shape of the hole is defined by interaction of the cluster with mass spins. If this interaction is estimated within the AWK model, then any mass configuration produces exponential wings of the hole $g(\Delta) \sim \exp(-\Delta / \mu)$, where Δ is detuning of the saturating field from the cluster transition frequency, and μ is defined by magnitude of the fields produced by other spins on the cluster, and by the rate of their fluctuations. After configurational averaging the wings become $g(\Delta) \sim \Delta^{-4}$, but averaged hole shape describes the transitions for small magnitude ω_1 of the saturating field only. In a general case the observable area of the hole σ should be averaged. For large ω_1 we have $\sigma \sim \ln\left(t_p \omega_1^2 / \mu\right)$, where t_p is the duration of the saturating pulse. Logarithm is a slow function, therefore $\langle \sigma \rangle_c \sim \ln\left(t_p \omega_1^2 / \langle \mu \rangle_c\right)$, which was observed in [60] and was interpreted as the exponential wing of Portis's packets. The described theory produces a microscopic picture of the phenomenon and indicates that Portis's packets have limited relation to the problem, because they are homogeneous by definition, i. e. $g(\Delta) = \langle g(\Delta) \rangle_c$, while in more microscopic cluster theory this relation isn't fulfilled.

10. Conclusion

Author believes that spin dynamics and magnetic resonance are most developed branches of nonequilibrium statistical physics. This position is based on such brilliant studies as investigation of phase transitions in nuclear magnetism [17] and inversion of evolution of spin systems [68]. Existence of ergodic theorem for realistic spin systems with nontrivial interspin interaction [69] is, probably, unique as well.

Preceding discussion indicates that spin dynamics of disordered systems exists now as developed section of the physics. It incorporates achievements of spin dynamics of dense systems and has its own problems, conceptions, methods and perspectives.

Modern state of understanding of the processes in disordered spin systems is satisfactory, because experimental studies in model spin system ^{8}Li-^{6}Li and successful numerical simulation of the process gave enough information for construction relatively full theoretical description. Nevertheless new measurements will be useful for better understanding of the nature of primary transition rates and their dependence on interspin distance. Of course, analytical solutions for long time and intermediate asymptotics of the processes still very desirable.

We have much less progress in description of magnetically diluted systems. New experimental investigations are necessary. First of all we need in precise measurement of free induction decay in diluted nuclear systems, which can be compared with classical work [46], fulfilled for dense nuclear paramagnets.

The work was supported by Russian Foundation for Basic Research (project # 11-02-00880).

References

[1] Dzheparov F S, Sorokina O N and Kovarski A L 2009 *JETP Letters* **89** 563
[2] Dzheparov F S 2007 *J. of Superconductivity and Novel Magnetism* **20** 161
[3] Dzheparov F S 2009 Spin dynamics in disordered solids *Encyclopedia of Complexity and Systems Science* Ed. Robert A Meyers (Berlin: Springer) pp. 8588-97
[4] Abov Yu G, Bulgakov M I, Borovlev S P, Gulko A D, Garochkin V M, Dzheparov F S, Stepanov S V, Trostin S S and Shestopal V E 1991 *Sov. Phys. JETP* **72** 534
[5] Dzheparov F S, Gulko A D, Ermakov O N Lyubarev A A, Stepanov S V and Trostin S S 2009

Appl. Magn. Reson. **35** 411
[6] Dzheparov F S 1991 *Sov. Phys. JETP* **72** 546
[7] Hoch M J R and Reynhardt E C 1988 *Phys. Rev.* B **37** 9222
[8] Verhulst A S, Maryenko D, Yamamoto Y and Itoh K M 2003 *Phys. Rev.* B **68**. 054105
[9] Hayashi H, Itoh K M and Vlasenko L S 2008 *Phys. Rev.* B **78**. 153201
[10] Li D, Dong Y, Ramos R G, Murray J D, MacLean K, Dementyev A E and Barrett S E 2008 *Phys. Rev.* B **77**. 214306
[11] Abragam A 1961 *The Principles of Nuclear Magnetism* (Oxford: Oxford Science)
[12] Dzheparov F S and Lundin A A 1978 *Sov. Phys. JETP* **48** 514
[13] Forster T 1949 *Z. Naturforsch.* (A) **4** 321
[14] Anderson P W 1951 *Phys. Rev.* **82** 342
[15] Zvyagin IP 1984 *Transport Phenomena in Disordered Semiconductors* (Moscow: MSU Publ)
[16] Scher H and Lax M 1973 *Phys Rev* B **7** 4491, 4502
[17] Abragam A and Goldman M 1982 *NuclearMagnetism: Order and Disorder* (Oxford: Oxford Science)
[18] Zobov V E, Lundin A A and Rodionova O E 2001 *JETP* **93** 542
[19] Dzheparov F S, Smelov V S and Shestopal V E 1980 *JETP Lett.* **32** 47
[20] Dzheparov F S, Lundin A A and Khazanovich T N 1987 *Sov. Phys. JETP* **65** 314
[21] Kubo R 1962 *J. Phys. Soc.* Jpn **17** 1100
[22] Shestopal V E 1981 *Preprint ITEP*-139 (ITEP, Moscow)
[23] Dzheparov F S and Kaganov I V 2002 *JETP Lett.* **75** 259
[24] Forster D 1975 *Hydrodynamic Fluctuations, Broken Symmetry, and Correlation Functions* (Massachusetts: W.A. Benjamin)
[25] Abov Yu G, Gul'ko A D, Dzheparov F S, Stepanov S V and Trostin S S 1995 *Phys. Part. Nucl.* **26** 692
[26] Dzheparov F, Gul'ko A, Heitjans P, L'vov D, Schirmer A, Shestopal V, Stepanov S and Trostin S 2001 *Physica* B **297** 288
[27] Dzheparov F S, Lvov D V, Nechaev K N and Shestopal V E 1995 *JETP Lett.* **62** 662
[28] Dzheparov F S, Lvov D V and Shestopal V E 1998 JETP 87, 1179
[29] Dzheparov F S, Lvov D V and Shestopal 2007 *J. of Superconduct. and Novel Magnetism* **20** 175
[30] Dzheparov F S and Stepanov S V 1996 *Hyperfine Interactions* C **1**, 609, 1996
[31] Dzheparov F S 2005 *JETP Lett.* **82** 521
[32] Gul'ko A D, Ermakov O N, Stepanov S V and Trostin S S 2007 *J. of Supercond. and Novel Magnetism* **20** 169
[33] Gul'ko A D, Dzheparov F S, Ermakov O N Lyubarev A A, Stepanov S V and Trostin S S 2008 *J. of Surface Investigation* **2**, 856
[34] Alexandrov I V 1975 *Theory of Magnetic Relaxation* (Moscow: Nauka).
[35] Atsarkin V A 1980 *Dynamical Polarization of Nuclei* (Moscow: Nauka).
[36] Khutsishvili G R 1966 *Sov. Phys. Usp.* **8** 743
[37] Malkin B Z, Vanyunin M V, Graf M J, Lago J, Borsa F, Lascialfari A, Tkachuk A M and Barbara B 2008 *Eur. Phys. J.* B **66** 155
[38] Tabti T, Goldman M, Jacquinot J-F, Fermon C and Le Goff G 1997 J. Chem. Phys. **107** 9239
[39] Dzheparov F S, Jacquinot J-F and Stepanov S V 2002 *Physics of Atomic Nuclei* **65** 2052
[40] Balagurov V Y and Vaks VT 1973 *Sov Phys JETP* **38** 968
[41] Pastur L A 1977 *Teor Mat Fiz* **32** 88
[42] Donsker M and Varadhan S 1975 *Commun Pure Appl Math* **28** 525
[43] Havlin S, Dishon M, Kiefer J E and Weiss G H 1984 *Phys. Rev. Lett.* **53** 407
[44] Renn S R 1986 *Nucl Phys* B**275**(FS**17**) 273
[45] Bulgakov M I, Gulko A D, Dzheparov F S, Stepanov S V and Trostin S S 1993 *JETP Lett.* **58** 592

[46] Engelsberg M and Lowe I J 1974 *Phys. Rev.* B **10** 822
[47] Drabolt D A and Fedders P A 1988 *Phys. Rev.* B **37** 3440
[48] Dzheparov F S and Shestopal V E 1987 *Russian Physics Journal* **30** 509
[49] Feynman R P 1972 *Statistical mechanics* (Massachusetts: W.A. Benjamin)
[50] Atsarkin V A and Rodak M I 1972 *Sov. Phys. Usp.* **15** 251
[51] Mims W 1972 *Electron Paramagnetic Resinance* **9** 263
[52] Altshuler S A and Mokeev A A 1969 *Fiz. Tverd. Tela* **11** 35
[53] Kochelaev B I and Nigmatulin R R 1972 *Fiz. Tverd. Tela* **14** 3414
[54] Maryasov A G, Dzuba S A and Salikhov K M 1982 *J. Magn. Res.* **50** 432
[55] Dzheparov F S and Henner E K 1986 *Fiz. Tverd. Tela* **28** 2182
[56] Dzheparov F S and Henner E K 1989 *Phys. Stat. Sol.* (b) **151** 663
[57] Dzheparov F S and Henner E K 1989 Zh. Eksp. Teor. Fiz. **96** 1884
[58] Atsarkin V A 1991 *Mag. Res. Rev.* **16** 1
[59] Sorokina O N, Kovarsky A L and Dzheparov F S 2010 *AIP Conf. Proc.* **1131** 96
[60] Atsarkin V A, Vasneva G A and Demidov V V 1986 Zh. Eksp. Teor. Fiz. **91** 1523
[61] Abov Yu G, Bulgakov M I, Borovlev S P, Gulko A D, Garochkin V M, Dzheparov F S, Stepanov S V, Trostin S S and Shestopal V E 1986 *Izv. Akad. Nauk SSSS Ser. Fiz.* **50** 2354
[62] Abov Yu G, Bulgakov M I, Borovlev S P, Gulko A D, Garochkin V M, Dzheparov F S, Stepanov S V, Trostin S S and Shestopal V E 1988 *Izv. Akad. Nauk SSSS Ser. Fiz.* **52** 460
[63] Abov Yu G, Bulgakov M I, Borovlev S P, Gulko A D, Garochkin V M, Dzheparov F S, Stepanov S V, Trostin S S and Shestopal V E 1988 *Izv. Akad. Nauk SSSS Ser. Fiz.* **52** 1699
[64] Dzheparov F S and Fel'dman E B 1988 *Izv. Akad. Nauk SSSS Ser. Fiz.* **52** 455
[65] Dzheparov F S and Henner E K 1993 *JETP* **77** 753
[66] Dzheparov F S, Kaganov I V and Henner E K 1997 *JETP* **85** 325
[67] Portis A M 1953 *Phys. Rev.* **91** 1071
[68] Rim W-K, Pines A and Waugh J S 1970 *Phys. Rev. Lett.* **25** 218
[69] Dzheparov F S 1999 *JETP* **89** 753

Journal of Physics: Conference Series **324** (2011) 012005 doi:10.1088/1742-6596/324/1/012005

The NMR line shape of magneto-active nanoclusters in moveable nano-containers with self-similar stochastic dynamics

A A Khamzin[1], R R Nigmatullin[1], A P Alekhin[1], A S Sitdikov[2], A S Nikitin[2]

[1]Kazan (Volga Region) Federal University, 420008, Kremlevskay 18, Kazan, Russia
[2]Kazan State Power Engineering University, 420066, Krasnosel'skay 51, Kazan, Russia

e-mail: airat.khamzin@rambler.ru

Abstract. NMR line shape of isolated spin clusters in moving nanocontainers with self-similar correlation law was calculated. It was shown that taking into account self-similarity in stochastic dynamics of pores leads to new shapes of NMR lines which differ from traditional shapes of Gauss and Lorentz types. Fractal dimension of spatial-temporal ensemble can serve as convenient fitting parameter for experimental data interpretation.

1. Introduction

In last decade one can see sharp increase of interest to study materials with nanometer pores and there are two reasons for it. In the first place synthesis of materials with pores with dimensions of several nanometers is the part of nanotechnology projects along with synthesis of nanoparticles. Laboratory research directed to obtaining homogeneous pores is conditioned by dependence of porous material properties from dimensions, shape and dimension distribution of pores [1]. For example, materials with ordered pores in the form of nanotubes (zeolite, aluminum silicates and so on) with constant section of tube mouth serve as molecular sieves for separation of mixture of gases molecules by their dimensions. Zeolite microcontainers are used as shells for dye molecules in dye lasers, as vessels for gases, electrolytes, as containers for toxic molecules of contrast medium for magnetic-resonance tomography and so on. In all cases nanocontainers affect the operation of the whole system container + filling. In the second place progress in the field of the nanotechnology have lead to the appearance of the spin physics of nanostructures and particularly to NMR spectroscopy of the gases in nanocavity [2]. High resolution of the NMR spectra and possibility of the precise analytic description of the NMR line shapes of the gases with molecules with nuclear spins in nanocavities [3] allow to get from experimental data the information about structure and dynamics of the nanocontainers investigated [3, 4].

Nanometer range of measurements opens the world of new material properties [5-9]. In comparison with bulk solid substance there are changes in parameters of crystal lattice and atomic dynamics, thermal and electronic properties; magnetic properties is changed, magnetic clusters become one-dimensional, for several metals increase and even appear magnetic moments on atom, abrupt magnetic phase transitions are observed and superparamagnetism occurs. All these effects have dimension character and strongly depend on nanocluster surface state, interactions between clusters and interaction of clusters with matrix. The properties of the isolated clusters differ from properties of the clusters that are forming nanosystems. Weakly interacting or isolated nanoclusters can be obtained in the form molecular clusters in gas-phase reactions by means of laser induced evaporation with consequent study in time-of-flight mass spectrometer and with application of photoelectron spectroscopy (see for example [11]), or by means of matrix isolation at solid-state and colloidal synthesis under condition of the weak interaction of clusters with matrix. Interactions between clusters and interactions of clusters with matrix allow not only change the properties of isolated clusters but also create complex crystal and supramolecular structures in which clusters play role of the atoms like regular crystal.

Success of analytic calculation of the NMR spectra of the gases and nanoclusters with fast configuration rebuilding in nanocavities was due to presence in system the small adiabatic parameter $\varepsilon = \tau_m / \tau_f << 1$, where τ_m - characteristic time of changing the spin-spin interactions of the particles as a result of the fast molecular motions in gas or fast configuration rebuilding of the cluster, τ_f - characteristic time of the spin magnetization flip of the particle as a result of the interaction with nearest neighborhood spin (coincide with flip-flop transition time). As a result with accuracy ε cluster spin dynamics with spin ½ in nanocavities on time interval

$$\tau_m << \Delta t << \tau_f \qquad (1)$$

is described by effective Hamiltonian H (see formula (2) of the first section) with space-independent effective spin-spin couplings. It is commonly known that main difficulty of the consecutive statistical calculation in big space-extended systems is the absence small adiabatic parameter. In such a case usually one have to use model Hamiltonians, the field of application of which is determined by comparison with results of sophisticated numerical calculations from first principles and also with results of the experiments. Contrary to that space-extended systems in which modeling linked with complicated accounting spatial-dependent interactions between spins at arbitrary values of coupling constants the spin dynamics and also thermodynamics with Hamiltonian (2) in limited nanocavities is actually precise and can be studied as much as possible in detail by analytic methods.

In papers [10, 11] with the help of the statistical Hamiltonian (2) the process of polarization transfer between nuclei in nanocavity is studied. The main result of this investigation was establishment of the fact that polarization transfer is nonergodic process. In paper [3] the method of the NMR line shape calculation of the gas of the spin bearing molecules in movable nanocontainers (described by Hamiltonian (2)) was developed. Taking into account fluctuation in time of the interdipolar bond constant the authors of this work for different time scales have obtained analytic expressions for line shape which differ from traditional Gauss or Lorentz dependencies. Particularly the line shape can have logarithmic singularity at $\omega \to 0$ and exponential decay on the tails. Singular behavior of the line shape can serve as recipe for finding the map of the angular distribution of the different ellipsoidal cavities.

Beginning with work [12] and then in works [13, 14] the authors offer to use model (2), which get the name of the model of the equal spin-spin interactions (ESSI), for description of the magnetoactive nanoclusters properties. For adequate description in the frame of the Hamiltonian (2) it is essential the presence in the system the small adiabatic parameter. Therefore it is necessary to consider clusters that are contained in containers (nanoclusters with matrix isolation) that can be presented by the pores with nanometer size. Under the influence of the container or due to any internal physicochemical features nanocluster can suffer fast configuration rebuilding (during time $\tau_m << \tau_f$). As an example one can give molecule bullvalene ($C_{10}H_{10}$) which (due to bond fluctuation) can suffer degenerate regrouping more than million times with the rate of the reorientation 10^{10} Hz ($\tau_m \simeq 10^{-10}$ sec) [15]. In paper [13] the thermodynamics of this model was studied in detail. Analysis of the thermodynamic properties of the model ESSI have revealed several specific features in them. Particularly it was shown that in finite spin system described by the model ESSI one can observe topological excitations (solitons) which firstly condition absence of the long-range order in the system and secondly form anomalous behavior of the heat capacity at low temperatures (appearance of the additional low temperature peak). Anomalous behavior of the heat capacity of the model is connected with appearance of the spin gap in system considered. The presence of the spin gap in the spectrum of the magnetic excitations leads also to specific features in magnetic properties of the model [14], notably nonmonotonic behavior of the susceptibility and presence fractional plateau in field dependence of the magnetization.

In this work authors set the problem of the theory construction for NMR line inhomogeneous broadening of the isolated spin clusters in moving nanocontainers with self-similar correlation law. Description of the hierarchical (self-similar) structure of the medium and processes occurring in it is the crucial task at present [see, for example, 16-18 and their references]. Taking into account the self-similar character of the kinetic processes leads to fractional power behavior of the many time characteristics of the system. Particularly taking into account self-similarity in dielectric relaxation processes leads to non-exponential (fractional power) character of the time dependence for relaxation function of the macroscopic dipole moment of the system [19, 20], that agree with experimental data for frequency dielectric spectra.

2. NMR line shape of clusters in nanocontainers with time-dependent volume

Let us consider systems (clusters) of n spins ½, that are enclosed in moving nanocontainers in external magnetic field B. On time interval Δt (1) the effective spin dynamics is described by time-dependent averaged Hamiltonian

$$\bar{H}(t) = -\omega \cdot I^z - \frac{1}{2}\left(A_1(t) - A_2(t)\right)\left(I^z\right)^2 - \frac{A_2(t)}{2} I^2 + \frac{n}{8}\left(A_1(t) + 2A_2(t)\right), \quad (2)$$

where $\omega = \gamma B$ (γ - gyromagnetic ratio for protons), nuclear spins are given by spin operators I_f^α, $f = 1,...,n$, ($\alpha = x, y, z$), operators $I^\alpha = \sum_{f=1}^{n} I_f^\alpha$ determine projections of total spin operator on x, y, z-axes, $A_{1,2}(t)$ - time-dependent averaged constants of longitudinal and transversal spin-spin interaction. Time dependence of spin-spin interaction constants is determined by dynamics of nanoccontainer's geometrical parameters (volume, shape, orientation with respect to external field and others). The square of total nuclear spin operator has the form $I^2 = \left(I^+ I^- + I^- I^+\right)/2 + (I^z)^2$.

Line shape is determined as Fourier transform of free induction decay (FID), which in high-temperature limit has the form [21, 22]

$$G(t) = \frac{\mathrm{Sp}\left(I^+(t) I^-\right)}{\mathrm{Sp}\left(I^+ I^-\right)}. \quad (3)$$

Time dependence of operator $I^+(t) = e^{it\bar{H}} I^+ e^{-it\bar{H}}$ in Heisenberg representation is determined easily from motion equation

$$\frac{dI^+(t)}{dt} = i[\bar{H}, I^+] = -i(A_1(t) - A_2(t))(I^z - 1/2) I^+ \quad (4)$$

and has the appearance

$$I^+(t) = \exp\left(-i\varphi(t)\left(I^z - 1/2\right)\right) I^+(0), \quad (5)$$

where

$$\varphi(t) = \int_0^t (A_1(t') - A_2(t')) dt'. \quad (6)$$

For calculation of FID $G(t)$ let's represent it in the form $G(t) = R(t)/R(0)$, where $R(t) = \mathrm{Sp}\left(e^{-i\varphi(t)\left(I^z - 1/2\right)} I^+ I^-\right)$. So long as Hamiltonian (2) eigenvalues are known

$$E_{ms}(t) = -\frac{1}{2}\left(A_1(t) - A_2(t)\right)m^2 - \frac{A_2(t)}{2}s(s+1) + \frac{n}{8}(A_1(t) + 2A_2(t)), \tag{7}$$

where $s = s_{\min}, ..., n/2$ ($s_{\min} = 1/2$ for odd particle quantity n and $s_{\min} = 0$ for even particle quantity n in system), $m = -s, ..., s$, the function $R(t)$ can be expressed in the form

$$R(t) = \sum_{s=s_{\min}}^{n/2} g(N,s) \sum_{m=-s}^{s} (s+m)(s-m+1)e^{i\varphi(t)(1/2-m)}, \tag{8}$$

where factor

$$g(n,s) = \frac{2s+1}{n+1} \cdot \binom{n+1}{n/2 - s} \tag{9}$$

determines the number of grouping ways of n spins ½ into total spin s. Taking into account formulas

$$\sum_{m=-s}^{s} (s+m)(s-m+1)e^{i\cdot\varphi(t)(1/2-m)} = \frac{1}{2}\frac{(s+1)\sin s\varphi(t) - s\sin(s+1)\varphi(t)}{\sin^3(\varphi(t)/2)}, \tag{10}$$

$$\frac{1}{2}\sum_{s=s_{\min}}^{n/2} g(n,s)\frac{(s+1)\sin s\varphi(t) - s\sin(s+1)\varphi(t)}{\sin^3(\varphi(t)/2)} = n2^n\left(\cos\frac{\varphi(t)}{2}\right)^{n-1}, \tag{11}$$

let us represent $R(t)$ in the form

$$R(t) = n2^n\left(\cos\frac{\varphi(t)}{2}\right)^{n-1}. \tag{12}$$

As a result the expression for FID (3) is determined as

$$G(t) = \left(\cos\frac{\varphi(t)}{2}\right)^{n-1}. \tag{13}$$

On time scale of NMR observation $t \le t_{\mathrm{nmr}} \sim 10^{-4}$ c we have $A_i t \le 10^{-2}$, then $\varphi(t) << 1$ and FID $G(t)$ for large numbers of spins n in nanocontainers transforms to the form

$$G(t) = e^{(n-1)\ln\cos(\varphi(t)/2)} \simeq e^{-n\varphi^2(t)/8}. \tag{14}$$

FID $G(t)$ (14) depends on constants $A_i(t)$ as internal parameters of function $\varphi(t)$ (6). Variation of the functions $A_i(t)$ leads to large variety of NMR line shape models, which mainly form two large groups: (a) homogeneous line broadening models and (b) inhomogeneous line broadening models [23, 24].

Nanosized containers are sensitive to the environment fluctuations, so constants $A_i(t)$ undergo fluctuations

$$A_i(t) = A_i + \delta A_i(t), \tag{15}$$

where fluctuation $\delta A_i(t)$ we will characterize by two moments

$$< \delta A_i(t) >= 0 \,, \quad < \delta A_i(t_1)\delta A_j(t_2) >=< \delta A_i \delta A_j > K_{ij}(|t_1 - t_2|) \,, \tag{16}$$

$K_{ij}(t)$ designates correlation function, which, for example, can have appearance $K(t) = \exp(-t/\tau_c)$, τ_c - correlation time. Averaging the function $G(t)$ with respect to Gauss fluctuations $\delta A_i(t)$ leads to the next result [3]

$$\langle G \rangle (t) = \frac{\exp\left(-\frac{(A_1 - A_2)^2 n t^2}{8(1 + nC(t)/2)}\right)}{\sqrt{1 + nC(t)/2}} \,, \tag{17}$$

where

$$C(t) =< \delta\varphi(0)\delta\varphi(t) >=< \delta A_1^2 > M_{11}(t) + < \delta A_2^2 > M_{22}(t) - 2 < \delta A_1 \delta A_2 > M_{12}(t) \,, \tag{18}$$

$$M_{ij}(t) = \int_0^t (t - t')K_{ij}(t')dt' \,. \tag{19}$$

Introducing designations

$$\frac{A_1}{A_2} - 1 = \delta \,, \quad \frac{< \delta A_i \delta A_j >}{A_i A_j} = \alpha_{ij} \,, \quad \nu = \sqrt{\frac{n}{2}} A_2 \tag{20}$$

and taking into account (18), we transform expression for FID (17) to the form

$$\langle G \rangle (t) = \frac{\exp\left(-\frac{\delta^2 \gamma^2 t^2}{4(1 + \nu^2 \widetilde{C}(t))}\right)}{\sqrt{1 + \nu^2 \widetilde{C}(t)}} \,, \tag{21}$$

where

$$\widetilde{C}(t) = (1+\delta)^2 \alpha_{11} M_{11}(t) + \alpha_{22} M_{22}(t) - 2(1+\delta)\alpha_{12} M_{12}(t) \,. \tag{22}$$

In the case of fluctuation absence ($\alpha_{ij} = 0$) the line shape

$$J(\omega) = \frac{1}{\pi} \int_0^\infty G(t) \cos \omega t dt \,, \tag{23}$$

has Gauss form $J(\omega) = \exp(-\omega^2 / \delta^2 \nu^2) / \delta \nu \sqrt{\pi}$. Fluctuations of nanocavity lead to broadening of the line shape. For demonstration of the precise calculations of the line shape $J(\omega)$ (following the work [3]) let's choose correlation function in the form $K_{ii}(t) = \exp(-t/\tau_c)$ ($K_{12}(t) = 0$). This choice of correlation function form leads to the next expression for $M_{ii}(t)$

$$M_{ii}(t) = \tau_c^2 \left(\exp(-t/\tau_c) + t/\tau_c - 1\right) \,. \tag{24}$$

Let's consider time fluctuations on two different time scales. When $\tau_c^2 < \delta A_i^2 > << 1$ function $M_{ii}(t)$ has the appearance $M_{ii}(t) = \tau_c t$ for time $t >> \tau_c$, which leads equation (21) to the form

$$G(t) = \exp(-\delta^2 t / (4\gamma\tau_c)) / \sqrt{1 + \gamma\tau_c \nu^2 t} \,, \tag{25}$$

where $\gamma = (1+\delta)^2\alpha_{11} + \alpha_{22}$, and Fourier transformation of this expression gives next presentation for line shape [25]

$$J(\omega) = \frac{1}{\gamma\sqrt{\pi\tau_c\nu^2}} \mathrm{Re}\left(\frac{e^z}{\sqrt{z}}\mathrm{erfc}(\sqrt{z})\right), \quad z = (2\gamma\tau_c\nu)^{-2} + i\omega(\gamma\tau_c\nu^2)^{-1}. \tag{26}$$

Function $J(\omega)$ (26) has bell-shaped profile with intermediate Lorentz asymptotic behavior $J(\omega) = \Gamma/\pi(\Gamma^2+\omega^2)$, where $\Gamma = (4\gamma\tau_c)^{-1}$ (upon $\tau_c \to 0$). On high frequencies ($\omega \to \infty$) the line shape has next asymptotic behavior

$$J(\omega) \sim \frac{9\delta^2 n^2 A_2\gamma}{16\pi\tau_c}\omega^{-4}. \tag{27}$$

In the case $\tau_c^2 < \delta A_i^2 >>> 1$ function $M_{ii}(t) = t^2/2$ for $0 \le t \le \tau_c \to \infty$. Therefore FID is represented in the form

$$G(t) = \exp(-\delta^2/2\gamma)/\sqrt{1+\gamma\nu^2 t^2/2}. \tag{28}$$

As a result for line shape we have next expression [3]

$$J(\omega) = \frac{\exp(-\delta^2/2\gamma)}{\pi\nu}\sqrt{\frac{2}{\gamma}}K_0\left(\frac{\omega\sqrt{2}}{\nu\sqrt{\gamma}}\right) = \begin{cases} \dfrac{\exp(-\delta^2/2\lambda)}{\pi\nu}\sqrt{\dfrac{2}{\gamma}}\left[-\ln\left(\dfrac{\omega\sqrt{2}}{\nu\sqrt{\gamma}}\right) + O(\omega)\right], & \omega \to 0, \\ \dfrac{\exp(-\delta^2/2\gamma)}{\sqrt{\pi\nu\sqrt{2\gamma}}}\dfrac{1}{\sqrt{\omega}}\exp\left(-\dfrac{\omega\sqrt{2}}{\nu\sqrt{\gamma}}\right), & \omega \to \infty. \end{cases} \tag{29}$$

In such a way this examples demonstratively proved wide applicability general formula (13), from which one can get different expressions for line shape under the assumption that random fluctuations have Gauss character.

3. The line shape of the clusters with self-similar correlation law of nanocontainers

In previous section we have shown calculation of the FID and line shape of spin clusters in fluctuating nanocavities with Gauss stochastic dynamics. Gauss ensemble of disordered containers is characterized by correlation time τ_c, changing of which leads to different line shapes.

According to recent ideas many experimental results can be described by entering conception about distribution of the correlation time. Taking into account this conception let us consider function of the correlation time distribution $g(\tau_c)$ and average function $M(t,\tau_c)$ (19) with this distribution. As a result we get average (effective) expression for this function

$$\overline{M}(t) = \int M(t,\tau_c)g(\tau_c)d\tau_c. \tag{30}$$

The simplest way to enter distribution function $g(\tau_c)$ is the next one. Let's divide set of nanocontainers (that form sample) on several groups, containing N_l nanocontainers with characteristic correlation time τ_{cl}. We assume that $M_{11}(t) = M_{22}(t) = M(t)$, $M_{12}(t) = 0$. The expression for $g(\tau_c)$ can be described by the formula

$$g(\tau_c) = \sum_l \frac{N_l}{N}\delta(\tau_c - \tau_{cl}), \tag{31}$$

where N – the number of pores in the sample. As a result the expression for averaged function $\overline{M}(t)$ of sample will accept the form

$$\overline{M}(t) = \sum_l \frac{N_l}{N} M(t, \tau_{cl}). \tag{32}$$

In this work we propose self-similar character of container's dynamics. Therefore distributions of numbers N_l and set of correlation times τ_{cl} will obey self-similarity conditions [19, 20]

$$N_l = N_0 b^l, \quad \tau_{cl} = \tau_{c0}\xi^l \quad (-\infty < l \le L, b > 1, 0 < \xi < 1). \tag{34}$$

From formula (24) one can see, that function $M(t,\tau_{cl})$ has structure $M(t,\tau_{cl}) = \tau_{cl}^2 f(t/\tau_{cl})$. As a result of the assumptions made the expression for effective memory function (32) will have the form

$$\overline{M}(\theta) = c\tau_{c0}^2 \sum_{l=-\infty}^{L} (b\xi^2)^l f(\theta\xi^{-l}), \tag{36}$$

where $c = N_0/N$, $\theta = t/\tau_{c0}$. For calculation of the sum (36) it is convenient to realize Mellin transform. Let's assume for definiteness that correlation function $K(t)$ has form $K(t) = \exp(-t/\tau_c)$, then $f(\theta) = e^{-\theta} + \theta - 1$. Mellin transform of the expression (36) leads to the next result

$$\mathfrak{M}(s) = c\tau_{c0}^2 \Gamma(s) \sum_{l=-\infty}^{L} \xi^{l(2-d+s)} = \frac{c\tau_{c0}^2 \Gamma(s) \xi^{(2-d+s)L}}{1-\xi^{-(2-d+s)}}, \quad -2 < \operatorname{Re} s < -(2-d). \tag{37}$$

Here $d = \ln b / \ln(1/\xi) > 0$, $\mathfrak{M}(s)$ – Mellin transform of the function $\overline{M}(\theta)$, $\Gamma(s)$ - gamma-function, and also we use formula $e^{-\theta} + \theta - 1 \overset{\text{MT}}{=} \Gamma(s), \quad -2 < \gamma = \operatorname{Re} s < -1$. Parameter d plays role of "dynamic" fractal dimension (i.e. dimension which connects geometrical b and dynamical ξ parameters of the system).

Making inverse Mellin transform in (37) we will get

$$\mathfrak{M}(s) \overset{\text{IMT}}{=} \overline{M}(\theta) = \frac{c\tau_{c0}^2}{2\pi i} \int_{\gamma-i\infty}^{\gamma+i\infty} \frac{\Gamma(s)\xi^{(s+2-d)L}\theta^{-s}}{1-\xi^{-(s+2-d)}} ds. \tag{38}$$

For calculation of the Mellin-Barnes integral in the right part of (38) we enclose integration line on contour by right half of circle (contour $L_{i\infty}$) and by Cauchy theorem we find

$$\overline{M}(\theta) = -c\tau_{c0}^2 \sum_{s_k \in L_{i\infty}} \operatorname*{Res}_{s_k} \left[\frac{\Gamma(s)\xi^{(s+2-d)L}\theta^{-s}}{1-\xi^{-(s+2-d)}} \right], \tag{39}$$

where s_k – poles of function (37). Function $\mathfrak{M}(s)$ (37) has two groups of poles

$$s'_k = -(2-d) + i\Omega k, \quad \Omega = \frac{2\pi}{\ln(1/\xi)}, \quad k = 0, \pm 1, \pm 2, \ldots, \tag{40a}$$

$$s''_k = -k, \quad k = 0, 1, 2, \ldots \tag{40б}$$

We take into account that inside of chosen contour there are all poles of the group (40a) and only poles $s''_0 = 0$, $s''_1 = -1$ of the group (40б) and $0 < d < 1$. As a result from (39) we will get

$$\overline{M}(\theta) = \frac{c\tau_{c0}^2 \theta^{2-d}}{\ln(1/\xi)} w_d\left(\frac{\ln\theta}{\ln(1/\xi)}\right) + c\tau_{c0}^2 \left(-\frac{\xi^{(2-d)L}}{1-\xi^{-(2-d)}} + \frac{\xi^{(1-d)L}\theta}{1-\xi^{-(1-d)}} \right), \tag{41}$$

where

$$w_d(z)=\sum_{k=-\infty}^{\infty}\Gamma(-(2-d)+i\Omega k)\exp[2\pi ikx], \tag{42}$$

- periodic function with period 1. In (41) we took into account, that $\operatorname{Res}_{-k}\Gamma(s)=(-1)^k/k!$. So long as we consider that $L\gg 1$ then expression in parentheses (41) is small and we neglect it. As a result expression for averaged function $\overline{M}(\theta)$ take the form

$$\overline{M}(\theta)=\frac{c\tau_{c0}^2\theta^{2-d}}{\ln(1/\xi)}w_d\left(\frac{\ln\theta}{\ln(1/\xi)}\right). \tag{43}$$

So long as $|\Gamma(-(2-d)+i\Omega k)|\ll\Gamma(-(2-d))$ at $k\neq 0$ then in present work we neglect the presence of log-periodic oscillations. Investigation of the influence of the log-periodic oscillations on line shape will be the problem of the next work. Thus function $\overline{M}(\theta)$ will take next final form

$$\overline{M}(\theta)=\frac{c\tau_{c0}^2\Gamma(d-2)\theta^{2-d}}{\ln(1/\xi)}. \tag{44}$$

As a result of substitution of the expression (45) in (21) we will get expression for FID with account of self-similar character of correlations in nanocavities motion

$$\langle\widetilde{G}\rangle(t)=\frac{\exp\left(-\dfrac{\delta^2\nu^2t^2}{4(1+\nu^2\gamma a(d)t^{2-d})}\right)}{\sqrt{1+\nu^2\gamma a(d)t^{2-d}}}, \tag{45}$$

where $a(d)=c\tau_{c0}^d\Gamma(d-2)/\ln(1/\xi)$. In case when $a(d)<\delta A_i^2>>>1$ expression (45) will take the form

$$\langle\widetilde{G}\rangle(t)\simeq\frac{\exp\left(-\dfrac{\delta^2t^d}{4\gamma a(d)}\right)}{\sqrt{1+\nu^2\gamma a(d)t^{2-d}}}. \tag{46}$$

On figure 1 there are time dependences of the FID for different values of fractal dimension d at next values of parameters: $\tau_{c0}=0{,}004$ c, $\alpha=100$, $\gamma\tau_{c0}=10$, $\delta=0{,}5$, $c=0{,}01$, $\xi=0{,}1$ (fig. 1a) and $\xi=0{,}8$ (fig. 1b). From this figure one can see that time dependence of the FID considerably depends on self-similarity parameters ξ and d, changing of which leads to change of the decay rate.

On figure 2 there are plots of line shape (23) for different values of fractal dimension d (parameter values corresponds to figure 1). From this figure one can see that line shape considerably changes at the edges of possible values interval of fractal dimension d.

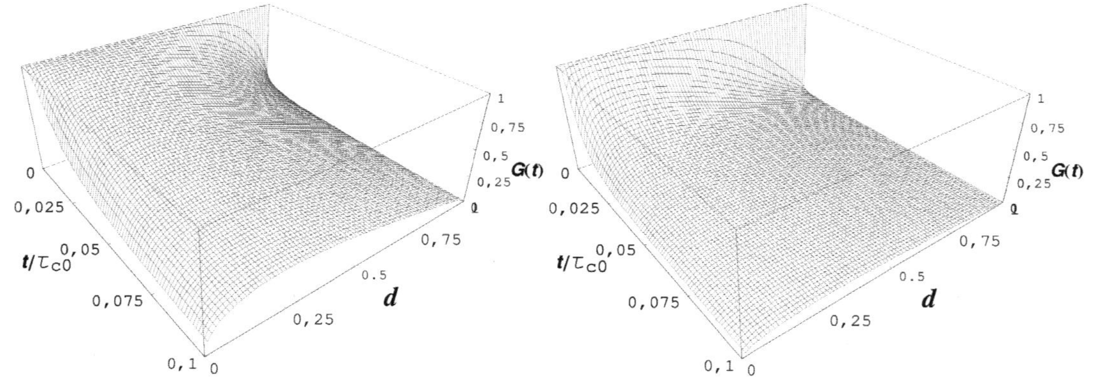

a b

Figure 1. The free induction decay at different values of the fractal dimension *d*. The 2D plots are constructed at the following values of the parameters: $\tau_{c0} = 0,004$ c, $\alpha = 100$, $\gamma\tau_{c0} = 10$, $\delta = 0,5$, $c = 0,01$, $\xi = 0,1$ (a) and $\xi = 0,8$ (b).

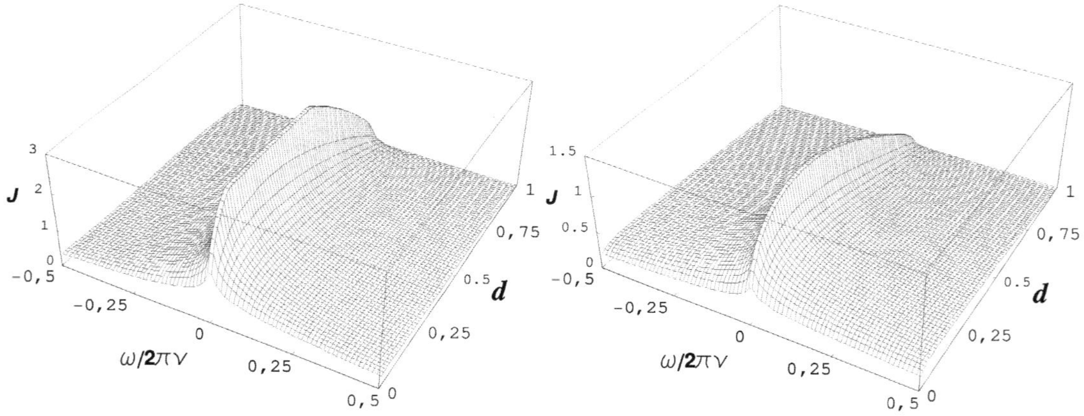

a b

Figure 2. The line shape at different values of the fractal dimension *d*. The plot is constructed at the following values of the parameters: $\tau_{c0} = 0,004$ c, $\alpha = 100$, $\gamma\tau_{c0} = 10$, $\delta = 0,5$, $c = 0,01$, $\xi = 0,1$ (a) and $\xi = 0,8$ (b).

4. Conclusion

The aim of the present work was development of the theory of NMR line shape of isolated in fluctuating nanopores magnetoactive clusters with self-similar Gauss stochastic dynamics. The possibility of precise analytic description of the NMR line shape of isolated nanoclusters with fast configuration rebuilding opens the way for subsequent development of the theory to bring it closer to real description of the physical systems, notably to take into account the self-similar character of pores dynamics.

In the frame of the model suggested and related to the self-similar hierarchy of the correlation times of fluctuations in nanocontainers it is shown how to take into account the scaling effects associated with collective dynamics of pores. The self-similar character of movement of nanocontainers leads to new analytical description of the NMR line shapes that are differed from the conventional Gaussian and Lorentz forms.The free decay and the corresponding line shape depends on the fractal dimension of the space-time self-similar ensemble and this important characteristic can serve as a convenient and clear understandable fitting parameter in interpretation of experimental data associated with NMR phenomena in disordered media.

Acknowledgments
Authors want to express their sincere thanks to Ministry of Higher Education and Science of the RF for their financial support (number of the federal grant 1.18.11).

References
[1] Devis M E 2002 *Nature* **417** 813
[2] Baugh J, Kleinhammes A, Han D, et al. 2001 *Science* **294** 1505
[3] Fel'dman E B and Rudavets M G 2004 *Chem. Phys. Lett.* **396** 458
[4] Inagaki S, Guan S, Ohsuna T and Terasaki O 2002 *Nature* **416** 304
[5] Haberland H 1994 *Clusters of atoms and molecules (Springer Series in Chemical Physics)* vol 52 (Berlin: Springer-Verlag)
[6] Prigogine I 1988 *Evolution of Size Effects in Chemical Dynamics* vol. 70 ed S Rice (New York: John Wiley & Sons) part 2
[7] Service R F 1996 *Science* **271** 920.
[8] Alivisatos A P 1996 *Science* **271** 933
[9] Shi J, Gider S, Babcock K and Awschlom D D 1996 *Science* **271** 937.
[10] Fel'dman E B and Rudavets M G 2002 *JETP Letters* **75** 635
[11] Fel'dman E B and Rudavets M G 2004 *JETP* **98** 207
[12] Kessel A R, Nigmatullin R R, Khamzin A A and Yakovleva N A 2005 *Theor. Math. Phys.* **145** 1727
[13] Khamzin A A and Nigmatullin R R 2010 *Theor. Math. Phys* **165** 1371
[14] Khamzin A A and Nigmatullin R R 2010 *JETP* **111** 1028
[15] Yannoni C S and Am J 1970 *Chem. Soc.* **92** 5237
[16] Olemskoi A I and Flat A Ya 1993 *Phys. Usp.* **36** 1087
[17] Zelenyi L M and Milovanov A V 2004 *Phys. Usp.* **47** 749
[18] Uchaikin V V 2003 *Phys. Usp.* **46** 821
[19] Nigmatullin R R 2005 *Physica B* **358** 201
[20] Nigmatullin R R 2006 *Physica A* **363** 282
[21] Abragam A, Goldman M 1982 *Nuclear magnetism: order and disorder* (Oxford: Clarendon Press) 626 p
[22] Goldman M 1970 *Spin temperature and nuclear magnetic resonance in solids* (Oxford: Clarendon Press)
[23] Abragam A 1961 *The principles of nuclear magnetism* (Oxford: Clarendon Press)
[24] Kubo R 1969 *Adv. Chem. Phys.* **15** 101
[25] Abramowitz M and Stegun I 1972 *Handbook of Mathematical Functions with Formulas, Graphs, and Mathematical Tables* (New York: Dover Publications)

Magnon Bose-Einstein condensation in $CsMnF_3$ and $MnCO_3$

E M Alakshin[a], Yu M Bunkov[b], R R Gazizulin[a], A V Klochkov[a], V V Kuzmin[a], A S Nizamutdinov[a], T R Safin[a] and M S Tagirov[a]

a) Kazan (Volga region) Federal University, 420008, Kremlyovskaja 18, Kazan, Russia
b) Institut Néel, CNRS et Université Joseph Fourier, F-38042 Grenoble, France

E-mail: `yuriy.bunkov@grenoble.cnrs.fr`

Abstract.
The Spin Supercurrent and Bose-Einstein condensation of magnons, similar to an atomic BEC, was observed in 1984 in superfluid ^{3}He-B. The same phenomena should exist in solid magnetic systems. We describe here the first observation of magnon BEC in solid easy plain antiferromagnets $CsMnF_3$ and $MnC0_3$. We have observed magnon BEC on a mode of coupled Nuclear-Electron precession. The dynamical properties of this mode have many similarities with NMR of superfluid ^{3}He-A. The frequency changes with deflection of nuclear magnetization. Furthermore, the involvement of electron ordered subsystem gives the magnon-magnon interaction, spin waves and spin supercurrent, while the nuclear subsystem gives the relatively long time of relaxation.

The conventional magnon BEC, the phase-coherent precession of magnetization was discovered for the first time in superfluid ^{3}He-B in 1984 [1]. It manifests itself by a domain with a fully phase-coherent precession of magnetization, known as the Homogeneously Precessing Domain (HPD). It exhibits all the properties of spin superfluidity: the supercurrent transports of magnetization and phase-slip processes at the critical current [2]; spin current Josephson effect [3]; spin current vortex [4] ets. The comprehensive review one can found in [5] and recent one in [6, 7]. The conventional magnon BEC has been also found in yttrium-iron garnet [8] for the magnons with non zero wave vector.

For a last 25 years there was found 5 different magnon BEC states in superfluid ^{3}He [6]. Particularly, the magnon BEC was observed and confirmed in superfluid ^{3}He-A immersed in aerogel squeezed along the magnetic field [9, 10]. In this case the aerogel reorients the orbital momentum of ^{3}He-A along the field [11, 12]. In this configuration the interaction of magnon is repulsive [13] the coherent precession is stable and the magnon BEC is possible. The same phenomena should exist in solid magnetic systems. Indeed, it was not observed up today. In this article we report the first observation of magnon BEC in solid easy plain antiferromagnets $CsMnF_3$ and $MnCO_3$. The preliminary results were published in [14]. We have observed this phenomena on a mode of coupled Nuclear-Electron precession, which shows the dynamical frequency shift (pulling). Magnetic systems with pulling were studied intensively in 70th of previous century [15]. This phenomenon takes place in the case when the frequencies of NMR and antiferromagnetic resonance are comparable. The typical system with pulling is the easy plain antiferromagnets with ions of Mn. The NMR frequency of Mn^{55} in hyperfine field is very high, about 600 MHz, and the antiferromagnetic resonance frequency in easy plain and cubic antiferromagnets can be very low. Due to coupling throw the hyperfine field, the two modes of resonance appear, low frequency quasi NMR mode (NEMR) and high frequency quasi antiferromagnetic resonance mode (ENMR). The properties of magnetic subsystems change significantly due to the couple precession. The hyperfine gap appears in a spectrum of antiferromagnetic spin waves. The quasinuclear spin waves appear, with the antiferromagnetic length of coherence. Other words, the nuclear magnetic system get some properties of magnetically ordered system.

The pulse or CW magnetic resonance can move the spin system out of equilibrium. This deviation can be described on the language of elementary excitation, the spin waves (magnons). Magnons can go fast to a new equilibrium state. The local magnetization in this case directed along the magnetic field and transverse components vanished. It means the reheating of the system and the saturation of the signal. Contrary the magnons can remains non-equilibrium for some time. Then they density matrix has a transverse terms. In this case one can see the signal of induction decay, radiates by a precessing spins. The signal can disappear in the case of thermalization or if the spins remains precessing, but its phase became different for the different place of the sample.

Indeed, there is one important question. The nuclear subsystem in $CsMnF_3$ is paramagnetic and after deflection should be dephased due to inhomogeneity of a superfine field. Furthermore, it should follows to Bloch equation. It means, that the nuclear system should thermalize for a time T_2 and the transverse component should vanish. In our experiments we have found that it is not a case! In CW NMR experiments the deflected nuclear magnetization precess homogeneously with the constant temperature and the NMR line does not saturated. The explanation can be found in the book [17]. This phenomena was also considered in[18]. There was considered two different scale for inhomogeneity. The microinhomogeneous broadening for a distance smaller then the length of coherence of Suhl-Nakamura interaction (about 10^4) of the lattice size, and the macroscopic broadening for a longer distances. The microscopic broadening can be suppressed in the case, when pulling is much bigger then a broadening. As a result the antiferromagnetic magnetization precess coherently at a short distance and force the nuclear subsystem precess coherently. For the longer distance it is not the case. It means that the nuclear subsystem shows the properties of ordered one, like the superfluid ^{3}He-A. The coherence length is even comparable. The magnetization follows the Landau-Lifshits scenario. It deflects but the module of magnetization does not change.

The density of non-equilibrium magnons depends from the angle of magnetization deflection β:

$$N = \frac{m - m_z}{\hbar} = \frac{\chi H}{\hbar\gamma}(1 - \cos\beta) \ . \tag{1}$$

The NMR frequency for ^{3}He-A in aerogel dependence on the magnon density (for $\Omega_L << \omega_L$):

$$\Omega = \omega_L - \frac{\Omega_L^2}{2\omega_L}\cos\beta, \tag{2}$$

where ω_L and Ω_L are Larmore and Leggett frequencies and β is the angle of magnetization deflection. The frequency dependence for NEMR mode in $CsMnF_3$ and $MnCO_3$ is very similar:

$$\Omega = \omega_n - \omega_p \frac{m_z}{m_0} = \omega_n - \omega_p \cos\beta, \tag{3}$$

where ω_n and ω_p are the non-shifted NMR frequency in hyperfine field and pulling, m_0 is the equilibrium nuclear magnetization and m_z its projection on a hyperfine field. It means that the both systems may have a very similar properties and fulfilled the condition of dynamic stability of coherent precession, reads [6, 7]:

$$\frac{\partial\Omega}{\partial\cos\beta} < 0, \tag{4}$$

Owing this similarity we have suggest to try to observe magnon BEC on a quasi NMR mode in $CsMnF_3$ and $CsMnF_3$[16]

We performed the BEC formation experiments in $CsMnF_3$ and $MnC0_3$ similar to one in ^{3}He-A described in Ref.[9, 10]. The frequency shift of Mn^{55} NMR depends from an external magnetic field H in a easy plain of antiferromagnet[18]:

$$\omega_p = \omega_n \frac{H_E H_n}{2H^2} \frac{m_0}{M_0}, \tag{5}$$

where H_E and H_n are exchange and hyperfine fields on an electron sublattice and m_0 and M_0 are the magnetization of the nuclear and electron sublattices. $MnC0_3$ is a week ferromagnet with the Dzyaloshinskii field about 440 mT. This gives the more complicate, but qualitatively the same equation for pulling.

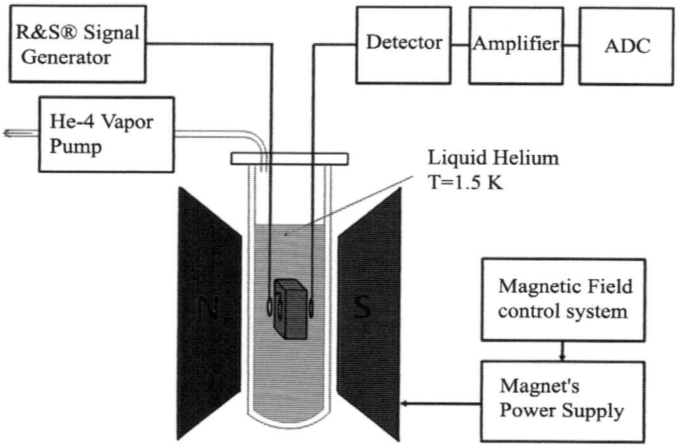

Figure 1. The schematic presentation of CW NMR spectrometr. The sample is placed inside the resonator. The two loops antennas used for exciting the resonator and to monitor the signal.

Our experiments with $CsMnF_3$ were done at the temperature of 1.5 K at a frequency of 566 MHz. The schema of spectrometer is shown in figure 1. The sample was displaced inside of the resonator. We have used the resonator described in [19]. The one coaxial cable with the loop antenna was used to excite the resonator, and other one to monitor the amplitude of the signal in resonator. The line of CW NMR at small RF excitation corresponds to 437 mT field. If we increase the excitation and sweep down the field, we can see the enormous growing of the signal. The frequency of NMR remains on a frequency of RF excitation even for a magnetic field far away from the resonance field. To explain this phenomena, we should suggest that the frequency shift (pulling) decreases. It can be done by two different way. The temperature may increase or the spins may be deflected on the angle β follow the equation:

$$\cos\beta = \frac{\omega_n - \Omega}{\omega_p}, \tag{6}$$

In the first case the NMR signal should saturated. In the second case the signal should grows. Let us suggest that we have a second case and that the all spins precess in phase. Then the amplitude of the signal should follows the equation

$$A = A_0 \sin\beta, \tag{7}$$

This dependence is shown in figure 2 by dashed line. The amplitude of the signal was calibrated

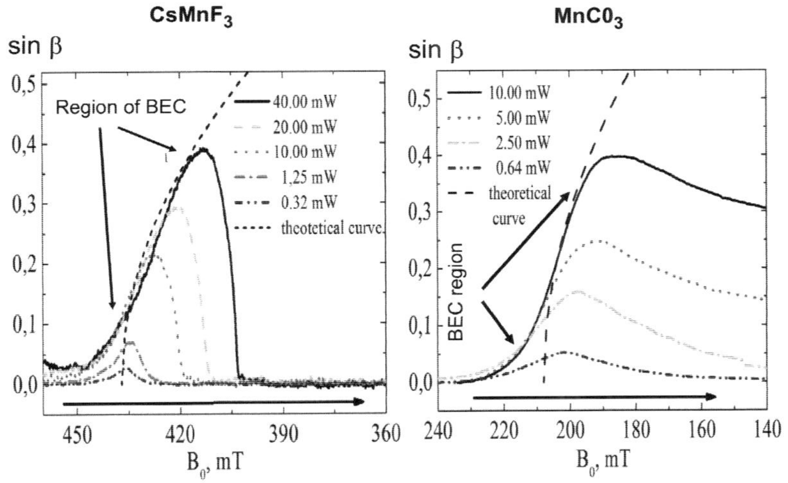

Figure 2. On a left side the amplitude of CW NMR signal from $CsMnF_3$ as a function of external magnetic field for different energy of excitation. The dashed line is a theoretical curve follows the equations (6,7). On the right side shown the similar results measured in $MnCO_3$

by the amplitude of induction after a short RF pulse. So we can present the signal in the units of angle of deflection β. There is a fields when all the magnetization precess homogeneously. This is a coherent BEC state. The BEC with deflection angle about 23^0 have been observed. At further sweep of the field down the RF power can not more compensate the processes of relaxation. The BEC in some part of the sample destroyed and the amplitude of signal decreased. Exactly the same behavior of NMR signal was early observed in superfluid ^{3}He-A in aerogel. Also similar behavior we have observed in $MnC0_3$. The results of these experiments are shown on a right side of figure 2. At the sweep back the coherent precession state creates at higher fields.

Finally, we have observed a coherent precession of a magnetization of all the sample. It is not a brute force coherence established by the RF pumping. The amplitude of RF field is only about 5 kHz, while the inhomogeneous broadening of the NMR line at small excitation is about 2 MHz. The external RF field gives the frequency of rotating frame in which the minimum of energy corresponds to a given angle of magnetization deflection (density of magnons). The magnon-magnon interaction and spin supercurrent establishing the coherent precession on a dimensions of the sample. All this phenomena for superfluid ^{3}He-A well explained in [9, 10] and reviews [6, 7]

It is important to point out that a similar experiments was described in Ref.[20]. The qualitative behavior of the signal was very similar. Indeed, it is difficult to figure out all parameters of the experiments, described there. To explain those results authors was used the scenario of saturation (heating) of the nuclear subsystem under a RF pumping. This scenario strongly contradicts to our experiments, described here and the other our experiments, which are under preparation for next publication.

This work is supported by the Ministry of Education and Science of the Russian Federation (FTP "Scientific and scientific-pedagogical personnel of the innovative Russia" contract

N02.740.11.5217)

1. References

[1] Borovik-Romanov A S, Bunkov Yu M, Dmitriev V V and Mukharskiy Yu M 1984 *JETP Lett.* **40** 1033; Fomin I A 1984 *JETP Lett.* **40** 1037
[2] Borovik-Romanov A S et al. 1989 *Phys. Rev. Lett.* **62** 1631
[3] Borovik-Romanov A S et al. 1988 *JETP Lett.***47** 478
[4] Borovik-Romanov A S et al. 1990 *Physica B* **165** 649
[5] Bunkov Yu M 2004 *J. Low Temp. Phys.* **135** 337; in: *Prog. Low Temp. Phys.* (Elsevier) **14** p. 69
[6] Bunkov Yu M and Volovik G E 2008 *J. Low Temp. Phys.* **150** 135; Bunkov Yu M and Volovik G E 2010 *J. Phys.: Condens. Matter* **22** 164210
[7] Bunkov Yu M 2009 *J. Phys.: Condens. Matter* **21** 164201
[8] Demokritov S O et al. 2006 *Nature* **443** 430
[9] Sato T et al. 2008 *Phys. Rev. Lett.* **101** 055301
[10] Hunger P et al. 2010 *J. Low Temp. Phys.* **158** 129.
[11] Kunimatsu T et al. 2007 *(JETP Lett.)* **86** *244*
[12] Elbs J et al. 2008 Phys. Rev. Lett. **100** *215304*
[13] Bunkov Yu M and Volovik G E 1993 Europhys. Lett. **21** *837.*
[14] Bunkov Yu M et al. 2011 JETP Lett. **94** *68.*
[15] Borovik-Romanov A S, Bunkov Yu M and Dumesh B S 1977 Physica **86** *1301*
[16] Bunkov Yu M 2010 Physics Uspekhi **53** *843*
[17] Kurkin M A and Turov E A 1990 NMR in magnetically ordered cristalls and its application (In Russian) (Moscow: Nauka)
[18] Borovik-Romanov A S et al. 1984 Sov. Phys. Uspekhi **142** *537*
[19] Hardy W N 1981 Rev. Sci. Instrum. **52** *213*
[20] Tulin V A 1969 Sov. Phys. JETP **28** *431*

Direct and two-phonon Orbach-Aminov type spin-lattice relaxation in molecular magnet V_{15}

Alex Tarantul[1] **and Boris Tsukerblat**[1,2]

[1]Department of Chemistry, Ben-Gurion University of the Negev, 84105 Beer-Sheva, Israel

E-mail: tsuker@bgu.ac.il

Abstract. In this article we propose a model of spin-phonon relaxation in $K_6[V^{IV}{}_{15}As_6O_{42}(H_2O)]\cdot 8H_2O$, the so called V_{15} cluster exhibiting the unique layered magnetic structure. The work is motivated by the recent observation of the Rabi oscillation [1] in this system and aimed to elucidate the role of spin-phonon interaction as a source of decoherence. The spin-phonon coupling is assumed to appear as a result of the modulation of the isotropic and antisymmetric (Dzyaloshinsky-Moriya) exchange interactions in the central triangular layer of vanadium ions by the acoustic lattice vibrations. The relaxation rates are estimated within the Debye model for the lattice vibrations. Within the pseudo-angular momentum representation the selection rules for the direct (one-phonon) transitions between Zeeman levels are derived and a special role of the antisymmetric exchange is underlined. The probabilities of the two-phonon Orbach-Aminov type processes are evaluated as well, while the Raman type relaxation is shown to have a negligible importance at low temperatures at which the Rabi oscillations have been detected.

Keywords: molecular magnetism, cluster V_{15}, exchange interaction, antisymmetric exchange, spin-phonon relaxation, spin frustration, coherence/decoherence

1. Introduction

During the last two decades quantum computing became one of the most promising disciplines lying on the border of computer science, quantum physics and theory of information. The main idea is that if some quantum system has (at least) two quantum states $|1\rangle$ and $|2\rangle$, not only these two pure states may be interpreted as information storage bits "0" and "1", but any superposition of them is also a carrier of information. Unlike its classical counterpart, this quantum information unit, or quantum bit (*qubit*) possesses intrinsic parallelism due to which quantum computation promises efficient algorithms which are not realizable on the classical computers [2]. However implementation of quantum computing is threatened by the decoherence, when the memory content of a qubit is lost due to its interaction with the environment (thermal bath, intercluster interaction or/and hyperfine spin-spin coupling). In order to successfully encode and process information the time of decoherence must be sufficiently long and, at least, not shorter than duration of the gate logical operation. This statement allows mentioning the central problem in the realization of quantum

[2] Author to whom any correspondence should be addressed

computing: to elaborate quantum system with a long coherence time. The first quantum logic gates have been already successfully realized on the trapped cold ions [3, 4]. The microscopic spins in quantum dots [5], semiconductors [6] and doped fullerenes [7] were also intensively studied as possible qubits.

A decade ago Leuenberger and Loss [8] proposed molecular magnets as promising candidates for the use as nanoscale qubits. In this respect it is to be mentioned that molecular magnets have a number of vitally important advantages: (a) unlike quantum dots the molecular magnets of a specified chemical composition are strictly identical and therefore have *identical physical characteristics* (energy levels, g-factors, magnetic exchange parameters, etc.); (b) by a proper chemical synthesis the magnetic clusters can be *engineered* to have desired physical characteristics that can be controlled by the due choice of the metal ions, ligands, etc. Moreover, in this way the main sources of decoherence (like the dipolar coupling and spin-nuclei interaction) can be suppressed by the chemical means. [1, 9, 10]; (c) Molecular magnets are relatively large (but still nanoscale!) objects, (as compare to single ions) and thus much easier for the *individual addressing* while processing the information; (d) magnetic molecules can be attached to *different types of platforms* by grafting individual clusters on solid surfaces [11-14], grafting monolayers [15-17] or embedding the isolated molecules in the amorphous media [1, 10]. This is expected to provide a firm stable structure in opposite to fragile and expensive apparatus for ion trapping.

The key question regarding molecular magnets as possible qubits was formulated by Winpenny et al. (see [9]): “will coherence times in molecular magnets permit quantum information processing?” A number of approaches for implementation of molecular magnets as qubits were proposed [8, 18-23] and the existence of the long living coherent states in different kinds of molecular magnets [1, 9, 23-27] have been predicted. The long phase-decoherence time T_2 in the heterometallic Cr_7M wheels was reported in [9]. The understanding of the significant importance of molecular magnets as possible qubits gave a strong impact on the study of the decoherence mechanisms, especially, spin-lattice and spin-spin interactions in molecular magnets [27-31]

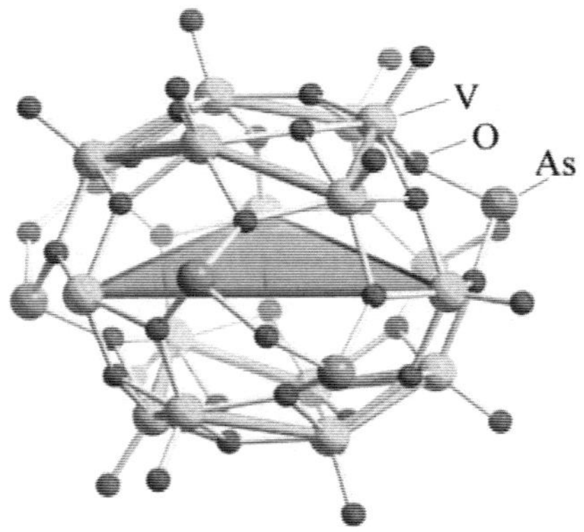

Figure 1. Ball-and-stick representation of the cluster anion $[V^{IV}{}_{15}As_6O_{42}(H_2O)]^{6-}$ emphasizing the V_3 triangle (the central water molecule is not indicated) [32].

A possibility of the long living coherent states in V_{15} have been theoretically predicted by Dobrovitski et al. [27] on the basis of the estimation of the characteristic times related to different mechanisms of decoherence. In fact, later on the existence and interpretation of the long-living coherent states was demonstrated by the first observation of the Rabi oscillations in the molecular magnet $K_6[V^{IV}{}_{15}As_6O_{42}(H_2O)]\cdot 8H_2O$, hereafter – V_{15} cluster [1].

The unique cluster V_{15} containing fifteen V^{IV} ions (S_i=1/2) and exhibiting layered structure was discovered more than two decades ago by Müller and Döring [32] and the first studies of this system date back to this period [32-35]. The metal skeleton of the V_{15} cluster has a distinct layered quasi-spherical structure [32, 33] in which fifteen magnetic V^{IV} ions ($s_i = 1/2$) are placed in a large central triangle sandwiched by two distorted hexagons possessing overall $\mathbf{D}_3$ symmetry (figure 1). The studies of the static magnetic susceptibility [34-37], energy pattern [38-43], *ab initio* electronic structure calculations [44-48] and inelastic neutron scattering [49, 50] showed that the low lying part of the energy spectrum is well isolated from the remaining spin levels and can be understood as a result of interaction between three moieties consisting of five strongly coupled spins (two almost paired spins from each hexagon coupled to a spin of the triangle) giving rise to spin S =1/2 of each moiety. The twenty-year studies of V_{15} cluster have been recently reviewed in [51].

In their study of the decoherence in molecular magnets the authors of [27] compared the dipole-dipole and spin-lattice relaxation times and according to their estimations the first was shown to be a dominant mechanism. The theory of spin-lattice relaxation in [27] had been based on a semiempirical model that led to a general estimations of some characteristic relaxation times . In this article we attempt to develop a microscopic model of spin-lattice interaction in V_{15} that takes into account isotropic and anisotropic exchange interactions and based on the accurate consideration of the spin states and Zeeman levels of V_{15} and on a simplified Debye model for the acoustic lattice vibrations. We consider direct (one-phonon) and two-phonon Raman and Orbach-Aminov type processes in order to estimate the relaxation rates and to reveal the importance of spin-phonon relaxation as a source of the decoherence in the V_{15} system.

2. Triangle model for V_{15} cluster

The model of spin triangle for the low lying spin excitations suggested in [34,35] includes the isotropic Heisenberg-Dirac-Van Vleck (HDVV) exchange interaction and the antisymmetric (AS) exchange proposed by Dzyaloshinskii [52] and Moriya [53] (see also [53-57]) as an origin of spin canting in magnetic materials. The last interaction was shown [57-62] to be especially important for the spin frustrated system possessing triangular structure. It was shown that the spin of hexagons are paired due to relatively strong antiferromagnetic interactions while the coupling inside the triangle is relatively small. That is why the model of an effective spin triangle of vanadium ions ($S = 1/2$) belonging to the central triangle provides an adequate description of the whole system at low temperatures. This triangular vanadium magnetic layer in the V_{15} structure is shaded in figure 1. Three spins of the central triangle are coupled through the antiferromagnetic isotropic exchange. The full Hamiltonian of the system in the external magnetic field looks as:

$$H \equiv H_0 + H_{AS} = 2J(\mathbf{S}_1\mathbf{S} + \mathbf{S}_2\mathbf{S}_3 + \mathbf{S}_3\mathbf{S}_1) + \sum_{i,j} \mathbf{D}_{ij}\left[\mathbf{S}_i \times \mathbf{S}_j\right] + g\mu\mathbf{SH}\,, \quad (1)$$

where for the sake of simplicity the Zeeman interaction is assumed to be isotropic. The eigen-values of the Hamiltonian H_0 with the antiferromagnetic ($J > 0$) coupling include two levels, namely, two "accidentally" degenerate spin doublets (ground manifold) and excited spin quadruplet separated by the gap $3J$. According to the overall point symmetry $\mathbf{D}_3$ the vector constants $\boldsymbol{D}_{ij}$ ($ij = 12,23,31$ numerate the sides) of the AS exchange have, in general, three independent components: along and perpendicular to the side (in plane of the triangle) and perpendicular to the plane component whose absolute values are D_l, D_t and D_n respectively. Consequently the Hamiltonian H_{AS} can be divided into two parts, $H_{AS}(\|)$ ("normal") and $H_{AS}(\perp)$ ("in-plane") which are defined as:

$$H_{AS}(\|) = D_n\left([\boldsymbol{S}_1 \times \boldsymbol{S}_2]_Z + [\boldsymbol{S}_2 \times \boldsymbol{S}_3]_Z + [\boldsymbol{S}_3 \times \boldsymbol{S}_1]_Z\right), \tag{2}$$

$$H_{AS}(\perp) = D_l\left([\boldsymbol{S}_1 \times \boldsymbol{S}_2]_X - \frac{1}{2}[\boldsymbol{S}_2 \times \boldsymbol{S}_3]_X + \frac{\sqrt{3}}{2}[\boldsymbol{S}_2 \times \boldsymbol{S}_3]_Y - \frac{1}{2}[\boldsymbol{S}_3 \times \boldsymbol{S}_1]_X - \frac{\sqrt{3}}{2}[\boldsymbol{S}_3 \times \boldsymbol{S}_1]_Y\right)$$
$$+ D_t\left([\boldsymbol{S}_1 \times \boldsymbol{S}_2]_Y - \frac{\sqrt{3}}{2}[\boldsymbol{S}_2 \times \boldsymbol{S}_3]_X - \frac{1}{2}[\boldsymbol{S}_2 \times \boldsymbol{S}_3]_Y + \frac{\sqrt{3}}{2}[\boldsymbol{S}_3 \times \boldsymbol{S}_1]_X - \frac{1}{2}[\boldsymbol{S}_3 \times \boldsymbol{S}_1]_Y\right) \tag{3}$$

where axes X and Y of the global coordinates system are in the plane (figure 3) while the axis Z is perpendicular to the plane. These two parts of the AS exchange are explicitly separated as they play quite different physical roles. The normal part of AS exchange splits the ground $S = 1/2$ doublets into two Kramers doublets and gives rise to a strong (first order) magnetic anisotropy while the in-plane part is responsible for the doublet-quadruplet mixing. This mixing leads to the second order zero-field splitting of the $S = 3/2$ level (that is $(D_l^2 + D_t^2)/8J$). It was shown [59] that D_l and D_t are combined into an effective parameter $D_\perp = \sqrt{D_t^2 + D_l^2}$ so the exchange model is fully specified by the three parameters J, D_n and $D_\perp$ (see Section 7 for the numerical estimations).

3. EPR transitions

The analysis of the energy pattern of the system shows that the ground term is represented by the orbital doublet ${}^2E$ (S=1/2) while the excited level is the orbital singlet ${}^4A_2(S = 3/2)$. This allows to find out the selection rule for the EPR transitions and to reveal the role of the AS exchange. The

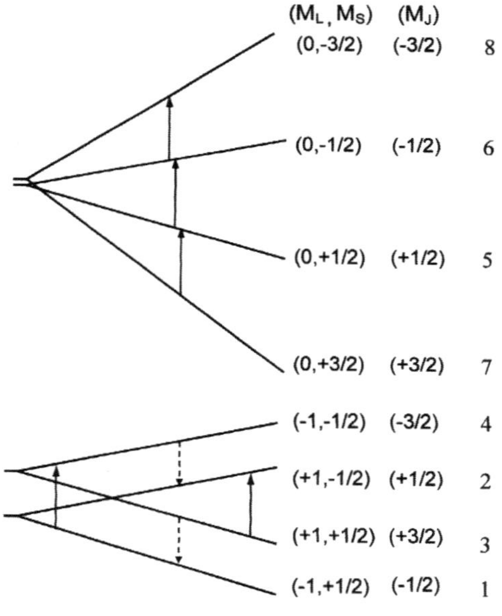

Figure 2. Allowed EPR transitions (low frequency range), $\boldsymbol{H} \,\|\, \boldsymbol{C}_3$ (solid arrows); allowed one-phonon transitions caused by the modulation of the isotropic exchange (dashed arrows), see Section 5.

component $H_{AS}(\|)$ of the AS exchange splits the ground manifold ^2E (S=1/2) while in-plane component $H_{AS}(\perp)$ of the AS exchange mixes the ground and excited levels and gives rise to a zero-field splitting of the excited $^4A_2(S=3/2)$ level. At low magnetic fields the effects of the in-plane component $H_{AS}(\perp)$ of the AS exchange are negligible due to the relatively large gap $3|J|$. Then, if the field is parallel to the $\boldsymbol{C}_3$ axis the energy levels may be labeled by following quantum numbers: the projection M_L of the orbital pseudo-angular momentum, the spin projection M_S and the full angular momentum projection $M_J = M_L + M_S$ [51,59,62]. Consequently, if $H_{AS}(\perp)$ is neglected, the EPR transitions satisfy the selection rules $\Delta M_L = 0, \Delta M_S = \pm 1, \Delta M_J = \pm 1$ [62]. There are five allowed transitions: two between the Kramers doublets and three within the excited quadruplet (figure 2). If the effects of $H_{AS}(\perp)$ are taken into account, then M_L and M_S are no longer "good" quantum numbers and the selection rule is now $\Delta M_J = \pm 1$. This allows for a number of additional EPR transitions which are, however, weak (their intensities are related to the degree of the intermultiplet mixing through the $H_{AS}(\perp)$) and may have importance only near the crossover points of the Zeeman sublevels. In terms of the quantum numbers M_L, M_S and M_J in Sec. 5 we will consider the selection rules for the one-phonon transitions.

4. Spin-phonon interaction

The interaction of spins with the lattice vibrations (heat reservoir) is assumed to arise from the modulation of the isotropic and AS exchange interactions by the molecular displacements X_i, Y_i, Z_i ($i = 1,2,3$ enumerates the ions) in course of the lattice vibrations [57, 58, 63]. It is convenient to deal with the symmetry adapted coordinates Q_α (α enumerates the vibrational mode) of the equilateral

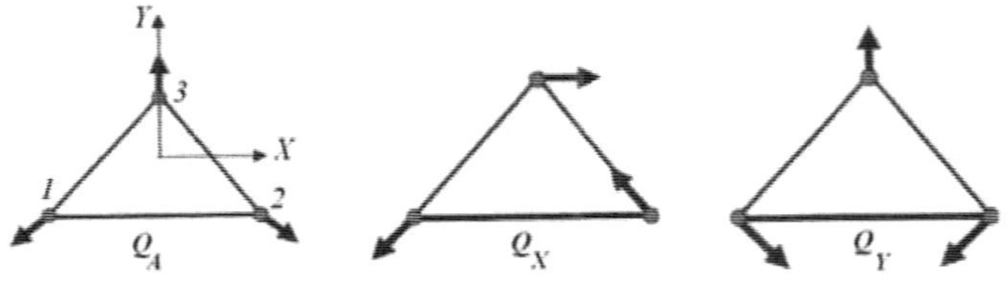

Figure 3. Vibrational coordinates of a symmetric triangular unit.

triangular unit: full symmetric $A_1(Q_{A_1} \equiv Q_A)$ and double degenerate E type $(Q_{Ex} \equiv Q_x, Q_{Ey} \equiv Q_y)$. These normal coordinates can be expressed in terms of the Cartesian displacements X_i, Y_i, Z_i of the constituent ions as following:

$$Q_A = \frac{1}{\sqrt{3}}\left[-\frac{1}{2}\left(\sqrt{3}X_1 + Y_1\right) + \frac{1}{2}\left(\sqrt{3}X_2 - Y_2\right) + Y_3\right]$$
$$Q_X = \frac{1}{\sqrt{3}}\left[\frac{1}{2}\left(X_1 + \sqrt{3}Y_1\right) + \frac{1}{2}\left(X_2 - \sqrt{3}Y_2\right) - X_3\right] \qquad (4)$$
$$Q_Y = \frac{1}{\sqrt{3}}\left[\frac{1}{2}\left(\sqrt{3}X_1 - Y_1\right) - \frac{1}{2}\left(\sqrt{3}X_2 + Y_2\right) + Y_3\right]$$

The displacements corresponding to these coordinates are shown in figure 3. Within the mentioned mechanism of spin-phonon interaction the exchange parameters are assumed to be the functions of the metal-metal distances R_{ij} so that the linear (with respect to the displacements) terms of the vibronic Hamiltonian H_{ev} can be represented as a sum of two contributions $H_{ev} = H'_{ev} + H''_{ev}$, where H'_{ev} and H''_{ev} are related to the isotropic and AS parts of the original Hamiltonian respectively:

$$H'_{ev} = \sum_{\alpha=A,X,Y} \left(\frac{\partial H_0}{\partial Q_\alpha} \right)_{Q_\alpha=0} Q_\alpha , \tag{5}$$

$$H''_{ev} = \sum_{\alpha=A,X,Y} \left(\frac{\partial H_{AS}}{\partial Q_\alpha} \right)_{Q_\alpha=0} Q_\alpha . \tag{6}$$

After substitution of the corresponding parts of the exchange Hamiltonians one obtains the following expressions for the spin-phonon coupling operators:

$$H'_{ev} = \sum_{ij} \sum_{\alpha=A,X,Y} 2\boldsymbol{S}_i \boldsymbol{S}_j \left(\frac{\partial J_{ij}(R_{ij})}{\partial R_{ij}} \right)_{\Delta R_{ij}=0} \cdot \frac{\partial R_{ij}}{\partial Q_\alpha} Q_\alpha \tag{7}$$

$$H''_{ev} = \sum_{ij} \sum_{\alpha=A,X,Y} [\boldsymbol{S}_i \times \boldsymbol{S}_j] \left(\frac{\partial \boldsymbol{D}_{ij}(R_{ij})}{\partial R_{ij}} \right)_{\Delta R_{ij}=0} \cdot \frac{\partial R_{ij}}{\partial Q_\alpha} Q_\alpha \tag{8}$$

Here R_{ij} are the instant metal-metal distances in course of the vibrations ($\Delta R_{ij} = 0$ corresponds to the equilibrium trigonal configuration and symmetric exchange network with the sides R_0). Modulation of the isotropic exchange can be described by the parameter λ which is defined as $\lambda \equiv \sqrt{6}\left(\partial J_{ij}(R_{ij})/\partial R_{ij}\right)_0$. For each side of the triangle there is also vector coupling parameter which related to AS exchange and defined as $\boldsymbol{\beta}_{ij} = \left(\partial \boldsymbol{D}_{ij}(R_{ij})/\partial R_{ij}\right)_0$. Due to trigonal symmetry of the system the absolute values of these three vector parameters have the same value β for each side of the triangle. One can define also the three components of $\boldsymbol{\beta}_{ij}$, namely, the normal part $\beta_n = \beta_{ijn}$ and the two perpendicular contributions $\beta_t = \beta_{ijt}$ and $\beta_l = \beta_{ijl}$ where the symbols *l* and *t* have the same meaning as in the definition of the AS exchange Hamiltonian. One can also define a combined vibronic parameter $\beta_\perp = \sqrt{\beta_t^2 + \beta_l^2}$ that appears in the final results.

After calculation of the derivatives $\partial R_{ij}/\partial Q_\alpha$ (using the relations between the Cartesian displacements X_i, Y_i, Z_i and normal coordinates Q_α, figure 3) one arrives at the following form of the spin-vibronic Hamiltonians:

$$H'_{ev} = \sum_{\Gamma\gamma} \lambda \hat{V}_{\Gamma\gamma} Q_{\Gamma\gamma} ,$$

$$H''_{ev} = \sum_{\Gamma\gamma} \beta \hat{W}_{\Gamma\gamma} Q_{\Gamma\gamma} ,. \tag{9}$$

Here $\Gamma\gamma = A_1, E_X, E_Y$ label the irreducible representations and basis functions of the $\boldsymbol{D}_3$ symmetry group and the operators $\hat{V}_{\Gamma\gamma}$ (related to the isotropic part of the Hamiltonian) are expressed in terms of the scalar products of spin operators:

$$\hat{V}_{A_1} = \sqrt{\frac{2}{3}}\left(\boldsymbol{S}_1\boldsymbol{S}_2 + \boldsymbol{S}_2\boldsymbol{S}_3 + \boldsymbol{S}_3\boldsymbol{S}_1\right),$$
$$\hat{V}_X = \frac{1}{\sqrt{6}}\left(\boldsymbol{S}_2\boldsymbol{S}_3 + \boldsymbol{S}_3\boldsymbol{S}_1 - 2\boldsymbol{S}_1\boldsymbol{S}\right), \tag{10}$$
$$\hat{V}_Y = \frac{1}{\sqrt{2}}\left(\boldsymbol{S}_2\boldsymbol{S}_3 - \boldsymbol{S}_3\boldsymbol{S}_1\right).$$

The operators $\hat{W}_{\Gamma\gamma}$ (related to the modulation of the AS exchange) are expressed in terms of the vector products of spin operators as follows:

$$\hat{W}_A = \boldsymbol{e}_{12}[\boldsymbol{S}_1 \times \boldsymbol{S}_2] + \boldsymbol{e}_{23}[\boldsymbol{S}_2 \times \boldsymbol{S}_3] + \boldsymbol{e}_{31}[\boldsymbol{S}_3 \times \boldsymbol{S}_1],$$
$$\hat{W}_X = \frac{1}{2}\left(\boldsymbol{e}_{12}[\boldsymbol{S}_1 \times \boldsymbol{S}_2] + \boldsymbol{e}_{23}[\boldsymbol{S}_2 \times \boldsymbol{S}_3] - 2\boldsymbol{e}_{31}[\boldsymbol{S}_3 \times \boldsymbol{S}_1]\right), \tag{11}$$
$$\hat{W}_Y = \frac{\sqrt{3}}{2}\left(\boldsymbol{e}_{23}[\boldsymbol{S}_2 \times \boldsymbol{S}_3] - \boldsymbol{e}_{31}[\boldsymbol{S}_3 \times \boldsymbol{S}_1]\right).$$

where $\boldsymbol{e}_{ij}$ are the unit vectors $\boldsymbol{\beta}_{ij}/\beta_{ij}$. The evaluation of the vibronic matrices was performed with the aid of the irreducible tensor operators approach [57, 58, 64-67].

5. One-phonon relaxation

In order to take into account the interaction of spins with the acoustic lattice vibrations, the symmetry adapted molecular displacement should be expanded into series of the longitudinal (l) and transverse (t) acoustic lattice vibrational modes $q_{\kappa p}$ specified by the wave vector $\boldsymbol{\kappa}$ and polarization $p = l, t$. We will also use a short notation $\xi = (\boldsymbol{\kappa}\, p)$ unless we have a special need to address any of them. Then one can obtain the following expression for the spin-phonon Hamiltonian:

$$H_{ev} = \sum_{\Gamma\gamma} \hat{G}_{\Gamma\gamma} \sum_{\xi} \left(\frac{\hbar}{M\omega_\xi}\right)^{1/2} a_\xi(\Gamma\gamma) q_\xi \tag{12}$$

Here the matrix $\hat{G}_{\Gamma\gamma}$ is defined as $\hat{G}_{\Gamma\gamma} = \lambda\hat{V}_{\Gamma\gamma} + \beta\hat{W}_{\Gamma\gamma}$, q_ξ are the dimensionless normal coordinates of the lattice, M is a mass of crystal and $a_\xi(\Gamma\gamma)$ are the so called Van Vleck coefficients introduced in his underlying theory of paramagnetic relaxation [68].

Since we are dealing with the transitions between the levels belonging to the ground manifold of V_{15} separated by the gaps of a few cm^{-1}, (Section 7) it is sufficient to use the long-wave approximation for the Van Vleck coefficients of the triangular unit:

$$\begin{aligned} a_{\kappa}(A_1) &= (1/2)R_0\left(p_x\kappa_x + p_y\kappa_y\right) \\ a_{\kappa}(E_X) &= -(1/2)R_0\left(p_x\kappa_x - p_y\kappa_y\right) \\ a_{\kappa}(E_Y) &= -(1/2)R_0\left(p_x\kappa_y - p_y\kappa_x\right) \end{aligned} \quad (13)$$

where κ_i, p_i are Cartesian components of the wave-vector and polarization vector and R_0 is V-V distance in the vanadium triangle. It is to be noted that the spin-phonon Hamiltonian, Eq. (12), is adapted to the triangle model for V_{15}. In fact, by definition the matrices $\hat{G}_{\Gamma\gamma}$ in this Hamiltonian act within the set of eight spin functions (two $S=1/2$ doublets and quadruplet $S=3/2$) of the vanadium triangle.

In the case of two non-degenerate electronic states $|m'\rangle$ and $|m\rangle$ separated by the energy gap $\Delta E = E_m - E_{m'} > 0$ the probability of the one-phonon transition $|m\rangle \rightarrow |m'\rangle$ is given by

$$w_{\sigma} = \frac{2\pi}{\hbar}\overline{\left|\langle m,n|H_{ev}|m',n'\rangle\right|^2}\delta(\hbar\omega - \Delta E) \quad (14)$$

where $|n\rangle = |....,n_{\sigma},.....\rangle$ and $|n'\rangle = |....,n_{\sigma} \pm 1,.....\rangle$. Using equation (12) for the spin-vibronic Hamiltonian one sees that the matrix element in (14) is the following:

$$\langle m,n|H_{ev}|m',n'\rangle = \sum_{\Gamma\gamma}\sum_{\xi}\left(\frac{\hbar}{M\omega_{\xi}}\right)^{1/2} a_{\xi}(\Gamma\gamma)\langle m|\hat{G}_{\Gamma\gamma}|m'\rangle\langle n_{\sigma}|q_{\xi}|n_{\sigma} \pm 1\rangle \quad (15)$$

The squared Van Vleck coefficients $b^2_{\kappa p}(\Gamma) = \overline{a^2_{\kappa p}(\Gamma\gamma)}$ averaged over polarization directions and directions of the wave-vectors are found as

$$\begin{aligned} b^2_{\kappa l}(A_1) &= \frac{2}{15}\kappa^2 R_0^2 \quad , \quad b^2_{\kappa l}(E) = \frac{1}{15}\kappa^2 R_0^2 \\ b^2_{\kappa t}(A_1) &= \frac{1}{60}\kappa^2 R_0^2 \quad , \quad b^2_{\kappa t}(E) = \frac{1}{20}\kappa^2 R_0^2 \end{aligned} \quad (16)$$

We use also a short notation:

$$b^2_{\kappa}(\Gamma) = b^2_{\kappa l}(\Gamma) + b^2_{\kappa t}(\Gamma) \quad (17)$$

Using the expression for the density of states $f(\omega) = \frac{V\omega^2}{(2\pi)^2 \upsilon^3}$ (V is the volume of crystal, υ is the mean sound velocity) and orthogonality of the Van Vleck coefficients one obtains the full probability $w^{(1)}_{m\rightarrow m'}$ of system's transition from $|m'\rangle$ to $|m\rangle$ with emission of a phonon of the resonance frequency $\omega = \Delta E/\hbar$:

$$w^{(1)}_{m\rightarrow m'} = \frac{\omega}{\pi\hbar\rho\upsilon^3}\frac{exp(\hbar\omega/kT)}{exp(\hbar\omega/kT)-1}\left(\sum_{\Gamma} b^2_{\kappa}(\Gamma)\sum_{\gamma}\left|\langle m|\hat{G}_{\Gamma\gamma}|m'\rangle\right|^2\right) \quad (18)$$

where ρ is a crystal density and $\kappa = \omega/\upsilon = \Delta E/\hbar\upsilon$ in the long-wave approximation so far employed. Using equations (16) and (17) one arrives at the following result which is specific just for a triangular unit:

$$w^{(1)}_{m\to m'} = \frac{R_0^2}{\pi\hbar\rho\upsilon^5}\frac{\exp(\hbar\omega/kT)}{\exp(\hbar\omega/kT)-1}\omega^3 B_{mm'}, \tag{19}$$

where

$$B_{mm'} = \frac{3}{20}\left|\langle m|\hat{G}_A|m'\rangle\right|^2 + \frac{7}{60}\left|\langle m|\hat{G}_X|m'\rangle\right|^2 + \frac{7}{60}\left|\langle m|\hat{G}_Y|m'\rangle\right|^2 \tag{20}$$

The one-phonon relaxation time is thus:

$$1/\tau_D = w_{m'\to m} + w_{m\to m'}.$$

If the in-plane part $H_{AS}(\perp)$ of AS interaction is neglected, the eigen-functions are given by the vectors $\boldsymbol{e}_i$, $i = 1,\cdots,8$, specified by the spin-projection M_S as well as by the pseudo-angular momentum quantum numbers M_L and M_J (Section 3). This vector set can be referred to as "pure" basis (Appendix). The $H_{AS}(\perp)$ part of the Hamiltonian causes mixing of the "pure" basis vectors that is quite negligible except of the levels crossover fields [51,59-62]. The matrix $\boldsymbol{B}$ in the "pure" basis is given in Appendix. The entries of matrix $\boldsymbol{B}$ in the "pure" basis refer to the strongest direct transitions which will be referred to as "first type" transitions.

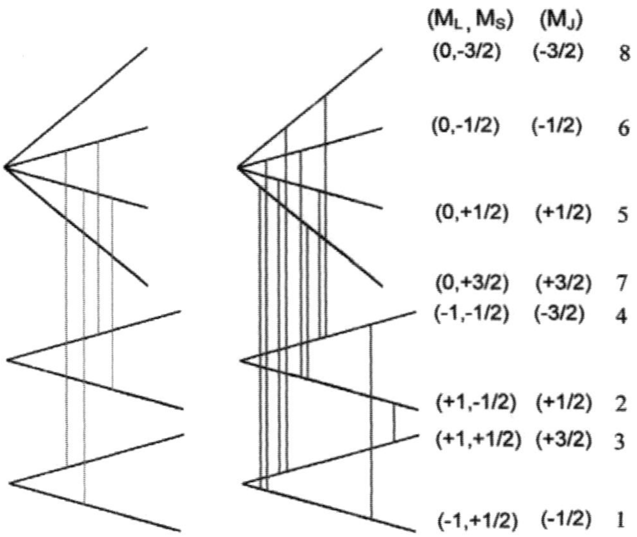

Figure 4. One phonon transitions caused by the phonon modulation of the normal part of the AS-exchange (left) and in-plane AS-exchange (right), $\boldsymbol{H} \| \boldsymbol{C}_3$.

Most of the first type transitions originate from the phonon-modulated AS exchange rather then from the modulated isotropic interaction. Indeed the modulation of the isotropic exchange produces only two first-type transitions within the $S = 1/2$ manifold (figure 2); these transitions obey the selection rules $\Delta M_S = 0, \Delta M_J = \pm 2$. Selection rules for the first-type transitions emerging from the acoustic modulation of the normal AS exchange component $H_{AS}(\|)$ are: $\Delta M_L = \pm 1,\ \Delta M_S = 0,\ \Delta M_J = \pm 1$ (figure 4, left). First-type transitions emerging from the acoustic modulation of the in-plane AS exchange component $H_{AS}(\perp)$ are shown in the figure 4, right. One can see that the most of the transitions originate from the spin modulation of the components of AS exchange rather then modulation of the isotropic exchange. Another important observation is that the first-type transitions within the $S = 3/2$ quadruplet are forbidden.

When the level mixing by the in-plane AS exchange Hamiltonian $H_{AS}(\perp)$ is taken into consideration, the M_L and M_S are no longer "good" quantum numbers and the selection rules should be formulated in terms of the total angular momentum projection M_J. Then the selection rules for the acoustically modulated isotropic exchange is now $\Delta M_J = \pm 2$ (quadrupole type of transitions) and for the modulated normal component $H_{AS}(\|)$ is $\Delta M_J = \pm 1$ (dipole type transitions). Selection rules for the modulated in-plane component $H_{AS}(\perp)$ become also less strict. This gives rise to emergence of additional very weak "second type" transitions whose probability is at least two orders smaller then that for the first-type relaxations.

6. Two phonon Orbach-Aminov relaxation

Since in the one-phonon transitions the density of the phonon states at the resonance frequencies is small, an important rule play the two-phonon relaxation processes. These are the Raman processes and the Orbach-Aminov type processes [69-73]. Due to the fact that the excited levels corresponding to the unpaired spins of the hexagons in V_{15} are higher then the Debye energy, the levels of the central vanadium triangle contribute significantly to the second order relaxation. That is why one can expect that only the resonant Orbach-Aminov type relaxation is actual for the V_{15} system at the relatively low temperatures corresponding to the actual EPR measurements. The Raman processes become important only at high temperatures. Later on we will derive general expressions adapted to the V_{15} system.

Let us consider now the two-step relaxation process in which the system passes from the initial state $|m,n\rangle$ to the final state $|m',n'\rangle$ through the intermediate state $|j,n''\rangle$. The probability of this two-stage transition is given in the framework of second-order perturbation theory (see, for example, [69]) by:

$$w_{\sigma_1,\sigma_2} = \frac{2\pi}{\hbar}\left|\sum_{j,n''}\frac{\langle m,n|H_{ev}|j,n''\rangle\langle j,n''|H_{ev}|m',n'\rangle}{E_m + E_n - E_j - E_{n''}}\right|^2 \delta\left(E_m + \hbar\omega_2 - E_{m'} - \hbar\omega_1\right) \tag{21}$$

Using the notation $\Delta_j = E_j - E_m$, the complex matrix element can be presented as:

$$\sum_{j,n} \frac{\langle m,n|H_{ev}|j,n''\rangle\langle j,n''|H_{ev}|m',n'\rangle}{E_m+E_n-E_j-E_{n''}} =$$

$$\sum_j \Big\{ \sum_{\Gamma\gamma,\xi} c(\xi)a_\xi(\Gamma\gamma)\langle m|G_{\Gamma\gamma}|j\rangle\langle n_{\sigma_1}|q_\xi|n_{\sigma_1}+1\rangle \frac{\sum_{\Gamma'\gamma'\xi'} c(\xi')a_{\xi'}(\Gamma'\gamma')\langle j|G_{\Gamma'\gamma'}|m'\rangle\langle n_{\sigma_2}|q_{\xi'}|n_{\sigma_2}-1\rangle}{-\hbar\omega_1-\Delta_j} \tag{22}$$

$$+\sum_{\Gamma\gamma,\xi} c(\xi)a_\xi(\Gamma\gamma)\langle m|G_{\Gamma\gamma}|i\rangle\langle n_{\sigma_2}|q_\xi|n_{\sigma_2}-1\rangle \frac{\sum_{\Gamma'\gamma'\xi'} c(\xi')a_{\xi'}(\Gamma'\gamma')\langle j|G_{\Gamma'\gamma'}|m'\rangle\langle n_{\sigma_1}|q_{\xi'}|n_{\sigma_1}+1\rangle}{\hbar\omega_2-\Delta_j}\Big\}$$

where $c(\xi)=(\hbar/M\omega_\xi)^{1/2}$. After deriving the expression for the squared matrix element, using the orthogonality properties of Van Vleck coefficients and performing thermal averaging one obtains:

$$\left|\sum_{j,n''} \frac{\langle m,n|H_{ev}|j,n''\rangle\langle j,n''|H_{ev}|m',n'\rangle}{E_m+E_n-E_j-E_{n''}}\right|^2$$

$$=\frac{\hbar^2}{M^2\omega_1\omega_2}\cdot\frac{\exp(\hbar\omega_1/kT)}{(\exp(\hbar\omega_1/kT)-1)(\exp(\hbar\omega_2/kT)-1)}$$

$$\times\sum_{j,j'}\Bigg\{\left(\sum_\Gamma b^2_{\kappa_1}(\Gamma)\sum_\gamma\langle m|G_{\Gamma\gamma}|j\rangle\langle m|G_{\Gamma\gamma}|j'\rangle\right)\cdot\frac{\left(\sum_\Gamma b^2_{k_2}(\Gamma)\sum_\gamma\langle j|G_{\Gamma\gamma}|m'\rangle\langle j'|G_{\Gamma\gamma}|m'\rangle\right)}{(\hbar\omega_1+\Delta_j)(\hbar\omega_1+\Delta_{j'})} \tag{23}$$

$$+\left(\sum_\Gamma b^2_{\kappa_1}(\Gamma)\sum_\gamma\langle j|G_{\Gamma\gamma}|m'\rangle\langle j'|G_{\Gamma\gamma}|m'\rangle\right)\cdot\frac{\left(\sum_\Gamma b^2_{k_2}(\Gamma)\sum_\gamma\langle m|G_{\Gamma\gamma}|j\rangle\langle m|G_{\Gamma\gamma}|j'\rangle\right)}{(\hbar\omega_2-\Delta_j)(\hbar\omega_2-\Delta_{j'})}$$

$$-\left(\sum_\Gamma b^2_{\kappa_1}(\Gamma)\sum_\gamma\langle m|G_{\Gamma\gamma}|j\rangle\langle j'|G_{\Gamma\gamma}|m'\rangle\right)\cdot\frac{\left(\sum_\Gamma b^2_{k_2}(\Gamma)\sum_\gamma\langle j|G_{\Gamma\gamma}|m'\rangle\langle m|G_{\Gamma\gamma}|j'\rangle\right)}{(\hbar\omega_1+\Delta_j)(\hbar\omega_2-\Delta_{j'})}$$

$$-\left(\sum_\Gamma b^2_{\kappa_1}(\Gamma)\sum_\gamma\langle j|G_{\Gamma\gamma}|m'\rangle\langle m|G_{\Gamma\gamma}|j'\rangle\right)\cdot\frac{\left(\sum_\Gamma b^2_{k_2}(\Gamma)\sum_\gamma\langle m|G_{\Gamma\gamma}|j\rangle\langle j'|G_{\Gamma\gamma}|m'\rangle\right)}{(\hbar\omega_1+\Delta_{j'})(\hbar\omega_2-\Delta_j)}\Bigg\}$$

Without lack of generality let us suppose that all intermediate levels are situated above the states $|m\rangle$, $|m'\rangle$ so that all parameters Δ_j are assumed to be positive. The probability of two-phonon transition should be considered in two different cases.

For the levels whose Δ_j is larger then $\hbar\omega_m$ (where ω_m is the maximal Debye frequency) we encounter the Raman process and the probability should be calculated by the integration of the matrix element given by equation (23) over all frequencies above ω_m. The resonance case when $\Delta_j < \hbar\omega_m$ requires a special consideration. This important advancement in the theory of paramagnetic relaxation was made within the frame of the general theory of resonant fluorescence [74] by Orbach [70] and Aminov [71-73] (see books of Altshuler and Kozyrev [69] and Aminov and Malkin [73]). In [73] the equation was given which allows to calculate the Orbach-Aminov relaxation times τ_R in a multilevel system in terms of the one-phonon relaxation probabilities. We adopt here this equation assuming that m and m' are the labels for the initial and final states while j numerates the intermediate levels :

$$\frac{1}{\tau_R} = \frac{1}{\tau_D} + \sum_j \frac{w^{(1)}_{m\to j} w^{(1)}_{j\to m'} + w^{(1)}_{m'\to j} w^{(1)}_{j\to m}}{w^{(1)}_{j\to m} + w^{(1)}_{j\to m'}} \tag{24}$$

Equation (24) contains contributions to the overall relaxation time arising from the direct one-phonon processes (first term) that have been evaluated in Section 5 as well as the contributions from the two-phonon resonance processes.

7. Numerical estimations

The parameter J was found to be about $0.85\,cm^{-1}$ [55] and the best fit to the low temperature stepwise magnetization data [55] provided the values of $D_\perp = 0.24\,cm^{-1}$, $D_n = 0.08\,cm^{-1}$ for the parameters of the AS exchange [60, 61]. The estimations for the spin-phonon coupling constants are: $\lambda = -3\,cm^{-1}/\text{Å}$, $\beta_n \approx -0.11\,cm^{-1}/\text{Å}$, $\beta_\perp \approx -0.34\,cm^{-1}/\text{Å}$ (see refs. [51, 75]). From the X-ray data one finds the parameters $R_0 = 7\,\overset{0}{A}$ for the side of the inner triangle in V_{15} and $\rho = 3\cdot 10^3\,kg/m^3$ for the density of V_{15} [76]. The sound velocity (both transversal and longitudinal) is assumed to be $\upsilon = 2\cdot 10^3\ m/s$.

The estimated one-phonon and Orbach-Aminov relaxation times for the five most intense EPR transitions (figure 2) at the frequency $h\nu = 0.3\ cm^{-1}$ $(\nu = 9\,GHz)$ in parallel field and at different temperatures are given in the Table 1.

Table 1. Relaxation times in the direct and Orbach-Aminov transitions at the frequency of 9 GHz and at different temperatures. Symbol * is related to almost forbidden one-phonon transitions (τ_D is of the order of tens of seconds).

T = 1 K				T = 5 K				T = 10 K			
m	m'	τ_D, s	τ_R, s	m	m'	τ_D, s	τ_R, s	m	m'	τ_D, s	τ_R, s
4	1	$1.3\cdot10^{-1}$	$1\cdot10^{-1}$	4	1	$3\cdot10^{-2}$	$8.3\cdot10^{-3}$	4	1	$1.5\cdot10^{-2}$	$4.1\cdot10^{-3}$
2	3	$1.5\cdot10^{-1}$	$1.1\cdot10^{-1}$	2	3	$2.7\cdot10^{-2}$	$7.8\cdot10^{-3}$	2	3	$1.4\cdot10^{-2}$	$3.4\cdot10^{-3}$
8	6	*	$2.2\cdot10^{-2}$	8	6	*	$3.8\cdot10^{-3}$	8	6	*	$3.3\cdot10^{-3}$
6	5	*	$3.5\cdot10^{-2}$	6	5	*	$5.1\cdot10^{-3}$	6	5	*	$3.9\cdot10^{-3}$
5	7	*	$2.2\cdot10^{-2}$	5	7	*	$4.6\cdot10^{-3}$	5	7	*	$2.7\cdot10^{-3}$

At $1K$ for both the one-phonon and Orbach-Aminov processes the relaxation times are of the order of $10^{-1}\,s$ within the $S = 1/2$ manifold while for the excited $S = 3/2$ manifold the direct processes show very long relaxation times of the order of tens of seconds (note, that this estimation in [75] contains an error). In this manifold Orbach-Aminov processes are much faster ($\tau_R \sim 10^{-2}\,s$). At higher temperatures Orbach-Aminov mechanism becomes the leading one in the ground manifold as well while the one-phonon processes are less significant. At these temperatures in both manifolds the Orbach-Aminov relaxation times are of the order of $10^{-3}\,s$. The relaxation times so far estimated are two orders longer then the coherence time of $1.8 \cdot 10^{-5}\ s$ measured in the Rabi oscillation experiment [1]. These estimations lead to the conclusion that the decoherence in this system can be probably attributed to the dipolar and hyperfine interactions.

8. Conclusions

In this paper we propose a model of spin-phonon relaxation in the V_{15} system that is the first molecular magnet exhibiting Rabi oscillations. It is assumed that spin-phonon interaction arises from the modulation of the isotropic and AS exchange interactions by the acoustic modes of the crystal lattice. Two mechanisms of the spin-lattice relaxations are considered, namely, the direct one-phonon relaxation and the two-phonon Orbach-Aminov mechanism within which the set of closely spaced levels of the system belonging to the central magnetic triangle act as intermediate levels. In terms of the pseudo-angular momentum representation we indicate the selection rules for the direct one-phonon processes. A special role of the different components of the AS exchange is emphasized. Most of the direct transitions arise from acoustic modulation of the AS exchange rather then from modulation of the isotropic exchange. Two phonon relaxation trajectories would not be realized if AS exchange was not present in the system.

The numerical estimations within the long-wave approximation for the acoustic vibrations show that at $1K$ for both the one-phonon and Orbach-Aminov processes the relaxation times are of the order of $10^{-1}\ s$ within the $S = 1/2$ manifold while for the excited $S = 3/2$ manifold the direct processes are inactive. In this manifold Orbach-Aminov processes are much faster ($\tau_R \sim 10^{-2}\ s$). At higher temperatures Orbach-Aminov mechanism becomes the leading one in the ground manifold as well while the one-phonon processes are less significant. At these temperatures in both manifolds the Orbach-Aminov relaxation times are of the order of $10^{-3}\ s$. The spin-phonon relaxation times prove to be two orders longer then the coherence time measured in the Rabi oscillation experiment for V_{15}.

To put the results into the perspective the following remarks are to be done. The second (after V_{15}) molecular magnet where coherent Rabi oscillations have been observed is the so called Fe_4 complex [10]. Spin-lattice relaxation time T_1 has been measured for this system and was found to be of the order of $10^{-6}\ s$. Due to its strong temperature dependence, T_1 in Fe_4 was attributed to the two-phonon processes, either Raman or Orbach-Aminov types. Since spin-lattice relaxation time has not been measured for V_{15}, and in this view it would be interesting to compare the results for two systems so far mentioned. One can rudely estimate that the isotropic spin-phonon coupling parameters λ in V_{15} and Fe_4 relate to each other as 1:10, as the isotropic constants J_0 relate ($0.85\ cm^{-1}$ in V_{15}, $8.2\ cm^{-1}$ in Fe_4 [77]). Spin-lattice relaxation times then should relate to each other as squared reciprocal values $(1/\lambda^2)$ and one may expect the order of $10^{-4}\ s$ for T_1 in V_{15}. However, since the ground state spins also relate as 1:10 ($1/2$ in V_{15}, 5 in Fe_4), relaxation rates in V_{15} may turn out to be significantly longer then even $10^{-4}\ s$ and reach the values estimated in this work.

Vanadium cluster is known for its very small spin-vibronic interaction. However, in the trinuclear low-spin systems similar to V_{15} but with strong spin-vibronic coupling (like copper cluster [78,79]) the phonon assisted relaxation is expected to be much faster, especially when the main decoherence mechanism, spin-to nuclei and dipole-dipole interactions, are substantially suppressed. According to [1] the coherence time may be increased up to $10^{-4}\ s$ when the main decoherence mechanisms are suppressed by the chemical means. The approach so far developed can be applied to more complicated spin-frustrated systems, like high-nuclearity magnetic polyoxometalates [80].

Acknowledgment

This work is done under the auspices of the Israel Science Foundation (ISF) whose financial support is gratefully acknowledged (grant no. 168/09).

Appendix

The vectors $\boldsymbol{e}_i$, $i = 1,\cdots,8$ are related to the eigen-vectors in the spin coupling scheme $\left|(S_{12})SM\right\rangle$ (see [51,59]) by the following relations:

$$\boldsymbol{e}_1 = \frac{1}{\sqrt{2}}\left|(0)\frac{1}{2},\frac{1}{2}\right\rangle - \frac{i}{\sqrt{2}}\left|(1)\frac{1}{2},\frac{1}{2}\right\rangle,$$

$$\boldsymbol{e}_2 = -\frac{i}{\sqrt{2}}\left|(0)\frac{1}{2},-\frac{1}{2}\right\rangle + \frac{1}{\sqrt{2}}\left|(1)\frac{1}{2},-\frac{1}{2}\right\rangle,$$

$$\boldsymbol{e}_3 = \frac{1}{\sqrt{2}}\left|(0)\frac{1}{2},\frac{1}{2}\right\rangle + \frac{i}{\sqrt{2}}\left|(1)\frac{1}{2},\frac{1}{2}\right\rangle,$$

$$\boldsymbol{e}_4 = \frac{i}{\sqrt{2}}\left|(0)\frac{1}{2},-\frac{1}{2}\right\rangle + \frac{1}{\sqrt{2}}\left|(1)\frac{1}{2},-\frac{1}{2}\right\rangle,$$

$$\boldsymbol{e}_5 = \left|(1)\frac{3}{2},\frac{1}{2}\right\rangle, \; \boldsymbol{e}_6 = \left|(1)\frac{3}{2},-\frac{1}{2}\right\rangle,$$

$$\boldsymbol{e}_7 = \left|(1)\frac{3}{2},\frac{3}{2}\right\rangle, \; \boldsymbol{e}_8 = \left|(1)\frac{3}{2},-\frac{3}{2}\right\rangle$$

The matrix $\boldsymbol{B}$ (Section 5) in the "pure" basis $\boldsymbol{e}_i$ ($i = 1,\cdots,8$) is the following:

$$\boldsymbol{B} = \frac{1}{60}\left(\begin{array}{cccc|cccc}
0 & 0 & \frac{21}{4}\lambda^2 & \frac{21}{8}\beta_\perp^2 & \frac{21}{8}\beta_n^2 & \frac{27}{16}\beta_\perp^2 & \frac{63}{32}\beta_\perp^2 & 0 \\
0 & 0 & \frac{21}{8}\beta_\perp^2 & \frac{21}{4}\lambda^2 & \frac{27}{16}\beta_\perp^2 & \frac{21}{8}\beta_n^2 & 0 & \frac{63}{32}\beta_\perp^2 \\
\frac{21}{4}\lambda^2 & \frac{21}{8}\beta_\perp^2 & 0 & 0 & \frac{21}{8}\beta_n^2 & \frac{21}{32}\beta_\perp^2 & \frac{81}{16}\beta_\perp^2 & 0 \\
\frac{21}{8}\beta_\perp^2 & \frac{21}{4}\lambda^2 & 0 & 0 & \frac{21}{32}\beta_\perp^2 & \frac{21}{8}\beta_n^2 & 0 & \frac{81}{16}\beta_\perp^2 \\
\hline
\frac{21}{8}\beta_n^2 & \frac{27}{16}\beta_\perp^2 & \frac{21}{8}\beta_n^2 & \frac{21}{32}\beta_\perp^2 & 0 & 0 & 0 & 0 \\
\frac{27}{16}\beta_\perp^2 & \frac{21}{8}\beta_n^2 & \frac{21}{32}\beta_\perp^2 & \frac{21}{8}\beta_n^2 & 0 & 0 & 0 & 0 \\
\frac{63}{32}\beta_\perp^2 & 0 & \frac{81}{16}\beta_\perp^2 & 0 & 0 & 0 & 0 & 0 \\
0 & \frac{63}{32}\beta_\perp^2 & 0 & \frac{81}{16}\beta_\perp^2 & 0 & 0 & 0 & 0
\end{array}\right)$$

References

[1] Bertaina S, Gambarelli S, Mitra T, Tsukerblat B, Müller A and Barbara B 2008 *Nature* **453**

203-207
[2] Nielsen M and Chuang I L 2000 *Quantum Computation and Quantum Information* (Cambridge: Cambridge University Press)
[3] Leibfried D, DeMarco B, Meyer V, Lucas D, Barrett M, Britton J, Itano W M, Jelenković B, Langer C, Rosenband T and Wineland D J 2003 *Nature* **422** 412-415
[4] Schmidt-Kaler F, Häffner H, Riebe M, Gulde S, Lancaster G P T, Deuschle T, Becher C, Roos C F, Eschner J and Blatt R 2003 *Nature* **422** 408-411
[5] Loss D and DiVincenzo D P 1998 *Phys. Rev. A* **57** 120-126
[6] Kane B E 1998 *Nature* **393** 133-137
[7] Morton J J L 2006 *Nature Phys.* **2** 365-366
[8] Leuenberger M N and Loss D 2001 *Nature* **410** 789-793
[9] Ardavan A, Rival O, Morton J J L, Blundell S, Tyryshkin A M, Timco G A and Winpenny R E P 2007 *Phys. Rev. Lett.* **98** 057201
[10] Schlegel C, van Slageren J, Manoli M, Brechin E K and Dressel M 2008 *Phys. Rev. Lett.* **101** 147203
[11] Zobbi L, Mannini M, Pacchioni M, Chastanet G, Bonacchi D, Zanardi C, Biagi R, del Pennino U, Gatteschi D, Cornia A and Sessoli R 2005 *Chem. Commun.* 1640-42
[12] Condorelli G G, Motta A, Fragalà I L, Giannazzo F, Raineri V, Caneschi A and Gatteschi D 2004 *Angew. Chem. Int. Ed.* **43** 4081-84
[13] Coronado E, Forment-Aliaga A, Romero F M, Corradini V, Biagi R, De Renzi V, Gambardella A and del Pennino **2005** *Inorg. Chem.* 7693-95
[14] Corradini V, Biagi R, del Pennino U, De Renzi V, Gambardella A, Affronte M, Muryn C A, Timco G A and Winpenny R E P 2007 *Inorg. Chem.* **46** 4937-43
[15] Cornia A, Fabretti A C, Pacchioni M, Zobbi L, Bonacchi D, Caneschi A, Gatteschi D, Biagi R, del Pennino U, De Renzi V, Gurevich L and Van der Zant H S J 2003 *Angew. Chem. Int. Ed.* **42** 1645-48
[16] Fleury B, Catala L, Huc V, David C, Zhong W Z, Jegou P, Baraton L, Albouy P-A and Mallah T 2005 *Chem. Commun.* 2020-22
[17] Salman Z, Chow K H, Miller R I, Morello A, Parolin T J, Hossain M D, Keeler T A, Levy C D P, MacFarlane W A, Morris G D, Saadaoui H, Wang D, Sessoli R, Condorelli GG and Kiefl R F 2007 *Nano Lett.* **7** 1551-55
[18] Troiani F, Ghirri A, Affronte M, Carretta S, Santini P, Amoretti G, Piligkos S, Timco G and Winpenny R E P 2005 *Phys. Rev. Lett.* **94** 207208
[19] Affronte M, Troiani F, Ghirri A, Carretta S, Santini P, Corradini V, Schuecker R, Muryn C, Timco G and Winpenny R E P 2006 *Dalton Trans.* 2810-17
[20] Affronte M, Troiani F, Ghirri A, Candini A, Evangelisti M, Corradini V, Carretta S, Santini P, Amoretti G, Tuna F, Timco G and Winpenny R E P 2007 *J.Phys.D: Appl.Phys* **40** 2999-3004
[21] Carretta S, Santini P, Amoretti G, Troiani F and Affronte M 2007 *Phys. Rev. B* 76 024408
[22] Stepanenko D, Trif M and Loss D 2008 *Inorg. Chim. Acta* **361** 3740-45
[23] Lehmann J, Gaita-Arino A, Coronado E and Loss D 2007 *Nature Nanotechnology* **2** 312-317
[24] Winpenny R E P 2008 *Angew. Chem. Int. Ed.* **47** 2–5
[25] Braun-Sand S B and Wiest O 2003 *J. Phys. Chem.*A **107** 285-291
[26] Morello A, Stamp P C E and Tupitsyn I S 2006 *Phys. Rev. Lett.* **97** 207206
[27] Dobrovitski V V, Katsnelson M I and Harmon B N 2000 *Phys. Rev. Lett.* **84** 3458-61
[28] Chudnovsky E M, Garanin D A and Schilling R 2005 *Phys. Rev.* B **72** 094426
[29] Mitrikas G, Sanakis Y, Raptopoulou C P, Kordas and G Papavassiliou G, 2008 *Phys. Chem. Chem. Phys.* **10** 743–748
[30] Georgopoulou A N, Sanakis Y and Boudalis A K 2011 *Dalton Trans.* **40** 6371-74
[31] Baibekov E I 2011 *JETP Lett.* **93** 292-297
[32] Müller A and Döring J 1988 *Angew. Chem. Int. Ed. Engl* **27** 1719 -21

[33] Gatteschi D, Pardi L, Barra A-L, Müller A and Döring J 1991 *Nature* **354** 463-465
[34] Barra A-L, Gatteschi D, Pardi L, Müller A and Döring J 1992 *J. Am. Chem. Soc.* **114** 8509-14
[35] Gatteschi D, Pardi L, Barra A-L and Müller A 1993 *Mol. Eng.* **3** 157
[36] Platonov V V, Tatsenko O M, Plis V I, Zvezdin A K and Barbara B 2002 *Physics of the Solid State (Rus. Fizika Tverdogo Tela)* **44** 2104-2106
[37] Kostyuchenko V V and Popov A I 2008 *J. Exp. Theor. Phys.* **107(4)** 595-602
[38] Raghu C, Rudra I, Sen D and Ramasesha S 2001 *Phys. Rev. B* **64** 064419
[39] Miyashita S, De Raedt H and Michielsen K 2003 *Prog. Theor. Phys.* **110** 889-899
[40] De Raedt H, Miyashita S, Michielsen K and Machida M 2004 *Phys. Rev. B* **70** 064401
[41] De Raedt H, Miyashita S, Michielsen K and Machida M 2004 *Phys. Stat. Solidi (b)* **241** 1180
[42] Konstantinidis N P and Coffey D 2002 *Phys. Rev. B* **66** 174426
[43] Machida M and Miyashita S 2005 *Physica E* **29** 538-540
[44] Kortus J, Pederson M R, Hellberg C S and Khanna S N 2001 *Eur. Phys. J.* D **16** 177-180
[45] Kortus J, Hellberg C S and Pederson M R 2001 *Phys. Rev. Lett.* **86** 3400
[46] Boukhvalov D W, Kurmaev E Z, Moiwes A, Zatsepin D A, Cherkashenko V M, Nemnonov S N, Finkelstein L D, Yarmoshenko Yu M, Neumann M, Dobrovitski V V, Katsnelson M I, Lichtenstein A I, Harmon B N and Koegerler P 2003 *Phys. Rev. B* **67** 134408
[47] Boukhvalov D W, Dobrovitski V V, Katsnelson M I, Lichtenstein A I, Harmon B N and Kögerler P 2004 *Phys. Rev. B* **70** 054417
[48] Barbour A, Luttrell R D, Choi J, Musfeldt J L, Zipse D, Dalal N S, Boukhvalov D W, Dobrovitski V V, Katsnelson M I, Lichtenstein A I, Harmon B N and Kögerler P 2006 *Phys. Rev. B* **74** 014411
[49] Chaboussant G, Basler R, Sieber A, Ochsenbein S T, Desmedt A, Lechner R E, Telling M T F, Kögerler P, Müller and Güdel H-U 2002 *Europhys. Lett.* **59** 291-297
[50] Chaboussant G, Ochsenbein S T, Sieber A, Güdel H-U, Mutka H, Müller A and Barbara B 2004 *Europhys. Lett.* **66** 423-429
[51] Tsukerblat B and Tarantul A 2011, The nanoscopic V_{15} cluster: a unique magnetic polyoxometalate, in *Molecular Cluster Magnets* , Ed: Winpenny R E P (Singapore: World Scientific Publishers) Ch. 3, pp. 106-180 (in press)
[52] Dzyaloshinskii I E 1957 *Sov. Phys. JETP* **5** 1259
[53] Moriya T 1960 *Phys. Rev. B* **120** 91–98
[54] Chiorescu I, Wernsdorfer W , Müller A and Bögge H 2000 *Phys. Rev. Lett.* **84** 3454
[55] Barbara B 2003 *J. Mol. Struct.* **656** 135-140
[56] Chiorescu I, Wernsdorfer W, Müller A, Bögge H and Barbara B 2000 *J.Magn. Magn. Mater.* **221** 103-109
[57] Tsukerblat B S and Belinskii M I 1983 *Magnetochemistry and Radiospectroscopy of Exchange Clusters (Russian)* (Kishinev: Stiintsa)
[58] Tsukerblat B S, Belinskii M I and Fainzilberg V E 1987 *Magnetochemistry and Spectroscopy of Transition Metal Exchange Clusters*, in *Soviet Sci. Rev. B*, M, Harwood Acad. Pub, New York, Vol. **9** pp 337- 481
[59] Tsukerblat B, Tarantul A and Müller A 2006 *Phys. Lett. A* **353** 48–59
[60] Tarantul A, Tsukerblat B and Müller A 2007 *Inorg. Chem.* **46** 161-169
[61] Tarantul A, Tsukerblat B and Müller A 2006 *Chem. Phys. Lett.* **428** 361–366
[62] Tsukerblat B, Tarantul A and Müller A 2006 *J. Chem. Phys* **125** 0547141
[63] Bates C A and Jasper R F, 1971 *J.Phys.C: Sol. State Phys.* **4** 2341 -53
[64] Varshalovich D A, Moskalev A N and Khersonskii V K 1988 *Quantum Theory of Angular Momentum* (Singapore: World Scientific)
[65] Gatteschi D, Sessoli R and Villain J 2006 *Molecular Nanomagnets* (Oxford: Oxford University Press)

[66] Bencini A and Gatteschi D 1990 *Electron Paramagnetic Resonance of Exchange Coupled Systems* (Berlin: Springer)
[67] Schnalle R and Schnack J 2010 *Int. Reviews in Physical Chemistry* **29** 403-452
[68] Van Vleck J H 1940 *Phys. Rev.* **57** 426-447
[69] Altshuler S A and Kozyrev B M 1974 *Electron Paramagnetic Resonance in Compounds of Transition Elements*, 2nd ed. (New York: Wiley)
[70] Orbach R 1961 *Proc. Roy. Soc.* **A254** (1319) 458-484
[71] Aminov L K 1962 *Sov. Phys. JETP* **15** 547
[72] Aminov L K 1974 *Zh. Exp. Teor. Phys.* **67** 79-83
[73] Aminov L K and Malkin B Z 2008 *Dynamics and Kinetics of Electronic and Spin Excitations in Paramagnetic Crystals (Russian)* (Kazan: Kazan State University)
[74] Heitler W 1957 *The Quantum Theory of Radiation* (Oxford: Clarendon Press)
[75] Tarantul A and Tsukerblat B 2010 *Inorg. Chim. Acta* **363** 4361-67
[76] Müller A and Döring J Private Communication
[77] Schlegel C, van Slageren J, Manoli M, Brechin E K and Dressel M 2009 *Polyhedron* **28** 1834-1837
[78] Belinsky M I 2008 *Inorg. Chem.* **47** 3521-31
[79] Belinsky M I 2008 *Inorg. Chem.* **47** 3532-39
[80] Kögerler P, Tsukerblat B and Müller A 2010 *Dalton Trans.* **39** 21–36

Journal of Physics: Conference Series **324** (2011) 012008 doi:10.1088/1742-6596/324/1/012008

Multi-photon Rabi oscillations in high spin paramagnetic impurity.

S. Bertaina[1], **N. Groll**[2,3], **L. Chen**[2,4] **and I. Chiorescu**[2]

[1] IM2NP-CNRS (UMR 6242) and Université Aix-Marseille, Faculté des Sciences et Techniques, Avenue Escadrille Normandie Niemen - Case 142, F-13397 Marseille Cedex, France.

[2] Department of Physics and The National High Magnetic Field Laboratory, Florida State University, Tallahassee, Florida 32310, USA

E-mail: `sylvain.bertaina@im2np.fr`

Abstract. We report on multiple photon monochromatic quantum oscillations (Rabi oscillations) observed by pulsed EPR (Electron Paramagnetic Resonance) of Mn^{2+} (S=5/2) impurities in MgO. We find that when the microwave magnetic field is similar or large than the anisotropy splitting, the Rabi oscillations have a spectrum made of many frequencies not predicted by the S=1/2 Rabi model. We show that these new frequencies come from multiple photon coherent manipulation of the multi-level spin impurity. We develop a model based on the crystal field theory and the rotating frame approximation, describing the observed phenomenon with a very good agreement.

1. Introduction

Quantum coherence of electron spin systems probed by pulsed EPR attracts a great interest due to its potential applications in quantum computation. Spin systems benefit from relatively large coherence and relaxation times, which makes them suitable as quantum bit implementations or as a new type of quantum random access memory [1, 2, 3, 4]. Recent studies show that the dilution of spins in diamagnetic hosts significantly reduces spin-spin dipolar interactions and allows spin quantum manipulations. Such situations are well exemplified by systems like N atoms in C_{60}[5], NV centers in diamonds [6, 7] , Er^{3+}[8, 9], Yb^{3+}[10] and Cr^{5+}[11] ions and molecular magnets[12, 13].

The majority of studies on electron spin qubits uses S=1/2 isotropic systems, which prevents strong static effects due to the crystal field or unwanted non linear effects. However, coherent manipulation of a multilevel system is needed as the first step towards the implementation of the Grover algorithm[14, 15, 16].

Although several quantum continuous wave (CW) EPR spectroscopy studies have been reported years ago[17], coherent multiple photon EPR (Rabi oscillations) have not been clearly demonstrated until recently [18].

[3] Present address: Materials Science Division, Argonne National Laboratory, 9700 S. Cass Avenue, IL 60439, USA.

[4] Present address: Department of Chemistry and Chemical Biology, Cornell University, Ithaca, NY 14853-1301, USA

In order to study quantum effects in multi level electronic spin systems, we chose Mn^{2+} (S=5/2) ion implanted inside the diamagnetic host MgO. The cubic symmetry of MgO induces a small crystal field anisotropy on Mn^{2+} and, as we will see in the following sections, the microwave magnetic energy can be brought to the same order of magnitude, an aspect shown to be instrumental in generating multi-photon Rabi oscillations.

2. Coherent dynamic of MgO:Mn^{2+}

2.1. *Spin Hamiltonian*

The spins S=5/2 of the Mn^{2+} ions are diluted in a MgO non-magnetic matrix of cubic symmetry $F_{m\bar{3}m}$ (lattice constant 4.216 Å). The Mn^{2+} ions are located in substitutional positions of Mg^{2+} ions. The high degree of symmetry ensures that the spins are seeing an almost isotropic crystalline environment and thus the fourth order magnetic anisotropy can be made much smaller than the Zeeman splittings. Interactions between Mn^{2+} ions and symmetry deformations are neglected. The spin Hamiltonian at resonance is given by [19, 20, 18]:

$$\begin{aligned} H &= a/6\left[S_x^4 + S_y^4 + S_z^4 - S(S+1)(3S^2-1)/5\right] \\ &\quad + \gamma\vec{H}_0\cdot\vec{S} - A\vec{S}\cdot\vec{I} + \gamma\vec{h}_{mw}\cdot\vec{S}\cos(2\pi f t) \end{aligned} \tag{1}$$

where $\gamma = g\mu_B/h$ is the gyromagnetic ratio ($g = 2.0014$ the g-factor, μ_B- Bohr's magneton and $h-$ Planck's constant), $a = 55.7$ MHz is the anisotropy constant, $A = 244$ MHz is the hyperfine constant of ^{55}Mn ($I = 5/2$), h_{mw} and f represent the microwave (MW) amplitude and frequency respectively, and $\vec{H}_0$ is the static field ($\vec{H}_0 \perp \vec{h}_{mw}$). In our experiments, the static field ensures a Zeeman splitting of $\gamma H_0 \approx f \sim$9 Ghz, much stronger than all the other interactions of Eq.(1). This implies that (i) $\vec{H}_0$'s direction can be approximated as the quantization axis and (ii) coherent MW driving is confined between levels of same nuclear spin projection m_I.

2.2. *Experimental procedure*

Rabi oscillations measurements have been performed in a Bruker Elexsys 680 pulse EPR spectrometer working at about $f = 9.6$ GHz (X-band). All data have been recorded at room temperature. Multiple photon resonance in continuous wave (CW) non coherent spectroscopy is explained in Fig. 1. In pulsed EPR both the static field and MW frequency are fixed. In our experiments the static magnetic field has been chosen to be in the middle of the transition line -1/2 → 1/2. A microwave pulse P_R start at $t = 0$ and coherently drive the magnetization. At the end of the pulse ($t = t_P$) the magnetization must be recorded in order to be compared with the present theory. Because of the dead time of the spectrometer ($\tau_{dt} \sim$ 80ns), it is impossible to directly measure the magnetization right after the pulse P_R. However if the line is sharp enough we can simply record the free induction decay (FID) emitted by the system when the microwave field is shut down. This method gives the value of the density of population at the end of the pulse (Fig. 1).

3. Results and discussions.

Examples of Rabi oscillations are given in Fig. 2. At low MW power only the 1-photon Rabi oscillation is observed. When the MW power is increased a second kind of oscillation superimposes with the one-photon process. The Fourier transform of the Rabi oscillation clearly show 2 well resolved frequencies.

In order to fully describe the experimental data we developed a model based on the Hamiltonian (1). Since the static field is much bigger than all the other interactions of (1),

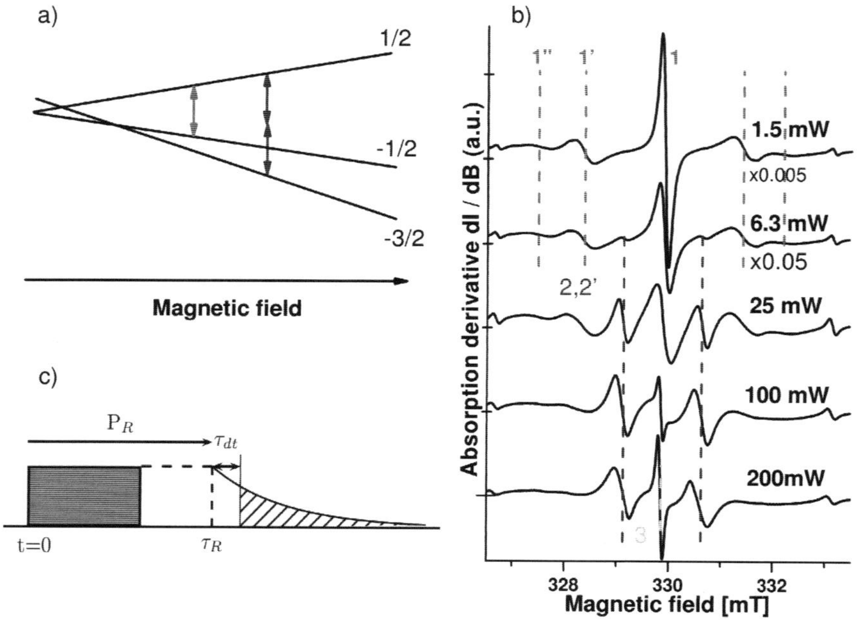

Figure 1. Multiple photon resonances in CW configuration. a) Schematic representation of the Zeeman levels (only 3 levels are shown) with increasing magnetic field. The red arrow is the conventional 1-photon transition. The blue arrows represent the 2-photon transition between $|-3/2\rangle \rightarrow |1/2\rangle$. b) CW spectra of MgO:$Mn^{2+}$. At low MW power, only the 1-photon transitions are induced (red dashed lines) but when the MW power is increased, multi-photon transitions appear. c) The pulse sequence used in the multi-photon Rabi oscillation measurements, starting with a strong excitation pulse P_R which induces the multi-photon nutation. Right after the P_R pulse, the FID signal gives the transverse magnetization state between -1/2 and 1/2.

the quantization axis is defined by $\vec{H}_0$. Once the static part of (1) is diagonalized, the full Hamiltonian (including the dynamical part) is transformed using the unitary transformation $U(t) = \exp(-i2\pi f\hat{S}_z t)$. Neglecting the counter rotation of the microwave radiation (rotating wave approximation) we get the Hamiltonian in the rotating frame :

$$\mathcal{H}_{RWA} = \frac{g\mu_B}{2h} h_{mw} m\hat{S}_x + \hat{E} - f\hat{S}_z \tag{2}$$

The dynamics of Hamiltonian (2) is described by the time evolution of the 6x6 density matrix $\rho(t)$:

$$i\frac{d\rho}{dt} = [\mathcal{H}_{RWA}, \rho]. \tag{3}$$

In the rotating frame (Dirac or interaction picture), the solution is $\rho(t) = U_p(t)\rho_0 U_p^{\dagger}(t)$ with ρ_0 the matrix density at thermal equilibrium :

$$\rho_0 = \frac{\exp(-\hat{E}/kT)}{Tr(\exp(-\hat{E}/kT))} \tag{4}$$

and $U_p(t)$ is the propagator operator: $U_p(t) = \exp(-i2\pi\mathcal{H}_{RWA}t)$.

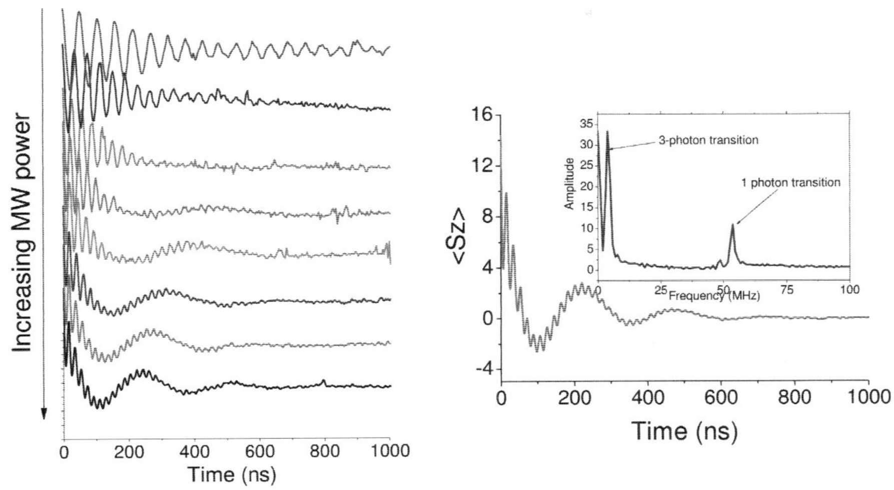

Figure 2. Rabi oscillations of the Mn^{2+} ions in MgO when the microwave power is increased, for $\vec{H}_0||[111]$. The FFT clearly shows 2 Rabi frequencies: the higher one is induced by a 1-photon process whereas the lower frequency is due to a 3-photon process.

When the Rabi pulse P_R is applied, the spin population is coherently manipulated. At the end of the sequence a weak $\pi/2$ pulse will selectively probe the difference of population between states -1/2 and +1/2. The signal out of the spectrometer is $S(t) = \sigma_{-1/2}(t) - \sigma_{1/2}(t)$.

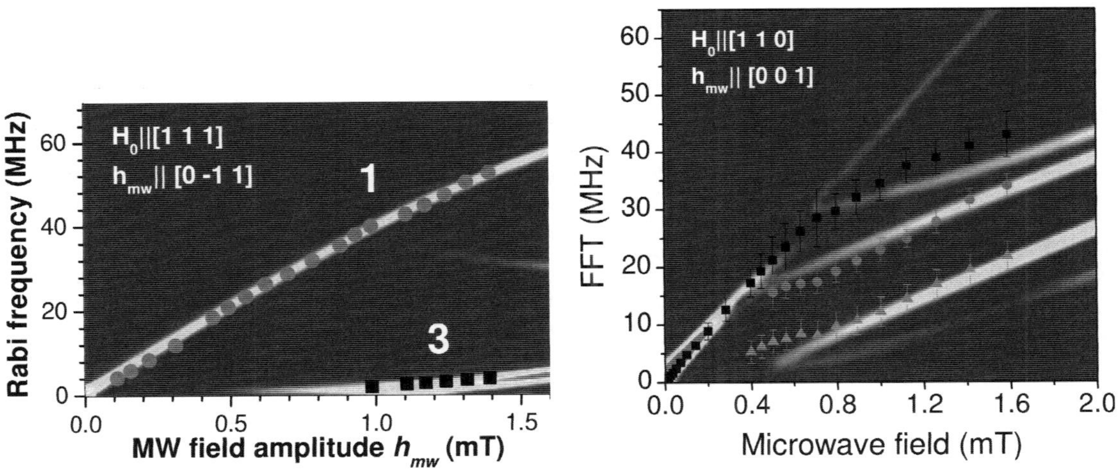

Figure 3. MW field dependence of the Rabi oscillations in MgO:Mn^{2+} for 2 orientations between the crystal and the static field $\vec{H}_0$. For both figures, the points are FFT peaks of the experimentally recorded Rabi oscillations. The contourplots have been computed numerically using the presented model.

Results of the presented model are presented in figure 3. The MW field dependence of the Rabi frequency of the experimental oscillations are in very good agreement with the presented model. Note that no fit parameter has been used and all parameters have been take from previous independent studies [19, 20]. For $\vec{H}_0||[111]$ the frequency distribution is quite simple showing 2 kinds of Rabi oscillations: 1-photon and 3-photon. For $\vec{H}_0||[110]$ the frequency distribution is more complicated but still well describes by the model. In this case, there are 1-photon (Black squares) and 2 kinds of 3-photon process (red circles and purple triangle), coupling resonantly different levels of the 6-level spin system.

We note that static field angular dependence studies [18, 21], not presented here, also shows also a very good agreement between this model and the experimental data.

4. Conclusion

We report on single and multi-photon coherent rotations in a $S = 5/2$ spin system featuring a cubic anisotropy. We introduce a model, using only the spin Hamiltonian, that describes the experimental data with a very good agreement. These result are an important step toward the implementation of the Grover algorithm which need the coherent manipulation of a multilevel system.

Acknowledgments

This work was supported by City of Marseille, NSF Cooperative Agreement Grant No. DMR-0654118, NSF grants No. DMR-0645408, Aix-Marseille University (BQR grant).

References

[1] Blencowe M 2010 *Nature* **468** 44–5 ISSN 1476-4687
[2] Chiorescu I, Groll N, Bertaina S, Mori T and Miyashita S 2010 *Phys. Rev. B* **82** 024413 ISSN 1098-0121 (*Preprint* 1004.3605)
[3] Kubo Y, Ong F, Bertet P, Vion D, Jacques V, Zheng D, Dréau A, Roch J F, Auffeves A, Jelezko F, Wrachtrup J, Barthe M, Bergonzo P and Esteve D 2010 *Phys. Rev. Lett.* **105** 1–4 ISSN 0031-9007
[4] Schuster D, Sears A, Ginossar E, DiCarlo L, Frunzio L, Morton J, Wu H, Briggs G, Buckley B, Awschalom D and Schoelkopf R 2010 *Phys. Rev. Lett.* **105** ISSN 0031-9007
[5] Morley G, van Tol J, Ardavan A, Porfyrakis K, Zhang J and Briggs G 2007 *Phys. Rev. Lett.* **98** 220501
[6] Hanson R, Dobrovitski V V, Feiguin A E, Gywat O and Awschalom D D 2008 *Science* **320** 352–5
[7] Dutt M V G, Childress L, Jiang L, Togan E, Maze J, Jelezko F, Zibrov A S, Hemmer P R and Lukin M D 2007 *Science (New York, N.Y.)* **316** 1312–6
[8] Bertaina S, Gambarelli S, Tkachuk A, Kurkin I N, Malkin B Z, Stepanov A and Barbara B 2007 *Nature nanotechnology* **2** 39–42 (*Preprint* 0703364)
[9] Bertaina S, Shim J H, Gambarelli S, Malkin B Z and Barbara B 2009 *Phys. Rev. Lett.* **103** 226402 (*Preprint* 0911.4818)
[10] Rakhmatullin R, Kurkin I N, Mamin G, Orlinskii S, Gafurov M, Baibekov E, Malkin B Z, Gambarelli S, Bertaina S and Barbara B 2009 *Phys. Rev. B* **79** 172404–172408
[11] Nellutla S, Choi K Y, Pati M, van Tol J, Chiorescu I and Dalal N S 2007 *Phys. Rev. Lett.* **99** 137601
[12] Bertaina S, Gambarelli S, Mitra T, Tsukerblat B, Müller A and Barbara B 2008 *Nature* **453** 203–6
[13] Takahashi S, van Tol J, Beedle C C, Hendrickson D N, Brunel L C and Sherwin M S 2009 *Phys. Rev. Lett.* **102** 4 (*Preprint* 0810.1254)
[14] Grover L 1997 *Phys. Rev. Lett.* **79** 325–328 ISSN 0031-9007
[15] Leuenberger M N and Loss D 2001 *Nature* **410** 789–93
[16] Leuenberger M N, Loss D, Poggio M and Awschalom D D 2002 *Phys. Rev. Lett.* **89** 1–4
[17] Sorokin P, Gelles I and Smith W 1958 *Phys. Rev.* **112** 1513–1515
[18] Bertaina S, Chen L, Groll N, Van Tol J, Dalal N S and Chiorescu I 2009 *Phys. Rev. Lett.* **102** 50501–50504 (*Preprint* 0812.4534)
[19] Low W 1957 *Phys. Rev.* **105** 793–800
[20] Smith S, Auzins P and Wertz J 1968 *Phys. Rev.* **166** 222–225
[21] Bertaina S, Chen L, Groll N and Chiorescu I *arXiv:1106.3365*

Journal of Physics: Conference Series **324** (2011) 012009 doi:10.1088/1742-6596/324/1/012009

Inhomogeneous broadening of ESR lines of rare earth impurities in scheelite $CaWO_4$ induced by internal electric field gradients

E I Baibekov[1] **and I N Kurkin**
Kazan Federal University, Kazan 420008, Russia

E-mail: edbaibek@gmail.com

Abstract. The inhomogeneous broadening of ESR line in erbium-doped $CaWO_4$ crystal is studied both theoretically and experimentally in the special case when external magnetic field is parallel to crystal *c* axis. Lorenzian shapes of ESR lines are observed and ascribed to the broadening from random electric field gradients produced by charged defects. The defect concentrations are estimated on the grounds of a simple statistical model and crystal-field calculations of impurity-defect coupling constants. Samples with different concentrations of erbium ions are used in order to find out how Er^{3+} impurities affect the linewidth.

1. Introduction

As is well-known, crystal lattice imperfections lead to the inhomogeneous broadening of ESR lines in crystals since the perturbation of the crystal field randomly shifts Larmor frequencies of paramagnetic ions. This occurs because energy levels and effective *g*-factrors of paramagnetic impurities (especially in case of rare earth ions) are extremely sensitive to variations of positions and charges of the neighboring ions. The possible lattice defects are various: point charges and consequent local strains, dislocations, mosaic structure [1]. Not taking into account random defects that are always introduced into the crystal during the process of its growth, the broadening is unavoidable in case when initially diamagnetic host matrix is activated with paramagnetic ions. Among the above-mentioned, the role of charged defects is appreciable since they produce electric field $E \sim 1/R^2$ slowly diminishing with distance R between the paramagnetic ion and the defect. These charged defects appear most naturally in case of heterovalent substitution, as in rare earth-doped scheelite crystals. In case when ESR line broadening is solely due to paramagnetic impurity, Larmor frequencies ω_1, ω_2 and the distance between the two given paramagnetic ions become correlated. Note that most statistical models of magnetic relaxation in magnetically diluted media do not take into account the above-mentioned correlations. These correlations become important when the corresponding transition probabilities depend on $|\omega_1 - \omega_2|$, as in the case of relaxation induced by interspin dipolar interactions [2, 3]. To summarize, in every theoretical study of magnetic relaxation that involves interspin interactions one must determine leading line broadening mechanisms that take place in the system.

These mechanisms can be revealed either in experimental or theoretical studies of the system. The former involve observations of shapes and widths of resonance lines; when magnetic anisotropy is

[1] Corresponding author.

present, one can also vary the direction of static magnetic field vector $\boldsymbol{H}_0$. The latter are direct calculations of the lineshapes based on the knowledge of relevant broadening mechanisms, impurity-defect coupling constants and defect concentration. In most cases this task is highly time-consuming, and only crude estimations are possible. In case of dislocations, the calculated linewidths are very approximate [1]. As for charged point defects, general model based on continuum approximation predicts Holtsmark and Lorenzian lineshapes in the presence of random electric fields and electric field gradients, respectively [1].

In the present paper we study both theoretically and experimentally the inhomogeneous broadening of ESR line in erbium-doped $CaWO_4$ crystal in the special case when external magnetic field is parallel to crystal c axis. We observe Lorenzian shapes of ESR lines and ascribe the broadening to random electric field gradients produced by charged defects. Based on a statistical model [1] and on our crystal-field calculations of impurity-defect coupling constants, we estimate the defect concentrations. A similar problem was studied before for Er^{3+} in MgO crystal [4]. Our study essentially differs from ref. [4] since we use samples with different concentrations of erbium ions in order to find out how Er^{3+} impurities affect the linewidth. The answer to this question is crucial since rare-earth-activated scheelites are frequently used in magnetic resonance experiments and were proposed as promising solid state media for the implementation of quantum computation [5, 6].

2. Theoretical background

We will briefly outline the procedure that enables to account for the random shifts of Larmor frequencies of rare earth (RE) ions due to their electrostatic interactions with charged point defects. We start with the electrostatic potential at the site of an electron of RE ion produced by the defect:

$$\varphi(\boldsymbol{R},\boldsymbol{r}) = \frac{qe}{|\boldsymbol{R}-\boldsymbol{r}|} = qe\sum_{p=0}^{\infty}\sum_{k=-p}^{p}\frac{r^p}{R^{p+1}}C_{pk}^{*}(\Theta,\Phi)C_{pk}(\vartheta,\varphi), \tag{1}$$

where vectors $\boldsymbol{R}(R,\Theta,\Phi)$ and $\boldsymbol{r}(r,\vartheta,\varphi)$ determine positions of the defect and the electron in spherical coordinates with respect to the nucleus of RE ion, q is the excessive charge of the defect in units of elementary charge e, and C_{pk} are spherical operators. Since $r \ll R$, one can leave only the first few terms of the multipolar expansion (1). The constant term with $p=0$ can be omitted. We will take only dipolar and quadrupolar terms $p=1$ and $p=2$ that correspond to electric field and electric field gradient of the defect, respectively.

If there are n electrons on the valency $4f$ shell of RE ion and N defects in the crystal, the perturbation of the crystal field at the site of RE ion will be as follows:

$$\delta\mathsf{H}_{CF} = -e\sum_{i=1}^{n}\sum_{j=1}^{N}\varphi(\boldsymbol{R}_j,\boldsymbol{r}_i). \tag{2}$$

The perturbation $\delta\mathsf{H}_{CF}$ alters eigenstates and energy splittings in Zeeman structure of the specter of RE ion. Since $p=1$ terms do not mix the states inside $4f$ shell, they are involved only in transitions via upper-lying configuration. Thus, despite the fact that $p=2$ terms are lesser in magnitude than $p=1$ terms, their final contributions into the frequency shift can be significant. Generally, it is hard to distinguish between dipolar and quadrupolar contributions from the experimentally observed lineshapes.

In $CaWO_4$ scheelite, trivalent RE ions occupy positions of Ca^{2+} with S_4 point symmetry. The ESR line broadening in that particular crystal due to random electric fields from charged point defects when the vector $\boldsymbol{H}_0$ is perpendicular to the crystal c axis was thoroughly investigated in ref. [7]. However, when $\boldsymbol{H}_0 \parallel c$, the terms with $p=1$ do not shift Larmor frequencies in the lowest order of perturbation theory due to symmetry selection rules [7, 8] and can be neglected. So we have here a special case when we are able to investigate solely $p=2$ contributions. Moreover, it is straightforward to show on the same symmetry grounds that only $C_{20}(\vartheta_i,\varphi_i)$ operators would alter the energy splittings, and the perturbation (2) takes the form

$$\delta \mathsf{H}_{CF} = -e^2 \sum_{i=1}^{n} \sum_{j=1}^{N} \frac{q_j r^2}{R_j^3} C_{20}(\Theta_j,\Phi_j) C_{20}(\vartheta_i,\varphi_i). \tag{3}$$

Since for rare earths the total angular moment J is a good quantum number, it is possible to project the operator (3) onto a basis $|JM\rangle$ of J_z eigenstates of the lowest multiplet ${}^{2S+1}L_J$:

$$\langle JM|\delta \mathsf{H}_{CF}|JM'\rangle = -\alpha_2 e^2 \langle r^2\rangle \langle JM|C_{20}|JM'\rangle \sum_{j=1}^{N} \frac{q_j C_{20}(\Theta_j,\Phi_j)}{R_j^3}, \tag{4}$$

were the coefficient α_p is the reduced matrix element [9] and $\langle r^2\rangle$ is the average value of r^2 of an electron in $4f$ shell. We will omit the brackets $|JM\rangle$ henceforth. Recalling that $C_{20} = O_{20}/2$ where O_{20} is Stevens operator, and that $C_{20}(\Theta_j,\Phi_j) = (3\cos^2\Theta_j - 1)/2$, we rewrite (4) as

$$\delta \mathsf{H}_{CF} = \frac{\alpha_2 e^2 (1-\sigma_2)\langle r^2\rangle}{4} \sum_{j=1}^{N} \frac{q_j(1-3\cos^2\Theta_j)}{R_j^3} \cdot O_{20} \equiv \alpha_2 \delta B_{20} \cdot O_{20}, \tag{5}$$

where the factor $(1-\sigma_2)$ is added to account for the shielding from $5s, 5p$ and $6s$ outer shells. The effective Hamiltonian of RE ion is now

$$\mathsf{H} = \sum_{pk} \alpha_p B_{pk} O_{pk} + g_J \mu_B H_0 J_z + \alpha_2 \delta B_{20} \cdot O_{20}, \tag{6}$$

where the first two terms represent interactions with crystal field and external magnetic field. Suppose that n is an odd number, and the energy levels of RE ion when $H_0 = 0$ are at least twice degenerate. The degeneracy is removed when $H_0 \neq 0$. Then it is possible to introduce an effective spin $s = 1/2$ and g-factor $g_{\parallel}$ for a given Kramers doublet. The last term in (6) shifts Larmor angular frequency ω by the value $\delta\omega$ so that

$$\delta\omega = \frac{\delta E_2 - \delta E_1}{\hbar} = \frac{2\alpha_2 g_J \omega \delta B_{20}}{g_{\parallel}} \sum_{m=3}^{2J+1} \frac{\langle 2|O_{20}|m\rangle\langle m|J_z|2\rangle + c.c.}{E_2 - E_m} \equiv G\omega \sum_{j=1}^{N} \frac{q_j(1-3\cos^2\Theta_j)}{R_j^3}, \tag{7}$$

where $E_1, E_2, \ldots, E_{2J+1}$ are energy levels of the multiplet when $H_0 = 0$, δE_m are their shifts due to electric field gradients in the presence of H_0, $\delta E_1 = -\delta E_2$, and indices 1,2 are those of the given Kramers doublet ($E_2 > E_1$ for nonzero H_0). The expression (7) is similar to the one obtained for the shift $\delta\omega_d$ due to dipolar interactions,

$$\delta\omega_d = \Delta\omega_{12} = \frac{g_{\parallel}^2 \mu_B^2}{2\hbar} \sum_j \frac{m_j(1-3\cos^2\Theta_j)}{R_j^3}, \tag{8}$$

where $m_j/2 = \pm 1/2$ accounts for the random orientation of j-th spin with respect to z axis. Under continuum approximation of statistical theory after the integration over all possible R_j the expression (8) results in Lorenzian lineshape with half-width $\Delta\omega_d = 4\pi^2 (9\sqrt{3}\hbar)^{-1} C g_{\parallel}^2 \mu_B^2$ [2], where C is the concentration of RE ions. The substitution $Ca^{2+} \to RE^{3+}$ in scheelite with excessive charge $q_+ = 1$ demands additional charge compensators with $q_- < 0$. In common case of Na^+ compensation $q_- = -1$, and the analogy with (8) is complete if we formally replace q_j with m_j. The resulting lineshape in case of random electric field gradients produced by defects of unit charge $\pm q$ and concentration C_q is Lorenzian with linewidth

$$\sigma = \frac{8\pi^2}{9\sqrt{3}} C_q \left|qG\right| \omega . \tag{9}$$

When there are several defect species with different $|q|$, one has to sum up the contributions (9) from all species to obtain the total half-width of ESR line. In the simplest case when the only charged defects in the crystal are RE^{3+} and Na^+ ions, one has $|q| = 1$ and $C_q = 2C_{RE}$.

3. Results and discussion

ESR spectra of two samples of $CaWO_4$:Er^{3+} crystals grown with natrium compensation and nominal erbium concentrations 0.05 at. % and 0.2 at. % (samples no. 1 and 2 henceforth) were acquired using Bruker ESP 300 X-band spectrometer at the frequency 9.4 GHz, temperature 10 K and $\boldsymbol{B}_0 \parallel c$. The ESR lines of different erbium isotopes proved to be Lorentzian, as expected in case of the broadening due to random electric field gradients. A part of ESR spectrum of the sample no. 2 corresponding to central line (zero nuclear spin) is presented in figure 1. The inhomogeneous half-widths σ of the central lines for both samples are presented in table 1.

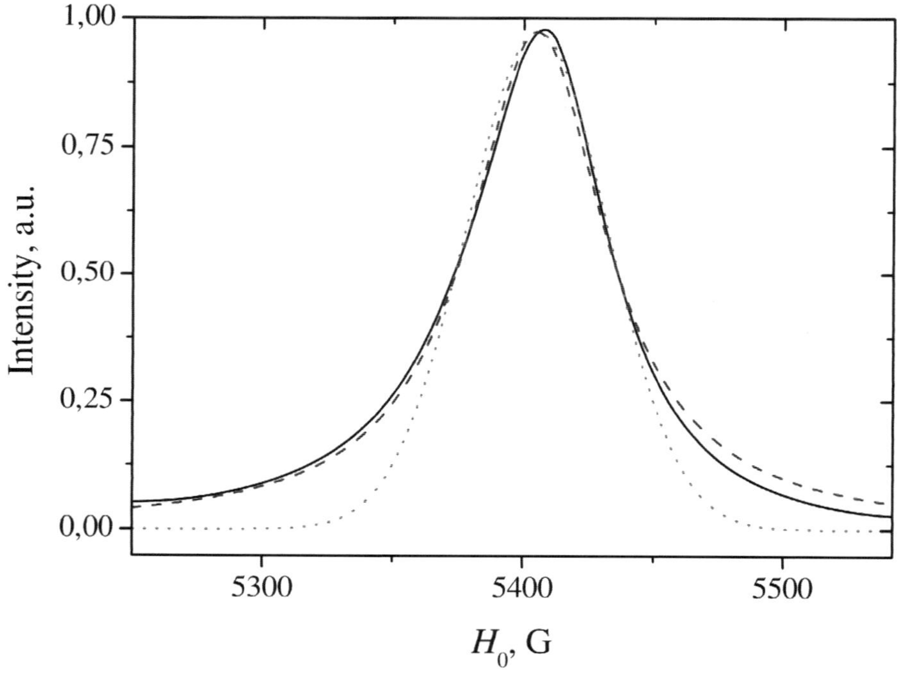

Figure 1. The central line (corresponding to erbium isotopes with zero nuclear spins) of ESR spectrum of the sample no. 2. Experimental data, fits to Lorenzian and Gaussian distributions are represented by black solid, blue dashed and red dotted lines, respectively.

The crystal field parameters B_{pk}, as well as the parameters $\sigma_2 = 0.620$, $\langle r^2 \rangle = 0.353$ Å, $g_J = 6/5$ and $\alpha_2 = 4/1575$ for the lowest multiplet $^4I_{15/2}$ of the ground configuration $4f^{11}$ of Er^{3+} ions were taken from literature data. The coupling coefficient $G = -3.2 \cdot 10^{-23} \mathrm{cm}^3$ in Eq. (9) was established for its lowest Kramers doublet with $g_{\parallel} = 1.25$ via direct diagonalization of the Hamiltonian (6). The defect concentrations C_q assuming $q = \pm 1$ were obtained from the fit to experimental σ values by means of Eq. (9).

In order to determine the role of Er^{3+} charged defects in the broadening of ESR lines we needed to know real Er^{3+} concentrations C_{RE} in the samples. They were obtained by means of comparison of erbium ESR spectra with those of standard samples with known impurity concentrations acquired under the same experimental conditions. Real concentrations turned out to be somewhat smaller: 0.021 at. % and 0.036 at. % for the samples no. 1 and 2, respectively. The relative error of concentration measurements was 30 %. Thus the values $2C_{RE}$ were less than C_q by nearly a factor of 4. Dipolar broadening was found negligible (the last column of table 1).

Table 1. Inhomogeneous half-widths, charged defect concentrations, erbium-induced excessive charge concentrations and dipolar half-widths obtained for the central ESR lines of the samples no. 1 and 2.

Sample no.	$\sigma / 2\pi$, MHz	C_q, 10^{19} cm^{-3}	$2C_{RE}$, 10^{19} cm^{-3}	$\Delta\omega_d / 2\pi$, MHz
1	35	2.3	0.52±0.16	0.07±0.02
2	56	3.7	0.91±0.27	0.12±0.04

4. Conclusions

Careful studies of ESR lines in $CaWO_4$:Er^{3+} crystal samples show that primary contributions to the observed linewidths do not come from impurity erbium ions. While there may be other charged defects with considerable concentractions, another possible broadening mechanisms like random strains and dislocation dipoles are known to produce the same Lorenzian lineshapes [1].

Acknowledgments

This work was supported by RFBR (grant N09-02-00930) and by Dynasty Foundation. Authors are grateful to A. A. Rodionov for the help in ESR measurements and to B. Z. Malkin for discussion of the results.

References

[1] Stoneham A M, 1969 *Rev. Mod. Phys.* **41** 82
[2] Mims W B 1968 *Phys. Rev.* **168** 370
[3] Baibekov E I 2011 *JETP Lett.* **93** 292
[4] Ayant Y and Belorizky E 1966 *J. Phys. France* **27** 24
[5] Bertaina S, Gambarelli S, Tkachuk A, Kurkin I N, Malkin B, Stepanov A and Barbara B 2007 *Nature Nanotechnol.* **2** 39
[6] Rakhmatullin R M, Kurkin I N, Mamin G V, Orlinskii S B, Gafurov M R, Baibekov E I, Malkin B Z, Gambarelli S, Bertaina S and Barbara B 2009 *Phys. Rev. B* **79** 172408
[7] Mims W B and Gillen R 1966 *Phys. Rev.* **148** 438
[8] Bagavantam S and Venkatarayudu T 1951 *Theory of groups and its application to physical problems* (Waltair: Andhra University Press)
[9] Abragam A and Bleaney B 1970 *Electron Paramagnetic Resonance of Transition Ions* (London: Oxford University Press)

Journal of Physics: Conference Series **324** (2011) 012010 doi:10.1088/1742-6596/324/1/012010

Effect of intramolecular interaction on time growth of second moment of multiple quantum NMR in solids

V E Zobov[1] and A A Lundin[2]

[1]L.V. Kirensky Institute of Physics, Siberian Branch, Russian Academy of Sciences, 660036, Krasnoyarsk, Russia,
[2]N.N. Semenov Institute of Chemical Physics, RAS, 117977, Moscow, Russia,

E-mail: [1]rsa@iph.krasn.ru, [2] andylun@orc.ru.

Abstract. The time dependence of the NMR multiple quantum spectrum in solids containing methyl groups is investigated. We studied both of types of the nuclear spin systems: namely linking by secular part of the mutual dipole-dipole interaction and linking by effective double quantum Hamiltonian. It is revealed for the first type of systems the rate with the time of the growing up of the second moment when intramolecular interaction is present becomes less in compare with situation when a strong intramolecular interaction is absent. On the contrary for the second type of Hamiltonian the above rate increases.

1. Introduction

Multiple quantum (MQ) NMR spectroscopy of solids [1-6] is unique and very suitable method of studying multiple spin correlation dynamics involving thousands spins [6]. In principle the field of applications of MQ spectroscopy extremely vast: from the medicine to superconductivity [3]. In particular, MQ spectroscopy presents the high-usage tool for the special study: what a type of particle distribution does explorer sees: uniform distribution or with clusters one (including the analysis of the type and structure of cluster)? Frequently this study provides by the means of exploring of the time dependence of the average number of correlated spins. Seems it has been done for the first time in ref. [2]. The experimental method evolved in ref. [2] has been working out for primarily using in molecular crystals and the theory developed there based on supposition that intra- and intermolecular dipole-dipole interactions are reciprocally independent. At the same time in conventional NMR for crystallohydrates the "principle 5/6 effect" is well known [7]. For example the width of Pake's doublet component stipulated by intermolecular interaction decreases by 5/6 factor.

In this paper we theoretically investigate the influence of a strong intramolecular interaction at the process of growing correlation.

2. Brief remarks on MQ NMR spectroscopy

The intensity of multiple-quantum coherences observed experimentally is determined by the time correlation function [1-6]:

$$\Gamma_{\varphi\alpha}(t,\tau) = Sp\{U^{+}(\tau)U_{\varphi\alpha}U(t)S_{\alpha}U^{+}(t)U^{+}_{\varphi\alpha}U(\tau)S_{\alpha}\}/Sp\{S_z^2\}\,. \quad (1)$$

Here we have, $U_{\varphi\alpha} = \exp(i\varphi S_\alpha)$ is the rotation operator by an angle φ about the z axis (if α=z) or x axis (if α=x); $S_\alpha = \sum_i S_{\alpha i}$, where S_α is the α component of the total spin of a nuclear spin system; $S_{\alpha i}$ is α – component of the spin S=1/2 vector operator in the lattice site "i" and $U(t)$ is the evolution operator with the secular part of the dipole-dipole interaction Hamiltonian H_d [8]

$$H_d = \sum_{i\neq j} b_{ij} S_{zi} S_{zj} + \sum_{i\neq j} a_{ij} S_{+i} S_{-j} = H_{zz} + H_{ff}, \tag{2}$$

or the evolution operator with this interaction but transformed by certain radiofrequency pulses into a new nonsecular effective Hamiltonian H_{eff} with double quantum interaction [1-6]:

$$H_{eff} = -\sum_{i<j} c_{ij} (S_{+i} S_{+j} + S_{-i} S_{-j}). \tag{3}$$

Here we have $b_{ij} = \gamma^2 \hbar (1 - 3\cos^2\theta_{ij}) / 2r_{ij}^3$, $a_{ij} = -b_{ij}/2$, $c_{ij} = b_{ij}/2$, $\vec{r}_{ij}$ - is the vector connecting spins i and j, θ_{ij} is the angle formed by vector $\vec{r}_{ij}$ and the external constant magnetic field, $S_\pm = S_x \pm iS_y$. For the use in the microscopic theory, we introduced the designation τ for the evolution with "time reversal." Experimental conditions $t = \tau$ will be fulfilled in the final equations.

As the consequence of the generalized form of the equation (1) we can describe as ordinary experiment with the Hamiltonian (3) and with rotation by angle φ around z-axis (let's name this one by z-basis [1]), as also the new type of experiment introduced recently [4] which contains rotation by angle φ around x-axis (let's name that one by x-basis). It is possible in the x-basis to observe multiple spin correlations appeared under influence ordinary secular dipole-dipole interaction in the process of the free induction decay (FID) [5].

We shall have the intensity of n order by Fourier transforming equation (1) (to say by other words by the using of the integration on variable $n\varphi$). Based on the simplest statistical model [1], a Gaussian shape is usually assumed in the experiment for the distribution of coherences of different orders in the MQ spectrum:

$$g_n(\tau) \sim exp(-n^2/N(\tau)). \tag{4}$$

The variance of the distribution in this model ($N(\tau)/2$) is determined by the number of spins $N(\tau)$ among which dynamical correlation due to the spin-spin interaction is settled during the preparation time τ. This number, called the number of correlated spins or the effective cluster size, grows as the preparation time increases.

Whether the spin system of many spins consists of groups with closed spins (e.g. molecules) it was proposed to describe MQ intensities by the sum of two independent Gaussian distributions [2]:

$$g_n(\tau) = m_c\, exp(-n^2/N_c(\tau)) + m_2\, exp(-n^2/N_2(\tau)). \tag{5}$$

Here $N_c(\tau)$ describes the increasing intramolecular correlations while $N_2(\tau)$ describes the intermolecular ones. The increasing of $N_c(\tau)$ restricts by the number of spins in molecule whereas $N_2(\tau)$ has no restriction.

3. Second moment of the intensity distribution for coherences of different orders in a multiple-quantum spectrum

As it has been shown in [9], the principle attribute of a number of correlated spins (quite the same is effective cluster size) what can be worked even for arbitrary form of a distribution function, is the second moment of the intensities of multiple quantum coherences of a different orders $<n^2(\tau)>$. For example, for Gaussian distribution this moment really coincides with the variance $N(\tau)/2$ in eq. (4) and there is no necessity to execute the Fourier- transformation because as it was shown in [9] actually anybody has the relation:

$$\begin{aligned} &< n^2(t) >= -d^2\Gamma_{\varphi\alpha}(t,t)/d\varphi^2\big|_{\varphi=0} = -Sp\{[S_\alpha, S_\alpha(t)]^2\}/Sp\{S_x^2\}, \\ &S_\alpha(t) = U(t) S_\alpha U^+(t) \quad (\alpha = z, x). \end{aligned} \tag{6}$$

Let us generalize this formula for situation $t \neq \tau$ (if $t=\tau$, $<< n^2(t,t) >>=< n^2(t,t) >=< n^2(t) >$)

$$<< n^2(t,\tau) >>=< n^2(t,\tau) > /2 + < n^2(\tau,t) > /2;$$

$$< n^2(t,\tau) >= 2 \sum_{i,j,f,q} \{Sp\{S_{\alpha j}S_{\alpha f}S_{\alpha i}(t)S_{\alpha q}(\tau)\} / Sp\{S_x^2\} - Sp\{S_{\alpha j}S_{\alpha i}(t)S_{\alpha f}S_{\alpha q}(\tau)\} / Sp\{S_x^2\}\}. \qquad (7)$$

Now we can rewrite the second summand from (7) in a form

$$Sp\{S_{\alpha j}(4S_{\alpha f}S_{\alpha f})S_{\alpha i}(t)S_{\alpha f}S_{\alpha q}(\tau)\} / Sp\{S_x^2\} = Sp\{S_{\alpha j}S_{\alpha f}\sigma_{\alpha f}S_{\alpha i}(t)\sigma_{\alpha f}S_{\alpha q}(\tau)\} / Sp\{S_x^2\} =$$

$$= Sp\{S_{\alpha j}S_{\alpha f}\overline{S_{\alpha i}(t)}^{(f)}S_{\alpha q}(\tau)\} / Sp\{S_x^2\},$$

because of $4S_{\alpha f}S_{\alpha f}=1$ for $S=1/2$. The notation $\overline{S_{\alpha j}(t)}^{(f)}$ means that in the time evolution operator we are using the Hamiltonian $\sigma_{\alpha f}H\sigma_{\alpha f}$ with the inverted spin f. So from (7) we get:

$$< n^2(t,\tau) >= \sum_{i,j,f,q} Sp\{S_{\alpha j}S_{\alpha f}\{S_{\alpha i}(t) - \overline{S_{\alpha i}(t)}^{(f)}\}S_{\alpha q}(\tau)\} / Sp\{S_x^2\}, \qquad (\alpha = z,\ x). \quad (8)$$

Whether we are starting to find the solution of our problem via expanding into series on the powers of a time the correlation function from (8), we shall see (as it has been shown in [10]), that the principle contributions into the second moment defined by equation (8) is carrying out by diagrams consisting of the chains with double connections starting with the lattice site i and ending with the site f. For the both of studying of interactions if we want to construct the require chains, we have to exploit two positions for the chain links (sections) with j and k sites (see figure 1):

1. Sequentially by the chain (see figure 1a)
2. Transversal relatively the chain (see figure 1b)

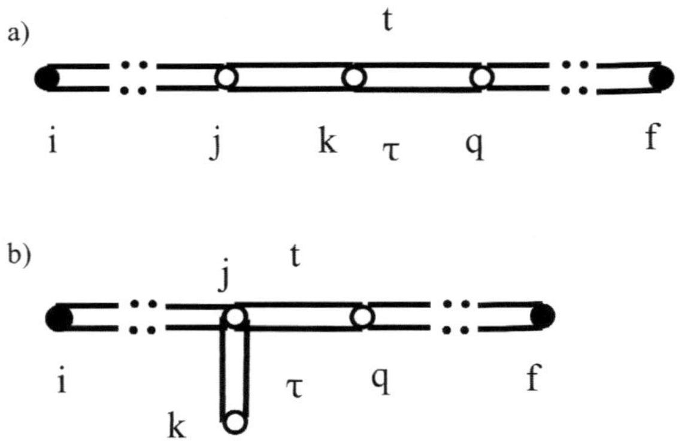

Figure 1. The location of the chain links at the lattice. The lattice points notates as i, j,k,q , f, correspond to the diagram vertexes and the lines at the figure present the interaction between spins occupied of these lattice points.

For the both of the above positions of the chain links (sections of the chain) (see figure 1) the diagrams with fixing locations of commutators and spin operators in the process of developing with the time are shown at figure 2 via chain with five links as example. Because of diagram construction laws we have to take one connection from the $\overline{S_{\alpha i}(t)}^{(f)}$ operator and the second one from the $S_{\alpha q}(\tau)$ ($q=i$) operator and we have to do so for the every double connection. As the consequence of integration by every of different variables t_m (τ_m) with the limits from 0 up to t_{m-1} (or τ_{m-1}) every skeleton diagram having the form of the chain with m links is giving us the time dependence as $(t^2\tau^2)^m$. We can take into account the interaction of spins belongs to chain with the rest of spins of a crystal by the dressing of the diagram lines. By the last process the decay with the time of the lines appears having quite sophisticated dependence. We have done the estimation of that dependence in [10]. In what follows we shall estimate the changing the account of the mentioned links of chain if the strong intramolecular interaction is taking place.

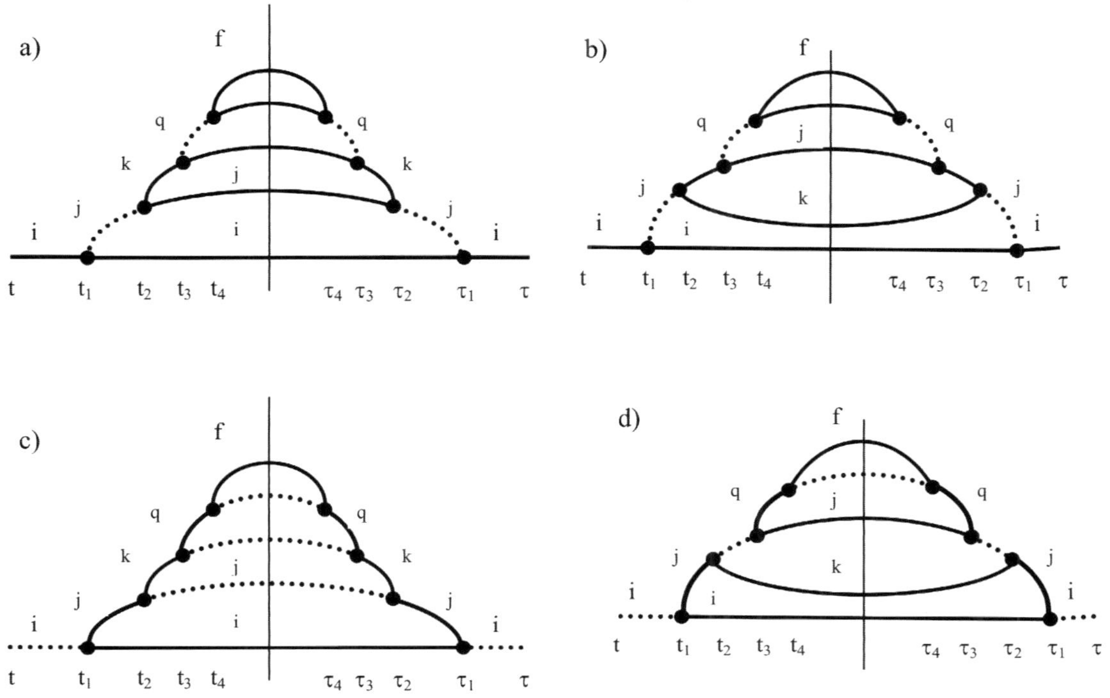

Figure 2. The examples of the diagram for the chains contain 5 spins. At the figures a) and b) it is shown the evolution with H_d interaction while figures c) and d) shows the occasion with H_{eff} interaction. The vertical line means the border between the forward and reversed evolution with the time (see (1) and (8)).The vertexes of diagram correspond to the acts of interaction between spins (actually it means the commutation of spin operator with the acting Hamiltonian) at the noted moments of time. Namely: t_m are on the left from the line mentioned above and τ_m are on the right from that line. Lines of the diagrams correspond to the $S_{\pm p}$ spin components (solid lines) or S_{zp} (dotted lines) at the indicated sites *i, j, f, k* etc.

4. Effect of intramolecular secular dipole-dipole interaction in methyl groups on the time growth of the second moment

Let us divide the Hamiltonian (2) on two divisions with the first of one reflects intramolecular interaction $H_M^{(d)}$ while the second one describes intermolecular interaction $H_{MM}^{(d)}$. Let us also mark that methyl groups usually expose oneself to rapid rotations about the axis of the third order. As a typical example we can point out at the Hexamethylbenzene [11] the MQ spectra of which were substantially studied in articles [1,2]. We shall describe the average intramolecular interaction by only one constant b_1 while the average intermolecular interaction depends only of the reference number of the methyl group. So we have:

$$H_M^{(d)} = \sum_i b_{i1}(S_{zi1}S_{zi2} + S_{zi1}S_{zi3} + S_{zi3}S_{zi2}), \tag{9}$$

$$H_{MM}^{(d)} = \sum_{i\neq j} b_{ij}S_{zi}S_{zj} + \sum_{i\neq j} a_{ij}S_{+i}S_{-j} = H_{zz} + H_{ff}, \tag{10}$$

$$S_{zi} = S_{zi1} + S_{zi2} + S_{zi3}, \quad S_{\pm i} = S_{\pm i1} + S_{\pm i2} + S_{\pm i3}.$$

Because of the permutation proton spins symmetry in the methyl group, anybody has to use only zz–part of intramolecular interaction $H_M^{(d)}$ as far as the exchange (or scalar) part of that interaction becomes unobservable [8]. The flip-flop processes between spins situated in a different methyl groups is getting registered only if the frequencies in the local field of the both of flipping spins coincide reciprocally. As the spectrum of alone methyl group is a triplet with frequencies $-b_1, 0, b_1$ and with the respective intensities $\frac{1}{4}, \frac{1}{2}, \frac{1}{4}$, for the probabilities of frequencies coincidence (if we suppose b_{11}=b_{21}) we can find

$$\left(\frac{1}{4}\right)^2 + \left(\frac{1}{2}\right)^2 + \left(\frac{1}{4}\right)^2 = \frac{3}{8}.$$

The calculated constant fixes the reduction of the square of flip-flop interaction between two different methyl groups: $A_1^2 = 3A^2/8$. Here $A^2 = 3\sum_j a_{ij}^2$ is the origin constant corresponds to situation b_1=0. Whether $b_{11} \neq b_{21}$ it can be found $A_1^2 = A^2/4$.

For the sake of simplicity we shall assuming in what follows below that all of a methyl groups has the same intramolecular interaction described by only one constant b_1 for the secular dipole-dipole interaction (see equation (9)).

If the secular dipole-dipole interaction produces the above described chains on diagrams, which combines the inverse spin with the number of {*f*} with the central (fixing) spin having the number of *i*, the chains has to be built with the help of flip-flop - processes connecting by zz –interaction (see figures 2 a and 2 b). Their contributions into skeleton diagrams are $2A^2B^2$ for the both of types chain links shown at figure 1 and these contributions has to be just sum. As it was revealed in [10] by summing the entire long chains the dependence of the second moment of MQ spectrum on a time takes a form

$$< n^2(t) >_d \sim \exp(a_d t)\frac{-b \pm \sqrt{b^2 - 4ac}}{2a},$$

where we got $a_d = 2(A^2 B\sqrt{\pi})^{1/3}$. Whether to take into account the flip-flop processes rate reduction, the exponent decreasing comes to $\sqrt[3]{\frac{8}{3}} \approx 1{,}39$ times if $b_1^2 >> M_{2MM}$ in compare with situation $b_1 = 0$ and it becomes

$$a_{Md} = (3A^2 B\sqrt{\pi})^{1/3}. \tag{11}$$

At last we should mark that $B^2 = 3\sum_j b_{ij}^2$ and $A^2 = 3\sum_j a_{ij}^2$. Also because of in the intermolecular contribution into the second moment of the conventional NMR spectrum one has to sum only by protons of the different methyl groups CH_3, (where 3 means the number of protons in one group) it can be got for the mentioned contribution:

$$M_{2MM} = 9 \cdot 3\sum_j b_{ij}^2.$$

5. Effect of intramolecular interaction describes double quantum effective Hamiltonian in methyl groups on the time growth of the second moment

As in the previous section let us divide the Hamiltonian (3) on two divisions with the first of one reflects intramolecular interaction $H_M^{(eff)}$ while the second one describes intermolecular interaction $H_{MM}^{(eff)}$:

$$H_M^{(eff)} = -\sum_i c_{i1}(S_{+i1}S_{+i2} + S_{+i1}S_{+i3} + S_{+i3}S_{+i2}) - \sum_i c_{i1}(S_{-i1}S_{-i2} + S_{-i1}S_{-i3} + S_{-i3}S_{-i2}), \quad (12)$$

$$H_{MM}^{(eff)} = -\sum_{i<j} c_{ij}(S_{+i}S_{+j} + S_{-i}S_{-j}) \leq, \quad (13)$$

As it was found in [12] the eigenvectors of isolated methyl group are:

$$\Psi_\pm = \pm\frac{1}{\sqrt{2}}|+++\rangle + \frac{1}{\sqrt{6}}(|+--\rangle + |-+-\rangle + |--+\rangle), \quad (14)$$

$$\varphi_1 = \frac{1}{\sqrt{2}}|-+-\rangle - \frac{1}{\sqrt{2}}|--+\rangle, \qquad \varphi_2 = \frac{2}{\sqrt{6}}|+--\rangle - \frac{1}{\sqrt{6}}|--+\rangle - \frac{1}{\sqrt{6}}|-+-\rangle. \quad (15)$$

with eigenvalues: $\mp\sqrt{3}c_1$, 0, 0 respectively. Four other functions $\overline{\Psi}_\pm$, $\overline{\varphi}_1$ and $\overline{\varphi}_2$ one can get by inversion of all spins. For eigenvectors of a single spin operator S_z in above formula notations $|\pm\rangle$ been used for spin projections $\pm 1/2$. Functions Ψ are symmetrical relatives to spin permutations while φ are antisymmetrical ones. So matrices of symmetric operators sums $S_{zi} = S_{zi1} + S_{zi2} + S_{zi3}$, $S_{\pm i} = S_{\pm i1} + S_{\pm i2} + S_{\pm i3}$ become as a block forms:

$$S_z = \begin{pmatrix} S_{z,\Psi} & 0 \\ 0 & S_{z,\varphi} \end{pmatrix}, \quad S_\pm = \begin{pmatrix} S_{\pm,\Psi} & 0 \\ 0 & S_{\pm,\varphi} \end{pmatrix}. \quad (16)$$

Here 4x4 matrices been introduced.

Let us construct four effective spins ½ with the help of basic functions set: $(\Psi_+, \overline{\Psi}_+)$ $(\Psi_-, \overline{\Psi}_-)$, $(\overline{\varphi}_1, \varphi_1)$ and $(\overline{\varphi}_2, \varphi_2)$. By the means of the Pauli matrices of these spins we can rewrite three matrices introduced above

$$S_{z,\varphi} = \begin{pmatrix} \sigma_1^z/2 & 0 \\ 0 & \sigma_2^z/2 \end{pmatrix}, \quad S_{\pm,\varphi} = \begin{pmatrix} -\sigma_1^\pm & 0 \\ 0 & -\sigma_2^\pm \end{pmatrix}, \quad (17)$$

and also the remaining of third other matrices secular parts in respect to strong intramolecular interaction.

$$S_{z,(sec)\Psi} = \begin{pmatrix} \sigma_{(+)}^z/2 & 0 \\ 0 & \sigma_{(-)}^z/2 \end{pmatrix}, \quad S_{\pm,(sec)\Psi} = \begin{pmatrix} \sigma_{(+)}^\mp + \sqrt{3}\sigma_{(+)}^\pm & 0 \\ 0 & \sigma_{(-)}^\mp - \sqrt{3}\sigma_{(-)}^\pm \end{pmatrix}. \quad (18)$$

To execute the calculations one needs to put (17) and (18) into (13). By exploiting of the well known features of the Pauli matrixes, commutators, what corresponds to vertexes of diagrams in new representation with $H_{MM}^{(eff)}$ operator can be found and also the traces of matrices for every diagram can be computed.

For the systems with effective Hamiltonian (3) the simplest chain between origin spin and "inverse" one "*f*" (see figures 1a and 2c) has to be constructed by the alternation of $c_{ij}S_{+i}S_{+j}$ and $c_{ij}S_{-i}S_{-j}$ interactions. The ch oice of the first connection linking to S_{zi} defines of all the other ones. As we have only two possibilities e.g.: $(+-+-+-\ldots$ or $-+-+-+\ldots)$ the combinatorial factor equal to number 2. Operators remain when commuting procedure is over has to be couple in the centre of diagram and the results of traces computations of a different methyl groups has to be mutually multiplicated. So we get that if a new section adds to a simplest chain one has to add a new factor $C_s^2 = 2,5\sum_j c_{ij}^2$ instead of $C^2 = 3\sum_j c_{ij}^2$ what is adequate when intramolecular interaction is absent. So the contribution diminishes by the factor 5/6.

At the same time we can build the chain by some other way (see figure1b and figure 2d). Really in section number j of the chain we can links $c_{ij}S_{+i}S_{+j}$ or $c_{ij}S_{-i}S_{-j}$ connections to S_{zj} operator and to begin new simple chain. Via this reason we are getting two novel possibilities. So by this way and in compliance with our calculations, if the strong intramolecular interaction is taking place and when all operations has to be done in the representation defined by functions from (14)-(15) the contribution from the diagram grows up by a factor $p = 3,4$ in comparing with the situation when intramolecular interaction is absent. In general situation for complete chain consisting of q simple parts the weight factor has to be taken as $(2p)^q$. If one needs in all weights N_{pm} of different chains with m section in each chain and with their weights it is necessary to review and then sum $(2p)^q$ analogues factors for all of the possible splitting of the section numbers between q simple segments and taking into account the condition $m = \sum_{i=1}^{q} m_i$. For doing the above operation let us use the generating functions respectively for the needs weights, the weights of simple chain (the ending parts of chains) and for their internal parts:

$$F = \sum_{m=1}^{\infty} N_{pm}\theta^m , \qquad F_1 = \sum_{m=1}^{\infty} 2\theta^m = \frac{2\theta}{1-\theta}, \qquad F_2 = p\sum_{m=2}^{\infty} 2\theta^m = \frac{2p\theta^2}{1-\theta} .$$

Where θ is a formal parameter. By exploit above described rules of the diagram construction we have the expression

$$F_p(\theta) = \sum_{q=0}^{\infty} \left(F_{2p}(\theta)\right)^q F_1(\theta) = \frac{F_1(\theta)}{1-F_2(\theta)} = \frac{\theta}{p(\theta_+ - \theta)(\theta - \theta_-)} \tag{19}$$

where we get

$$\theta_\pm = \frac{1}{4p}\left(-1 \pm \sqrt{1+8p}\right).$$

If $p = 3,4$ we get $\theta_+ \cong 1/3,155$ instead of 0.5 when p=1. In principle multiplying series by θ powers in (19) one can find the needing coefficient N_{pm} in front of θ^m . But because of we are interesting in the exponent constant for the asymptotically long times we shall not do it. The fact is that the changing of the exponent constant what we actually need defines by changing of the amount θ_+ . So for the chains with the large number of sections the average number of the weighting factor for one section is $1/\theta_+ \cong 3,16$ and not 2 when p=1.

Thereby if the intramolecular interaction averages up to zero we are getting with the help of theory from [10] the estimation for the MQ second moment growing up:

$$< n^2(t) >_e \sim \exp(a_e t) ,$$

The exponent above is $a_e \approx 1,5C \approx 0,5\sqrt{M_{2MM}}$ and it was formed by the constant $C^2 = 3\sum_j c_{ij}^2$ where index "j" runs all of the methyl groups. Strong intramolecular interaction involves the changing of this constant to other amount

$$C_M^2 = \frac{5}{6} \cdot \frac{3,16}{2} \cdot C^2 = 1,32C^2 . \tag{20}$$

It means that the exponent a_e grows up to $\sqrt{1,32} = 1.15$ if the exponent a_e is taken by amount $\sqrt{M_{2MM}}$.

6. Summary

Using the methyl groups contains three nuclear spins as example the estimation of the influence of a strong intramolecular interaction on the rate of the growing up of a number of correlated spins has been done. This rate defines the growing up of the second moment of MQ spectra (quite the same is the parameter $N_2(\tau)$ in equation (5)). As one can see the rate reduces in ~1.4 times in occasions when the dipole-dipole interaction H_d governs by the evolution of a system, and increases in ~1.2 times when H_{eff} interaction controls our system. The trends marked above reveals in compare with situation when intramolecular interaction is absent.

Certainly mentioned effects realizes in all different systems consist of molecules or clusters of spins and this effect has to be taken into account for any exploiting of the MQ NMR spectroscopy in solids contains quite isolated groups of spins.

References

[1] Baum J, Munovitz M, Garroway A N and Pines A 1985 *J. Chem. Phys.* **83** 2015
[2] Baum J and Pines A 1986 *J. Am. Chem. Soc.* **108** 7447
[3] Ernst R, Bodenhausen G and Wokaun A 1987 *Principles of Nuclear Magnetic Resonance in One and Two Dimensions* (Oxford: Clarendon Press)
[4] Ramanathan C, Cho H, Cappellaro P et al. 2003 *Chem. Phys. Lett.* **369** 311
[5] Cho H, Ladd T D, Baugh J, Cory D G and Ramanathan C 2005 *Phys. Rev. B* **72** 054427
[6] Krojanski H G and Suter D 2006 *Phys. Rev. Lett.* **97** 150503
[7] Pedersen B 1963 *J. Chem. Phys.* **39** 720
[8] Abragam A 1961 *The Principles of Nuclear Magnetism* (Oxford: Clarendon Press)
[9] Khitrin A K 1997 *Chem. Phys. Lett.* **274** 217
[10] Zobov V E and Lundin A A 2006 *J. Exp. Theor. Phys.* **103** 904
[11] Andrew E R 1950 *J. Chem. Phys.* **18** 607
[12] Roy A K and Gleason K K 1996 *J. Magn. Reson. A* **120** 139

Ferromagnetic resonance in nanostructures, rediscovering its roots in paramagnetic resonance

Klaus Baberschke
Institut für Experimentalphysik, Freie Universität Berlin, Arnimallee 14, D-14195 Berlin, Germany
E-mail: bab@physik.fu-berlin.de

Abstract. Both techniques went different routes: The EPR explored an enormous variety of paramagnets in solids, liquids, and gas phase. The focus was to determine orbital- and spin-magnetic moments (g-tensor), hyperfine interactions, and from the linewidth the spin dynamics (T_1, T_2 relaxation). In FMR most of the experiments and theory assumed the total value M to be constant in the equation of motion and used only one effective damping parameter (Gilbert). This is an enormous, unnecessary limitation for today's analysis of magnetism in nanostructures and ultrathin films. To assume M = const ignores spin wave excitations, scattering between longitudinal and transverse components of M. Moreover, in the framework of itinerant ferromagnetism, the magnetic moment/atom μ was assumed to be isotropic with $g \approx 2$! That ignores the anisotropy of μ in nanostructures and the importance of the orbital magnetic moments with $\mu_L/\mu_S = (g-2)/2$. Without finite μ_L we would have no magnetic anisotropy energy (MAE), no hard magnets, no magnetic storage media. Only recently the "language" of EPR was adapted to FMR in ultrathin films. A g-tensor is discussed and its interrelation with the MAE is pointed out. Also recent theory points out, that "*there is no reason to assume a fixed magnetization length for nanoelements*". This allows a detailed discussion of magnon-magnon scattering, spin-spin, and spin-lattice relaxation - useful, for example, for fs spin dynamics. Recent FMR experiments using frequencies from 1 GHz up to several hundred GHz, will allow measuring the proper g-factor components and μ_L, μ_S. From the frequency dependent linewidth magnon-magnon scattering can be separated from dissipative spin-lattice damping.

1. Historical reminiscences

Shortly after the discovery of EPR, here in Kazan, Altshuler and Kosyrew published a comprehensive book. Immediately after, this book was translated into German [1], it is one of the first EPR-textbooks, others appeared more recently (e.g.[2]). The main focus of EPR experiments with transition metal ions in solid state physics was to determine orbital- and spin-magnetic moments, the chemical bonding of the magnetic ion, its coupling to the nuclei, and its spin dynamic. In figure 1 this is visualized in short, illustrating the difference of the development in the ferromagnetic resonance (FMR), as reviewed in the next paragraph. Low symmetry and spin-orbit coupling (SOC) splits the e_g and t_{2g} degeneracy and mixes excited states into the ground state. In second order perturbation theory the Λ_{ij} matrix elements enter and modify the energy levels and EPR spectrum in to ways: They modify the magnetic moment of the ion, the g-factor components, and they produce an additional energy splitting, the so-called *zero field splitting* (ZFS), the D or B_2 parameters. This is textbook knowledge and was tabulated and did characterize numerous paramagnetic crystals and solids. Important for EPR and FMR

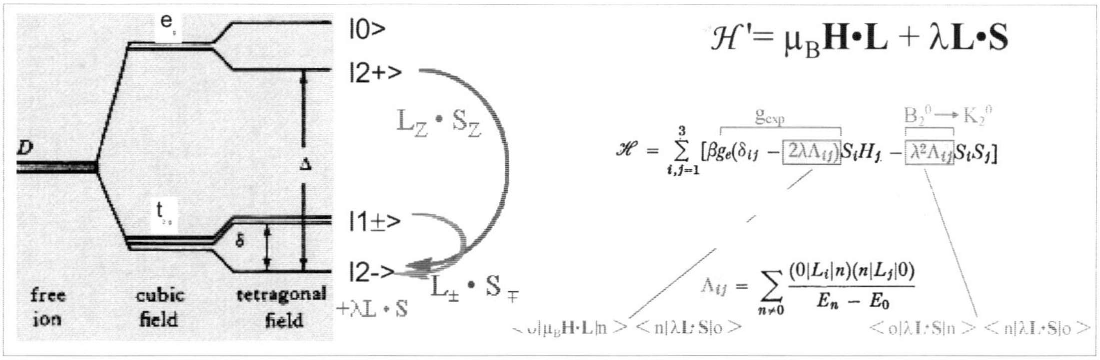

Figure 1. Schematic sketch of d-level energy splitting, the SOC, and the perturbative part of the Spin Hamiltonian $\mathcal{H}'$. For details see text and [1, 2].

students is, that Fig.1 visualizes the same origin for anisotropic magnetic moments and the ZFS, namely the admixture of excited states into the ground state by SOC. As exercise for students it is interesting to note that the off diagonal part of the SOC mixes also different spin states (orientations), "−" goes to "+" and vice versa.

The FMR went a complete different route. The resonant microwave absorption in ferromagnetic metals for a given frequency, let's say X-band, is controlled by an effective field H_{eff} (see figure 2), which is a vector sum of an external applied field and various contributions of internal magnetization. Consequently it is almost impossible to disentangle the product of the torque term in the Landau-Lifshitz equation (LL) (figure 2). In addition, a larger fraction of the literature, treats 3*d* ferromagnets (FM), like Fe and Co, in the framework of *itinerant ferromagnets* with delocalized d-band electrons with an isotropic value of $g \approx 2$. Very little is

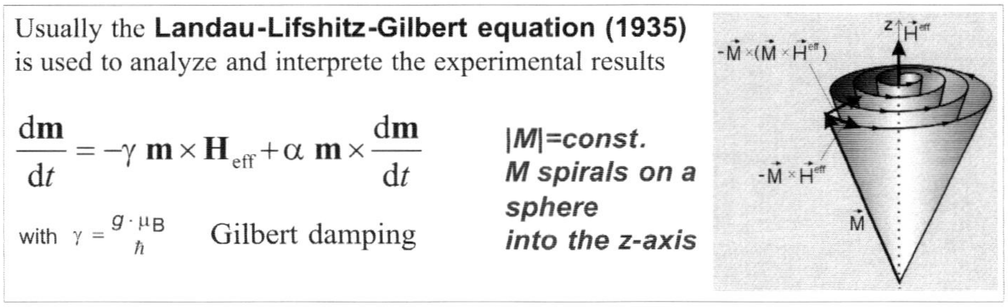

Figure 2. Standard equation of motion for a constant magnetization M in an external field.

said about an anisotropic magnetic moment per ion, for example in Fe, Co, Ni in the standard literature of the *fundamentals of nanomagnetism* [3, 4]. However, this is important for a proper Hamiltonian $\mathcal{H}'$ (figure 1). An isotropic γ or g value, as given in the LL-equation (figure 2), ignores this. For example, the magnetic moment μ per atom in Ni single crystals is larger along the [111] direction (see figure3 and [5]). Clearly the effect is small ($< 10^{-3}$) but all important. Otherwise we would not understand hard and easy axis in FM. We will come back to this in Section 2.

Figure 2 illustrates another handicap or "incompleteness" in frequently reasoning of the spin-dynamics in FM nanostructures [4]: It is assumed that the length of the magnetization vector M is constant. If so, than it is clear that with some constant α, the Gilbert damping in the LLG equation, the vector spirals on a sphere into the z-axis. This is energy dissipation, an

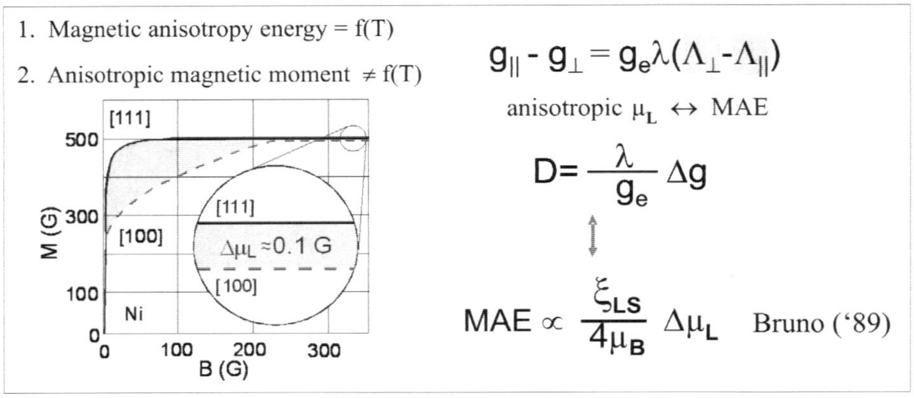

Figure 3. Magnetic Anisotropy Energy (MAE) and orbital magnetic moment μ_L and its interrelation via 2nd order perturbation theory. For details see [5].

irreversible process. In the early days of magnetism this may have been justified, experiments were carried out at ambient temperature or below and the Curie temperatures were at $T_C \approx 1000$ or 1300 K. A $T = 0$ approach was reasonable, but for ultrathin films and nanostructures with

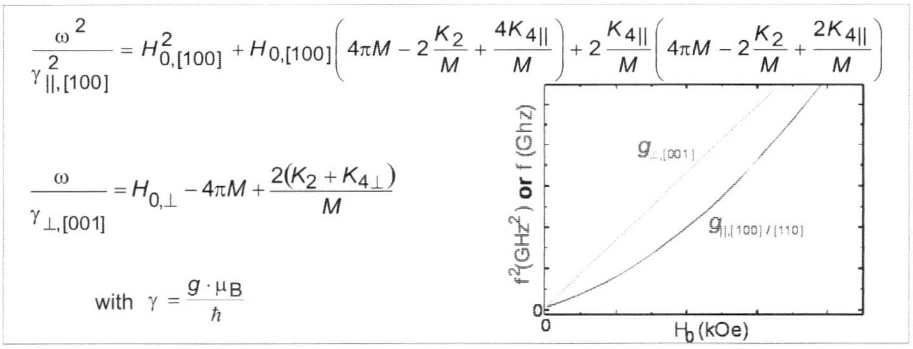

Figure 4. FMR resonance equation for in-plane and out-of plane orientation of the external field H_0 with respect to the crystallographic film plane. For details and notation, see[6, 7].

low T_C, this ignores important information, which is contained in the FMR linewidth. Not to be misunderstood, there exists classical literature discussing spin-wave excitations at finite temperatures and more than one relaxation channel, e.g. Harry Suhl in [8] and the nice book about *Magnetic Oscillations and Waves* [9]. But this literature is cited and used very little by the FMR community. The majority of FMR papers, review article, and textbooks use as an equation of motion the LLG equation (figure 2) with **one relaxation channel**, only. This is an unnecessary limitation, it ignores spin-wave excitation at finite T. Historically it may be understandable, because in the early days there was a heavy debate whether spin-waves could exist in ferromagnetic metals. One opinion (Peter Wohlfarth, et al.) said, no spin waves only Stoner excitations are important in metallic FM. Others, like David Edwards, could show that long lived spin waves indeed exist in metals. In Sec.3 we show how magnon-magnon scattering can be extracted from the FMR linewidth [7, 10].

2. Orbital- and spin-magnetic moments μ_L, μ_S in ferromagnetic monolayers

Is it possible to measure the g-tensor components and energy splitting (ZFS or MAE) in magnetic resonance experiments separately? If yes, this will be a great success, because both parameters are related, as indicated in figures1 and 3. In figure 4 the solution of the LL-equation is given for two orientations of the external field (for a more general solution, see [7]). For in-plane orientation we have a quadratic dependence of $\omega(H)$, and a linear dependence for out-of-plane. But in both cases there are internal contributions to the local field [6]. These contributions make it very difficult to determine γ or g on the left hand side of the equation, if only **one** frequency is used. Consequently most publications in FMR of metallic FM assume $g \approx 2$ or keep $g = \text{const}$ and determine only the MAE and K_i parameters in the analysis of angular dependent FMR experiments. In Sec.1 and figure 1 we mentioned the interrelation between ZFS, D, B_2 and the Δg. This is a standard knowledge for EPR. In figure 3 we show the corresponding case for FMR: MAE $\propto \Delta\mu_L$ and the EPR Bleaney parameters $\mathbf{B}_n^m$ can be related to the MAE parameters $\mathbf{K}_n^m$ in FMR. Surprisingly, this standard treatment (figure 1) in EPR was transfered to itinerant electrons in the language of band structure very late, only in 1989 by Patric Bruno (figure 3) [11]. With this pioneering work one bridge was established between localized and itinerant pictures of EPR and FMR in metals.

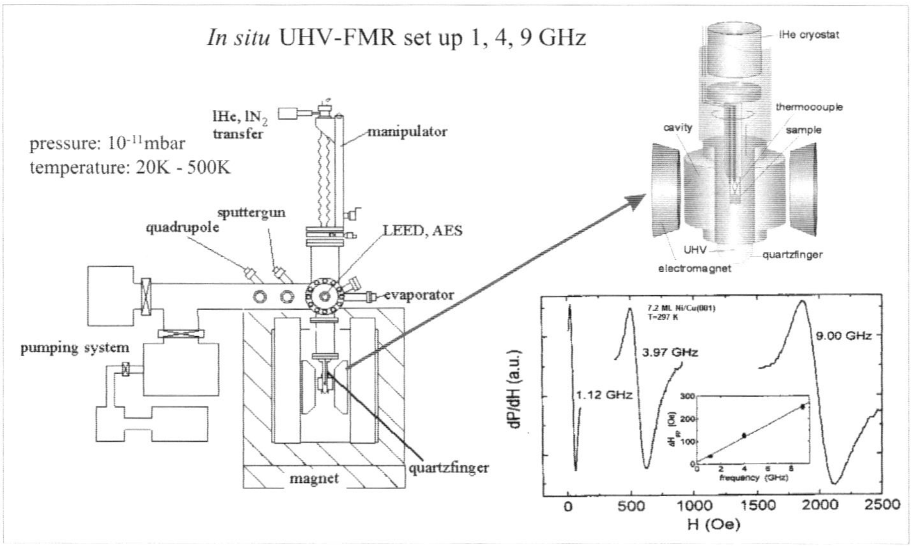

Figure 5. EPR, FMR spectrometer in combination with an UHV-chamber. The sample is prepared and characterized with LEED, AES, etc in the upper level. The long manipulator moves the specimen down into the quarzfinger and rotates it with respect to H_0 [12, 13].

If however, multifrequency FMR is used, the first term at the right hand side of the resonance equation in figure 4 depends only on the external field, H_0, which is know very precisely. Today multifrequency EPR and FMR is no problem, commercial equipment for low frequencies 1, 4, 9 GHz is available, as well as frequency analyzer and high frequencies up to THz. This allows to measure in addition to the K_i parameters also γ or g, as indicated in a sketch of figure 4. This new development of GHz technology makes the full potential of the FMR technique accessible.

In Fig.5 a set-up of an UHV-FMR spectrometer is shown. In the bottom part of the UHV chamber a quartzfinger is mounted with an one inch diameter. This allows very easily to replace standard cavities with "large excess hole" of 9, 4, 1 GHz (Varian, Bruker). One example is given for 7 ML of Ni on a Cu single crystal surface (inset, figure 5). The ultrathin film stays in perfect

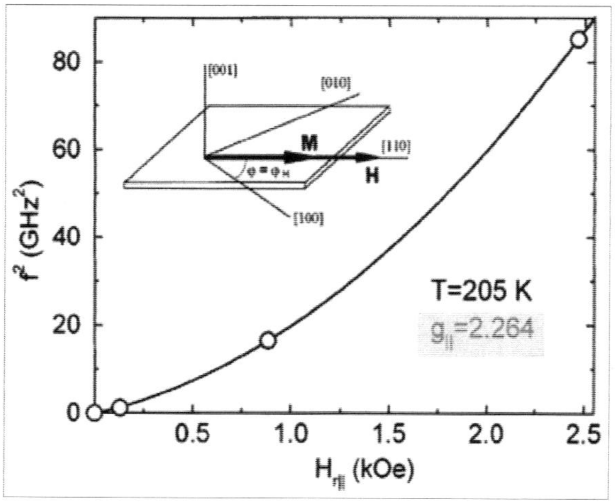

Figure 6. Analysis of $g_{\|}$ for Fe_2V_5 according to the equation given in figure 4.[14].

UHV, no protecting cap layer is needed. The microwave cavity operates in normal laboratory air. Within 30 min. cavity and frequency can be changed and the identical sample is measured at different microwave frequencies. Note the narrow linewidth at 1 GHz, the frequency dependent plot of $\Delta H(\omega)$ ends for $\omega = 0$ almost at zero on the y-axis, no apparent residual linewidth at $\omega = 0$ appears [3]. Details of the set-up, its sensitivity, the preparation of ultrathin films, and the adsorption of molecules, are given in [7, 12, 13].

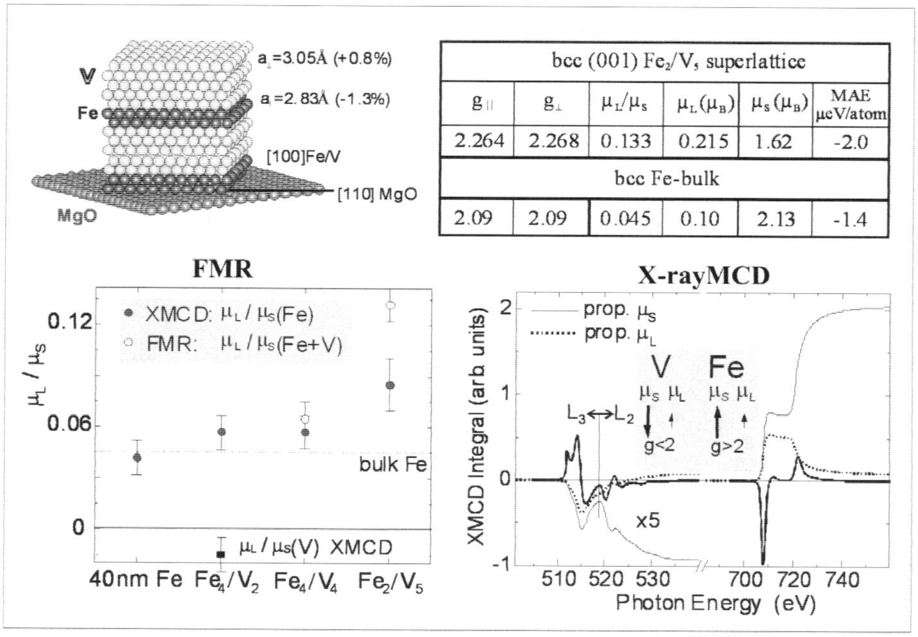

bcc (001) Fe_2/V_5 superlattice					
$g_\parallel$	$g_\perp$	μ_L/μ_S	$\mu_L(\mu_B)$	$\mu_S(\mu_B)$	MAE µeV/atom
2.264	2.268	0.133	0.215	1.62	-2.0
bcc Fe-bulk					
2.09	2.09	0.045	0.10	2.13	-1.4

Figure 7. Schematic sketch for epitaxial grown Fe_nV_m superlattice structure, FMR and XMCD results, table for g-factors, μ_L, and μ_S [7, 14].

In figures 6 and 7 FMR results for a Fe_2V_5 multilayer or a Fe_2/V_5 superlattice is shown[14].

The data follow perfectly a parabola of $f^2(H_r)$ with one g-factor component of $g_{\|} = 2.26$. Note, that the spectroscopic g-factor is a fundamental proportionality factor for the magnetic moment μ. In some FMR literature g is used as a simple fitting parameter, yielding different values for 9 and 24 GHz, for example. Other publications report on temperature dependent g-factors at low T. This should not be mixed up with the discussion of orbital- and spin-magnetic moments, as defined in EPR. It indicates only the difficulties to extract proper g-values in FMR. The table in figure 7 (top right) and diagram (bottom left) show the results. From the g-factor for Fe_2V_5 we determine an orbital moment of 13 % for μ_L/μ_S. This is 3 to 4 times larger than in bulk Fe. The diagram also shows the change of the μ_L/μ_S ratio for thicker Fe_4 superlattice films and for bulk Fe. Another experimental technique, the X-ray Magnetic Circular Dichroism (XMCD), also measures orbital- and spin-moments, by applying the sum rules (for details of XMCD see [15].) The right-hand diagram in figure 7 shows the element selective XMCD results, measured at the $L_{3,2}$ absorption edges of V and Fe. This is of interest, because it tells us, that at interfaces of FM films and its nonmagnetic substrate or cap material, a magnetic moment may be induced at the nonmagnetic element. Here at the Fe/V interface a magnetic moment at V is induced with $\mu_L/\mu_S < 0$. This is in agreement with $g < 2$ for the beginning of the $3d$ series. For Fe it is opposite, $g > 2$. Therefore it is not a surprise, that the measured μ_L/μ_S value is larger in FMR than in XMCD. FMR is not element specific, it measures the total moment (Fe+V, see the arrows in the X-ray MCD figure). This result shows some warning, when discussing FMR results in nanostructures. An enhanced μ_L/μ_S ratio in FMR does not necesserily mean that μ_L is larger. On the other hand, FMR is in favor, it can measure both, the MAE and the g-tensor, as indicated in Sec.1 and figure 3.

In summary: The handicap and misinterpretation in the past of FMR in nanostructures , was caused because only X-band (or at most K- and Q-band) were used [16]. Now with a multifrequency technique the full potential and a proper interpretation of FMR is available.

3. Spin-phonon, spin-spin dynamics

In Fig.8 two most popular equations of motion are given for the dynamics in magnetism. Most of the FMR community uses the LLG-equation, as discussed in Sec.1. In EPR and NMR very clearly a spin-spin and a spin-lattice relaxation rate are used. This can be formulated by the Bloch-Bloembergen (BB) equation. There exists some literature in which different relaxation channels in metallic FM are discussed [8, 9]. But the FMR community made very little use out of it. The resonant microwave absorption in a FM starts with a uniform motion of a $k = 0$ mode (figure 8b), but this uniform mode can scatter into spin waves. In the cartoon of figure 8 this is indicated by a variation of the transverse components of the magnetization. In figure 8b path 2 indicates this. It is in principle a reversible process, the energy stays in the magnetic subsystem. In contrast, path 1, a scatter to the lattice, is irreversible with energy dissipation [8]. Path 3 in figure 8b indicates a magnon-phonon scattering. This is a vital research topic, currently, for femtosecond LASER pump-probe experiments. This community discusses very well transverse spin-spin relaxation, a T_2 process in the BB-model. Indeed, a *dephasing* in the spin wave dynamic is important. Details for metallic FM can be seen in [17].

So the question arises: Can FMR contribute to disentangle scattering within the magnetic subsystem from dissipation and energy flow to the thermal bath? The standard FMR literature uses the LLG equation and concludes *that the FMR linewidth is strictly linearly dependent on the microwave frequency* ω...[4] (left part of figure 9). On the other hand, already Suhl stated that *spin wave excitations need to be taken into account...and...the LLG damping no longer apply....* Similar reasoning can be found in [9]. 1999 D. L. Mills and coauthors calculated magnon-magnon scattering in ultrathin films and its contribution to the FMR linewidth (right part of figure 9)[18]. As they discuss: *Two magnon scattering is a dephasing event*, contributes to the transverse relaxation time T_2, but leaves the longitudinal relaxation time T_1 unaffected

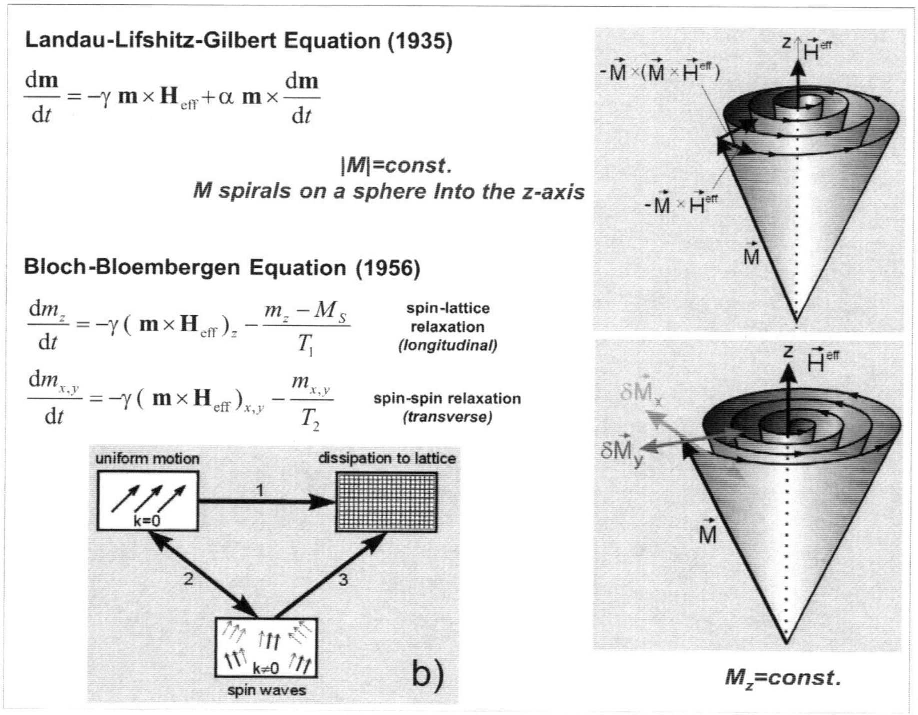

Figure 8. Two models for Larmor precision [1, 2, 9]. The arrows $\delta H_{x,y}$ indicate transverse spin scattering, dephasing, see text. Figure 8b is taken from [8].

[19]. For such a process, which describes the time evolution of the total magnetization including magnon-magnon scattering, the LL equation is inappropriate. For ultrathin films, some given surface structure, and anisotropy, it appears that dipolar interaction between spins becomes important. As a consequence, the spin-wave dispersion relation is not anymore a parabola, but a superposition of linear and quadratic terms. This is indicated in the sketch $\omega(k)$ in figure 9. Now it is easy to understand that the excitation of the uniform FMR mode is degenerated with a finite wave vector mode. These authors even, derive an analytical function for the contribution to the FMR linewidth due to magnon-magnon scattering, as given in figure 9. Now we have two contributions to the FMR linewidth $\Delta H(\omega)$, the linear Gilbert damping with the prefactor G and an *arcsin* function for the 2-magnon scattering with prefactor Γ. All other parameters (M_S, K_2, ω_0) are known from the static resonance parameters.

The left part of figure 10 shows the result, a strong curvature at low frequencies and a saturation at high ω. In other words: For realistic parameters of ultrathin films magnon-magnon scattering strongly increases in the range of zero to 30 GHz, whereas the Gilbert damping increases slowly linear with the microwave excitation frequency. Here we demonstrate the importance of low frequency FMR. In former FMR analysis a residual linewidth ΔH_0 at $\omega = 0$ was discussed [3, 4]. This may be an erroneous misinterpretation resulting from the fact, if only few frequencies (10, 24, 36 GHz) have been measured - one always can fit a linear line through two or three data points. If however at 1 GHz a very narrow linewidth is measured - like in figure 5 and right-hand diagram of figure 10 - it becomes evident that the measured ΔH is a relaxation and no inhomogeneous broadening! The data and diagram for Fe_4V_2 and Fe_4V_4 samples show clearly the two relaxation processes. For $H \perp$ to the film plane no magnon scattering is possible, consequently we see only a linear frequency dependence (green dots). For

Journal of Physics: Conference Series **324** (2011) 012011 doi:10.1088/1742-6596/324/1/012011

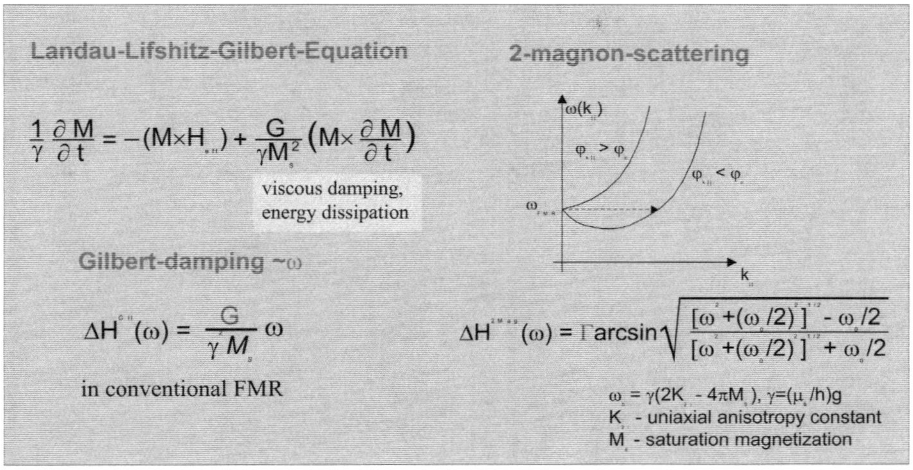

Figure 9. FMR linewidth in the LLG-model [3, 4] and for the 2-magnon-scattering [18, 19]. See text

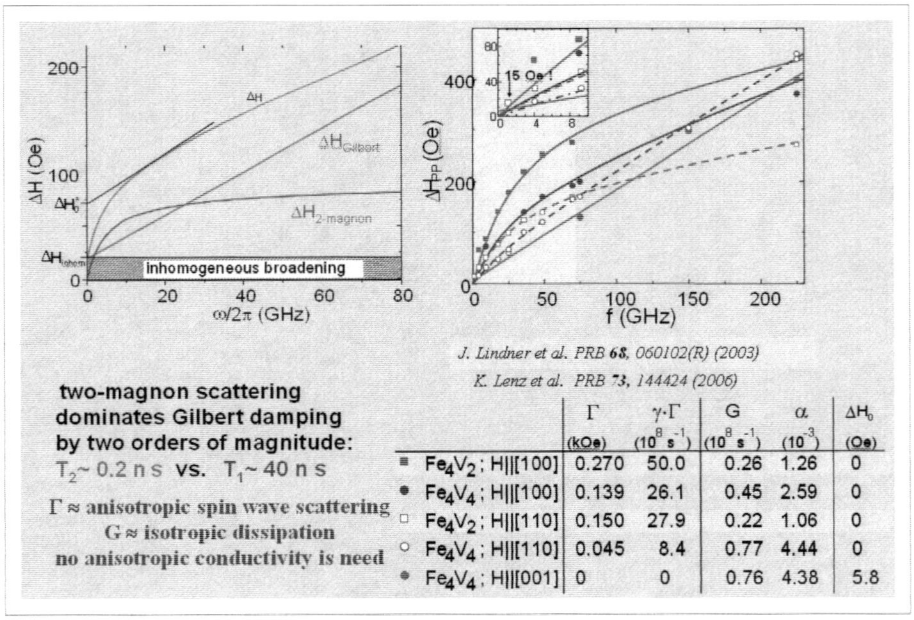

Figure 10. FMR linewidth of Fe_nV_m superlattices as function of microwave frequencies from 1 to 225 GHz. Left: as model; right: experimental results taken from [20, 21]. Note the very narrow ΔH for 1 GHz in the inset.

H in-plane strong magnon-magnon scattering is observed. Moreover, we observe that the Gilbert damping G is more or less isotropic for different orientations of H in the film plane, whereas the magnon-magnon $\gamma\Gamma$ is anisotropic by a factor of 2 or 3 for different in-plane orientation (figure 10 Table) [19, 20, 21]. No residual linewidth was observed, the experimental width is purely a relaxation process. Finally, the fitting procedure yields numbers for the relaxation times. The spin-spin relaxation is about two orders of magnitude faster than the spin-lattice rate - a very plausible result. In the table of figure 10 we give the Gilbert relaxation rates in dimensionless units of α and G in units of s^{-1}. For the 2-magnon scattering Γ rates/second are given as $\gamma\Gamma$ and as linewidth in kOe. Field units (Oe) and dimensionless α parameters are very little informative. For a better understanding of fundamental resonance physics, we strongly recommend to return to the notation of EPR and give units of relaxation rates in s^{-1}. This is not a simple question of style and common language, it facilitates the communication with theory of femtosecond dynamics in magnetism.

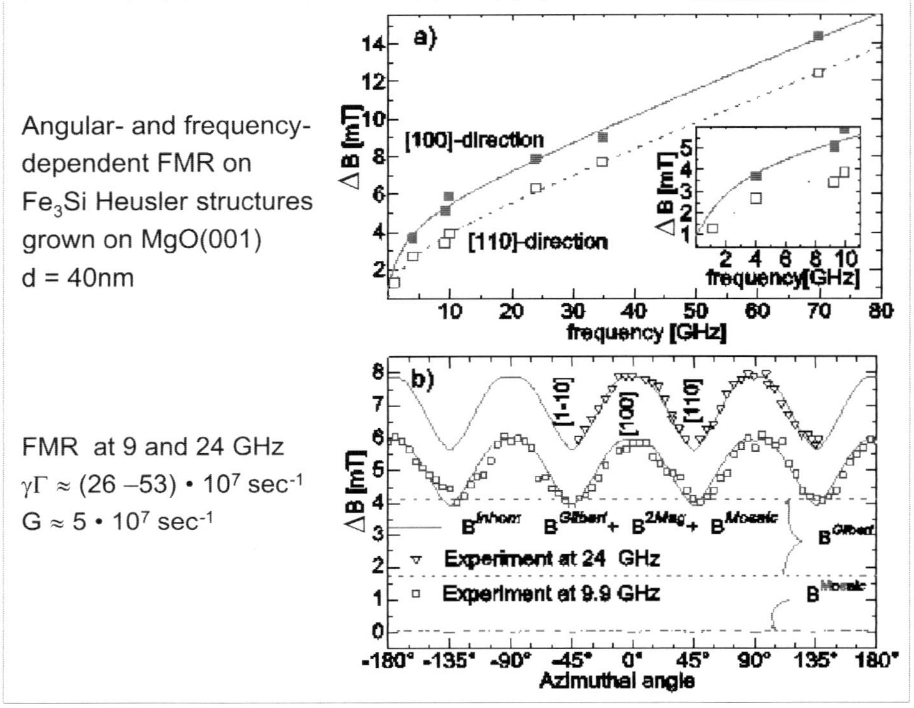

Figure 11. FMR linewidth for Heusler alloys as function of frequency taken from [23]. Note the narrow line at 1 GHz. Figure 11b shows the angular dependent 2-magnon scattering and isotropic Gilbert damping.

The results in [20, 21] were the only experimental FMR measurement, which did check in a quantitative way the theoretical prediction for the magnon-magnon scattering [18, 19]. All the other FMR literature for more than 10 years from 1999 on, used only the LLG-equation [22]. Certainly the *arcsin* function may not be the ultimate solution, but experimentalists should proof or disproof exciting model. Or even batter, they should not stick on old models, but publish experimental data free of bias.

Recently similar results have been observed for ultrathin Fe_3Si Heusler structures [23]. The film was grown epitaxially on a MgO single crystal surface. This is an interesting case because Heusler ultrathin films are used in many experiments of nanomagnetism, and if FMR can help to determine specific parameters for the spin dynamic, this will influence other interpretations of spin transport measurements. Figure 11 shows the result for the FMR linewidth measured from 1 to 70 GHz. Again, the non-linear frequency dependence of ΔB is seen, note the very narrow line at 1 GHz (inset of figure 11a). More important is the frequency and angular dependent (azimuthal and polar) measurement. In figure 11b the results are shown for two(!) frequencies, 9 and 24 GHz. With this complete set of FMR data the different contributions to the linewidth are separated: A strong angular dependent 2-magnon scattering with minimum along the [001] direction, an isotropic Gilbert damping, and a vanishing inhomogeneous line broadening was observed. The detailed analysis and fitting procedure by Zakeri et al. [23] used only the linewidth parameters for the fitting, all other input parameters have been determined before from the static resonance parameter (e. g. $g = 2.075$). As a results the authors find an isotropic Gilbert damping of $G \approx 5 \cdot 10^7$ sec^{-1}, a value which agrees very well with other Fe damping parameters. However the magnon-magnon scattering is not as fast as in the Fe/V superlattice, it is only a factor of 10 faster, than the Gilbert damping.

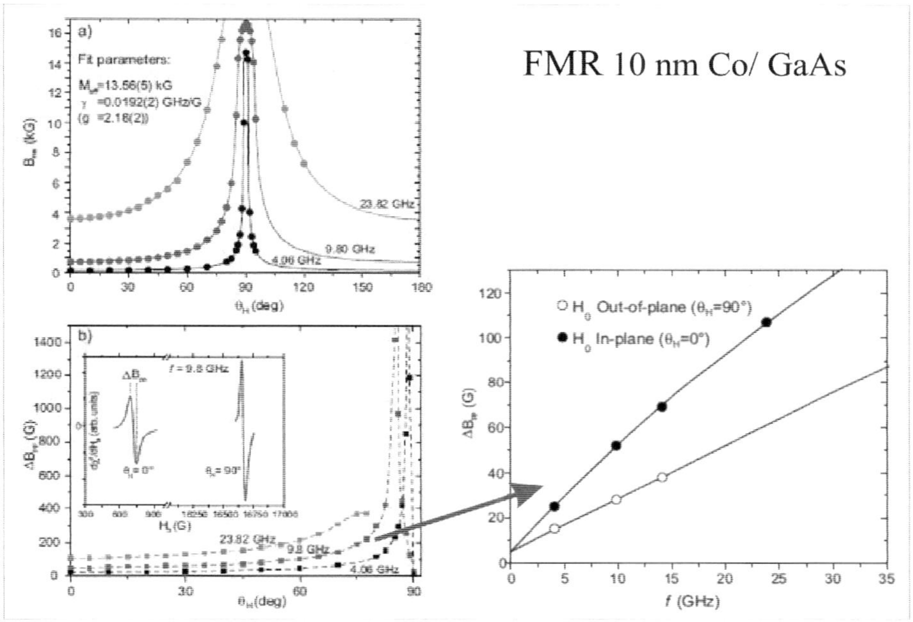

Figure 12. FMR resonance field and linewidth for a Co ultrathin film, measured at 4 frequencies, taken from [24].

For a long time the first experimental evidence for magnon-magnon scattering, determined from FMR has been ignored or criticized, to be a singularity and special feature of the Fe/V superlattice. The previous results at the Heusler structure gave some more confidence. As last and most recent case, we will shortly discus the FMR experiments on pure Co films by Lindner et al. [24]. To measure the dynamical response of a prototype film of polycrystalline hcp Co film of about 10 nm thickness on GaAs is of general interest for spin dynamics in nanostructures. In figure 12 the results are given. Figure 12a shows the field for resonance as function of the polar angel for 3(!) frequencies. As discussed in Sec.2, for a proper interpretation all data together need to be fitted with one set of parameters. As a result the g-factor of $g = 2.18$ is very close to bulk Co. The error bar of ± 0.02 is too large to see the anisotropy in hcp Co. With these numbers the frequency dependent linewidth (figure 12b+c) can be fitted. The Gilbert damping of $G \approx 0.16 \cdot 10^9$ sec^{-1} is about a factor of 10 faster for polycrystalline specimen, than in high purity epitaxially grown films. The magnon-magnon scattering is in the same order of magnitude. The authors analyzed some residual linewidth at $\omega = 0$, however experiments at 1 GHz are missing. Our experience is, that in cases with 1 GHz FMR data the line narrows again, surprisingly. In other words: The remaining width may still be caused by scattering of some magnons in the range of few GHz.

So far we have discussed the LLG- and BB- model and the magnon-magnon scattering in the FMR linewidth by Arias and Mills [18]. This theory assumes a certain model structure on the surface of ultrathin FMs. Not to be misunderstood, this calculation and the classical models of spin dynamics are based on idealistic structures, which may be far from reality. More and better theory is needed for a better understanding of the FMR linewidth. But this is not the point. In Sec.3 we wanted to demonstrate the full potential of FMR linewidth analysis. Frequency dependent experiments will give the FMR the same full potential in linewidth analysis as it is a standard procedure in EPR.

4. Summary and future

Two aspects have been discussed for the future development of FMR in nanomagnetism: The static resonance conditions and the spin dynamics in the linewidth. The message of Sec.2 will be: The old-fashion standard procedure to discuss magnetism of 3d-metallic FM in the framework of e_g and t_{2g} levels with eigenstates like $d_{z^2}, d_{x^2-y^2}, d_{xy}, d_{xz}, d_{yz}$ is insufficient. It may explain some MOKE, STM, or photoemission experiments, but it misses orbital magnetism and MAE. For the above unperturbed 3d-eigenstates $\langle L_z \rangle = 0$. In reality the groundstate is an admixture via SOC with $\langle L_z \rangle \neq 0$, see figure 1. This produces a g-tensor, to be measured by FMR, and an MAE, see figure 4. The new development in multifrequency- and UHV-technology opens a huge field for FMR. Old-fashion experiments and review papers, for X-band only, etc. used the g-factor as a fitting parameter, but not as a fundamental property of the magnetic moment (different g values for X- and K-band were published).

The spin wave dynamics in magnetic nanostructures is a very active and broad research field, currently. Not only 2-magnon scattering, as discussed in Sec.3, is of interest. 4-magnon scattering, for example, has been discussed in a high condensate of magnons in YIG thin films. Microwave photons split into pairs of correlated magnons [25]. In the same material X-ray detected magnetic resonance was performed in the nonlinear regime of spin waves [26]. These authors discuss also the redistribution of energy within internal degrees of freedom of the spin system, before it is transferred to the lattice. They introduce also a T_1^{-1} and a T_2^{-1}, longitudinal and transverse, relaxation rates. In LASER pump-probe experiments [17] the authors discuss magnon induced damping and magnon-phonon scattering (path3 in figure 8b). In particular these experiments initiated new theories [27], the Landau-Lifshitz-Bloch equation (LLB). Two relaxation rates are introduced, $\alpha_\perp$ and $\alpha_\parallel$. Note, these rates are not identical to the phenomenological T_1^{-1} and T_2^{-1} rates in the BB model. Sec.3 demonstrates that todays FMR is able to measure separately scattering within the spin wave system and energy dissipation. Multifrequency experiments seem to be essential. This is possible, starting at ≈ 1 GHz up to THz [28].

Now, on the occasion of the 100th anniversary of S. A. Altshuler it seems to be time that the young generation of FMR leaves the old-fashion path of LLG-equation, assuming an isotopic $g \approx 2$, fitting Gilbert parameter with an apparent linear frequency dependence. All this is not completely wrong, but it gives very little insight on the fundamental properties of nanomagnetism in 3d metallic FMs. Many of the more recent aspects in the interpretation of FMR of nanostructures, as discussed in Sec.2 and 3, are given in [29].

Acknowledgment

Intensive and ongoing discussions with D. L. Mills, M. Farle, J. Lindner, and K. Lenz are acknowledged. E. Kosubek and J. Kurde assisted in preparing this manuscript.

References

[1] Altshuler S A and Kosyrew B M *Paramagnetische Elektronenresonanz* 1963 (Leipzig: Teubner)

[2] Abragam A and Bleaney B *Electron Paramagnetic Resonance of Transition Ions* W. Marshall, D. H. Wilkinson, Eds. Clarendon Press, Oxford, (1970)
[3] Heinrich B and Bland J A C Eds. 1994 *Ultrathin Magnetic Structures II*, Springer, Berlin
[4] Bland J A C and Heinrich B Eds.2005 *Ultrathin Magnetic Structures III*, Springer, Berlin
[5] Baberschke K 2001 *Lecture Notes in Physics* Springer **580** 27
[6] for a detailed discussion of K_2, K_4, etc., see Farle M 1998 *Rep. Prog. Phys.* **61** 755
[7] Baberschke K 2007 in H. Kronmüller and S. S. Parkin, Eds. *Handbook of Magnetism and Advanced Magnetic Materials* Vol. **3** 1627 John Wiley & Sons, Ltd
[8] Suhl Harry 1998 *IEEE Transactions on Magnetics* **34** 1834
[9] Gurevich A G and Melkov G A 1996 *Magnetic Oscillations and Waves* (CRC Press, Boca Raton)
[10] Baberschke K 2008 *Physica status solidi* (b) **245** 174
[11] Bruno P 1995 *Phys. Rev.* B **52** 411
[12] Zomack M and Baberschke K 1986 *Surf. Sci.* **178** 618
[13] Lindner J and Baberschke K 2003 *J. Phys.: Condens. Matter* **15** R193
[14] Anisimov A N, Farle M, Poulopoulos P et al. 1999 *Phys. Rev. Lett.* **82** 2390 and Scherz A, Wende H, Poulopoulos P et al. 2001 *Phys. Rev.* B **64** 180407
[15] Wende H 2004 *Rep. Prog. Phys.* **67** 2105
[16] In some FMR literature g-factors are fitted from angular dependent measurements at one frequency. This may lead to erroneous results. We believe only frequency dependent fits, as in figure 6, will lead to unambiguous results.
[17] Melnikov A et al. 2008 *Phys. Rev. Lett.* **100** 247401 and Bovensiepen U 2007 *J. Phys. Condens. Matter* **19** 083201
[18] Arias R and Mills D L 1999 *Phys. Rev.* B **60** 7395)
[19] Mills D L and Rezende S M in B. Hillebrands, K. Ounadjela (Eds.) *Spin Dynamics in Confined Magnetic Structures II*, Springer, Berlin (2003)
[20] Lindner J, Lenz K, Kosubek E, Baberschke K et al. 2003 *Phys. Rev.* B **68** 060102(R)
[21] Lenz K, Wende H, Kuch W, Baberschke K et al. 2006 *Phys. Rev.* B **73** 144424
[22] Note that in [21] also earlier results by Woltersdorf and Heinrich have been reanalyzed.
[23] Zakeri Kh, Lindner J, Barsukov I, Meckenstock R et al. 2007 *Phys. Rev.* B **76** 104416 and 2009 (*Phys. Rev.) B **80** 059901*
*[24] Lindner J, Barsukov I, Raeder C, Hassel C et al. 2009 Phys. Rev. B **80** 224421*
*[25] Demokritov S O et al. 2006 Nature **443** 430*
*[26] Goulon J et al. 2010 J. Mag. Mag. Mat. **322** 2308*
*[27] Chubykalo-Fesenko O et al. 2006 Phys. Rev. B **74** 094436 and Garanin D A et al. 2004 Phys. Rev. B **70** 212409*
*[28] A Schnegg et al. 2009 Phys. Chem. Chem. Phys. **11** 6820*
*[29] J Stöhr and H C Siegmann 2006 Magnetism Solid-State Sciences **152** (Berlin: Springer)*

The magnetic-resonance properties study of nanostructures for spintronics by FMR

G Kupriyanova, A Zyubin, A Astashonok, A Orlova and E Prokhorenko

Immanuel Kant Baltic Federal University ,236041, A.Nevskogo 14, Kaliningrad, Russia

e-mail: galkupr@yandex.ru

Abstract. In this work we report the study of the magnetic-resonance properties such as magnetic anisotropy, magnetic damping, and interlayer exchange coupling between ferromagnetic layers separated by a nonmagnetic spacer by FMR to assess their applicability in a functional magnetic tunnel junction.

1. Introduction

One of the important tasks today is to create new materials and combinations for the applications of spintronics. The magnetic tunnel junctions (MTJ) are of a very special interest. The basic element of MTJ consists of the ferromagnetic – insulator – ferromagnetic structures, which are the main element of the random access memory of novel generation. The magnitude of the tunnel magnetoresistance (TMR), which determines the functional properties of the memory cell based on MTJ, is defined as the relative current resistance for parallel and antiparallel orientations of magnetizations in FM electrodes, and according to the Julliere model [1] critically depends on the spin polarization of both electrodes. The magnetic property of MTJ ferromagnetic layers differs considerably, as it involves the possibility of an independent magnetization switching in two ferromagnetic electrodes. It was found that the materials combination for the MTJ must satisfy a number of requirements, which are the following: a) high conductivity electrons polarization in a ferromagnetic electrode, b) the absence of paramagnetic phase on the interface; c) the thickness of the tunnel insulator layer should be within 2-3 nm; d) homogeneity of insulator; e) the roughness of the insulator within the ranges of monolayer thicknesses; f) defectlessness [2]. Many of these properties, including magnetic anisotropy and interlayer exchange coupling parameters which play the most important role in MTJ effect may be studied by ferromagnetic resonance.

The perspective class of materials for the application as a magnetic stiff layer for magnetic tunneling junctions is the ferromagnetic oxides. The common ferromagnet Fe_3O_4 (magnetite) has been predicted by band structure calculations to be half metal with the minority spin polarization at the Fermi level [3]. According to the latest experimental investigations, thin film Fe_3O_4 shows spin polarization P=-80% [4]. Furthermore, Fe_3O_4 meets important prerequisites for the compatibility with a memory cells industry, like a Curie temperature (T_C) = 860 K and relatively high (~100-400 Oe) coercivity [5]. However, to use ferrite as a magnetic stiff layer we should master several anomalous properties which make it impossible to reach the saturation magnetization by fields up to 10 kOe and flat hysteresis curve. With the aim of choosing of an optimum structure for the magnetic stiff layer, we have analyzed structures with different consequences of layers Fe_3O_4/Fe and Fe/Fe_3O_4. The other applicable magnetic material, which might be used as the second ferromagnetic layer is Fe_3Si. It is a ferromagnetic binary Heusler alloy and was found to have a spin polarization of 43% and a high Curie temperature of 820 K [6] and coercive force 50Oe. It was obtained that for thin film structures Fe_3Si with the Fe upper layer a 100 % spin polarization is inherent. As samples to be analyzed some structures were grown on Si/SiO_2, which differ by the position of layers and the nature of the isolating layer and a growth temperature such as $Fe/Fe_3O_4/MgO/Fe_3Si$, $Fe_3/SiO_2/Fe_3O_4/Fe$, $Fe_3Si/SiO_2/Fe/Fe_3O_4$.

2. Experimental details and results

At first we analyzed the magnetic properties of polycrystalline sample $Si/SiO_2/Fe_3Si$ and the magnetic film properties of magnetite during the introduction of a Fe sublayer which were investigated with regard to the position of a Fe sublayer and the annealing temperature by the FMR method and by using vibrating sample magnetometer (VSM). Polycrystalline sample $Si/SiO_2/Fe_3Si$ was grown by the method of pulse laser deposit method by evaporation (PLD) of alloyed target Fe_3Si. After depositing the sample was annealed at in situ in the growth camera at T=600 °C for 30 minutes. The samples Fe_3O_4/Fe and Fe/Fe_3O_4 structure on Si/SiO_2 substrate were grown at room temperature by means of PLD technique, and then Fe/Fe_3O_4 structure divided in four parts, three of which were vacuum-annealed at T=200 °C, 350 °C and 500 °C respectively. The structure $Fe/Fe_3O_4/MgO/Fe_3Si$ was grown on the amorphous surface Si/SiO_2 substrate by pulsed laser deposition technique. A Fe sublayer of 10 nm thick was deposited on the amorphous surface of SiO_2, then a Fe_3O_4 layer 20 nm thick was deposited from the stechiometrical alloyed target Fe_3O_4 at room temperature in ultrahigh vacuum (P = 10^{-6} Pa) in one vacuum cycle. The formation of the upper polycrystalline FM layer Fe_3Si was performed by means of deposition from the stechiometrical alloyed target Fe_3Si with the following vacuum annealing up at temperature 400 °C.

The obtained thicknesses results are inherent to all layers except MgO (3nm).The POP data analysis and ion profiling analysis by electronic spectroscopy evidence the formation of MgO layer of 3-5 nm.

The ferromagnetic resonance spectra of the thin film samples were analyzed by the Radiopan electron spin resonance spectrometer operating at the X-band frequency 9.5 GHz and the modulation frequency 100 kHz. The first field derivative of the microwave power absorption was registered as a function of applied magnetic field at room temperature. The samples were mounted on the side of a special holder passing through the center of the TE_{102} microwave cavity. The orientation of the film was controlled by a two axial goniometer with the accuracy by 0.5^0. The out-of –plane dependencies of FMR spectra and line widths have been measured for the thin magnetic films made at the different annealing temperatures. The magnetic parameters such as the effective magnetization, magnetic anisotropy and line widths were extracted for the samples with different positions of layers.

Table 1. Structures, synthesized on Si/SiO_2 Fe(10 nm)/Fe_3O_4(20 nm).

№ Fe/Fe_3O_4	Annealing T,°C	H_c, Oe	M_s emu/cm^3	The results of modelling M_1/M_2 emu/ cm^3	Interrelation constants (of the first and the second order) J(erg/cm^2)	Anisotropy constants, erg/cm^3
1	20	10	490	M_1=500 M_2=500	J_1=+0.2	K_2= -2*10^5 K_4= -2.3*10^5

2	200	10	620	M_1=640 M_2=640	J_1=+0. 24	K_2= -5.1*10^5 K_4 = -2.9*10^5
3	350	95	670	M_1=670 M_2=670	J_1=+0.1	K_2= -5.1*10^5 K_4=-3.1*10^5
4	500	110 (85)	760 (780)	M_1=780 M_2=780	J_1=+0.09	K_2= -5.6*10^5 K_4= -2.9*10^5

For the sample with structure Si/SiO_2/Fe_3O_4/Fe two signals were observed: the narrow intense signal in area H_{res} = 336 mT with the line width ΔH=18 mT, which doesn't depend on the sample orientation, and the wide weak signal which is lower by one order of magnitude, its resonance field and line width depend on the θ angle. For each of the three samples [7] with the reverse sequence of layers Si/SiO_2/Fe/Fe_3O_4 one integrated signal is observed. Its position and the line widths depend on the sample orientation. Moreover, for the unannealing sample, along direction of external magnetic field, a very narrow line H =4,1 mT is observed, if compared in width to the line width of the pure Fe H =1,8 mT. Magnetic anisotropy of Fe/GaAs(001) ultrathin films was investigated by in situ ferromagnetic resonance [8]. It was obtained that the absolute values of magnetization in systems Si/SiO_2 /Fe/Fe_3O_4 is considerably at higher annealing temperature 500 ^{0}C, Ms ~ 760 emu/cm^3, than in Fe_3O_4 (Ms = 480 emu/cm^3 for the volume sample and comparable to the values obtained in epitaxial films Fe/Fe_3O_4 on the MgO substrate. In work [9] it is given that the maximum magnetization of saturation Ms = 970 emu/cm^3 is found at the sample with Fe layer width, equal to 8 nm in the field with 10 kOe. In the quoted work it was obtained that the difference of the sample Fe_3O_4, the absolute value of magnetization saturation in samples of Fe/Fe_3O_4 depends on the annealing temperature. The magnetization rises together with the rise of annealing temperature. This fact was derived from the analysis of the hysteresis curve and from the FMR experiments analysis. With increasing of annealing temperature the FMR line width is broadening. It is important, that the value of coercive force of the analyzed samples (Table 1) is increased several times if compared to the coercive force of the thin film of the magnetite and the coercive force of film Fe (10÷15 Oe), and reaches the values relevant for the realization of MTJ (100÷200 Oe). When this happens the hysteresis loop form becomes rectangular in comparison to the initial flat hysteresis curve of the thin film Fe_3O_4 .

Table 2. Different samples and its propertiees.

Sample #	The sample structure (in brackets the layers' thicknesses are given in A) on Si/SiO_2 substrate	Magnetis ation (M), emu/cm^3	Sampl e thickn ess L, A	Coerciv e force (H_C), Oe	The value of resonance fields (line width) mT
2669	Fe_3Si(190)/SiO_2/Fe_3O_4(200)Fe (80) /Pt(1000) RT	1029	350	15	Br_1=58 (7,5) Br_2=111 (9,3)
2671	Fe_3Si(190)/SiO_2/Fe_3O_4(200)Fe (80)/ Pt(1000) the larger growth temperature T of Fe_3O_4 и Fe,	1014	400	135	Br_1=73(11) Br_2=97 (6,5) (low angular dependence)

	annealing temperature ~ 450 C.				
2673	Fe_3Si(190A)/SiO_2/Fe57(60)/Fe_3O_4(160)	683	400	60	Br_1=93,9(20,3) Br_2=120 (6,0)
2642	Fe57(40)Fe_3O_4(100)/MgO(30)/Fe_3Si(130)	1399,2	270	Hc_1=115 Hc_2=850	Br_1=98 (29) Br_2=131(11)

The preliminary study of samples 2669 and 2671, which have the similar structure but different growth temperature was carried out by using vibrating sample magnetometer (VSM), showed that, unlike one-layer structures, increased annealing temperature result in the formation of structures with smaller magnetization value but higher coercive force (Table 2)

With the aim of anisotropic properties analysis, the in-plane and out-of–plane dependencies of FMR spectra and line widths have been measured for thin magnetic films. Two distinct resonance peaks were observed in out-of-plane FMR spectra for all the samples. For in-plane spectra the positions of the two resonance fields don't change with the azimuthal angle of the applied field. However, only minor changes of resonance peak intensities were observed. Picture 1 shows angular dependencies of the resonance fields of the both modes for samples 2671 and 2669. The strong angular dependence is attributed mainly shape anisotropy. The comparison of pictures shows, that the position of resonance peaks of the sample 2671 undergoing the larger growth temperature T of Fe_3O_4 и Fe grow closer if compared to sample 2669. This effect is the same for decreasing of spacer thickness effect, which is equivalent to increase of magnetic coupling between ferromagnetic layers. Figure 2 shows the angular dependencies of the line widths. It is seen that the two narrow signals symmetrical to the direction when the field is applied to the film normal.

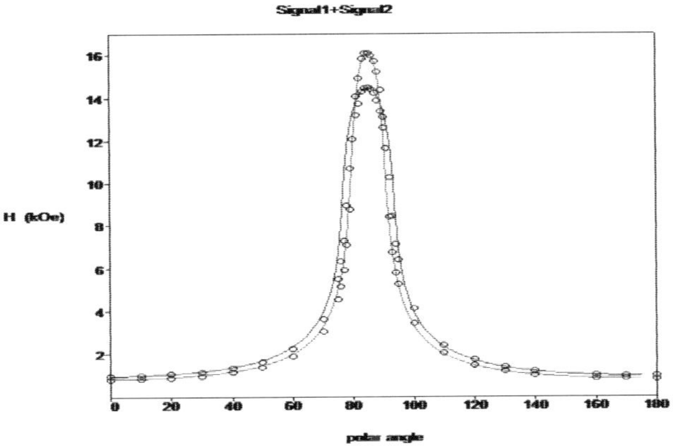

Figure 1. Dependence of the experimental and simulated resonance field value as a function of θ_H for sample 2671

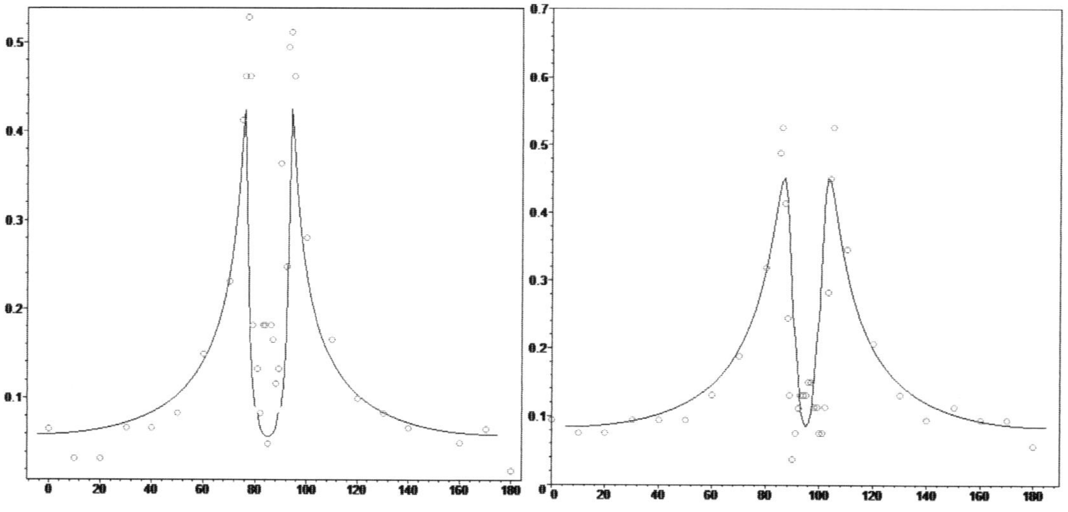

a b

Figure 2. Dependence of the experimental and simulated linewidth as a function of θ_H for sample a) 2671 and b)2669

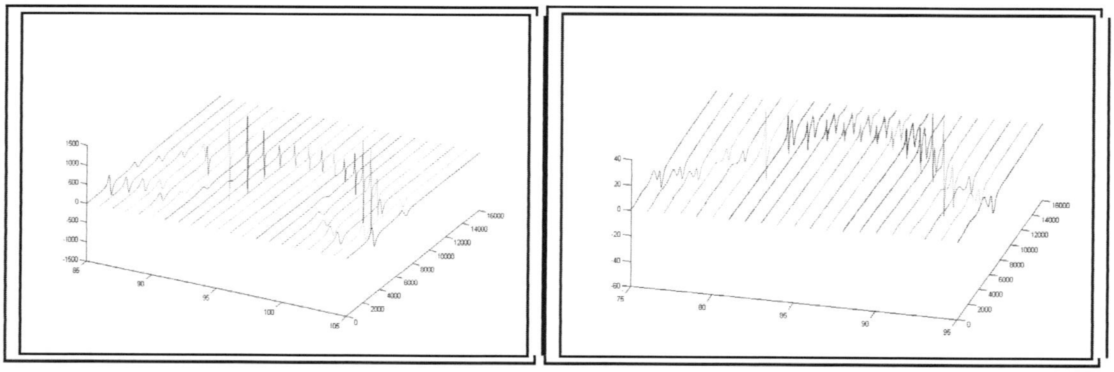

a b

Figure 3. The behaviour of the resonance signals for a) sample 2669 and b) sample 2671 close to the position where the magnetic field is perpendicular to the film

The particular interest is represented by the sample Fe/FeOx/MgO/Fe_3Si. The analysis a hysteresis curve shows that in the structure (2642) Fe/FeOx/MgO/Fe_3Si there is an independent switching of magnetization ferromagnetic layers, switching soft layer Fe_3Si is distinctly observed at fields H ~ (-100÷100 Oe), while the hard layer Fe/Fe_3O_4 keeps the magnetization in a range of magnetic fields H ~ (-500÷500) [10].

3. Theoretical aspects

The experimental data for a system consisting two magnetic layers with saturation magnetization $M_{1,2}$ and layer thickness $d_{1,2}$ in external field H can be described by following theoretical model. Magnetic energy surface density of such system is

$$E = d_1F_1 + d_2F_2 - J_\ell(\sin\theta_1 \sin\theta_2 \cos(\varphi_1 - \varphi_2) + \cos\theta_1 \cos\theta_2) - \\ - J_q(\sin\theta_1 \sin\theta_2 \cos(\varphi_1 - \varphi_2) + \cos\theta_1 \cos\theta_2)^2 \qquad (1)$$

Here F_1 and F_2 are free energy density for first and second layer. In general case free energy F consist of Zeeman energy F_z of structure, magnetostatic F_{ms} and magnetocrystalline anisotropy energy F_c:

$$F = F_z + F_{ms} + F_c, F_z = -M \cdot H(\sin\theta \sin\theta_H \cos(\varphi - \varphi_H) + \cos\theta\cos\theta_H), F_{ms} = -2\pi M^2 \sin^2\theta, \\ F_c = -K_2 \cos^2\theta - \frac{K_4}{2}\cos^4\theta - \frac{K_4}{8}(3 + \cos 4\varphi)\sin^4\theta, \qquad (2)$$

where K_2 is the uniaxial out-of-plane anisotropy constant, and K_4 is fourfold anisotropy constant. The angles (θ_i, θ_H) and (ϕ_i, ϕ_H) are, respectively, the polar and azimuthal angles for magnetization vector M and external field vector H with respect to the film normal. J_l and J_q are, respectively, the bilinear and biquadratic coupling parameters. The value of J_l can be either positive or negative (ferromagnetic or antiferromagnetic coupling respectively). The large negative value of J_q corresponds 90°-type coupling. The biquadratic term is smaller compared to the bilinear term and can be neglected.

Strictly speaking the Equations (1), (2) are valid in a case of monocrystalline film. For polycrystal the accounting of domain structure is necessary. If distribution of easy axis magnetization is isotropic in film plane the effective magnetocrystalline energy is

$$\overline{F}_c = -K_2 \cos^2\theta - \frac{K_4}{2}\cos^4\theta - \frac{3K_4}{8}\sin^4\theta = -\widetilde{K}_2 \cos^2\theta - \frac{\widetilde{K}_4}{2}\cos^4\theta + const, \qquad (3)$$

where $\widetilde{K}_2 = K_2 - 3K_4/4, \widetilde{K}_4 = 7K_4/4$. Further constants are omitted. Equation (3) is obtained from (2) by averaging on azimuthal angle. The expressions for Zeeman and interaction energy do not change if (ϕ_1, ϕ_2) are understood as the azimuthal angles of resulting vector $M_{1,2}$ of layers as a whole.

The equilibrium conditions for azimuthal angles (ϕ_1, ϕ_2) give

$$\frac{\partial E}{\partial \varphi_{1,2}} = 0 \Rightarrow \\ d_1M_1H \sin\theta_1 \sin\theta_H \sin(\varphi_1 - \varphi_H) + J_\ell \sin\theta_1 \sin\theta_2 \sin(\varphi_1 - \varphi_2) + \\ 2J_q \sin\theta_1 \sin\theta_2 \sin(\varphi_1 - \varphi_2)(\sin\theta_1 \sin\theta_2 \cos(\varphi_1 - \varphi_2) + \cos\theta_1 \cos\theta_2) = 0; \qquad (4) \\ d_2M_2H \sin\theta_2 \sin\theta_H \sin(\varphi_2 - \varphi_H) + J_\ell \sin\theta_1 \sin\theta_2 \sin(\varphi_2 - \varphi_1) + \\ 2J_q \sin\theta_1 \sin\theta_2 \sin(\varphi_2 - \varphi_1)(\sin\theta_1 \sin\theta_2 \cos(\varphi_1 - \varphi_2) + \cos\theta_1 \cos\theta_2) = 0.$$

For sustainable equilibrium condition $\partial^2 E/\partial\varphi_i^2 > 0$ must be hold also. From this and Eqs.(4) follows that $\varphi_1 = \varphi_2 = \varphi_H$ for positive J_l. If $J_l<0$ and $J_q \sim 0$ there are three possible variants:

(i) $\varphi_1 = \varphi_2 = \varphi_H$ (for $H > H_{crit}^{(1)}, H > H_{crit}^{(2)}$)

(ii) $\varphi_1 = \varphi_H, \varphi_2 = \varphi_H + \pi$ (for $H < H_{crit}^{(2)}$)

(iii) $\varphi_1 = \varphi_H + \pi, \varphi_2 = \varphi_H$ (for $H < H_{crit}^{(1)}$)

Critical fields are $H_{crit}^{(1)} = \dfrac{|J_\ell|}{d_1 M_1 \sin\theta_H}\sin\theta_2$, $H_{crit}^{(2)} = \dfrac{|J_\ell|}{d_2 M_2 \sin\theta_H}\sin\theta_1$. For large resonance fields ($H > H_c$) therefore $\varphi_1 = \varphi_2 = \varphi_H$.

The resonance frequency ω is given by the following fourth-order equation:

$$\omega^4 - \left[\left(\frac{\gamma_1}{M_1\sin\theta_1}\right)^2 \det(E''_{11}) + \left(\frac{\gamma_2}{M_2\sin\theta_2}\right)^2 \det(F''_{22}) + \frac{2\gamma_1\gamma_2}{M_1M_2\sin\theta_1\sin\theta_2}\det(F''_{12})\right]\omega^2 + \\ + \left(\frac{\gamma_1\gamma_2}{M_1M_2\sin\theta_1\sin\theta_2}\right)^2 \det(F'') = 0, \tag{5}$$

where γ_i is a gyromagnetic ratio for each layer. In Equation (5) matrices F''_{ij}, F_{ij} (i, j=1,2), are introduced. Each of them consists of second derivatives of free energy on angles in equilibrium:

$$F''_{ij} = \begin{vmatrix} F_{\theta_i\theta_j} & F_{\theta_i\varphi_j} \\ F_{\theta_j\varphi_i} & F_{\varphi_i\varphi_j} \end{vmatrix}_0 \tag{6}$$

$$F'' = \begin{vmatrix} F_{\theta_1\theta_1} & F_{\theta_1\varphi_1} & F_{\theta_1\theta_2} & F_{\theta_1\varphi_2} \\ F_{\theta_1\varphi_1} & F_{\varphi_1\varphi_1} & F_{\varphi_1\theta_2} & F_{\varphi_1\varphi_2} \\ F_{\theta_1\theta_2} & F_{\theta_2\varphi_1} & F_{\theta_2\theta_2} & F_{\theta_2\varphi_2} \\ F_{\theta_1\varphi_2} & F_{\varphi_1\varphi_2} & F_{\theta_2\varphi_2} & F_{\varphi_2\varphi_2} \end{vmatrix}_0 \tag{7}$$

The equilibrium polar angles (θ_1, θ_2) can be derived from equations:

$$M_1 H\sin(\theta_1 - \theta_H) - (2\pi M_1^2 - K_2^{(1)})\sin 2\theta_1 + 2K_4^{(1)}\sin\theta_1\cos^3\theta_1 - d_1^{-1}J_\ell\sin(\theta_1 - \theta_2) = 0, \\ M_2 H\sin(\theta_2 - \theta_H) - (2\pi M_2^2 - K_2^{(2)})\sin 2\theta_2 + 2K_4^{(2)}\sin\theta_2\cos^3\theta_2 + d_2^{-1}J_\ell\sin(\theta_1 - \theta_2) = 0. \tag{8}$$

We obtain from Equations. (5), (8) the following equation for polar dependence of resonance field $H(\theta_H)$:

$$\omega^4 - (\gamma_1^2 f_1 g_1 + \gamma_2^2 f_2 g_2 + 2\gamma_1\gamma_2 fg)\omega^2 + \gamma_1^2\gamma_2^2(f_1f_2 - f^2)(g_1g_2 - g^2) = 0, \\ f_i = \sigma_i\left(H\cos(\theta_i - \theta_H) + \left(-4\pi M_i + \frac{2K_2^{(i)}}{M_i} + \frac{K_4^{(i)}}{M_i}\right)\cos 2\theta + \frac{K_4^{(i)}}{M_i}\cos 4\theta_i\right) + \frac{J_\ell}{M_i}(d_1 + d_2)^{-1}\cos(\theta_1 - \theta_2), \\ g_i = \sigma_i\left(H\cos(\theta_i - \theta_H) + \left(-4\pi M_i + \frac{2K_2^{(i)}}{M_i}\right)\cos^2\theta_i + \frac{2K_4^{(i)}}{M_i}\cos^4\theta_i\right) + \frac{J_\ell}{M_i}(d_1 + d_2)^{-1}\cos(\theta_1 - \theta_2), \\ f = -\frac{J_\ell\cos(\theta_1 - \theta_2)}{(M_1M_2)^{1/2}(d_1 + d_2)}, g = -\frac{J_\ell}{(M_1M_2)^{1/2}(d_1 + d_2)}, \sigma_i = d_i(d_1 + d_2)^{-1}. \tag{9}$$

If interlayer coupling is absent Equation (9) reduces to two independent equations:

$$(\omega^2 - \gamma_1^2 f_1 g_1)(\omega^2 - \gamma_2^2 f_2 g_2) = 0 \tag{10}$$

Solutions of Eq. (10) simply describe resonant modes for first and second layer. For example if external field is applied in film plane (i.e. $\theta_1 = \theta_2 = \theta_H = 90°$) we have from (10) the well-known relation for a polycrystalline film:

$$\omega^2 = \gamma^2 H_{res}(H_{res} - 4\pi M_{eff}), 4\pi M_{eff} = -4\pi M + 2K_2 / M.$$

In general case for fixed frequency Equantion (9) with Equation (8) gives two solutions for resonance field. This corresponds to two mode FMR-signals. Polar dependence of resonance field provides information on magnetization density, magnetic anisotropy and interlayer coupling.

The line width for frequency may be evaluated by the follow expression:

$$\Gamma = \omega_0^2 \frac{\omega_0^2 \left(p_1 b_1 \Delta_{\theta_1\varphi_1} F + p_2 b_2 \Delta_{\theta_2\varphi_2} F\right) - G}{\omega_0^4 - \dfrac{p_1^2 p_2^2}{\sin^2(\theta_{10})\sin^2(\theta_{20})} \det(\mathbf{F}'')} \quad (11)$$

Then the peak – to –peak line width is given as follows:

$$\mathrm{D}H = \frac{dH}{dw}\frac{gb\mathrm{D}_{q,j} F}{M}. \quad (12)$$

where b is the dimensionless intrinsic damping factor in the Landau-Lifshitz-Gilbert equation, $\mathrm{D}_{q,j}$ is the angle part of Laplace operator.

The the peak – to –peak line width along magnetic field is $\mathrm{D}_0 = \dfrac{2bw}{g}$

We can see modeling data for # 2671 and 2669 samples in the table 3.

Table 3. Modeling data for # 2671 and 2669 samples

	M_1, emu/cc	M_2, emu/cc	$K_2^{(1)}$	$K_2^{(2)}$	$K_4^{(1)}$	$K_4^{(2)}$	J_ℓ, erg/cm^2
			10^5 erg/cc		10^5 erg/cc		
2669	~700	~1270	-1.6	-5.6	-1.2	-1.4	+0.20
2671	~960	~1040	3.3	2.05	2.1	-0.18	+0.05

4. Summary

The modeling experimental data for 2470 showed that the rise in annealing temperature, the layers magnetization increases and tends to equalize. As supplementary to this, the constant which describes the interaction between the layers, has a ferromagnetic character and decreases. The possible underlying factor is that during the annealing process the consistent ferromagnetic phase Fe_3O_4 with increased concentration of ferrum is formed, which causes the merger of the border between the ferrum and the magnetite layers. In addition the linewidth broadening is observed, caused by diminishing of the antiphase boundarys ' density and growing of grain size. It was derived that the optimum annealing temperature is $400^0 - 450^0$C. This data was used at obtaining multilayer structures.

An additional point is that the anisotropy constants depend on the annealing temperature.

It should be noted that the preliminary investigations of magnetite structures which contain ferrous substrates showed that the rising annealing temperature change the magnetization level and the coercive force and anisotropy constants. The modelling allowed follow the magnetization changes in each of the layer. The analysis showed that there is the optimum temperature at which the structure has the relevant properties. At 350C the magnetization of the second layer rises, as well as the interrelation constant, which determines the nonferromagnetic character of the ferrous substrate and magnetite interrelation. At a higher temperature the magnetization rises, however, the interrelation constant does not change. The comparison of spectral data of FMR samples 2669 and 2740 shows the similarity of one of the signals of sample 2669 to the signal of annealing sample 2740, which allow make a conclusion of the weak interchanges between the layers.

From (12) Gilbert constant was evaluated . For sample 2669 b_1=0.012, b_2 =0.010, its give possibility to find the Gilbert constants $G_1=1.4\times10^8$ s^{-1}, $G_2=2.2\times10^8$ s^{-1} for each layer. Gilbert constant for the first layer some decreases, but increasing for second layer of annealing sample 277. $G_1=1.3\times10^8$ s^{-1}, $G_2=4.0\times10^8$ s^{-1}

It is necessary to notice that the magnetic properties differences for samples made with same structure but with different annealing temperature. The process of №2671 sample annealing initiate cooptation of resonance lines and intensity. Annealing process also leads to coercive force values grow. It equivalent to exchange coupling value rising process between film layers through interface. Inversing layers order (for samples №2669 and №2673) leads to intensity change and initiate cooptation of resonance lines. The change of coercive force parameter had been registered caused by Fe sublayer bringing in film structure before magnetite deposition process. The same picture had been observed for 2740 series.

For sample with independence switching we observed resonance field shift into higher fields and leveling of resonance modes signals intensity with its cooptation.

The present work was supported by Russian Ministry of Education and Science within the frames of the Federal Special-Purpose program of Education at the Contract No.02.740.11.0550

References

[1] Julliere M 1975 *Phys. Lett. 54* A **3**, 225
[2] Goihman A, Kupriyanova G, Prokhorenko E and Chernenkov A 2010 *Vestnik I. Kant SU* vol **4** 65-71
[3] Butler W and Gupta A 2004 *Nature Mater* **3** 845
[4] Fonin et al. 2007 *J. Phys. Cond. Matt* **19**, 315217
[5] Parames M, Mariano J, Viskadourakis Z, Popovici N, Rogalski M, Giapintzakis J and Conde O 2006 *App. Surf. Sci* **252** 4610
[6] Niculescu A, Burch T and Budnick J 1983 *J. Magn. Magn.Mater.* **39** 223
[7] Zubin A, Orlova A, Goikman A, Kupriyanova G and Nevolin V 2010 The magnetic properties study of nanostructures with Fe3O4 by ferromagnetic resonance *Proc. of Int. Conf.* p 116 – 119
[8] Zakeri Kh, Kebe Th, Lindner J and Farle J 2006 *Magn. Magn. Mater.* 299
[9] Magen C, Snoeck E, Luders U and Bobo J 2008 *J. of Appl. Phys.* **104** 013913
[10] Goikhman A 2010 Magnetic and structure propeties of nanomagnetic ferromagnetic-isolator-ferromagnetic layers on base $FeSi_x$ и FeO_y PhD Dissertation of Physics and Mathematics MIFI Moscow.

Fe/Ni thin films temperature investigation with MgO and SiO_2 interfaces by ferromagnetic resonance

A Zyubin, A Orlova, A Astashonok, G Kupriyanova and V Nevolin
Immanuel Kant Baltic Federal University, 236041, A. Nevskogo 14, Kaliningrad, Russia

e-mail: azubin@mail.ru

Abstract. In this work the temperature study of magnetic – resonance properties of the structures such as Fe/MgO/Ni, Fe/SiO2/Ni differing thickness of spacer and of method of preparation was carried out by FMR. These systems are investigated to estimate their applicability in model creation experiments for a spintronics devices research [1-4]. The special attention was given to the temperature dependence research of three layer films linewidths. The out-of-plane temperature dependences of FMR signal position and linewidths have been measured for Fe/Ni samples with MgO and Si/SiO2 interfaces in static position of 0 and 90 degrees rotation angle to the external static magnetic field. The extracted magnetic parameters such as linewidths and resonance field position were studied.

1. Introduction

The study of ferromagnetic resonance (FMR) linewidth of multilayer structure is a useful diagnostic of the quality of the structures developed for a variety of applications [5] and exchange coupling between ferromagnetic layers across the nonmagnetic spacer [5] for different structures with Fe and Si content. One of the important problems today is a development of new materials and combinations for applications in field of spintronics. A lot of alloys and structures were investigated today for applicability of spintronics and microelectronics devices research. $Fe_{1-x}Mn_x$ alloys had been studied for «spin valve» devices application in [6] and suggest about foundation of different phases in nanoscale particles with various magnetic properties. Temperature dependent magnetization dynamics in thin FeNi layers in contact with an antiferromagnetic NiO were observed by J. Dubovik [7]. The anomalous broadening of linewidth and negative lineshift were observed for in-plane and out-of-plane geometry at ~150K temperature.

At the other hand, there is a great interest in the magnetization dynamics of magnetic multilayers because of their potential applications non-volatile magnetic random access memory and microwave devices. In the field of MRAM, much effort has been devoted to decreasing power consumption through the use of current-induced magnetization. [8]. Experimentally it was shown, that current-induced magnetization was observed as the current perpendicular to planetype giant magnetoresistivity of a nanopillar, in which the spin-polarized current injected from the fixed layer exerts a torque on the magnetization of the free layer. The torque induced by the spin current is utilized to generate microwaves.

The Gilbert damping constant is an important parameter for spin electronics since the critical current density of CIMR is proportional to the Gilbert damping constant [9, 10] and fast-switching time magnetization reversal is achieved for a large Gilbert damping constant [11]. Several mechanisms intrinsic to ferromagnetic materials, such as phonon drag [9] and spinorbit coupling [10], have been proposed to account for the origin of the Gilbert damping constant. In addition to these intrinsic mechanisms, Mizukami et al.[11, 12] and Tserkovnyak et al.[13, 14] showed that the Gilbert damping constant in a non-magnet (N) / ferromagnet (F) non-magnet (N) trilayer system is enhanced due to spin pumping. Tserkovnyak et al. [16] also studied spin pumping in a collinear F/N/F trilayer system and showed that enhancement of the Gilbert damping constant depends on the precession angle of the magnetization of the free layer [8].

To analyze magnetic and transport properties it was suggested to use different interface layers at the formation process of Fe/Ni thin films, e.g. SiO_2 and MgO. In this work we report FMR study of the magnetic properties of the Fe/MgO/Ni, Fe/SiO_2/Ni systems realized in a *polycrystalline* 3-layer structures to assess their applicability in a functional magnetic tunnel junction.

2. Experimental details

The samples in this work such as Fe/SiO2/Ni were synthesized by ion-plasma deposition technique, deposited on substrate with high energy ion beam, and Fe/MgO/Ni formatted by pulsed laser deposition on also RT amorphous Si/SiO2 substrate. The spacer layer thickness were varied from 1 to 50 nm, making possible establishment exchange interaction properties between magnetic Fe and Ni layers by FMR.

Ferromagnetic resonance spectra of the films were taken using Bruker electron spin resonance spectrometer with temperature module. Experiment was held at the X-band frequency 9.5 GHz and the modulation frequency 100 kHz. The first field derivative of the microwave power absorption was registered as a function of applied steady magnetic field at the different temperature.

The out-of -plane dependences of FMR spectra and linewidths have been measured for two samples. Samples were investigated at 80K – 273K for Fe/SiO_2/Ni and at 10K – 273K for Fe/MgO/Ni. The extracted magnetic parameters such as the linewidths, effective magnetization, resonance field position were studied.

3. Results

FMR signals were analyzed at different samples orientation relatively to the external static magnetic field at different temperature in position of 0 and 90 Fe/SiO2/Ni and in position of 90 for Fe/MgO/Ni degrees rotation angle to the external static magnetic field at 80K – 273K for Fe/SiO2/Ni and at 10K – 273K for Fe/MgO/Ni. The experimental results are presented at the table 1 and figures 1, 2.

Table 1. Samples structure, with FMR linewidth ΔB and Bres values.

№	Sample structure, synthesized on Si/SiO2	ΔB, G	B res, G
1	Fe/SiO2/Ni	143	8885,5
		143	8400,5
2	Fe/MgO/Ni	1854	9399

The FMR research of Si/SiO_2/Fe/SiO_2/Ni and Si/SiO2/Fe/MgO/Ni structures (sample №1 in table 1) showed that the spectrum parameters depends on temperature and rotation angle to the external static magnetic field . Two signals were observed for Si/SiO_2/Fe/SiO_2/Ni: the narrow high intense signal in area B_{res} = 8885,5 G with the line width ΔB=143 G, which depend on the temperature, and the narrow weak signal with the intense lower by one order of magnitude, with resonance field and line width depends on the temperature. Such signal behavior points at the presence of two magnetic phases in this sample. Temperature dependence of the FMR signal position data we obtained correlate with J. Kienert et al. [16] data for FMR signal position and number of modes for Ni thin films.

For the sample (table 1, sample №2) with MgO interface $Si/SiO_2/Fe/MgO/Ni$ sample one ferromagnetic signal was observed, the position and linewidths of which depend on the temperature. Dramatically changes in temperature (from 273 K to 80 K) of the sample with MgO interface are confirmed by the significantly FMR signal position changes. It was suggested to $Si/SiO_2/Fe/SiO_2/Ni$ sample as a material for model experiments for magnetic tunneling junctions development, because of coincidence with [16] FMR resonance field behavior.

On figure 1. FMR signal position (a) and linewidth (b) for sample with $Si/SiO_2/Fe/SiO_2/Ni$ structure are shown (sample №1 (table 1). FMR modes 1 and 2 [5] (figure 1a) are interested as they had steady shift into lower fields parallel to each other. It may be used as sample quality evidence for $Si/SiO_2/Fe/SiO_2/Ni$ structure. On figure 1b we obtained linewidth temperature dependence for both modes of signal. It has a symmetric behavior. We have coincidence of temperatures at 160 K to 220 K for modes 1 and 2. We suggest that it caused by thickness of $Si/SiO_2/Fe/SiO_2/Ni$ layers structure. We have the same gradual behavior of FMR signal position as at [5] for Ni structures, but for linewidth behavior is different.

On figure 2. FMR signal position (a) and linewidth (b) for sample with Si/SiO2/Fe/MgO/Ni structure for FMR signal mode. We have 273 K – 10 K temperature diapason for FMR signal position (figure 1a) and linewidth (figure 2a). We have intensive shift of signal position from 220K to 100K into lower fields and slow anomalous shift from 100K to 70K. Then we have signal position shift into the lower fields until 10 K. Linewidth also begun to rise after the point of 100 K temperature. The same behavior for FMR Signal position from temperature we have in [6] for 300 – 500 K temperature range, but in [7] we have gradual lowered temperature dependence for NiO thin films, without any shape shifts. Linewidth dependence results correlates with [6] for sample heating process.

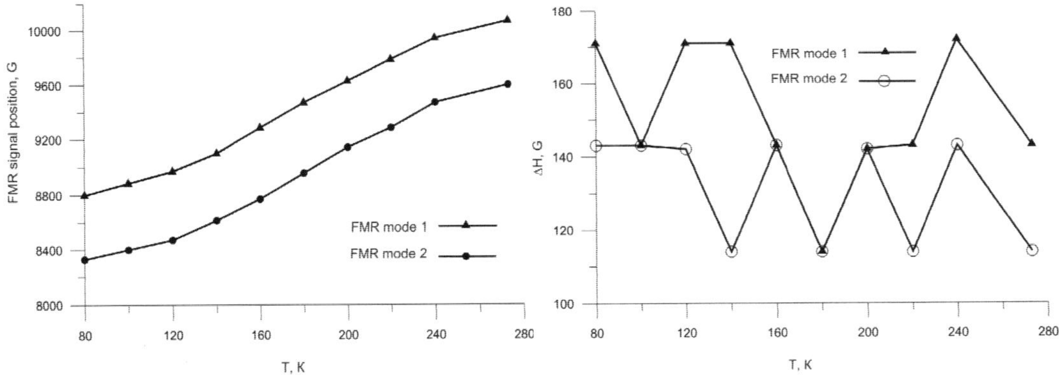

(a) (b)

Figure 1. FMR signal position (a) and linewidth (b) for sample with $Si/SiO_2/Fe/SiO_2/Ni$ structure: FMR signal mode 1 and signal mode 2 are shown

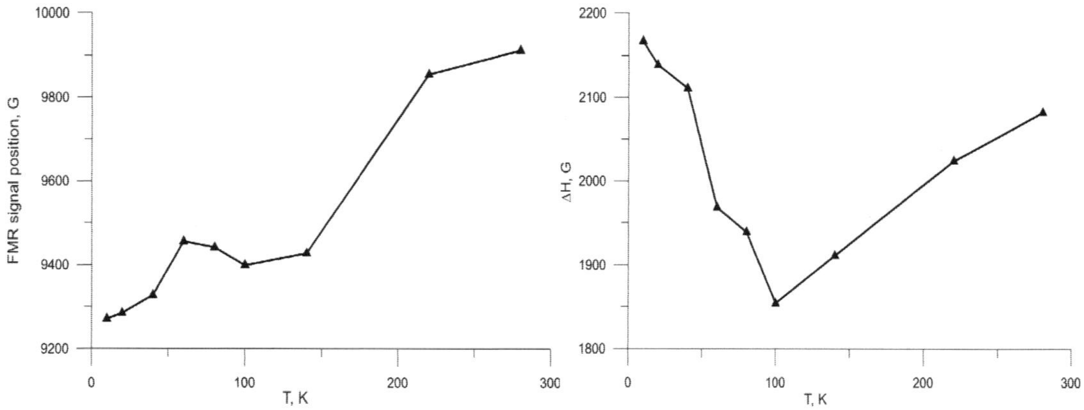

(a) (b)

Figure 2. FMR signal position (a) and linewidth (b) for sample with Si/SiO2/Fe/MgO/Ni structure: FMR signal mode is shown

4. Summary

In this work the experimental temperature study of magnetic properties of the $Si/SiO_2/Fe/SiO_2/Ni$ and $Si/SiO_2/Fe/MgO/Ni$ systems realized in a polycrystalline structures was made by ferromagnetic resonance (FMR) technique at different temperature. The investigations of magnetic properties of different multilayer structures, grown up on Si/SiO_2 substrate by FMR, were done. The analysis showed that parameters of FMR spectra of multilayer structure are very sensitively to the arrangement of temperature and nature of the interface material. Such systems are interesting because of behavior of spectrum parameters dependent of temperature become more clear. Work [15] illustrates the same data behavior of FMR parameters and correlate with our data.

The present work was performed within the frames of the State contract No. 02.740.11.0550 and No. P148 dated from April 15th, 2010.

Acknowledgments

Authors wishing to acknowledge for encouragement from Brucker Corporation.

References

[1] Comstock R L 2002 J. of Mater. Science: Materials in Electronics 13 509
[2] Berkowitz A E, Takano K 1999 J. Magn. Magn. Mater. 200 552
[3] Nogu´es J, Schuller I K 1999 J. Magn. Magn. Mater. 192 203
[4] Anhøj T A, Jacobsen C S, Mørup S 2004 J. Appl. Phys. 95 3649
[5] Kalarickal S S, Xu X Y, Lenz K, Kuch W and Baberschke K 2007 Phys. Rev. 75 224429
[6] Koksharov Y A, Yurkov G Y, Baranov D A and Malaxo A P Spectri electronnogo magnitnogo rezonansa v nanochasticax [EMR spectra in nanoscale particles Fe1-xMnx] 2006 J. of Solid State Physics 48 [in Russian]
[7] Dubowik J, Goscianska I, Paetzold A, Roll K 2003 Molecular physics reports 38 p 64-69
[8] Taniguchi T and Imamura H 2007 Phys. Rev. B 76 092402
[9] Suhl H 1998 IEEE Trans. Magn. 34 1834
[10] Kambersk´y V, Can J 1970 Phys. 48 2906
[11] Mizukami S, Ando Y and Miyazaki T 2002 J. Magn. Magn. Mater. 42 239
[12] Mizukami S, Ando Y and Miyazaki T 2002 Phys. Rev. B 66 104413
[13] Tserkovnyak Y and Brataas A and Bauer G E W 2002 Phys. Rev. Lett. 88 117601
[14] Tserkovnyak Y, Brataas and Bauer G E W 2002 Phys. Rev. B 66 224403
[15] Tserkovnyak Y, Brataas A and Bauer G E W 2003 Phys. Rev. B 67 140404(R)

[16] Kienert J, Schwieger S, Lenz K, Lindner J, Baberschke K and Nolting W 2007 Phys. Rev. Lett. 98 057205

Relaxation function theory for spin dynamics of strongly correlated layered copper oxide superconductors

Igor A Larionov

Magnetic Radiospectroscopy Laboratory, Institute of Physics, Kazan Federal University, 420008 Kazan, Russia

E-mail: Igor.Larionov@ksu.ru

Abstract. We present the relaxation function theory for dynamic spin susceptibility for doped two-dimensional $S = 1/2$ Heisenberg antiferromagnetic (AF) system in the *paramagnetic* state as obtained by means of the Mori-Zwanzig projection operator procedure to the t–J model. The results of the calculations are discussed in connection with the peculiar properties of layered copper oxide high-temperature superconductors (high-T_c). These include the neutron resonance peak, pseudogap properties, ω/T scaling, and the temperature and doping dependence of plane copper and oxygen nuclear spin-lattice relaxation rates (NSLRR). Particularly, the role of AF short range order, its evolution with doping and saturation of AF correlation length at low temperatures, is highlighted in view of the dynamic spin response of high-T_c up to optimal (maximum T_c) doping. The contribution from spin diffusion to relaxation rates is evaluated and is shown to play a dominant role in plane oxygen NSLRR of lightly doped ($\sim$3% holes per copper site) high-T_c at low temperatures. It is shown that the spin-wave-like theory is able to reproduce the main features of spin dynamics in high-T_c as observed experimentally.

1. Introduction

Dynamic phenomena in strongly correlated electron systems (SCES) continues to be in the focus of attention by condenced matter physicists due to rich variety of properties ranging from high temperature superconductivity (high-T_c) in layered copper oxides and exotic phases in other low-dimensional materials [1, 2, 3, 4, 5, 6]. Despite of considerable both the experimental and theoretical effort these systems still hide a host of undisclosed properties. Theoretical studies of dynamical properties [7, 8, 9, 10] make use of auxiliary quantities, viz. the Green's and relaxation function methods which have been brilliantly and exhaustively reviewed by Balucani, Lee, and Tognetti [11] in connection with both the quantum (e.g., described by Heisenberg Hamiltonian) and classical many-body systems.

Here we will discuss the spin dynamics of doped by holes a two dimensional S=1/2 Heisenberg antiferromagnetic (2DHAF) system in the paramagnetic state and compare the theoretical results obtained by Mori-Zwanzig projection operator procedure with magnetic resonance and inelastic neutron scattering data in layered copper high-T_c. The relation between the Kubo's relaxation function and the Green's function is discussed in [11].

Inelastic neutron scattering (NS) is a powerful tool in the high frequency studies of wave vector $\mathbf{k}$ and frequency ω dependent dynamic spin susceptibility $\chi''(\mathbf{k},\omega)$. Emergent observations of the resonance peak [12] at AF wave vector $\mathbf{Q} = (\pi, \pi)$ and frequency $\omega_r \approx$ 40 meV in optimally doped $YBa_2Cu_3O_7$, and the ω/T scaling of the averaged over the Brillouin zone the imaginary part of dynamic spin susceptibility, $\chi''(\omega, T) = \int \chi''(\mathbf{q}, \omega, T)\mathrm{d}^2\mathbf{q} \approx \chi''(\omega, T \to 0)f(\omega/T)$, in the underdoped high-T_c compounds [2, 13] are awaiting for theoretical understanding.

Soon after the discovery of high-T_c in $La_{2-x}(Ba,Sr)_xCuO_4$ family compounds it has been established that the parent, carrier free La_2CuO_4 compound is the 2DHAF insulator in which the magnetic correlation length ξ is described, e.g., by a quantum nonlinear σ model [14, 15] and by the isotropic spin-wave theory [16] in accord with NS experiments [17]. The majority of theories for resonance peak predominatingly treat it as a feedback of superconductivity (SC) which arises in the d-wave channel [18, 19, 20, 21, 22, 23]. However, the spin-wave like features of spin susceptibility in the underdoped $YBa_2Cu_3O_{6+x}$ above T_c have been emphasized [24, 25, 26, 27], suggesting that the resonant features may be caused *not only* by the emergence of the SC state. This conclusion has been encouraged by the absence of the isotope effect in the resonance peak frequency [28], contrary to the prediction within the RPA approach with d-wave SC [22]. In addition, Hwang, Timusk, and Gu [29] have recently shown by means of infrared spectroscopy that the resonance peak disappears completely in the overdoped $Bi_2Sr_2CaCu_2O_{8+y}$ sample with T_c = 55 K, thus suggesting the magnetic origin of the resonance peak. Thus the spin-wave concept appears to provide a natural elementary excitation in doped high-T_c. For a decade the resonant feature was attributed solely to the double layered cuprates, meanwhile, the observation of the resonance peak in the single layered $Tl_2Ba_2CuO_{8+y}$ [34] compound and similar features in $La_{2-x}Sr_xCuO_4$ [35] designate it as a generic feature of the layered copper oxides.

The ω/T scaling of $\chi''(\omega, T)$ above T_c is referred to a nearby quantum phase transition [30, 31], nevertheless, the theories [18, 32] used the temperature dependence of ξ that *disagrees* with the experimental data [13]. Nuclear magnetic/quadrupole resonance (NMR/NQR) studies [33] revealed the extension of the universal behavior of $\chi''(\omega, T)$ down to the MHz frequency range.

In this report, using the Mori-Zwanzig projection operator procedure with a *three* pole approximation for the relaxation function [11, 37, 38, 39], we will show that the resonant features in doped 2DHAF, as well as the nearly universal ω/T dependence of $\chi''(\omega, T)$ may be explained within a spin-wave-like theory [40, 41], where the correlation length, its doping dependence and saturation with lowering temperature, governs the main features of $\chi''(\mathbf{k}, \omega)$ as observed experimentally and *no* SC will be presupposed in the present calculations.

2. Basic relations

2.1. The t-J model Hamiltonian

We employ the $t - J$ Hamiltonian[42] since it is the minimal model for the electronic properties of high-T_c cuprates:

$$H_{t-J} = \sum_{i,j,\sigma} t_{ij} X_i^{\sigma 0} X_j^{0\sigma} + J \sum_{i>j} (\mathbf{S}_i \mathbf{S}_j - \frac{1}{4} n_i n_j), \qquad (1)$$

written in terms of the Hubbard operators $X_i^{\sigma 0}$ that create an electron with spin σ at site i and $\mathbf{S}_i$ are spin-1/2 operators. Here, the hopping integral $t_{ij} = t$ between the nearest neighbors (NN) describes the motion of electrons causing a change in their spins and $J = 0.12$ eV is the NN AF coupling constant. The spin and density operators are defined as follows:

$$S_i^{\sigma} = X_i^{\sigma\bar{\sigma}}, \qquad S_i^z = \frac{1}{2} \sum_{\sigma} \sigma X_i^{\sigma\sigma}, \qquad n_i = \sum_{\sigma} X_i^{\sigma\sigma}, \qquad (\sigma = -\tilde{\sigma}), \qquad (2)$$

with the standard normalization $X_i^{00} + X_i^{++} + X_i^{--} = 1$.

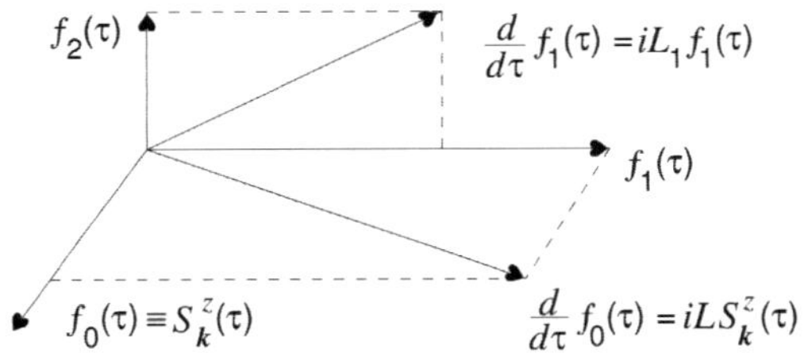

Figure 1. Schematic relation between random forces $f_1(\tau)$ and $f_2(\tau)$, and their effect on the fluctuating variable $f_0(\tau) \equiv S^z_{\mathbf{k}}(\tau)$.

2.2. Mori's projection operator procedure and a three pole approximation for the dynamic relaxation function

We formulate our study of the spin fluctuations following Mori [37], who showed it's efficiency for both the classical (and essential equivalence to Brownian motion) and quantum (e.g. Heisenberg systems of arbitrary dimension) many body systems [11]. The time evolution of a dynamical variable $S^z_{\mathbf{k}}(\tau)$, say, is given by

$$\dot{S}^z_{\mathbf{k}}(\tau) \equiv \frac{dS^z_{\mathbf{k}}(\tau)}{d\tau} = iLS^z_{\mathbf{k}}(\tau). \tag{3}$$

The Liouville superoperator L represents the Poisson bracket in the classical case and in the quantal case, which is the case of the present study, it corresponds to the commutator with the Hamiltonian (1): $iLS^z_{\mathbf{k}}(\tau) \to [H_{t-J}, S^z_{\mathbf{k}}(\tau)]$. The projection of the vector $S^z_{\mathbf{k}}(\tau)$ onto the $S^z_{\mathbf{k}} \equiv S^z_{\mathbf{k}}(\tau = 0)$ axis is given by:

$$\mathcal{P}_0 S^z_{\mathbf{k}}(\tau) = R(\mathbf{k}, \tau) \cdot S^z_{\mathbf{k}}, \tag{4}$$

and defines the linear projection Hermitian operator $\mathcal{P}_0$. One may separate $S^z_{\mathbf{k}}(\tau)$ into the projective and vertical components with respect to the $S^z_{\mathbf{k}}$ axis:

$$S^z_{\mathbf{k}}(\tau) = R(\mathbf{k}, \tau) \cdot S^z_{\mathbf{k}} + (1 - \mathcal{P}_0) S^z_{\mathbf{k}}(\tau), \tag{5}$$

where

$$R(\mathbf{k}, \tau) \equiv (S^z_{\mathbf{k}}(\tau), (S^z_{-\mathbf{k}})^*) \cdot (S^z_{\mathbf{k}}, (S^z_{-\mathbf{k}})^*)^{-1} \tag{6}$$

is the relaxation function in the inner-product bracket notation:

$$(S^z_{\mathbf{k}}(\tau), (S^z_{-\mathbf{k}})^*) \equiv k_B T \int_0^{1/k_B T} d\varrho \langle e^{\varrho H} S^z_{\mathbf{k}}(\tau) e^{-\varrho H} (S^z_{-\mathbf{k}})^* \rangle, \tag{7}$$

and the angular brackets denote the thermal average.

For future evaluations, it is convenient to introduce a set of quantities $f_0(\tau)$, $f_1(\tau)$, …, $f_j(\tau)$, … defined by equations: $f_j(\tau) \equiv \exp(iL_j\tau) f_j \equiv \exp(iL_j\tau) iL_j f_{j-1}$, where $L_0 \equiv L$, $f_0(\tau) \equiv S^z_{\mathbf{k}}(\tau)$, $L_j \equiv (1 - \mathcal{P}_{j-1}) L_{j-1}$, and $\Delta_j^2 \equiv (f_j, f_j^*) \cdot (f_{j-1}, f_{j-1}^*)^{-1}$ for $j \geq 1$. The set $\{f_j\}$ forms an orthogonal set. The *larger* number of f_j is used, the *finer* description of $S^z_{\mathbf{k}}(\tau)$ is obtained. The last quantity from this set f_n, affected by evolution operator $\exp(iL_n\tau)$, resulting in $f_n(\tau)$, is called the "n-th order random force" [37] acting on the variable $S^z_{\mathbf{k}}(\tau)$ and is responsible for fluctuation from its average motion.

In terms of Laplace transform of the relaxation function, $R(\mathbf{k},\tau)$, one may construct a continued fraction representation for $R^L(\mathbf{k},s)$, for which Lovesey and Meserve [39, 11] suggested a three pole approximation,

$$R^L(\mathbf{k},s)=\int_0^\infty d\tau\, e^{-s\tau}R(\mathbf{k},\tau)\approx 1/\{s+\Delta^2_{1\mathbf{k}}/[s+\Delta^2_{2\mathbf{k}}/(s+1/\tau_{\mathbf{k}})]\}, \tag{8}$$

with a cutoff characteristic time $\tau_{\mathbf{k}}=\sqrt{2/(\pi\Delta^2_{2\mathbf{k}})}$, by arguing that $S^z_{\mathbf{k}}(\tau)$ fluctuations are weakly affected by the higher order random forces. For the relaxation shape function $F(\mathbf{k},\omega)=Re[R^L(\mathbf{k},i\omega)]/\pi$, this gives

$$F(\mathbf{k},\omega)=\frac{\tau_{\mathbf{k}}\Delta^2_{1\mathbf{k}}\Delta^2_{2\mathbf{k}}/\pi}{[\omega\tau_{\mathbf{k}}(\omega^2-\Delta^2_{1\mathbf{k}}-\Delta^2_{2\mathbf{k}})]^2+(\omega^2-\Delta^2_{1\mathbf{k}})^2}, \tag{9}$$

where $\Delta^2_{1\mathbf{k}}$ and $\Delta^2_{2\mathbf{k}}$ are related to the frequency moments

$$\langle\omega^n_{\mathbf{k}}\rangle=\int_{-\infty}^\infty d\omega\,\omega^n F(\mathbf{k},\omega)=\frac{1}{i^n}\left[\frac{d^nR(\mathbf{k},\tau)}{d\tau^n}\right]_{\tau=0}, \tag{10}$$

of $R(\mathbf{k},\tau)$ as $\Delta^2_{1\mathbf{k}}=\langle\omega^2_{\mathbf{k}}\rangle$, $\Delta^2_{2\mathbf{k}}=(\langle\omega^4_{\mathbf{k}}\rangle\,/\,\langle\omega^2_{\mathbf{k}}\rangle)-\langle\omega^2_{\mathbf{k}}\rangle$ for $\tau>\tau_{\mathbf{k}}$. Note that $F(\mathbf{k},\omega)$ is real, even in both $\mathbf{k}$ and ω, and normalized to unity $\int_{-\infty}^{\infty}d\omega F(\mathbf{k},\omega)=1$.

Here we will discuss two approximations for the imaginary part of the dynamic spin susceptibility $\chi''(\mathbf{k},\omega)$. The first one is in the undamped spin-wave approximation, where $F(\mathbf{k},\omega)$ is related to $\chi''(\mathbf{k},\omega)$ as

$$\chi''_F(\mathbf{k},\omega)=\omega\chi(\mathbf{k})F(\mathbf{k},\omega). \tag{11}$$

Within the second one, since the relaxation function can be understood within the spin-wave framework,[11] the temperature and doping dependence of the *damping* of the spin-wavelike excitations may be studied further. The spin-wavelike dispersion, renormalized by interactions, is given by the relaxation function,[11]

$$\omega^{sw}_{\mathbf{k}}=2\int_0^\infty d\omega\,\omega F(\mathbf{k},\omega), \tag{12}$$

where the integration over ω in Eq. (12) has been performed analytically and exactly. One may assume the Lorentzian form for the imaginary part of the dynamic spin susceptibility,

$$\chi''_L(\mathbf{k},\omega)=\frac{\chi(\mathbf{k})\,\omega\Gamma_{\mathbf{k}}}{[\omega-\omega^{sw}_{\mathbf{k}}]^2+\Gamma^2_{\mathbf{k}}}+\frac{\chi(\mathbf{k})\,\omega\Gamma_{\mathbf{k}}}{[\omega+\omega^{sw}_{\mathbf{k}}]^2+\Gamma^2_{\mathbf{k}}}, \tag{13}$$

for $\mathbf{k}$ around the AF wave vector (π,π), where the wave vector dependence of the damping is given by $\Gamma=\sqrt{\langle\omega^2_{\mathbf{k}}\rangle-(\omega^{sw}_{\mathbf{k}})^2}$.

The expression for the second moment is straightforward,

$$\left\langle\omega^2_{\mathbf{k}}\right\rangle=i\langle[\dot S^z_{\mathbf{k}},S^z_{-\mathbf{k}}]\rangle/\chi(\mathbf{k})=-\left(8Jc_1-4t_{eff}T_1\right)\left(1-\gamma_{\mathbf{k}}\right)/\chi(\mathbf{k}), \tag{14}$$

where $\gamma_{\mathbf{k}}=\frac{1}{z}\sum_\rho\exp(i\mathbf{k}\rho)=\frac{1}{2}(\cos k_xa+\cos k_ya)$, where a=3.8 Åis a lattice unit and $z=4$ is the number of nearest neighbours on the square lattice, while it is rather cumbersome for the fourth moment $\langle\omega^4_{\mathbf{k}}\rangle=i\langle[\ddot S^z_{\mathbf{k}},\dot S^z_{-\mathbf{k}}]\rangle/\chi(\mathbf{k})$ and has been calculated with the approximations using the decoupling procedures for thermal averages as will be described in following SubSection. The final result is

$$
\begin{aligned}
\langle\omega_{\mathbf{k}}^4\rangle \simeq -\Big\{&128J^3[c_2\left(1-\gamma_{\mathbf{k}}^2\right)\left(\zeta c_2\left(\gamma_{\mathbf{k}}-\tfrac{3}{4}\right)-\tfrac{1}{4}c_0\right)+c_0c_1\left(\tfrac{7}{4}-\tfrac{5}{2}\gamma_{\mathbf{k}}+\tfrac{3}{4}\gamma_{\mathbf{k}}^2\right)\\
&+\zeta c_1c_2\left(\tfrac{13}{4}-\tfrac{15}{2}\gamma_{\mathbf{k}}+\tfrac{17}{4}\gamma_{\mathbf{k}}^2\right)+\zeta c_1^2\left(\tfrac{3}{2}-\tfrac{43}{8}\gamma_{\mathbf{k}}+\tfrac{21}{4}\gamma_{\mathbf{k}}^2+\tfrac{5}{8}\cos k_xa\cos k_ya-2\gamma_{\mathbf{k}}^3\right)]\\
&+16T_1t_{eff}^3[c_1\left(3-2\gamma_{\mathbf{k}}^2-\cos k_xa\cos k_ya\right)+\zeta T_2\left(7-12\gamma_{\mathbf{k}}+5\gamma_{\mathbf{k}}^2\right)\\
&+\tfrac{1-\delta}{2}\left(1-4\gamma_{\mathbf{k}}+3\gamma_{\mathbf{k}}^2\right)+(\delta+\lambda)\left(-\tfrac{9}{2}+9\gamma_{\mathbf{k}}-3\gamma_{\mathbf{k}}^2-\tfrac{3}{2}\cos k_xa\cos k_ya\right)]\\
&+16t_{eff}J^2T_1[c_0\left(-\tfrac{39}{8}+\tfrac{31}{4}\gamma_{\mathbf{k}}-\tfrac{23}{8}\gamma_{\mathbf{k}}^2\right)+c_2\left(-\tfrac{85}{8}+\tfrac{93}{4}\gamma_{\mathbf{k}}-\tfrac{101}{8}\gamma_{\mathbf{k}}^2\right)\\
&+c_1\left(16\gamma_{\mathbf{k}}^3-35\gamma_{\mathbf{k}}^2+25\gamma_{\mathbf{k}}-\tfrac{9}{2}-\tfrac{3}{2}\cos k_xa\cos k_ya\right)+\tfrac{9}{16}\cdot\tfrac{1-\delta}{2}\left(\gamma_{\mathbf{k}}-1\right)]\\
&+16t_{eff}^2J[c_1\left(\tfrac{3}{4}\gamma_{\mathbf{k}}-\tfrac{3}{4}\right)+T_2\left(\tfrac{1}{4}T_0+\tfrac{3}{4}T_0^2-\lambda\right)\left(2\gamma_{\mathbf{k}}^2-3\gamma_{\mathbf{k}}+1\right)\\
&+\tfrac{1+\delta}{2}c_1\left(6\gamma_{\mathbf{k}}^2-\tfrac{45}{4}\gamma_{\mathbf{k}}+\tfrac{21}{4}\right)+\left(\tfrac{3}{4}\lambda^{+-}\left(1-T_0\right)-T_0^2+\lambda T_0\right)\left(\gamma_{\mathbf{k}}-1\right)\\
&+\tfrac{1+\delta}{2}c_2\left(2\gamma_{\mathbf{k}}^2-\tfrac{9}{2}\gamma_{\mathbf{k}}+\tfrac{5}{2}\right)+c_1T_0\left(\tfrac{9}{4}\gamma_{\mathbf{k}}^2-\tfrac{5}{2}\gamma_{\mathbf{k}}+\tfrac{1}{4}\right)+c_1T_2\left(\tfrac{11}{4}\gamma_{\mathbf{k}}^2-\tfrac{15}{2}\gamma_{\mathbf{k}}+\tfrac{19}{4}\right)\\
&+c_2T_0\left(-2\gamma_{\mathbf{k}}^2+\tfrac{9}{2}\gamma_{\mathbf{k}}-\tfrac{5}{2}\right)+\zeta T_1^2\left(-4\gamma_{\mathbf{k}}^3+6\gamma_{\mathbf{k}}^2+\tfrac{11}{4}\gamma_{\mathbf{k}}-\cos k_xa\cos k_ya-\tfrac{15}{4}\right)\\
&+T_2c_2\left(16\gamma_{\mathbf{k}}^3-21\gamma_{\mathbf{k}}^2-\tfrac{5}{2}\gamma_{\mathbf{k}}+\tfrac{15}{2}\right)+T_0T_2\left(2\gamma_{\mathbf{k}}^2-\tfrac{9}{2}\gamma_{\mathbf{k}}+\tfrac{5}{2}\right)\\
&+T_2c_0\left(-5\gamma_{\mathbf{k}}^2+\tfrac{9}{2}\gamma_{\mathbf{k}}+\tfrac{1}{2}\right)+\zeta T_2^2\left(-2\gamma_{\mathbf{k}}^3+6\gamma_{\mathbf{k}}^2-\tfrac{19}{4}\gamma_{\mathbf{k}}+\tfrac{3}{4}\right)]\Big\}/\chi(\mathbf{k})
\end{aligned}
\tag{15}
$$

(see Reference [40] for details). Note that in the expression for $\langle\omega_{\mathbf{k}}^4\rangle$ the decoupling procedures were employed for the thermodynamic averages in spirit of papers by Hubbard and Jain [43], and by Kondo and Yamaji [44]. The averages with four operators are approximated, as usually, by products of two-operator correlation functions [39], however, multiplied now with the decoupling parameter ζ, e.g.: $\langle S_i^{\sigma}S_r^{\tilde{\sigma}}S_m^{\sigma}S_j^{\tilde{\sigma}}\rangle \to \zeta\langle S_i^{\sigma}S_r^{\tilde{\sigma}}\rangle\langle S_m^{\sigma}S_j^{\tilde{\sigma}}\rangle$ and so on. This parameter may be fixed from the total moment sum rule, however, the uncertainty in the correlation length and the destruction of fraction of the Cu^{2+} moments by holes makes this restriction less rigorous and we fix ζ from the comparison with experimental data.

2.3. Thermodynamic averages

To calculate the thermodynamic averages, we use the retarded Green's functions formalism. The equation of motion for a retarded Green's function $\langle\langle A|B\rangle\rangle_{\omega}$ takes the form

$$\omega\langle\langle A|B\rangle\rangle_{\omega} = \langle[A,B]_+\rangle + \langle\langle[A,H]|B\rangle\rangle_{\omega}, \tag{16}$$

where $\langle...\rangle$ denotes the thermal average. The standard relationship between correlation and Green's function may be written as

$$\langle BA\rangle = \frac{1}{2\pi i}\oint d\omega f(\omega)\langle\langle A|B\rangle\rangle_{\omega}, \tag{17}$$

where $f(\omega) = [\exp(\omega/k_BT)+1]^{-1}$ is the Fermi function; the contour encircles the real axis without enclosing any poles of $f(\omega)$.

In general, Equation (16) cannot be solved exactly and one needs some sort of approximation. To evaluate the Green's function $\langle\langle [A,H]|B\rangle\rangle_\omega$ in Equation (16), one uses a decoupling scheme originally proposed by Roth [46] for calculations on the Hubbard model. It can be shown that Roth's method is essentially equivalent to the Mori-Zwanzig projection technique [47, 48] and is strongly related to the moments method as applied to the evaluation of the spectral density of the Green's functions [49, 50]. Roth's method has been studied by many authors [48, 51], and became a general method to treat approximately the quasiparticle spectrum of an interacting system. The reliability of the method has been demonstrated by comparison with the exact diagonalization results [51].

Roth's method [46] implies that we seek a set of operators A_n, which are believed to be the most relevant to describe the one-particle excitations of the system of interest. Also, it is assumed that, in some approximation, these operators obey the relations [46]

$$[A_n, H] = \sum_m K_{nm} A_m, \tag{18}$$

where the parameters K_{nm} are derived through a set of linear equations

$$\langle [[A_n, H], A_l]_+ \rangle = \sum_m K_{nm} \langle [A_m, A_l^+]_+ \rangle. \tag{19}$$

Thus, it remains to define the operators A_n. Because, in the framework of the $t-J$ model, the quasiparticles are described by the Hubbard operators $X_{\mathbf{k}}^{0\sigma}$, a set of operators A_n contains only one operator $A = X_{\mathbf{k}}^{0\sigma}$. Hence, the matrix K_{nm} is diagonal and also contains one element $K = E_{\mathbf{k}}^{\sigma}$, where $E_{\mathbf{k}}^{\sigma}$ is the energy of of an electron with wave vector $\mathbf{k}$ and spin projection σ. Consequently, Equations (18) and (19) become

$$[X_{\mathbf{k}}^{0\sigma}, H] = E_{\mathbf{k}}^{\sigma} X_{\mathbf{k}}^{0\sigma}, \tag{20}$$

$$\langle [[X_{\mathbf{k}}^{0\sigma}, H], X_{\mathbf{k}}^{\sigma 0}]_+ \rangle = E_{\mathbf{k}}^{\sigma} \langle [X_{\mathbf{k}}^{0\sigma}, X_{\mathbf{k}}^{\sigma 0}]_+ \rangle. \tag{21}$$

In the 2D $t-J$ model, long-range order is absent at any finite temperature and hence, $E_{\mathbf{k}}^{\sigma}$ does not depend on σ. Thus, we can replace $E_{\mathbf{k}}^{+}$ and $E_{\mathbf{k}}^{-}$ by $E_{\mathbf{k}}$.

For our evaluations we need the thermal averages of the following types: $\left\langle X_i^{\sigma 0} X_j^{0\sigma} \right\rangle$ and $\langle X_i^{\sigma\sigma} X_j^{\sigma'\sigma'} \rangle$. First, one should note, that in the absence of long-range order, $\langle X_i^{\tilde{\sigma}\tilde{\sigma}} \rangle$ does not depend on the site index and hence, $T_0 = \langle X_i^{\tilde{\sigma}\tilde{\sigma}} \rangle = \langle X_i^{\sigma\sigma} \rangle = (1-\delta)/2$ and $c_0 = \langle S_r^z S_r^z \rangle = 1/4$.

The transfer amplitude between the first neighbours $T_1 = pI_1$ is given by

$$T_1 = pI_1 = -\frac{1}{z} \sum_\rho \left\langle X_i^{\sigma 0} X_{i+\rho}^{0\sigma} \right\rangle \tag{22}$$

and may be calculated using the spectral theorem

$$I_1 = -\sum_{\mathbf{k}} \frac{\gamma_{\mathbf{k}}}{\exp\left[(E_{\mathbf{k}} - \mu)/(k_B T)\right] + 1} \equiv \sum_{\mathbf{k}} \gamma_{\mathbf{k}} f_{\mathbf{k}}^h. \tag{23}$$

The latter equivalence has been obtained with the help of the identity $\sum_{\mathbf{k}} \gamma_{\mathbf{k}} = 0$. The sum (integral) over the wave vectors $\mathbf{k}$ in the 2D Brillouin zone is normalized by its area, $(2\pi)^2$, which is omitted for brevity. The parameter I_1 in Equation (23) has been estimated in [52],

$$I_1 \approx \frac{4}{\pi}\left(1 - e^{-\pi\tilde{\delta}}\right) - 2\tilde{\delta}, \qquad \tilde{\delta} = \frac{\delta}{1+\delta}, \tag{24}$$

with an accuracy of a few percent over the whole region of δ from 0 to 1. Here one should note that for very small δ and low temperatures, $I_1 \approx 2\delta$. Similarly, the transfer amplitude between the second neighbours, T_2, is given by

$$\begin{aligned} T_2 &= \frac{1}{z(z-1)} \sum_{\rho \neq \rho'} \left\langle X_i^{\sigma 0} X_{i+\rho-\rho'}^{0\sigma} \right\rangle, \\ T_2 &= \frac{p}{z(z-1)} \sum_{\mathbf{k}} \frac{16\gamma_{\mathbf{k}}^2 - 4\cos k_x a \cos k_y a - 4}{\exp[(E_{\mathbf{k}} - \mu)/(k_B T)] + 1} \\ &\equiv -\frac{p}{z(z-1)} \sum_{\mathbf{k}} \left(16\gamma_{\mathbf{k}}^2 - 4\cos k_x a \cos k_y a - 4\right) f_{\mathbf{k}}^h. \end{aligned} \tag{25}$$

$$c_1 = \frac{1}{z} \sum_{\rho} \langle S_i^z S_{i+\rho}^z \rangle, \qquad c_2 = \frac{1}{z(z-1)} \sum_{\rho \neq \rho'} \langle S_i^z S_{i+\rho-\rho'}^z \rangle, \tag{26}$$

are the first and second neighbour spin-spin correlation functions respectively, the index ρ runs over nearest neighbours. The numerical values of c_1, c_2 have been calculated following the expressions as described in Reference [54].

For p we have

$$p = (1 + \delta)/2, \tag{27}$$

where δ is the number of *extra* holes, due to doping, per one plane Cu^{2+}, which can be identified with the Sr content x in $La_{2-x}Sr_xCuO_4$. The excitation spectrum of holes is given by

$$E_{\mathbf{k}} = 4t_{eff}\gamma_{\mathbf{k}}, \tag{28}$$

where the hoppings, t, are affected by electronic and AF spin-spin correlations c_1, resulting in *effective* values [42, 52, 53], for which we set

$$t_{eff} = \delta J/0.2 \tag{29}$$

to match the insulator-metal transition. The chemical potential μ is related to δ by

$$\delta = p \sum_{\mathbf{k}} f_{\mathbf{k}}^h. \tag{30}$$

where $f_k^h = [\exp(-E_{\mathbf{k}} + \mu)/k_B T + 1]^{-1}$ is the Fermi function of holes.

To obtain the thermodynamic averages of the type $\langle X_i^{\sigma\sigma} X_{i+\rho}^{\sigma'\sigma'} \rangle$ it is convenient to make the following definitions

$$\lambda = \lambda_{\tilde{\sigma}\tilde{\sigma}} = \frac{1}{z} \sum_{\rho} \langle X_i^{\tilde{\sigma}\tilde{\sigma}} X_{i+\rho}^{\tilde{\sigma}\tilde{\sigma}} \rangle, \tag{31}$$

and

$$\lambda_{\sigma\sigma'} = \frac{1}{z} \sum_{\rho} \langle X_i^{\sigma\sigma} X_{i+\rho}^{\sigma'\sigma'} \rangle. \tag{32}$$

To obtain λ and $\lambda_{\sigma\tilde{\sigma}}$ we use the two Green's functions [52]

$$G_{\mathbf{k}}^{(1)}(\omega) = \frac{1}{z} \sum_{\rho} \langle\langle X_{\mathbf{k}}^{0\tilde{\sigma}} | X_i^{\tilde{\sigma}0} X_{i+\rho}^{\tilde{\sigma}\tilde{\sigma}} \rangle\rangle_\omega, \tag{33}$$

$$G_{\mathbf{k}}^{(2)}(\omega) = \frac{1}{z} \sum_{\rho} \langle\langle X_{\mathbf{k}}^{0\tilde{\sigma}} | X_i^{\tilde{\sigma}0} X_{i+\rho}^{\sigma\sigma} \rangle\rangle_\omega. \tag{34}$$

Note that in the paramagnetic state, $\lambda_{\sigma\sigma} = \lambda_{\bar{\sigma}\bar{\sigma}}$ and $\lambda_{\sigma\bar{\sigma}} = \lambda_{\bar{\sigma}\sigma}$.

According to Equations (16) and (20), the equation of motion for $G_{\mathbf{k}}^{(1)}(\omega)$ and $G_{\mathbf{k}}^{(2)}(\omega)$ can be written as

$$(\omega - E_{\mathbf{k}})G_{\mathbf{k}}^{(1)}(\omega) = \frac{e^{i\mathbf{k}r_i}}{\sqrt{N}}(1 - p - \lambda_{\bar{\sigma}\sigma} + pI_1\gamma_{\mathbf{k}}), \tag{35}$$

$$(\omega - E_{\mathbf{k}})G_{\mathbf{k}}^{(2)}(\omega) = \frac{e^{i\mathbf{k}r_i}}{\sqrt{N}}(1 - p - \lambda_{\bar{\sigma}\bar{\sigma}}), \tag{36}$$

where N is the number of sites. According to Equation (17):

$$\lambda_{\bar{\sigma}\bar{\sigma}} = \frac{1}{2\pi i}\sum_{\mathbf{k}} \frac{e^{-i\mathbf{k}r_i}}{\sqrt{N}} \oint d\omega f(\omega) G_{\mathbf{k}}^{(1)}(\omega), \tag{37}$$

$$\lambda_{\bar{\sigma}\sigma} = \frac{1}{2\pi i}\sum_{\mathbf{k}} \frac{e^{-i\mathbf{k}r_i}}{\sqrt{N}} \oint d\omega f(\omega) G_{\mathbf{k}}^{(2)}(\omega). \tag{38}$$

Consequently, Equations (35) and (36) lead to a system of linear equations for $\lambda_{\bar{\sigma}\bar{\sigma}}$ and $\lambda_{\bar{\sigma}\sigma}$ with the trivial solution

$$\lambda = \lambda_{\bar{\sigma}\bar{\sigma}} = (1-p)^2 - \frac{p^3}{2p-1}I_1^2, \tag{39}$$

$$\lambda_{\sigma\bar{\sigma}} = (1 - p - \lambda)\frac{1-\delta}{1+\delta} = (1-p)^2 + \frac{(1-p)p^2}{2p-1}I_1^2. \tag{40}$$

2.4. Decoupling procedures

We now describe the decoupling procedures for the thermodynamic averages performed following the papers by Hubbard and Jain [43] and by Kondo and Yamaji [44], and are performed in spirit of the self-consistent Born approximation (noncrossing approximation) [45].

The averages of the type $\langle X_i^{\sigma 0} X_l^{0\sigma} X_m^{\bar{\sigma}0} X_j^{0\bar{\sigma}}\rangle$ are decoupled resulting in products of transfer amplitudes and the decoupling parameter ζ,

$$\langle X_i^{\sigma 0} X_l^{0\sigma} X_m^{\bar{\sigma}0} X_j^{0\bar{\sigma}}\rangle \to \zeta \langle X_i^{\sigma 0} X_l^{0\sigma}\rangle \langle X_m^{\bar{\sigma}0} X_j^{0\bar{\sigma}}\rangle. \tag{41}$$

The four-spin correlation functions are approximated, as usually, by products of two-spin correlation functions [39], however, multiplied now with the decoupling parameter ζ. Thus, we employ the decoupling procedures

$$\langle S_i^{\sigma} S_r^{\bar{\sigma}} S_m^{\sigma} S_j^{\bar{\sigma}}\rangle \to \zeta \langle S_i^{\sigma} S_r^{\bar{\sigma}}\rangle \langle S_m^{\sigma} S_j^{\bar{\sigma}}\rangle, \tag{42}$$

and

$$\langle S_i^{z} S_r^{z} S_m^{\sigma} S_j^{\bar{\sigma}}\rangle \to \zeta \langle S_i^{z} S_r^{z}\rangle \langle S_m^{\sigma} S_j^{\bar{\sigma}}\rangle, \tag{43}$$

for $i \neq r$ and $m \neq j$, whereas

$$\langle S_r^{\sigma} S_r^{\bar{\sigma}} S_m^{\sigma} S_j^{\bar{\sigma}}\rangle \to 2c_0 \langle S_m^{\sigma} S_j^{\bar{\sigma}}\rangle. \tag{44}$$

The averages with the products of operators $X_i^{\sigma 0} X_r^{0\sigma}$ between the nearest(next-nearest) neighbours and $(1 - X_m^{\bar{\sigma}\bar{\sigma}})(1 - X_j^{\sigma'\sigma'})$ are decoupled as follows:

$$\langle X_i^{\sigma 0} X_r^{0\sigma}(1 - X_m^{\bar{\sigma}\bar{\sigma}})(1 - X_j^{\sigma'\sigma'})\rangle \to \langle X_i^{\sigma 0} X_r^{0\sigma}\rangle \langle 1 - X_m^{\bar{\sigma}\bar{\sigma}} - X_j^{\sigma'\sigma'} + X_m^{\bar{\sigma}\bar{\sigma}} X_j^{\sigma'\sigma'}\rangle \tag{45}$$

and so on.

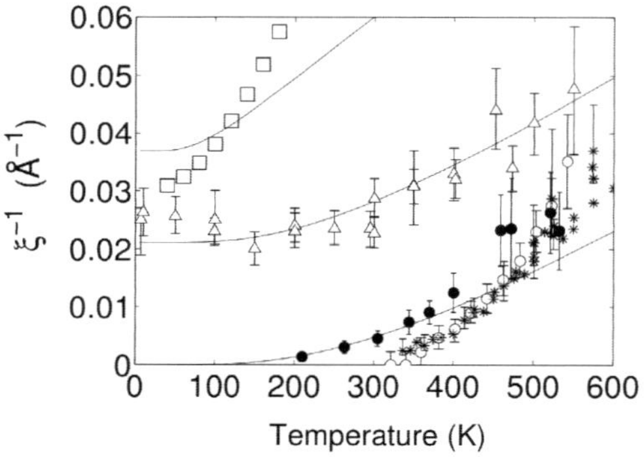

Figure 2. Inverse correlation length ξ_{eff} *vs* temperature fitted (solid lines) to the experimental data as obtained from neutron scattering experiments. For carrier free La_2CuO_4: filled circles from [17] (fitted data), asterisks from [57] and open circles from [13]. For doped $La_{2-x}Sr_xCuO_4$: up triangles for $x = 0.04$ from [13] and open squares for $x = 0.14$, modeled by $\xi^{-1}_{x=0.14} = \sqrt{k_o^2 + a^{-2}(T/E)^{2/Z_A}}$, with $k_o = 0.03$ Å^{-1}, $E = 690$ K, and $Z_A = 0.8$, following Reference [31]).

The averages with spin and Hubbard operators are decoupled as follows:

$$\langle X_i^{\sigma 0} X_j^{0\sigma} S_l^{\bar\sigma} S_r^{\sigma} \rangle \rightarrow \langle X_i^{\sigma 0} X_j^{0\sigma} \rangle \langle S_l^{\bar\sigma} S_r^{\sigma} \rangle, \tag{46}$$

and with spin and density operators:

$$\langle X_i^{\sigma\sigma} S_m^{\bar\sigma} S_r^{\sigma} \rangle \rightarrow \langle X_i^{\sigma\sigma} \rangle \langle S_m^{\bar\sigma} S_r^{\sigma} \rangle. \tag{47}$$

The averages $\langle X_i^{\sigma\sigma} X_j^{\sigma'\sigma'} \rangle$ between the second neighbouring operators are decoupled simply by

$$\langle X_i^{\sigma\sigma} X_{i+2}^{\sigma'\sigma'} \rangle \rightarrow \langle X_i^{\sigma\sigma} \rangle \langle X_{i+2}^{\sigma'\sigma'} \rangle, \tag{48}$$

because an inspection of Equations (39) and (40) shows that the values of averages of these type between the first neighbours differ only slightly from $\langle X_i^{\sigma\sigma} \rangle \langle X_{i+\rho}^{\sigma\sigma} \rangle$. Therefore, the averages between second neighbours in Equation (48) are thought as independent. In addition, because the averages $\langle X_i^{\sigma\sigma} X_j^{\sigma'\sigma'} \rangle$ between the first, in contrast with next-nearest, neighbours are calculated exactly, the averages like $\langle X_r^{\sigma\sigma} X_m^{\sigma\sigma} X_j^{\sigma'\sigma'} \rangle$ are decoupled in a way to avoid, where possible, the averages of the type as given in Equation (48).

2.5. Static susceptibility

In the present work we employ the static quantities that has been *derived* for both the carrier free and doped by charge carriers 2DHAF systems and work in the overall temperature range. The expression for static spin susceptibility is given by [54],

$$\chi(\mathbf{k}) = \frac{4|c_1|}{Jg_-(g_+ + \gamma_\mathbf{k})}, \tag{49}$$

and its structure is the same as in the isotropic spin-wave theory [16]. The meaning of g_+ is clear: it is related to ξ via the expression $\xi/a = 1/(2\sqrt{g_+ - 1})$. Here we will use the doping and temperature dependence of ξ following the explicit formulation given in Reference [54].

To mimic the low temperature behavior of the correlation length we use the expression, as in Reference [52], resulting in *effective* correlation length ξ_{eff}, given by,

$$\xi_{eff}^{-1} = \xi_0^{-1} + \xi^{-1}. \tag{50}$$

Here, in Equation (50), ξ is affected by doped holes, in contrast with the Keimer *et al.* [13] empirical equation, where ξ is given by the Hasenfratz-Niedermayer formula [15] and there was no influence of the hole subsystem on ξ. For strongly doped systems the expression for ξ is much more complicated compared with simple relation $\xi/a \simeq \frac{J\sqrt{g_-}}{k_BT}\exp(2\pi\rho_S/k_BT)$, which is valid for carrier free or lightly doped systems [40, 54]. Thus from now on we replace ξ by ξ_{eff}. In the best fit of ξ_{eff} to NS data [13, 31] (see Figure 2) we use $\xi_0 = a/n_\xi\delta$, where n_ξ is given in Table I. Whether its value follows from stripe ordering [55] or more exotic states [56] remains to be shown. The parameter g_- in Equation (49) has been introduced in Reference [54] and its numerical values with doping are listed in Table I. The second neighbour spin-spin correlation function c_2 is related to g_- as [54]: $g_- = \frac{4}{3}(1 + 30c_2)$.

3. Comparison with experiments and discussion

The results of the calculations are summarized in Table I. For brevity we consider here the cases: $La_{2-x}Sr_xCuO_4$ with $x = 0$, $x = 0.025$, $x = 0.035$, $x = 0.04$ and $x = 0.14$, and $YBa_2Cu_3O_{6.5}$ for which we accept [26] $\delta = 0.09$, and, particularly, $p = (1 + \delta/2)/2$, due to the bilayered structure that affects also $\langle\omega_{\mathbf{k}}^2\rangle$, $\langle\omega_{\mathbf{k}}^4\rangle$, and ξ.

Table 1. The calculated NN AF spin-spin correlation function $c_1 = \frac{1}{z}\sum_\rho\langle S_i^z S_{i+\rho}^z\rangle$, the parameter g_-, and the spin stiffness constant ρ_S using the expressions and the procedure as described in References [54] and [52] in the $T \to 0$ limit together with the spin diffusion constant D as calculated following References [40, 41] and the n_ξ and ζ values as extracted from comparison with NS data.

δ	c_1	g_-	$2\pi\rho_S/J$	D/J	n_ξ	ζ
0.0	−0.115215	4.1448	0.38	2.66	-	1.8
0.04	−0.1055	3.913	0.30	2.48	2	1.0
0.09	−0.0851	3.46	0.24	3.5	∼1.5	2.8
0.14	−0.0657	3.034	0.15	6.5	1	4.0

Our results for $\chi''(\mathbf{k},\omega)$ agree with the basic relations known from general physical grounds for small wave vectors $\mathbf{q}$ and frequency ω [8],

$$\chi''(\mathbf{q} \sim 0, \omega \sim 0) \simeq \frac{2\chi_S\omega D\mathbf{q}^2}{\omega^2 + (D\mathbf{q}^2)^2}, \tag{51}$$

where

$$D = \lim_{q\to 0}\frac{1}{\pi q^2 F(\mathbf{q},0)} = \lim_{q\to 0}\frac{1}{q^2}\sqrt{\pi < \omega_{\mathbf{q}}^2 >^3 /(2 < \omega_{\mathbf{q}}^4 >)}$$

is the spin diffusion constant. The Equation (51) for diffusive spin dynamics may be obtained from linear response theory, hydrodynamics and fluctuation-dissipation theorem. Using Equation (51) (or, equivalently, from Equation (9)) one may easy estimate the value of $\mathbf{q}_0$ in the local maximum of the imaginary part of dynamic spin susceptibility $\chi''(\mathbf{q},\omega)$, which is given by

$$q_0^2 \simeq \omega/D. \tag{52}$$

Journal of Physics: Conference Series **324** (2011) 012014 doi:10.1088/1742-6596/324/1/012014

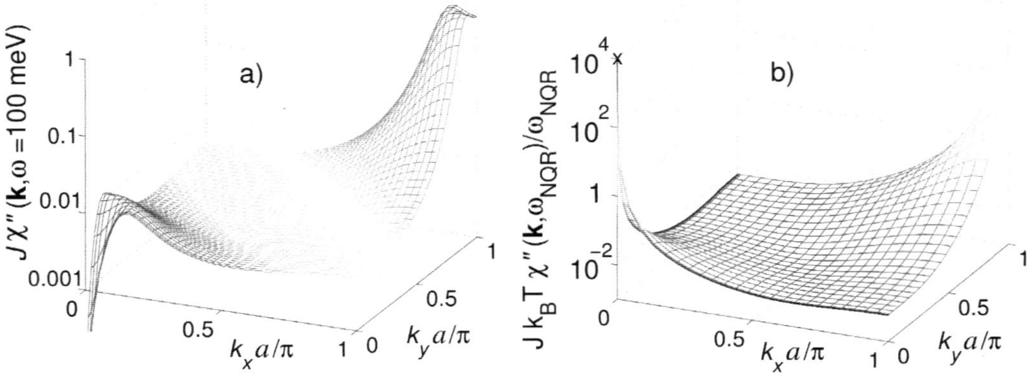

Figure 3. (Color online) Semilog-scale mesh of the calculated imaginary part of dynamic spin susceptibility $\chi''(\mathbf{k}, \omega)$ in the Brillouin zone for (a) $T = 90$ K and $\delta = 0.09$ and (b) $T = 300$ K and $x = 0.04$. The cross on the vertical axis marks the value of $\chi''(\mathbf{q_0}, \omega_{\rm NQR} = 34$ MHz) in its maximum at small wave vectors.

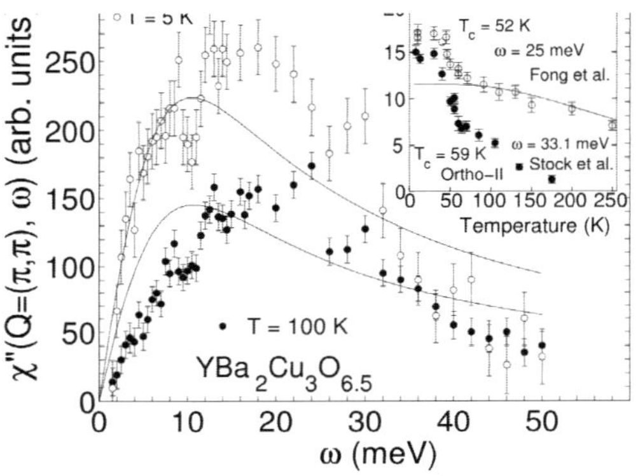

Figure 4. Imaginary part of the odd spin susceptibility $\chi''(\mathbf{Q}, \omega)$ from NS studies [24, 25, 26] of $\mathrm{YBa_2Cu_3O}_{y\approx 6.5}$ samples *versus* frequency ω. The lower solid line shows the calculated $\chi''_F(\mathbf{Q}, \omega)$ in the undamped approximation for $T = 100$ K and the upper solid line for $T = 5$ K scaled up by a factor of 1.5. The inset shows $\chi''_F(\mathbf{Q}, \omega)$ *versus* T in arb. units for each data set. The solid line shows the calculated and scaled to fit the temperature dependence of $\chi''_F(\mathbf{Q} = (\pi, \pi), \omega = 10$ meV) data above T_c.

For typical value of the measuring NMR frequency, $\omega \approx 1$ mK, $q_0 a \approx \pi \times 10^{-4}$. For extremely small $\mathbf{q} \ll \mathbf{q_0}$ with finite ω the imaginary part of dynamic spin susceptibility $\chi''(\mathbf{q}, \omega)$ approaches zero: $\chi''(\mathbf{q_0} \gg \mathbf{q} \to 0, \omega) \to 0$.

3.1. Inelastic neutron scattering

Figures 3-7 show the wave vector, frequency, doping and temperature dependence of $\chi''(\mathbf{k}, \omega)$. We note that for all temperatures the form of $F(\mathbf{q}, \omega)$ gives the elastic peak at $q = 0$ and $\omega = 0$. Figure 3 shows that for large ω the diffusive (small $\mathbf{k}$) dynamics is negligible, the calculated $\chi''(\mathbf{k}, \omega)$ for $\delta = 0.09$ is peaked at $\mathbf{Q} = (\pi, \pi)$ for $\omega < 55$ meV and becomes incommensurate with a spin-wave like cone (symmetric ring of scattering) for $\omega > 55$ meV in agreement with high-energy NS studies [27].

Figure 4 shows $\chi''_F(\mathbf{Q}, \omega)$ in the undamped approximation *versus* frequency and temperature.

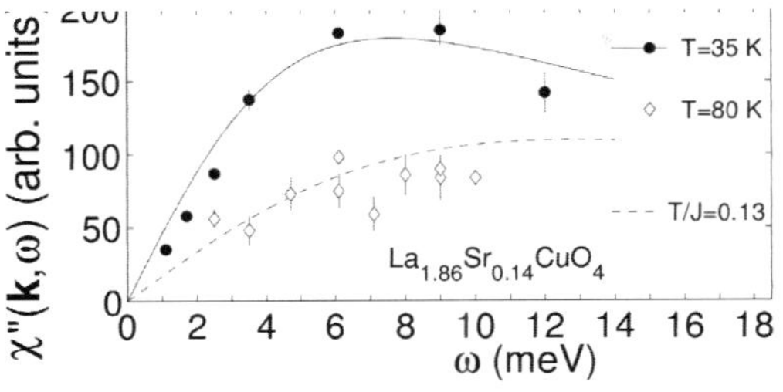

Figure 5. Imaginary part of dynamic spin susceptibility $\chi''(\mathbf{k},\omega)$ *versus* ω (symbols: NS data for $La_{1.86}Sr_{0.14}CuO_4$ of the incommensurate peak from Reference [31]. The lines show the calculated $\chi''_F(\mathbf{Q}=(\pi,\pi),\omega)$) in the undamped approximation.

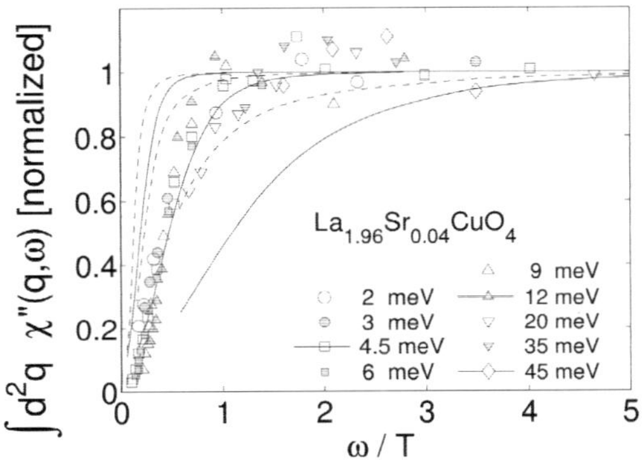

Figure 6. The averaged over the Brillouin zone imaginary part of dynamic spin susceptibility $\chi''(\omega,T) = \int \chi''(\mathbf{q},\omega,T)d^2\mathbf{q}$ *versus* ω/T (symbols: NS data for $La_{1.96}Sr_{0.04}CuO_4$ from Reference [13]. Solid lines show the calculated $\chi''_L(\omega,T)$) with Lorentzian, dashed lines show the calculated $\chi''_F(\omega,T)$) in the undamped spin-wave approximation.

The inset shows that $\chi''_F(\mathbf{Q},\omega)$ may not exhibit the sharp increase below T_c, in contrast with the predictions within the weak coupling theories [21, 22, 23]. Indeed, the more underdoped $YBa_2Cu_3O_{y\approx 6.5}$ sample (controlled by T_c) with the smaller resonance frequency shows the smaller increase of $\chi''_F(\mathbf{Q},\omega)$ below T_c. Figure 5 shows the results of our calculations in spirit of undamped spin-wave picture of Kondo and Yamaji [44] and suggests that the damping of spin-wave like excitations affects $\chi''(\mathbf{k},\omega)$ noticeably in doped 2DHAF even at low temperatures. Noting that the relaxation shape function $F(\mathbf{k},\omega)$ can be understood within the spin-wave like [11] framework, $\omega_{\mathbf{k}}^{SW} = 2\int_0^\infty d\omega\, \omega F(\mathbf{k},\omega)$, the temperature and doping dependence of the damping of the spin-wave-like excitations may be studied further. Our results suggest that, in contrast with [18], the damping of spin-wave like excitations is, however, *does not qualitatively affects* $\chi''(\mathbf{k},\omega)$ even in the normal state of optimally doped high-T_c cuprates. This may be caused by oversimplifications in [18] in the expression for susceptibility and simultaneous use of the temperature independent correlation length parameter, as indeed observed [13] *only* at $T <$400 K in the *lightly doped* regime together with the numerical results that are valid solely in the $T > J/2 \approx$700 K limit.

Figure 6 shows the averaged over the Brillouin zone and normalized imaginary part of dynamic spin susceptibility $\chi''(\omega,T)$ *versus* ω/T. Both the undamped approximation and the Lorentzian form with damping for the imaginary spin susceptibility suggest the ω/T scaling for underdoped

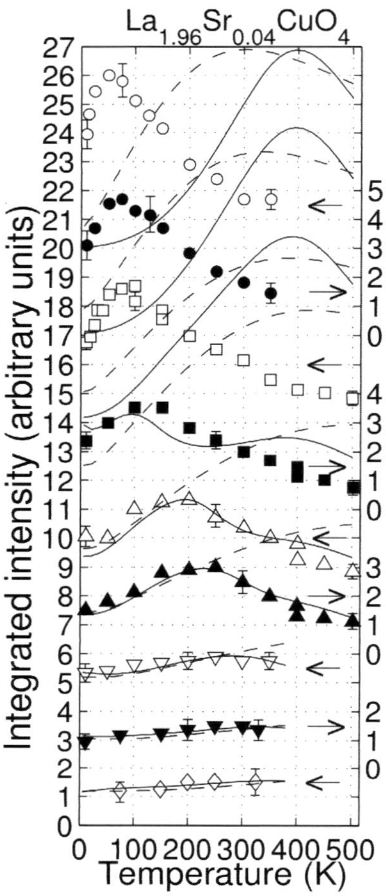

Figure 7. Temperature dependence of the dynamic structure factor $S(\mathbf{k},\omega)$, space and time Fourier transform of the spin-spin correlation function, as measured by neutron scattering in $La_{1.96}Sr_{0.04}CuO_4$ from top to bottom ω = 2, 3, 4.5, 6, 9, 12, 20, 35 and 45 meV, respectively. The results of the calculations are given for $x = 0.04$ in both approaches: $\chi''_L(\mathbf{q},\omega,T)$ with damping (Lorentzian form) (solid lines) and in the approximation for undamped paramagnon-like excitaions (dashed lines).

high-T_c layered cuprates with a deviations in qualitative agreement with NS data [13].

Figure 7 shows the dynamic structure factor $S(\mathbf{k},\omega)$, space and time Fourier transform of the spin-spin correlation function, as measured by neutron scattering in $La_{1.96}Sr_{0.04}CuO_4$. At large and medium frequencies the agreement between theory and experiment is very good. At small ω both theoretical approaches have valuable deviations form experimental data. It was already mentioned in the original experimental reports that the behaviour at small ω in NS experiments deviates from universal curves and we therefore will compare the results of our calculations with NMR data in order to check also the absolute values of $\chi''(\mathbf{k},\omega)$ with considerably smaller ω.

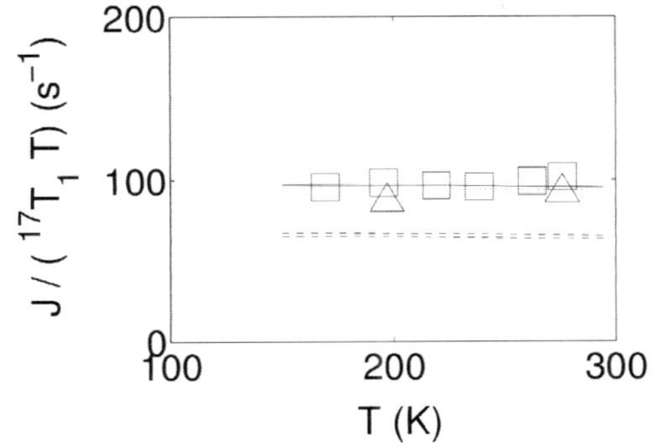

Figure 8. The calculated plane oxygen nuclear spin-lattice relaxation rates $^{17}(1/T_1)$ (lines) and the experimental data for $La_{2-x}Sr_xCuO_4$ as measured by NMR with $x = 0.025$ (triangles) and $x = 0.035$ (squares) from [60]. The experimental points have been rearranged with $J = 1393$ K. The results of the calculations in the undamped paramagnon-like excitations approxiamtion with $\omega = 2\pi \times 52$ MHz (9 T) are given for $x = 0.035$ by solid line and for $x = 0.025$ by dotted line. The contribution to $^{17}(1/T_1)$ from spin diffusion for $x = 0.035$ with $\omega = 2\pi \times 52$ MHz is shown by upper dashed line and with $\omega = 2\pi \times 81.4$ MHz (14.1 T) by lower dashed line.

3.2. Plane copper and oxygen nuclear spin-lattice relaxation rates

The nuclear spin-lattice relaxation rate $1/T_1$ is given by

$$^{\alpha}(1/T_1) = \frac{2k_BT}{\omega_0} \sum_{\mathbf{k}} {}^{\alpha}F(\mathbf{k})^2 \chi''(\mathbf{k}, \omega_0), \tag{53}$$

where ω_0 ($\ll T, J$) is the measuring NMR/NQR frequency. The quantization axis of the electric field gradient coincides with the crystal axis c which is perpendicular to CuO_2 planes defined by a and b. The wave vector dependent hyperfine formfactor for plane ^{63}Cu sites[58, 59] is given by, $^{63}F(\mathbf{k})^2 = (A_{ab} + 4\gamma_{\mathbf{k}}B)^2$, where $A_{ab} = 1.7 \cdot 10^{-7}$ eV and $B = (1 + 4\delta) \cdot 3.8 \cdot 10^{-7}$ eV are the Cu on-site and transferred hyperfine couplings, respectively. The relation for B is used to match the weak changes with Sr doping [62]. For plane oxygen sites we use $^{17}F(\mathbf{k})^2 = 2C^2\,(1 + \gamma_{\mathbf{k}})$, with $C = 2.8 \times 10^{-7}$ eV.

We first estimate the value of contribution to $1/T_1$ from small $\mathbf{q}$. A direct numerical integration over $\mathbf{q}$ is difficult, because $^{\alpha}F(\mathbf{q})^2\chi''(\mathbf{q}, \omega)$ has an extremely sharp peak at very small $\mathbf{q}_0$. This requires an unattainably large number of points in numerical integration over the Brillouin zone. Expanding $\chi''(\mathbf{q}, \omega)$ around q_0 we obtain,

$$^{\alpha}(1/T_1)_{Diff} = \frac{^{\alpha}F(0)^2 k_B T a^2 \chi_S}{\pi\hbar D}\Lambda, \tag{54}$$

where Λ depends on frequency through q_0. A simple and rough estimate gives

$$\Lambda \sim \ln(1/q_0^2) \sim \ln(const \times J/\omega). \tag{55}$$

This result explains the reason of the negligible shift of the oxygen $^{17}(1/T_1)$ relaxation rate as measured by NMR at 9 Tesla ($\omega_0 = 2\pi\times$ 52 MHz) and 14.1 Tesla ($\omega_0 = 2\pi\times$ 81.4 MHz) that lies within the experimental accuracy [60]. One should note that ω is much less than $J = 1.8\times10^8$ MHz, hence $\ln(J/52\text{ MHz})/\ln(J/81.4\text{ MHz}) \approx 1.03$. A sophisticated calculation

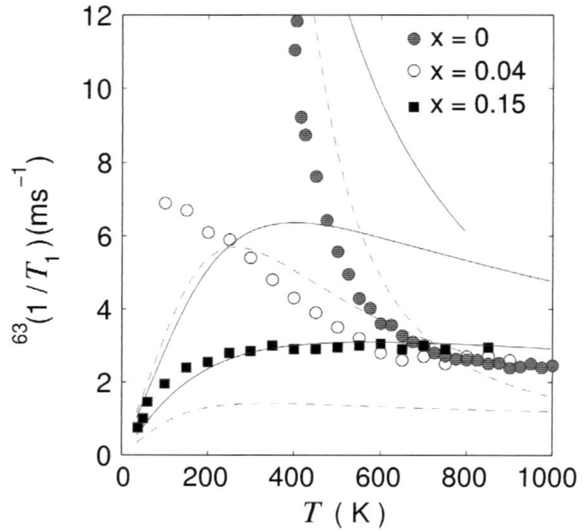

Figure 9. The calculated temperature dependencies of plane copper nuclear spin-lattice relaxation rate $^{63}(1/T_1) = 2W$ in $La_{2-x}Sr_xCuO_4$ (data from [33]) for $x = 0.0$, $x = 0.04$ and for $x = 0.14$ by solid lines for $\chi''_L(\mathbf{q}, \omega, T)$ with damping (Lorentzian form) and in the approximation for undamped paramagnon-like excitaions (dashed lines).

gives $\Lambda(35\text{ MHz}) = 2.52$, $\Lambda(52\text{ MHz}) = 2.44$, and $\Lambda(81.4\text{ MHz}) = 2.37$. In view of the result that the spin diffusive contribution is 70 %, the relative shift of the measured $^{17}(1/T_1)$ will be ≈ 2 %, that lies within the experimental error (see Figure 8).

Figures 2 and 9 show the temperature dependencies of inverse correlation length and plane copper nuclear spin-lattice relaxation rate, $^{63}(1/T_1)$. Equations (11), (53), and Figures show that the temperature dependence of $^{63}(1/T_1)$ is governed by the temperature dependence of the correlation length and by the factor k_BT. At low T, where $\xi_{eff} \simeq const$, the plane copper $^{63}(1/T_1) \propto T$, as it should. At high T, the correlation length shows the weak doping dependence and behaves similarly to that of carrier free La_2CuO_4 and $^{63}(1/T_1)$ of doped samples behaves similarly to that of La_2CuO_4. Thus our results shown in Figures 2 and 8 suggest that the "pseudogap" effect seen with NMR in the high-T_c cuprates is *hidden* in the correlation length that affects the observable quantities, and, generally, is in agreement with the conclusion based on the nearly AF Fermi liquid concept [61, 62] about the leading role of the correlation length in temperature and doping dependence of $1/T_1$.

4. Conclusion

In conclusion, we have presented a relaxation function theory for dynamic spin properties of layered copper high-T_c in the normal state. The resonant feature, ω/T scaling, and spin-lattice relaxation at planar sites may be explained within the undamped spin-wave-like theory. We conclude that the resonant feature seen by NS may be caused not only by the feedback effect of the d-wave superconducting state. We discuss the influence of AF short range order (affected by stripes or, possibly, more exotic phases) on the *dynamic* quantities is discussed, possessing a reasonable agreement with the observations by means of neutron scattering and magnetic

Journal of Physics: Conference Series **324** (2011) 012014 doi:10.1088/1742-6596/324/1/012014

resonance experiments in the underdoped high-T_c layered copper oxides. At the same time, the future model for magnetic properties should match also a crossover from the localized spins picture (paramagnon-like excitation) to itinerant weak coupling theory with doping.

Acknowledgments
It is a pleasure to thank Mikhail V. Eremin for discussions. This work was supported in part by Dynasty foundation.

References
[1] Lee P A, Nagaosa N and Wen X G 2006 *Rev. Mod. Phys.* **78** 17
[2] Kastner M A, Birgeneau R J, Shirane G and Endoh Y 1998 *Rev. Mod. Phys.* **70** 897
[3] Dagotto E 1999 *Rep. Prog. Phys.* **62** 1525
[4] Izyumov Yu A 1997 *Usp. Fiz. Nauk* **167** 465 (in Russian) [1997 (*Physics-Uspekhi*) **40** 445]
[5] Izyumov Yu A 1999 *Usp. Fiz. Nauk* **169** 225 (in Russian) [1999 (*Physics-Uspekhi*) **42** 215]
[6] Ovchinnikov S G 1997 *Usp. Fiz. Nauk* **167** 1043 (in Russian) [1997 (*Physics-Uspekhi*) **40** 993]
[7] Zubarev D N 1971 (*Nonequilibrium Statistical Thermodynamics*) Nauka, Moscow (in Russian) [Consultants Bureau, New York, 1974]; Zwanzig R 1965 *Annual Rev. Phys. Chem.***16** 67
[8] Forster D 1975 *Hydrodynamic Fluctuations, Broken Symmetry, and Correlation Functions* Frontiers in Physics, Vol. **47** (Benjamin, Reading, MA); [Atomizdat, Moscow, 1980 (in Russian)]
[9] Marshall W and Lowde R D 1968 *Rep. Prog. Phys.* **31** 705
[10] Alexandrov I V 1975 *Theory of Magnetic Relaxation. Relaxation in Liquids and Solid Nonmetallic Paramagnets* (Moscow: Nauka) (in Russian)
[11] Balucani U, Lee M H and Tognetti V 2003 *Phys. Rep.* **373** 409
[12] Rossat-Mignod J *et al.* 1991 *Physica C* **185-189** 86
[13] Keimer B *et al.* 1992 *Phys. Rev.* B **46** 14034
[14] Chakravarty S, Halperin B I and Nelson D R 1988 *Phys. Rev. Lett.* **60** 1057; 1989 *Phys. Rev.* B **39** 2344
[15] Hasenfratz P and Niedermayer F 1991 *Phys. Lett.* B **268** 231; 1993 *Z. Phys.* B **92** 91
[16] Sokol A, Singh R R P and Elstner N 1996 *Phys. Rev. Lett.* **76** 4416
[17] Endoh Y *et al.* 1988 *Phys. Rev.* B **37** 7443
[18] Prelovsek P, Sega I and Bonca J 2004 *Phys. Rev. Lett.* **92** 027002; 2003 *Phys. Rev.* B **68** 054524
[19] Morr D K and Pines D 1998 *Phys. Rev. Lett.* **81** 1086
[20] Abanov A and Chubukov A 1999 *Phys. Rev. Lett.* **83** 1652
[21] Liu D, Zha Y and Levin K 1995 *Phys. Rev. Lett.* **75** 4130
[22] Eremin I, Kamaev O and Eremin M V 2004 *Phys. Rev.* B **69** 094517
[23] Norman M R 2001 *Phys. Rev.* B **61** 14751
[24] Bourges P 1999 in *The Gap Symmetry and Fluctuations in High Temperature Superconductors* edited by Bok J., Deutscher G., Pavuna D., Wolf S.A. (Plenum Press, New York, 1998), p. 349 (Proceedings of NATO ASI summer school held September 1-13, 1997 in Cargese, France); also in arXiv xxx.lanl.gov/cond-mat/9901333, 1999
[25] Fong H F *et al.* 2000 *Phys. Rev.* B **61** 14773
[26] Stock C *et al.* 2004 *Phys. Rev.* B **69** 014502
[27] Stock C *et al.* 2005 *Phys. Rev.* B **71** 024522
[28] Pailhes S *et al.* 2005 *Phys. Rev.* B **71** 220507(R)
[29] Hwang J, Timusk T, Gu G D 2004 *Nature (London)* **427** 714; Norman M *Nature (London)* 2004 **427** 692
[30] Vojta M 2003 *Rep. Prog. Phys.* **66** 2069
[31] Aeppli G*et al.* 1997 *Science* **278** 1432
[32] Varma C M *et al.* 1989 *Phys. Rev. Lett.* **63** 1996
[33] Imai T *et al.* 1993 *Phys. Rev. Lett.* **70** 1002
[34] He H *et al.* 2002 *Science* **295** 1045
[35] Tranquada J M *et al.* 2004 *Phys. Rev.* B **69** 174507
[36] Sidis Y *et al.* 2000 *Phys. Rev. Lett.* **92** 5900
[37] Mori H 1965 *Prog. Theor. Phys.* **34** 399
[38] Zwanzig R 1961 *Phys. Rev.* **124** 983; Zwanzig R, Nordholm K S J and Mitchell W C 1972 *Phys. Rev.* A **5** 2680
[39] Lovesey S W and Meserve R A 1973 *J. Phys. C* **6** 79
[40] Larionov I A 2004 *Phys. Rev.* B **69** 214525
[41] Larionov I A 2005 *Phys. Rev.* B **72** 094505

[42] Anderson P W 1987 *Science* **235** 1196; Baskaran G, Zou Z and Anderson P W 1987 *Solid State Commun.* **63** 973
[43] Hubbard J and Jain K P 1968 *J. Phys. C* **1** 1650
[44] Kondo J and Yamaji K 1972 *Prog. Theor. Phys.* **47** 807
[45] Plakida N M and Oudovenko V S 1999 *Phys. Rev.* B **59** 11949
[46] Roth L M 1969 *Phys. Rev.* **184** 451
[47] Unger P and Fulde P 1993 *Phys. Rev.* B **48** 16607
[48] Mehlig B, Eskes H, Hayn R and Meinders M B J 1995 *Phys. Rev.* B **52** 2463
[49] Bowen S P 1975 *J. Math. Phys.* **16** 620
[50] Harris A B and Lange R V 1967 *Phys. Rev.* **157** 295
[51] Beenen J and Edwards D M 1995 *Phys. Rev.* B **52** 13636
[52] Zavidonov A Yu, Larionov I A and Brinkmann D 2000 *Phys. Rev.* B **61** 15462
[53] Plakida N M, Hayn R and Richard J L 1995 *Phys. Rev.* B **51** 16599
[54] Zavidonov A Yu and Brinkmann D 1998 *Phys. Rev.* B **58** 12486
[55] Tranquada J M *et al.* 1995 *Nature (London)* **375** 561; 1996 *Phys. Rev.* B **54** 7489; 1997 *Phys. Rev. Lett.* **78** 338
[56] Kivelson S A, Fradkin E and Emery V J 1998 *Nature (London)* **393** 550; Hayden S M *et al.*, 2004 *ibid.* **429** 531; Tranquada J M *et al. ibid.* 2004 **429** 534; Hinkov V 2004 *et al. ibid.***430** 650
[57] Birgeneau R J *et al.* 1999 *Phys. Rev.* B **59** 13788
[58] Mila F and Rice T M 1989 *Physica C* **157** 561
[59] Shastry B S 1989 *Phys. Rev. Lett.* **63** 1288
[60] Thurber K R, Hunt A W, Imai T, Chou F C and Lee Y S 1997 *Phys. Rev. Lett.* **79** 171
[61] Millis A J, Monien H and Pines D 1990 *Phys. Rev.* B **42** 167
[62] Zha Y, Barzykin V and Pines D 1996 *Phys. Rev.* B **54** 7561

Features of Magnetic Phase Transitions in $EuB_{6-x}C_x$

T S Altshuler[1], Yu V Gorjunov[1], A V Levchenko[2] and V B Filippov[2]

[1]Zavoisky Physical-Technical Institute, Sibirsky trakt 10/7, 420029 Kazan, Russia
[2]Institute for problems of Materials Science of NASU, 03680, Kiev, Ukraine

E-mail: tatiana@kfti.knc.ru

Abstract. The ESR measurements were performed on the single crystals $EuB_{6-x}C_x$ with x = 0, 0.02 and 0.07 at X band in the temperature range T = 10 - 300 K. It was observed the magnetic phase separation in spin system of Eu^{2+} ions The reason of the separation is accompanied by the self-localization of free charge carriers and the formation of magnetic polarons of two types. With decreasing temperatures the 2 phase transitions take place. The first transition is connected with the effect of magnetic ordering of the most intensive subsystem. The second transition is obliged to the delocalization of charge carriers and the system is converted from heterogeneous to homogeneous state.

1. Introduction

Europium hexaboride is part of the large and heterogeneous class of materials that exhibit colossal magnetoresistance (CMR). EuB_6 has a cubic unit cell with Eu-ions at its vertices and a boron octahedron at its center and because of its simple lattice it is very convenient for the researchers of CMR effect. However, despite numerous experimental and theoretical studies of the pure and doped EuB_6, understanding of the electronic properties and magnetic coupling still provides challenges. The substance shows two magnetic transitions: at T_M = 15.3K and at T_C = 12.5K. It orders ferromagnetically at 15,3K, which is accompanied by a huge decrease of resistivity and a significant blue shift of the reflectivity plasma edge. At 12.7 K another phase transition takes place, which is observed as a broad peak in the specific heat or an anomaly in the resistivity [1-3]. The origin of this transition was still unclear.

Until now the study of EuB_6 was carried out, in essence, by the methods, which reveal only physical properties of an entire sample. Meanwhile the electron paramagnetic resonance gives the possibility to observe such local phenomenon as magnetic phase separation [see, for example, 4-5], to investigate its temperature dependence, and hence also the features of the magnetic phase transitions.

2. Experimental

The electron spin resonance measurements were performed on the single crystals $EuB_{6-x}C_x$ with x = 0, 0.02 and 0.07 in X band in the temperature range T = 10 - 300 K. Carbon ion substitutes the boron ion and brings one electron into the conduction band of EuB_6. The single crystals of the pure and doped by carbon europium hexaboride were grown by inductive floating zone melting. The starting rods for melting were prepared by sintering of two types of the europium hexaboride powders with carbon and without it. EuB_6 crystals were synthesized by borothermal reduction of Eu_2O_3. Eu_2O_3

Published under licence by IOP Publishing Ltd

purity was 4N, natural boron – 3N. The final compositions of three types of single crystals may be estimated as EuB_6, $EuB_{5.98}C_{0.02}$ and $EuB_{5.93}C_{0.07}$, according to chemical analyses and hydrostatic density measurements. According to an optical spectral analysis the total content of the impurities apart from carbon was no more than 10^{-3} wt % for single crystals. The $EuB_{5.93}C_{0.07}$ and EuB_6 lattice parameters are 4.176161±0.000036 Å and 4.184679±0.000015 Å correspondingly in agreement with published data [6]. The samples are the polished single crystal plates with 0.5 mm thickness cut from central part of the corresponding single crystals. One single crystal was grown by flux method also.

The ESR measurements were performed on frequency of 9.25 GHz in TE_{102} rectangular resonator with the flux helium cryostat, allowed to measure in the temperatures range from 7K to 300K. The measurements of the temperature dependences of the ESR spectra for thin single crystal plates were performed for perpendicular ($\theta = 90^0$) orientations of the plate with respect to the magnetic field in the carbon content europium hexaboribes and for both ($\theta = 0^0$ and $\theta = 90^0$) orientations in clean europium hexaboribe.

3. Results and Discussion

The single line from Eu^{2+} was observed at 300 K in all single crystals by ESR method. The lineshape is Lorentzian with Dyson distortion. The Eu^{2+} ion has the spin state with the electron spin S = 7/2 and two stable isotopes with a nuclear spin of 5/2; therefore 84 resonance lines should be observed. The absence of the fine and hyperfine structures is associated with the averaging of the lines due to strong exchange interactions (fig. 1). The Lorentzian lineshape can also indicate the strong exchange. The position and ESR lineshape are independent from the orientation of the sample plane to the static magnetic field.

3.1. ESR in EuB_6

Below 50K the ESR spectrum depends on the orientation of the sample plane to the static magnetic field. The position of the resonance field for the parallel and perpendicular orientation to EuB_6 plate is pointed in fig2 at T=22K. Such shift of the resonance lines is mainly determined by the demagnetizing fields. In our simplest case, where the external magnetic field is parallel or perpendicular to the sample plane, the position of the resonance line is calculated by the following well known formulas. The position of the magnetic resonance in parallel geometry at $\theta = 0$ is given by the expression

$$H_r(T) = \sqrt{H_{r0}^2 + 4\pi^2 M_{eff}^2(T)} - 2\pi M_{eff}(T) \qquad (1)$$

where H_{r0} is the resonance field at high temperatures (paramagnetic state) and $M_{eff}(T)$ is the effective magnetization. For a thin-plate sample in the magnetic field perpendicular to its plane (θ= 90°),

$$H_r(T) = H_{r0} + 4\pi M_{eff}(T). \qquad (2)$$

Although these formulas are incompletely correct for a well-ordered ferromagnet near the Curie point, we use them for the rough description of the positions of the resonance lines.

Further we will examine the case, when external magnetic field is directed perpendicularly to the plane of the sample.

At T = 50K the ESR line is split into two lines with g factors of 1.95 and 2.06. (fig 1 and 2) and shifts of the g factors are $\Delta g = -0.05$ and +0.06, respectively, The intensity and temperature behaviour of the split lines are different. The ratio of the less intensive line (close circles in fig. 2) to the more intensive line (open squares) was 1:4. or 1.3 .

The number of works [3, 7-9], directly points that magnetic and transport properties of EuB_6 can be explained by the existence of magnetic polarons. According to the magnetic phase diagram obtained in [7] for EuB_6, the region of polaron states begins at temperatures 40–30 K. Their appearance is accompanied by the self-localization of the charge carriers; this self-localization is a cause for the increase in the resistivity of europium hexaboride observed at 15 -40 K [11]. According to numerous investigations, no other features (e.g., structure changes, etc.) are observed in EuB_6 in this

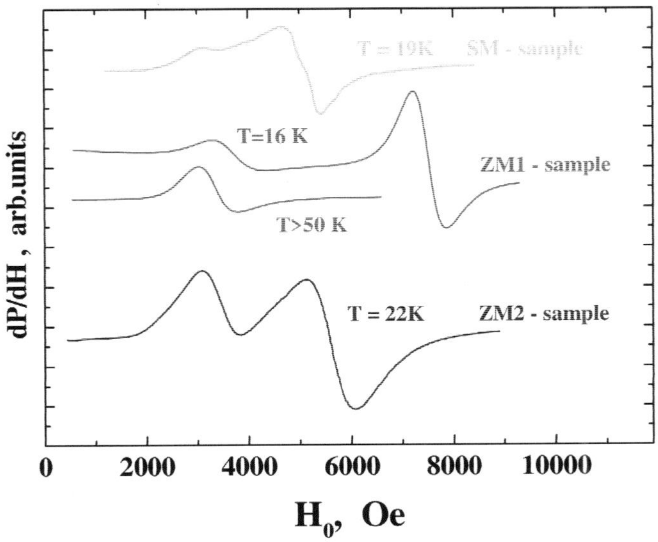

Figure 1. ESR spectra of europium hexaboride at different temperatures and the perpendicular orientation of the single crystal plane to the magnetic field. SM is the single crystal grown by flux melting, ZM1 and ZM2 are the single crystals grown by zone melting.

temperature range. Therefore, the splitting of the EPR line into two lines is due to only the formation of the polaron. Since the temperature dependences of these two lines do not correlate with each other, we can conclude that there are two different regions of polaron states. Since g = 2.00, the more and less intense lines have the negative and positive shifts of the g factor, Δ g = –0.05 and +0.06, respectively. We believe that the sign and value of the shift of the g factor from its value g = 2 for the free electron is obviously determined by the presence and character of the charge carriers interacting with a localized magnetic moment of the Eu^{2+} ion. In order for the shift Δg to be negative or positive, it is necessary to increase or decrease the local magnetic field acting on the Eu^{2+} ion, respectively. To this end, the magnetic moment of the localized spin must be parallel or antiparallel to the spins of the charge carriers, respectively. Kunes and Pickett [10] theoretically described the physical properties of EuB_6 in the model of the two-band: the antiferromagnetic (Kondo) coupling of the valence electrons with the local 4f magnetic moments and the ferromagnetic one (anti-Kondo) of the conduction electrons with the local 4f magnetic moments.

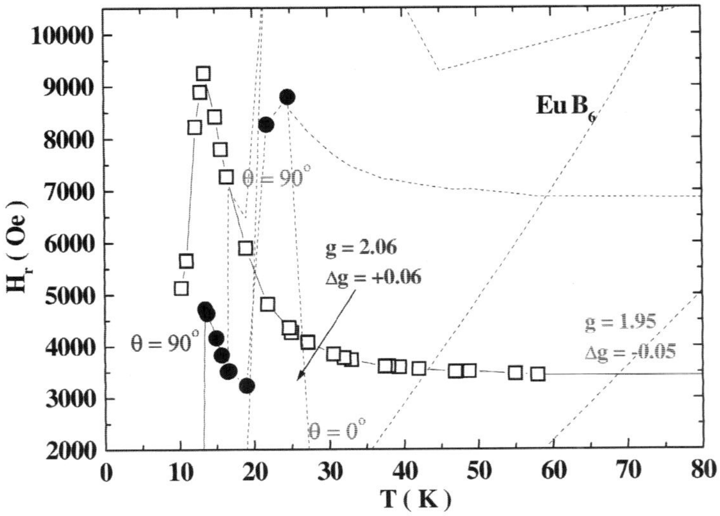

Figure 2. The temperature dependences of the resonance field H_r of the ESR-lines at the perpendicular orientation of the "zone melt" EuB_6 single crystal plate to the magnetic field. The magnetic resonance field H_r of the ESR signal for the parallel orientation at T= 22K.

Thus, our data are experimental evidence that two types of coupling of free charge carriers–with Kondo and anti- Kondo couplings are formed in EuB_6. With a temperature decrease ferromagnetic ordering is observed in both (Kondo and anti-Kondo) phases, but ordering temperature T_c and value of the magnetization M of these phases are different. It is clear that Kondo phase has the lower values T_c and M than anti- Kondo phase. It is known that the value of the shift of ESR line into the high fields is proportional to the magnetization of the system.

At temperatures below 40 K the most intense line of EuB_6 shifts towards higher fields (see Figs. 1, 2). reaching 10 kG at a temperature of 15K. This is the first phase transition. It is connected with the ferromagnetic exchange of the magnetic polarons in the anti-Kondo phase. At T = 13.5T this ESR line is shifted towards lower fields and the second line is shifted to the high field on towards the first. This is the region of the second phase transition. It is connected with exchange interaction between two spin subsystems (Kondo and anti - Kondo phases). The meeting of two ESR lines in the lower fields indicates that the exchange between two subsystems has an antiferromagnetic nature.
The sharp decrease in the resistivity observed in this temperature range [11] can be attributed to the delocalization of the charge carriers.

3.2. ESR in $EuB_{5.98}C_{0.02}$ and $EuB_{5.93}C_{0.07}$

In $EuB_{5.98}C_{0.02}$ we observed also the magnetic phase separation at T = 45-50K as in EuB_6. Then the more intensive ESR line is shifted into the large fields, reached only 4 kG, and the second line is shifted to the lower fields, i.e. the magnetization of this sample is lower then in EuB_6. Carbon ion substitutes the boron ion and brings one electron into the conduction band of EuB_6. The transport and magnetization investigations showed that carbon, doping in EuB_6,.increases conductivity and decreases magnetization of the samples [11 -13]. Even insignificant doping with carbon changes picture.

More stronger changes are observed into $EuB_{5.93}C_{0.07}$. At high temperatures the position and the shape of the ESR line in $EuB_{5.93}C_{0.07}$ were not differed practically from the position and the shape in EuB_6

At T = 135K the ESR line is split into two lines as well as in the case of a pure EuB_6 (fig 3, 4). One of the lines has the negative g-shift and another one has the positive g-shift. We attribute this phenomenon to the formation of the magnetic polarons of two types: Kondo- and anti-Kondo coupling of europium localized magnetic moments with the charge carriers. However the formation of

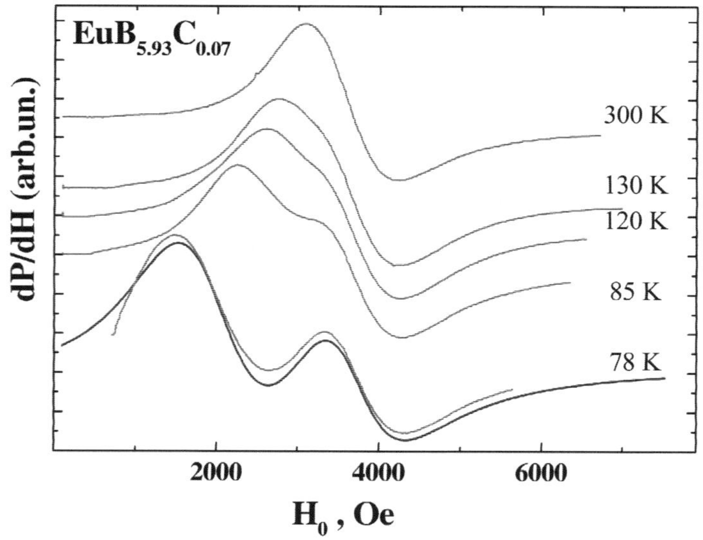

Figure 3. ESR spectra of the $EuB_{5.93}C_{0.07}$ at the different temperatures

magnetic polarons occurs at the more higher temperature then in EuB_6 (T=40K). A tetravalent carbon substitutes a trivalent boron and acts as a single electron donor. The higher density of conduction electrons in $EuB_{5.93}C_{0.07}$ increases the RKKY exchange interaction between localize europium moments and therefore the magnetic structure is generating more rapidly. We observed another than in EuB_6 the temperature dependence of the two split lines. Now more intensive line belongs to Kondo-phase (with the antiparallel i.e.anti-ferromagnetic connection of valence electrons with the f- electrons of europium). It is shifted into the lower field. The less intensive line (anti-Kondo-phase) had a slight displacement to the high fields. It is clear that doping of the europium hexaboride by carbon reduces magnetization of the both phases. This means that the exchange interaction between additional conduction electrons carried by carbon and each of Kondo or anti-Kondo phases has antiferromagnetic nature. However, we see signal ESR i.e., the compensation for magnetic moments of both phases incomplete. It is possible that here as in $EuB_{5.95}C_{0.05}$ [13 -15] the magnetic structure can be presented as helimagnetic domain. Actually, Kondo phase, described by most intensive ESR line, consists of the complex ferro- and antiferromagnetic exchanges. $EuB_{5.93}C_{0.07}$ was investigated by ESR up to 74K. We saw a good ESR signals from Eu^2 at 76K, but already at 74K the ESR signals from both phases disappeared. This fact can be interpreted, in our opinion, by the antiferromagnetic exchange of two phases. System passed into the antiferromagnetic state with the complete compensation for magnetic moments. The reason of this process is probably a delocalization of the charge carriers, i.e. a destruction of the polarons of both phases.

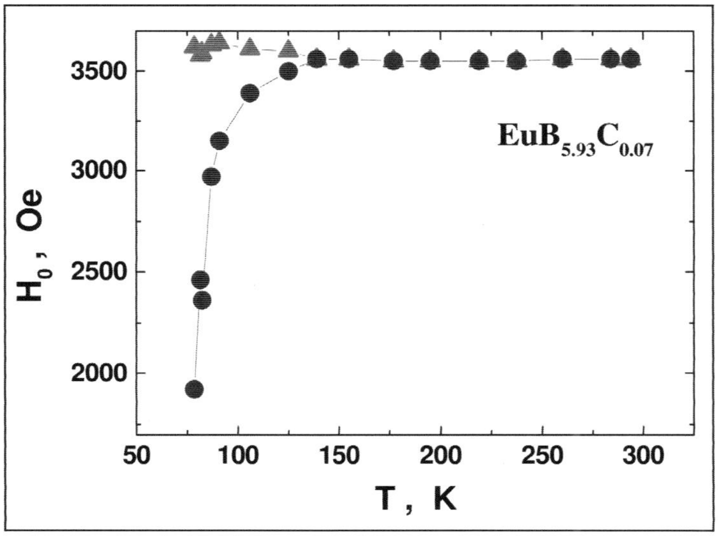

Figure 4. The temperature dependences of the resonance field of the ESR-lines at the perpendicular orientation of the $EuB_{5.93}C_{0.07}$ single crystal plate to the magnetic field. Lines are drawn for eye

Our investigations showed that there are in $EuB_{5.93}C_{0.07}$ two phase transitions. The first transition is connected with the helimagnetic ordering of the most intensive Kondo- phase, the second transition is connected with the delocalization of the charge carriers of both phases and the passage of entire system in the uniform antiferromagnetic state.

One of the most interesting and requiring further studies is the fact of the very high temperatures of magnetic phase separation - 130K and the phase transition of 74K in $EuB_{5.93}C_{0.07}$.

Until now phase transitions in hexaboride of europium, clean and alloyed by different admixtures were observed only at low temperatures; in the carbonized samples, for example, at 7 -18 [11-13].

4. Conclusion

The $EuB_{6-x}C_x$,where x = 0, 0.02 and 0.07, have been investigated in the temperature range 10–300 K. A single ESR line from Eu^{2+} is observed in the room temperature. With a further decrease in the temperature, it is split into two lines: at 40-50K in EuB_6 and $EuB_{5.98}C_{0.02}$ and at 130K in $EuB_{5.93}C_{0.07}$. An analysis of the ESR spectra indicates that the magnetic phase separation of $EuB_{6-x}C_x$ are observed. It is due to the formation of the bound states of the magnetic moments of Eu^{2+} ions with the conduction electrons and valence electrons (holes). We have observed two different types of coupling between the localized europium spins and charge carriers: Kondo coupling with holes (or valent electrons) and anti-Kondo coupling with conduction electrons. The interaction of the charge carriers with localized magnetic moments in europium hexaboride is likely accompanied by the self-localization of free charge carriers and the formation of magnetic polarons of two types. It was observed 2 phase transitions in all single crystals. The first phase transition is connected with ferromagnetic ordering of the more intensive anti-Kondo phase in EuB_6 and $EuB_{5.98}C_{0.02}$ at 15-16K and the helimagnetic ordering of the more intensive Kondo phase in $EuB_{5.93}C_{0.07}$ at 76K. The second phase transition is connected with antiferromagnetic interaction between Kondo and anti- Kondo phases. In this case the delocalization of the charge carriers occurs and the system passes on a homogeneous state

Acknowledgments
The work was supported by grants from RFBR and Presidium of RAS.

References

[1] Degiorgi L, Felder E, Ott H R, Sarrao J L and Fisk Z 1997 *Phys. Rev. Lett.* **79** 5134
[2] Sulow S, Prasad I, Aronson M C, Sarrao J L, Fisk Z, Hristova D, Lacerda A H, Hundley M F, Vigliante A and Gibbs D 1998 *Phys. Rev.* B **57** 5860
[3] Sulow S, Prasad I, Aronson M C, Bogdanovich S, Sarrao J L and Fisk Z 2000 *Phys. Rev.* B **62** 11626
[4] Deisenhofer J, Braak D, Krug von Nidda H-A and Eremina R M 2005 *Phys.Rev. Lett.* **95** 257202
[5] Badrutdinov A O, Zarubezhnova E M, Talanov Yu. I. et al 2007 *JETP*, **105** 79
[6] Kasaya M, Tarascon J M, Etourneau J and Hagenmuller P 1978 *Mater. Res. Bull.*, **13** 751
[7] Snow S, Cooper S L, Young D P, Fisk Z, Comment A and Ansermet J P 2001 *Phys.Rev.* B **64** 174412
[8] Calderon M J, Wegener L.G.L and Littlewood P B 2004 *Phys.Rev.* B **70** 092408
[9] Yu U and. Min B I 2006 *Phys.Rev.* B **74** 094413
[10] Kunes J and Picket W E 2004 *Phys.Rev.* B **69** 165111
[11] Tarascon J M, Etourneau J, Dordor P, Hagenmuller P, Kasaya M and Coey J M D 1980 *J. Appl. Phys.* **51**(1) 574
[12] von Molnar S, Tarascon J.M and Etourneau J 1981 *J. Appl. Phys.* **52** 2158

Journal of Physics: Conference Series **324** (2011) 012016 doi:10.1088/1742-6596/324/1/012016

Theoretical investigation of NMR spectra in rare-earth manganites

L E Gonchar[1,2], J V Leskova[2] and A E Nikiforov[2]

[1] Ural State Railway Transport University, 620034, 66 Kolmogorov St., Ekaterinburg, Russia
[2] Ural Federal University, 620000, 51 Lenin Av., Ekaterinburg, Russia

lyudmila.gonchar@usu.ru

Abstract. The work is aimed to theoretical investigation of orbital and magnetic structure of manganites. The orbitally-dependent exchange interaction model was used in order to describe the superexchange interaction in all range of $RMnO_3$ (R=La, Pr, Nd, Tb, Dy, Ho) orthorhombic compounds. The model of nearest-neighbour and next-nearest neighbour exchange was used to describe the magnetic structures of manganites with small rare-earth ion sublattice (R=Dy, Tb, Ho). We investigate different types –*A*, *E*, non-collinear – of magnetic structures. The models of NMR spectra calculation of Mn^{3+} ion in $RMnO_3$ compounds are proposed. The theoretical study of the magnetic structure could distinguish The NMR-spectra of antiferromagnetic (*A* or *E*) or strongly non-collinear ordering type.

1. Introduction

The investigation of manganite crystals with general formula $RMnO_3$ (R is a trivalent rare-earth ion) and orthorhombic crystal structure (denoted as *Pbnm* or *Pnma* space symmetry group) was very intensive last two decades because of colossal magnetoresistance effect taking place in alkaline-earth doped crystals. These compounds are the striking examples of a strong correlation between crystal, charge, orbital, and magnetic subsystems. The modern studies pay now their attention to one's more intriguing feature of orthorhombic manganites, i.e. multiferroicity [1, 2].

The reason of the multiferroical behavior of orthorhombic manganites with small rare-earth ions sublattice is proposed to be strongly canted magnetic structure [2]. The type of magnetic structure could be non-commensurate sinusoidal, *bc*-cycloidal and *ac*-cycloidal (in *Pbnm*-notation) with different angles between magnetic moments [3].

Thus, the observed types of magnetic structures in orthorhombic rare-earth manganites are antiferromegnetic *A*-type (R=La, Pr, Nd) [4] non-commensurate or cycloidal (R=Dy, Tb)[2, 3] and *E*-type (R=Ho, Er) [1]. The mechanism of forming *A*-type magnetic structure is the orbital structure [5]. The causes of cycloidal, sinusoidal and *E*-type structures are named to be the Dzyaloshinsky-Moria antisymmetric exchange interaction and second-neighbor [3] or third neighbor superexchange interaction [6] in addition to orbitally-dependent nearest-neighbor superexchange.

This paper is devoted to the theoretical investigation of the microscopic origin of interrelation between crystal, orbital and magnetic structures over all the rare-earth ions range in manganite compounds. We predict the nuclear magnetic resonance spectra of different magnetic structures.

2. Crystal structure and superexchange interaction model of orthorhombic rare -earth manganites

The crystal structure of orthorhombic manganites is described by a space group with two denominations – *Pnma* or *Pbnm*.

The distortions are of two kinds: the rotational distortions of oxygen octahedral and lanthanum ions shift like in orthoferrite crystal and the Jahn-Teller distortions of oxygen octahedra of *e*-type [7]. The strong electron-lattice interaction causes the presence of *e*-type distortions due to degenerated trivalent manganese ion ground state and helps the orbital structure to be established.

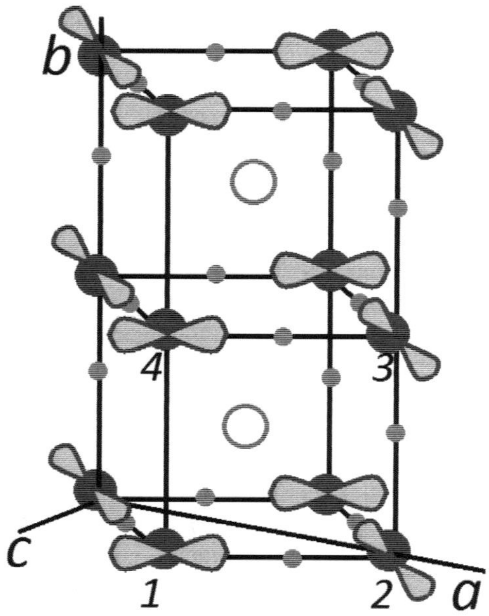

Figure 1. Crystal and orbital structures. Lanthanum ions are omitted. Manganese ions are drawn as *e*-orbitals. Numbers enumerate orbital and magnetic (*A*-type) sublattices. Axes are given in *Pnma* orientation.

The wave function of the ground state of each manganese ion in distorted perovskite crystal is [7]:

$$\psi_n = \sin(\phi_n / 2)\varphi_{n\theta} + \cos(\phi_n / 2)\varphi_{n\varepsilon}, \tag{1}$$

where $\varphi_{n\theta}$ and $\varphi_{n\varepsilon}$ – are the eigenfunctions of degenerated ^{5}E term. The values of ϕ_n could be derived from experimental data, using parameters of lattice distortions [8] within the framework of strong electron-lattice coupling. The orbital structure is described by correlation of orbital angles:

$$\phi_1=-\phi_2=-\phi_3=\phi_4=\phi, \tag{2}$$

where ϕ characterize the orbital state of Mn^{3+} sublattice according to equation (2).

Within the framework of the strong electron-lattice correlation [7], the parameter of orbital function's mixing could be calculated as function of symmetrized crystal lattice distortion

$$\cos\phi_n = \frac{Q_{n\theta}}{\sqrt{Q_{n\theta}^2 + Q_{n\varepsilon}^2}}; \cos\phi_n = \frac{Q_{n\varepsilon}}{\sqrt{Q_{n\theta}^2 + Q_{n\varepsilon}^2}} \tag{3}$$

2.1. Orbitally dependent superexchange interaction

In order to predict the magnetic structure, we use following spin-Hamiltonian:

$$\hat{H} = J_b \sum_{n>m} (\mathbf{S}_n \cdot \mathbf{S}_m) + J_{ac} \sum_{n>m} (\mathbf{S}_n \cdot \mathbf{S}_m) + J_2 \sum_{n>m} (\mathbf{S}_n \cdot \mathbf{S}_m) + \hat{H}_n^{anis} \tag{4}$$

where

$$J_b = J_0 \frac{\cos^2 \varphi_b}{r_b^{10}} \left[1 + 2\alpha\cos\phi + \beta\cos^2\phi\right],$$

$$J_{ac} = J_0 \frac{\cos^2 \varphi_{ac}}{r_{ac}^{10}} \left[1 - \alpha \cos\phi + \beta\left(\cos^2 \phi - 3/4\right)\right], \tag{5}$$

the parameters of dependencies are [5] J_0=1.69×104 K·Å10, α =1.0, β =4.5, and the experimental structural parameters are φ_b, φ_{ac} (Mn–O–Mn bond angles), r_b=1.97 Å, r_{ac}=2.05 Å (mean Mn–O distances), ϕ (angle of orbital mixing). These parameters one could find in the table 1. J_2 is included into Hamiltonian by authors of papers [2, 3] second neighbour superexchange interaction.

The single-ion anisotropy of manganese ions in octahedral oxygen environment [5]:

$$\widehat{H}_n^{anis} = D_n \cdot S_{nzl}^2 + E_n \cdot \left(S_{nxl}^2 - S_{nyl}^2\right), \quad D_n = 3P\cos\phi,\ E_n = \sqrt{3}P\sin\phi \tag{6}$$

where P= –1.15 K [5], n enumerate manganese ions, l denote local axes of octahedra, nearly corresponding to cubic axes. In order to take into account the rotational distortions, we have to transform the Hamiltonian (6) into *Pnma* axes.

Table 1. Lattice parameters of rare-earth manganites. The parameters of the crystal structure are calculated using experimental data of paper [8]

R	ϕ, deg	φ, deg	ψ,deg	φ_b, deg	φ_{ac}, deg	r_b, Å	r_{ac}, Å
La	107.1	12.1	9.9	155.5	155.1	1.97	2.04
Pr	110.9	13.5	12.6	152.4	150.5	1.95	2.06
Nd	111.5	14.1	12.9	150.9	149.6	1.95	2.06
Tb	115.5	16.5	14.9	145.0	145.4	1.94	2.06
Dy	115.7	17.3	15.2	143.2	144.7	1.94	2.06
Ho	116.3	17.6	15.5	142.5	144.1	1.94	2.06
Er	116.2	17.6	15.3	142.9	143.9	1.94	2.05
Y	116.3	17.3	15.3	143.5	144.6	1.94	2.05

The nearest-neighbor superexchange parameters and Neel temperatures of the manganite compounds are placed in the table 2. This table is a result of our calculation taking into account only orbitally-dependent nearest-neighbor exchange interaction.

Table 2. Calculated exchange parameters and Neel temperature of the rare-earth manganites

R	J_b, K	J_{ac}, K	T_N, K – theory	T_N, K – exper. [2]
La	13,1	-18,9	135	140
Pr	14,0	-13,7	112	100
Nd	13,9	-12,9	107	85
Tb	14,5	-9,2	89	~40*
Dy	13,7	-9,0	86	~40*
Ho	13,7	-8,5	83	~40*
Er	14,2	-9,0	87	~40*
Y	14,4	-9,0	88	~40* [1]

* The temperature of transition from incommensurate to paramagnetic phase. The temperature of transition from *E*-type to incommensurate phase is lower [2].

As it was proposed in the papers [2, 3], the second-neighbor exchange interaction is acting in the manganites with a small rare-earth radii. That type of exchange as the paper [2] proposes arises due to

Journal of Physics: Conference Series **324** (2011) 012016 doi:10.1088/1742-6596/324/1/012016

interaction of orbital structure with strong rotational distortion. Not considering the microscopic reasons of the exchange interaction, we try to describe the magnetic structure and resonance spectra.

3. Possible magnetic structures of manganite

The magnetic structure of manganite is determined by exchange interaction and single ion anisotropy. Within the framework of the model of paper [2], the magnetic structure is a result of competing exchange interaction between nearest neighbour magnetic ions and second-neighbour magnetic ions. We have studied 8-sublattice commensurate model. (see figure 2).

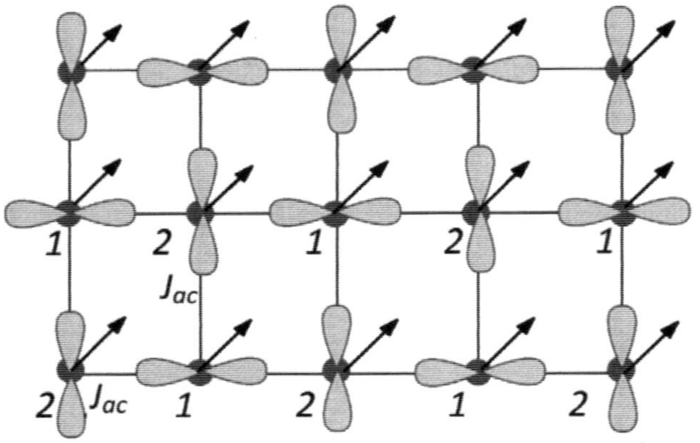

Figure 2a. One *ac*-plane of *A*-type magnetic structure (magnetic sublattices 1,2). Orbital structure of manganese ions is shown. The directions of magnetic moments of the neighbouring *ac*-plane (sublattices 3,4) are opposite. The orbitals directions are the same.

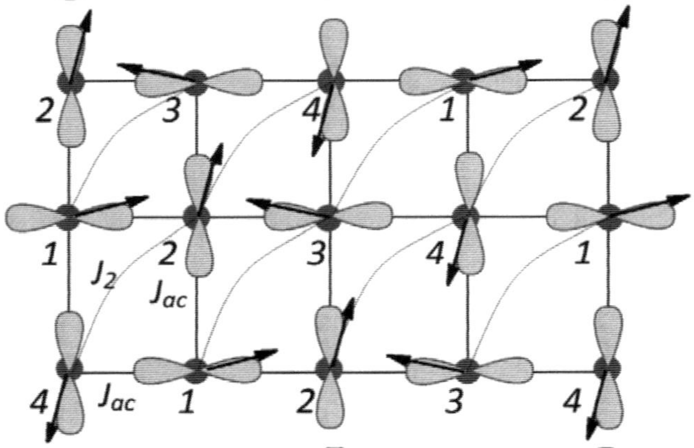

Figure 2b. One *ac*-plane of non-collinear magnetic structure (magnetic sublattices 1–4). Orbital structure of manganese ions is shown. The directions of magnetic moments of the neighbouring *ac*-plane (sublattices 5–8) are opposite. The orbitals directions are the same.

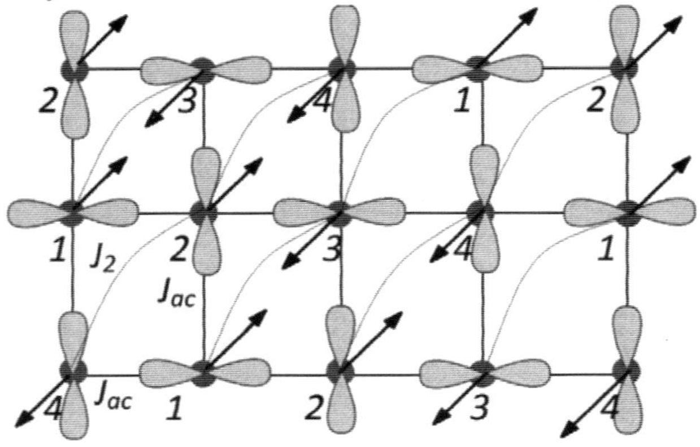

Figure 2c. One *ac*-plane of *E*-type magnetic structure (magnetic sublattices 1–4). Orbital structure of manganese ions is shown. The directions of magnetic moments of the neighbouring *ac*-plane (sublattices 5–8) are opposite. The orbitals directions are the same.

This model results two cases.

I. $J_2 < J_{ac}$: *A*-type magnetic structure. Large rare-earth ion leads to small rotation of MnO_6 – octahedra. The second-neighbour exchange interaction is small [2, 3]. The type of magnetic structure is $(A_X F_Y G_Z)$. This case takes place if R=La, Pr, Nd. The single-ion anisotropy described by Hamiltonian (6) is sufficient to mark out an easy magnetic axis of the crystal (see figure 2a).

II. $J_2 \sim J_{ac}$: strongly non-collinear structure. Small rare-earth ion leads to large rotation of MnO_6 – octahedra. The second-neighbour exchange interaction is of the same order as ferromagnetic exchange in *ac*-plane. [2, 3]. This case take place if R=Dy, Tb, Er, Ho. The single-ion anisotropy described by Hamiltonian (6) causes strongly canted structure (see figure 2b). For description of *E*-type structure neither the single-ion anisotropy (6) nor antisymmetric exchange interaction [3] are not sufficient using commensurate model. The single-ion anisotropy like DS_{xi}^2 helps to establish collinear *E*-type magnetic structure (see figure 2c).

4. Nuclear magnetic resonance model

In order to investigate the magnetic subsystem, we propose the model of NMR on ^{55}Mn ion. The spin-Hamiltonian of the model is

$$\hat{H} = \hat{H}_Q - \gamma_N \hbar \left(\mathbf{H}_{dip} + \mathbf{H}_{hf} \right) \cdot \mathbf{I}, \tag{7}$$

where hyperfine interaction

$$\mathrm{H}_{hf} = \mathrm{H}_{hf}(\phi) = A_s(\mathbf{S}\cdot\mathbf{I}) + A_{II}\left([3S_z I_z - (\mathbf{S}\cdot\mathbf{I})]\cos\phi + \sqrt{3}[S_x I_x - S_y I_y]\sin\phi\right) \tag{8}$$

is dependent upon orbital mixing angle ϕ and could be divided into two parts – anisotropic (with A_s constant) and anisotropic (with A_{II} constant);

$$\hat{H}_Q = \nu_Q \cdot \left[\left(3 \cdot I_z^2 - I(I-1)\right) + \eta \cdot \left(I_x^2 - I_y^2\right) \right] \tag{9}$$

is nuclear quadrupole interaction; and the values of the fields are approximately equal to H_{dip}~0.1 T, H_{hf}~30 T. The nuclear spin of ^{55}Mn ion is $I=5/2$, and γ_N=10.46 MHz/T – is nuclear hyromagnetic factor. Considering the magnetic ion, the magnetic hyperfine interaction is much stronger than quadrupole interaction, thus, it determines the NMR spectra completely.

The *ab initio* calculation of hyperfine interaction (HFI) was carried out within the framework of unlimited MO LCAO SCF Hartree-Fock-Ruthaan method using GAMESS programs [9].

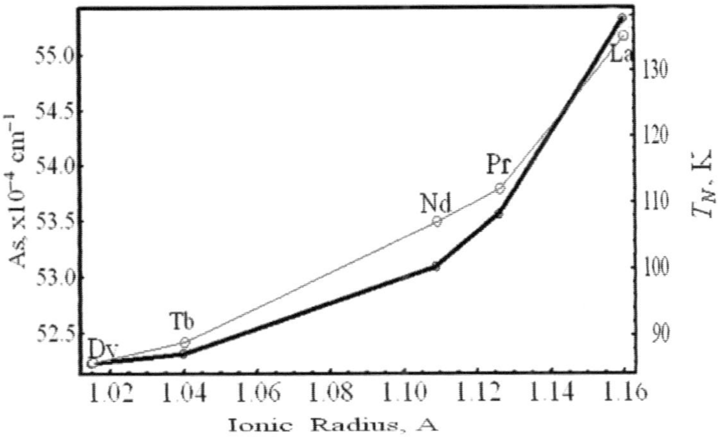

Figure 3. Dependency of magnetic isotropic HFI (solid circles) of Mn^{3+} and Neel temperature (open circles) in manganites (considered *A*-type magnetic structure)

We calculate the line position of NMR spectra of trivalent manganese ion of $RMnO_3$ compounds (R=La, Pr, Nd, Tb, Dy, Ho).

The changes of rare-earth sublattice lead to changes of Mn-O bond configuration. The rotational and shift distortion change their magnitude. The magnetic structure changes its parameters and even the type. In our work, the hyperfine parameters are calculated for different rare-earth ions sublattice. The results of calculation are presented on figure 3.

The isotropic term of HFI (7) is dependent upon the Mn-O bond length. If the ionic radius of rare-earth element becomes smaller, the bond length decreases and the covalence effect become stronger. The anisotropic part of HFI (7) is dependent on orbital and magnetic structure. After taking into account the anisotropic term of HFI, the direction of HF field on the nucleus changes and depends upon the orbital structure. In the dependence of figure 3, only *A*-type of magnetic structure is used. For small rare-earth ions, the other types of magnetic structures could be used. The results of prediction of the NMR line position are presented in the table 3.

The changes of the magnetic structure type cause the shift of the NMR line. In this work, the NMR spectrum of $HoMnO_3$ was taken as an example.

Table 3 NMR line position, MHz

La	Pr	Nd	Tb	Ho
360	336	333	314	312 (*A*) 324 (*E*) 442 (non-collinear)

The results of model for $HoMnO_3$ show that one could not reliably distinguish *A* or *E* magnetic structure using NMR method. Nevertheless, the non-collinear structure is appreciably different. The cycloidal magnetic structure, proposed in paper [3] gives three lines instead one as in presented model.

5. Summary

Thus, we have carried out the theoretical study of different rare-earth manganites. The type of rare-earth sublattice effects not only on the crystal structure but also upon orbital and magnetic orderings. The orbital dependence of superexchange interaction is very important microscopic mechanism of interplay between crystal, orbital, and magnetic structures. Nevertheless, for the small-size rare-earth ion in R^{3+} sublattice, the rotational distortions have much stronger effect on exchange interaction comparing with lanthanum manganite. These two mechanisms form *A*, *E*, or non-collinear magnetic structure of manganites. We propose the NMR method as a local probe of magnetic structure.

References

[1] Zhou J-S *et al.* 2006 *Phys. Rev. B* **74** 014422
[2] Kimura T *et al.* 2003 *Phys. Rev. B* **68** 060403(R)
[3] Mochizuki M and Furukawa N 2009 *Phys. Rev. B* **80** 134416
[4] Jirak Z *et al.* 1985 *J. Magn. Magn. Mater.* **53** 153
[5] Gontchar L E and Nikiforov A E 2002 *Phys. Rev. B* **66** 014437
[6] Soloviev I V 2011 *Phys. Rev. B* **83** 054404
[7] Nikiforov A E, Popov S E and Shashkin S Yu 1999 *Phys.Met. Met.* **87** 16
[8] Alonso J A *et al.* 2000 *Inorg. Chem.* **39** 917
[9] Schmidt M W *et al.* 1993 *Comput. Chem.* **14** 1347.

Spin relaxation in Kondo lattices

S I Belov, A S Kutuzov and B I Kochelaev

Kazan Federal University, Kremlevskaya, 18, Kazan 420008, Russian Federation

E-mail: Sergei.Belov@ksu.ru

Abstract. A model of spin relaxation in Kondo lattices is proposed to explain the presence of an electron spin resonance (ESR) signal in the heavy fermion compounds $Yb_2Rh_2Si_2$ and $Yb_2Ir_2Si_2$. Coupled equations for dynamical susceptibilities of Kondo ions and conduction electrons are derived by means of the functional derivative method. The perturbational scaling approach reveals the collective spin motion of Yb-ions with conduction electrons in the bottleneck regime. A common energy scale due to the Kondo effect regulates the temperature dependence of the different kinetic coefficients and results in a mutual cancelation of all divergent parts in a collective spin mode. The angular dependence of the ESR linewidth is shown to be in a qualitative agreement with experimental data on $Yb_2Rh_2Si_2$ and $Yb_2Ir_2Si_2$. Linewidth contributions other than the Kondo interaction are also discussed.

1. Introduction

The discovery of electron spin resonance (ESR) in the Kondo lattice $Yb_2Rh_2Si_2$ below the Kondo temperature $T_K = 25$ K [1] (T_K revealed by specific heat data) was a great surprise for the condensed matter physics community and stimulated different approaches for its explanation [2, 3, 4]. According to the picture based on the single ion Kondo effect, magnetic moments of Yb-ions are screened by conduction electrons at $T < T_K$ and the ESR linewidth is of the order $\Delta\nu = k_B T_K / 2\pi\hbar = 500$ GHz. The experimental results were completely opposite: at X-band (9.4 GHz) and $T = 0.7$ K a linewidth of 0.3 GHz was observed. The angular dependence of the ESR g-factor and linewidth reflects the local properties of Yb-ion in a crystal electric field (CEF). Similar results were obtained later for $Yb_2Ir_2Si_2$ [5]. It seems that many properties of heavy fermion systems in the non-Fermi-liquid state can be described in terms of quasi-localized f-electrons. Recently we successfully studied the static magnetic susceptibility of $Yb_2Rh_2Si_2$ and $Yb_2Ir_2Si_2$ at temperatures below T_K [6], based on entirely local properties of Yb-ions in a CEF. In another work [4] a theoretical model of spin relaxation was proposed to explain the ESR signal existence in Kondo lattice systems with heavy fermions. We showed that the collective spin motion of quasi-localized f-electrons and wide-band conduction electrons is the key ingredient for understanding this phenomenon. Our model successfully explained the temperature and magnetic field dependence of the ESR linewidth and resonance frequency in the case of the static magnetic field oriented perpendicular to the crystal symmetry axis.

This work gives a more detailed account of our approach developed in [4] and extends the earlier results to the case of an arbitrary angle between the static magnetic field and the crystal symmetry axis. In section 2 we analyze the anisotropy of the Kondo interaction in the heavy fermion systems $Yb_2Rh_2Si_2$ and $Yb_2Ir_2Si_2$ and obtain the basic Hamiltonian of our model. Section 3 presents the model of spin relaxation in a Kondo lattice. Equations of spin motion are derived up to the third order in the Kondo exchange constants. In section 4 the perturbation expansion is improved by the renormalization

Published under licence by IOP Publishing Ltd

of the Kondo couplings at low temperatures on the basis of the "Poor Man's Scaling" approach, and section 5 concludes the article. The mathematical basis of our approach is the functional derivative method of Kadanoff and Baym [7]. We outlined it briefly in the appendix.

2. Theoretical model

Our basic theoretical model includes the kinetic energy of conduction electrons, the Zeeman energy, the Kondo interaction of Yb-ions with conduction electrons, and the coupling between the Yb-ions via conduction electrons (RKKY interaction).

A free Yb^{3+} ion has a $4f^{13}$ configuration with one term ^2F. The spin-orbit interaction splits the ^2F term into two multiplets: $^2F_{7/2}$ with $J = 7/2$ and $^2F_{5/2}$ with $J = 5/2$, where J is the value of the total momentum $\mathbf{J} = \mathbf{L} + \mathbf{S}$ with $\mathbf{L}$ and $\mathbf{S}$ as the orbital and spin momentum of the ion. The excited multiplet $^2F_{5/2}$ is separated from the ground state $^2F_{7/2}$ by about 1 eV. Since this value is much larger than the energy of the crystal electric field (CEF), we consider in the following the ground multiplet only. It is reasonable to express the Zeeman energy of the i-th Yb-ion H_{ZJ} for the lowest multiplet $J = 7/2$ via the total electronic momentum of the ion. Using the Lande g-factor g_J, we have

$$H_{ZJ} = (\mathbf{L}_i + 2\mathbf{S}_i)\mathbf{B} = g_J \mathbf{J}_i \mathbf{B}, \tag{1}$$

where $\mathbf{B}$ is the external magnetic field multiplied by the Bohr magneton.

The Kondo exchange coupling of the rare earth ion with the conduction electrons occurs due to the hybridization of their wave functions at the ion site. The exchange integral can be written in the form (see, e.g. [8])

$$A(\mathbf{k}, \mathbf{k}') = \sum_i \int \psi^*(\mathbf{r}', \mathbf{k}) \Psi_{4f}(\mathbf{r}_1, ..., \mathbf{r}_i, ..., \mathbf{r}_n) \frac{e^2}{|\mathbf{r}_i - \mathbf{r}'|} \psi(\mathbf{r}_i, \mathbf{k}') \Psi_{4f}(\mathbf{r}_1, ..., \mathbf{r}', ..., \mathbf{r}_n) d\mathbf{r}' d\mathbf{r}_1 ... d\mathbf{r}_n . \tag{2}$$

Here $\psi(\mathbf{r}, \mathbf{k})$ is the Bloch wave function of the conduction electrons. The wave function of the $4f$-electrons Ψ_{4f} is represented by the determinant constructed from the one-electron wave functions of the type $R_{4f}(r_i) Y_3^m(\mathbf{r}_i)$. Expanding the Bloch functions and $|\mathbf{r}_i - \mathbf{r}'|^{-1}$ in spherical harmonics, one can obtain $A(\mathbf{k}, \mathbf{k}')$ as an expansion in multipoles. As a matter of fact the small parameter of this expansion is the value $k_F \langle r_{4f} \rangle \ll 1$, the product of the wave vector of the conduction electron at the Fermi surface and the average radius of the $4f$-electron. The Kondo interaction at the i-th site corresponding to the zero order of this expansion is isotropic and can be written in the form [9, 10]

$$H_{KJ} = A_0 \mathbf{S}_i \boldsymbol{\sigma}_i = A_0 (g_J - 1) \mathbf{J}_i \boldsymbol{\sigma}_i , \tag{3}$$

where $\boldsymbol{\sigma}_i$ is the spin density of the conduction electrons at the ion site. The next terms of the expansion in multipoles are less and anisotropic (detailed calculation of them can be found in [11, 12] and especially in [13]). In the following we neglect these terms, since the CEF gives a very strong anisotropy for the zero order of the Kondo exchange interaction (3).

The tetragonal CEF splits the ground multiplet into four Kramers doublets, each one described by the wave functions of the type $\psi_\pm = \sum_M c_{\pm M} |\pm M\rangle$. Within the every Kramers doublet the total Zeeman energy can be represented by the effective spin Hamiltonian with an effective spin $S = 1/2$:

$$H_{ZS} = \sum_i \left[g_\perp \left(S_i^x B_i^x + S_i^y B_i^y \right) + g_\parallel S_i^z B_i^z \right], \tag{4}$$

where $g_\perp = g_J \langle \psi_+ | J_x + iJ_y | \psi_- \rangle$ and $g_\parallel = 2g_J \langle \psi_+ | J_z | \psi_+ \rangle$ (details see in [6, 14, 15]).

According to neutron scattering experiments [16, 17] the excited energy levels are located at 17, 25, 43 meV (197, 290, 499 K) in the case $YbRh_2Si_2$ and at 18, 25, 36 meV (209, 290, 418 K) for $YbIr_2Si_2$. It means that the physics of low energy spin excitations at temperatures $T \ll 200\,\text{K}$ can be described by the ground Kramers doublet. In the following we shall relate the Zeeman Hamiltonian (4) to the ground Kramers doublet; in other words, we have projected to this state the starting Zeeman energy with the isotropic Lande g-factor.

It is evident that the projection of the isotropic Kondo interaction (3) involves the same matrix elements of the total momentum **J** as in the case of the Zeeman energy. Hence, after projection onto the ground Kramers state the total Kondo interaction can be expressed via the g-factors $g_\perp$, $g_\parallel$ given above:

$$H_{s\sigma} = \sum_i \left[J_\perp \left(S_i^x \sigma_i^x + S_i^y \sigma_i^y \right) + J_\parallel S_i^z \sigma_i^z \right] \tag{5}$$

with $J_\parallel = J g_\parallel$, $J_\perp = J g_\perp$, $J = A_0 \dfrac{g_J - 1}{g_J}$. As a matter of fact all these results are simple consequences of the well known the Wigner-Eckart theorem.

The same arguments can be used to reveal the anisotropy of the RKKY interaction between the Kondo ions. Although this interaction appears in the second order of the Kondo interaction (3) it is convenient to consider it independently. The result is given, in particular, in [8]. Starting with the isotropic exchange Hamiltonian for two Kondo ions $H_{ex}^{ij} = I_{ij}^{RKKY} \mathbf{S}_i \mathbf{S}_j = I_{ij}^{RKKY} (g_J - 1)^2 \mathbf{J}_i \mathbf{J}_j$ the authors of [8] arrive to the effective anisotropic interaction

$$H_{RKKY} = \sum_{ij} I_{ij} \left[g_\perp^2 \left(S_i^x S_j^x + S_i^y S_j^y \right) + g_\parallel^2 S_i^z S_j^z \right], \tag{6}$$

where $I_{ij} = I_{ij}^{RKKY} \left(\dfrac{g_J - 1}{g_J} \right)^2$. It is evident that the same procedure was used as in deriving the equation (5).

The kinetic energy of the conduction electrons and their Zeeman energy can be written as

$$H_c = \sum_{ij\lambda} t_{ij} c_{i\lambda}^+ c_{j\lambda} - \mu \sum_{i\lambda} c_{i\lambda}^+ c_{i\lambda}, \tag{7}$$

$$H_{z\sigma} = g_\sigma \sum_i \mathbf{B} \boldsymbol{\sigma}_i . \tag{8}$$

Here $\lambda = \pm 1$ labels the orientation of the conduction electron spin, μ is the chemical potential, and g_σ denotes the g-factor of conduction electrons. The conduction electron density is expressed in terms of the creation and annihilation operators

$$\boldsymbol{\sigma}_i = \sum_{\lambda\lambda'} \mathbf{s}_{\lambda\lambda'} c_{i\lambda}^+ c_{i\lambda'}, \tag{9}$$

where $\mathbf{s}_{\lambda\lambda'}$ are the matrix elements of the spin operators $s = 1/2$.

Collecting all terms together we obtain an effective Hamiltonian

$$H = H_c + H_Z + H_{S\sigma} + H_{RKKY} . \tag{10}$$

Here H_c, $H_{s\sigma}$, H_{RKKY} are defined by (7), (5), (6), respectively, H_Z is the total Zeeman energy of conduction electrons and Yb-ions:

$$H_Z = H_{ZS} + H_{Z\sigma} = B \sum_i \left[g_\sigma (\sin\theta\, \sigma_i^y + \cos\theta\, \sigma_i^z) + g_\perp \sin\theta\, S_i^y + g_\parallel \cos\theta\, S_i^z \right], \tag{11}$$

where θ is the angle between the static magnetic field and the crystal symmetry axis.

The next step is to diagonalize, for convenience of further calculations, the Zeeman part of the effective Hamiltonian. After an unitary transformation to a new quantization axis the Hamiltonian (10) takes its final form $H = H_0 + H_{S\sigma} + H_{RKKY}$ with

$$H_0 = \sum_{ij\lambda} t_{ij} c^{+}_{i\lambda} c_{j\lambda} + \sum_{i\lambda} \varepsilon^{\lambda} c^{+}_{i\lambda} c_{i\lambda} + g_s B \sum_i S_i^z , \tag{12}$$

$$H_{s\sigma} = \sum_{i,\alpha\beta} J_{\alpha\beta} \sigma_i^{\alpha} S_i^{\beta} , \tag{13}$$

$$H_{RKKY} = \sum_{ij,\alpha\beta} I_{ij}^{\alpha\beta} S_i^{\alpha} S_j^{\beta} . \tag{14}$$

Here $\varepsilon^{\lambda} = -\mu + \lambda g_{\sigma} B / 2$, $\alpha, \beta = x, y, z$,

$$g_s = \sqrt{g_{\|}^2 \cos^2\theta + g_{\perp}^2 \sin^2\theta} , \tag{15}$$

$J_{\alpha\beta}$, $I_{ij}^{\alpha\beta}$ represent Kondo- and RKKY couplings after rotation of the quantization axis:

$$\begin{aligned} &J_{xx} = J g_{\perp},\ J_{yy} = J g_{\|} g_{\perp} / g_s,\ J_{zz} = J g_s, \\ &J_{yz} = J \frac{g_{\perp}^2 - g_{\|}^2}{g_s} \sin\theta\cos\theta, \\ &J_{zy} = J_{xy} = J_{yx} = J_{xz} = J_{zx} = 0, \end{aligned} \tag{16}$$

$$\begin{aligned} &I^{xx} = I g_{\perp}^2,\ I^{yy} = I \frac{g_{\|}^2 g_{\perp}^2}{g_s^2},\ I^{zz} = I\left(g_{\|}^2 + g_{\perp}^2 - \frac{g_{\|}^2 g_{\perp}^2}{g_s^2} \right), \\ &I^{yz} = I^{zy} = I g_{\|} g_{\perp} \frac{g_{\perp}^2 - g_{\|}^2}{g_s} \sin\theta\cos\theta, \\ &I^{xy} = I^{yx} = I^{xz} = I^{zx} = 0. \end{aligned} \tag{17}$$

The angle ψ between the new quantization axis and the crystal symmetry axis is the one given by the relation

$$\cot\psi = \frac{g_{\|}}{g_{\perp}} \cot\theta . \tag{18}$$

3. Spin relaxation

The ESR response is given by the total transverse dynamical susceptibility

$$\chi(\omega) = \sum_{\alpha\beta} \chi_{\alpha\beta}(\omega); \quad \alpha, \beta = s, \sigma \tag{19}$$

with partial susceptibilities $\chi_{\alpha\beta}(\omega)$:

$$\begin{aligned} &\chi_{ss} = -g_{\perp}^2 \langle\langle S^- \mid S^+ \rangle\rangle, \qquad \chi_{s\sigma} = -g_{\perp} g_{\sigma} \langle\langle S^- \mid \sigma^+ \rangle\rangle, \\ &\chi_{\sigma s} = -g_{\perp} g_{\sigma} \langle\langle \sigma^- \mid S^+ \rangle\rangle, \quad \chi_{\sigma\sigma} = -g_{\sigma}^2 \langle\langle \sigma^- \mid \sigma^+ \rangle\rangle. \end{aligned} \tag{20}$$

Here $\langle\langle A \mid B \rangle\rangle$ is the Fourier transform of a retarded Green function

$$\langle\langle A \mid B \rangle\rangle = -i \int_0^{\infty} dt \exp(i\omega t) \langle [A(t), B] \rangle , \tag{21}$$

S, **σ** are the total spin operators of Yb-ions and conduction electrons, respectively, $S^{\pm} = (S^x \pm i S^y)/\sqrt{2}$, $\sigma^{\pm} = (\sigma^x \pm i\sigma^y)/\sqrt{2}$, $\langle \ldots \rangle$ means the statistical average at temperature *T*. The definition (20), (21) imply the symmetry relation

$$\chi^{*}_{\alpha\beta}(\omega) = \chi_{\alpha\beta}(\omega^{*}) . \tag{22}$$

$\chi_{\alpha\beta}$ can be represented as the analytical continuation of the temperature Green functions A, B, D (see appendix, (A.8)) onto the real axis:

$$\begin{aligned}
\chi_{ss}(\omega+i0) &= -g_\perp^2 D^{-+}(\mathbf{k}=0, i\omega_n \to \omega+i0),\\
\chi_{\sigma\sigma}(\omega+i0) &= -g_\sigma^2 A^{-+}(\mathbf{k}=0, i\omega_n \to \omega+i0),\\
\chi_{\sigma s}(\omega+i0) &= -g_\sigma g_\perp B^{-+}(\mathbf{k}=0, i\omega_n \to \omega+i0),\\
\chi_{s\sigma}(\omega+i0) &= \chi_{\sigma s}^*(\omega-i0).
\end{aligned} \tag{23}$$

The definitions, main properties and calculation technique for the temperature Green functions are given in appendix.

As a first step we find $\chi_{\alpha\beta}$ for the high temperature region $T > B$ up to the third order in the Kondo coupling. The RKKY interaction being treated in molecular field approximation only. The result is convenient to be presented as a matrix equation

$$\begin{pmatrix} a_{ss} & a_{s\sigma} \\ a_{\sigma s} & a_{\sigma\sigma} \end{pmatrix} \begin{pmatrix} \chi_{ss} & \chi_{s\sigma} \\ \chi_{\sigma s} & \chi_{\sigma\sigma} \end{pmatrix} = \begin{pmatrix} g_\perp^2 \langle S^z \rangle & 0 \\ 0 & g_\sigma^2 \langle \sigma^z \rangle \end{pmatrix} \tag{24}$$

with

$$\begin{aligned}
a_{ss} &= \omega - \omega_s + \Sigma_{ss}, & a_{s\sigma} &= g_\perp^2 \langle S^z \rangle \lambda - \frac{g_\perp}{g_\sigma} \Sigma_{s\sigma},\\
a_{\sigma s} &= g_\sigma^2 \langle \sigma^z \rangle \lambda - \frac{g_\sigma}{g_\perp} \Sigma_{\sigma s}, & a_{\sigma\sigma} &= \omega - \omega_\sigma + \Sigma_{\sigma\sigma}.
\end{aligned} \tag{25}$$

Here $\lambda = J(g_\parallel + g_\perp)/2g_s g_\sigma$, $\langle S^z \rangle = -g_s B/4(T+T_W) + O(J)$, $\langle \sigma^z \rangle = -g_\sigma \rho B/2 + O(J)$, ω_s, ω_σ are resonance frequencies of Yb-ions and conduction electrons, respectively, containing the first order Knight shifts due to the Kondo- and RKKY interaction:

$$\begin{aligned}
\omega_s &= g_s B + J g_s \langle \sigma^z \rangle + 4T_W \langle S^z \rangle,\\
\omega_\sigma &= g_\sigma B + J g_s \langle S^z \rangle.
\end{aligned} \tag{26}$$

T_W is the Weiss temperature which originates from the RKKY interaction in a molecular field approximation:

$$T_W = \frac{1}{4}\left(g_\parallel^2 + g_\perp^2 - \frac{g_\parallel^2 g_\perp^2}{g_s^2} \right) \sum_i I_{ij}, \tag{27}$$

ρ is the conduction electron density at the Fermi surface.

The coefficients $\Sigma_{\alpha\beta}$ describe the spin kinetics in the system of Yb-ions and conduction electrons. Their imaginary parts give the relaxation rates, their real parts give the corresponding resonance frequencies. We calculate $\Gamma_{\alpha\beta} \equiv \mathrm{Im}(\Sigma_{\alpha\beta})$ up to the third order in the Kondo interaction at high temperatures $T > B$:

$$\begin{aligned}
\Gamma_{ss}(\omega) &= \frac{\pi}{2}\frac{\omega}{\omega_s}(\rho J)^2 T \left[\frac{3}{2} g_\perp^2 + g_\parallel^2 \left(1 - \frac{g_\perp^2}{2g_s^2} \right) - 4\rho J g_\parallel g_\perp^2 \ln(T/W) \right],\\
\Gamma_{\sigma\sigma}(\omega) &= \frac{\pi}{4}\frac{\omega}{\omega_s}\frac{g_\sigma}{g_s} \rho J^2 \frac{T}{T+T_W} \left[g_\perp^2 + \frac{g_\parallel^2 + g_s^2}{2} - 4\rho J g_\parallel g_\perp^2 \ln(T/W) \right],
\end{aligned} \tag{28}$$

$$\Gamma_{\sigma s}(\omega)=\frac{\pi}{2}\frac{\omega}{\omega_s}(\rho J)^2 g_\perp(g_\parallel+g_s)T\left[1-\frac{g_\perp^2+g_\parallel g_s}{g_s}\rho J\ln(T/W)\right],$$

$$\Gamma_{s\sigma}(\omega)=\Gamma_{\sigma s}(\omega)\frac{g_s}{2g_\sigma\rho(T+T_W)}.$$

Here ω is the frequency of the alternating transverse magnetic field, W is the bandwidth of the conduction electrons and g_s is defined by (15).

The kinetic coefficient $\Gamma_{ss}(\omega=\omega_s)$ represents the relaxation rate of the transverse magnetic moment of Yb-ions toward conduction electrons which remain in a thermodynamical equilibrium state (Korringa relaxation). For an isotropic system with $g_\parallel=g_\perp=g$ we have the well known result in the second order in J: $\Gamma_{ss}=\pi(\rho Jg)^2T$. The third order reveals logarithmic Kondo anomalies which can be important at low temperatures. Similarly, substituting $\omega=\omega_\sigma$ into $\Gamma_{\sigma\sigma}(\omega)$ we obtain the Overhauser relaxation rate (conduction electrons relax toward the Yb spin system, being in the equilibrium with the thermal bath). Note that the strong anisotropy of the Kondo interaction breaks the detailed balance relation for $\Gamma_{ss}(\omega_s)$, $\Gamma_{\sigma\sigma}(\omega_\sigma)$ except for the cases of parallel ($g_s=g_\parallel$) and perpendicular ($g_s=g_\perp$) orientation of the static magnetic field to the crystal symmetry axis.

However, an equilibrium state approximation for the conduction electron spin system is not valid to study the ESR response of the samples with a high concentration of Kondo ions [18] which is especially the case for a Kondo lattice. Instead, one has to treat the coupled kinetic equations (24) for both magnetic moments of Kondo ions and conduction electrons. Hence two additional kinetic coefficients $\Gamma_{s\sigma}$, $\Gamma_{\sigma s}$ couple the kinetic equations of motion of the transverse magnetizations of localized moments and conduction electrons. Besides, for a correct analysis of a stationary solution one has to take into account the spin relaxation of Kondo ions and conduction electrons spin toward the thermal bath ("lattice"). Correspondingly, the kinetic coefficients Σ_{ss}, $\Sigma_{\sigma\sigma}$ in equations (24) should be replaced with $\Sigma_{ss}+\Sigma_{sL}$, $\Sigma_{\sigma\sigma}+\Sigma_{\sigma L}$, respectively. To the second order in J our results coincides with equations derived by means of the non-equilibrium statistical operator [19]. To study the ESR response of the total system we have to find solutions of the system (24), where the coefficients $a_{\alpha\beta}$ include the spin lattice relaxation terms. The poles of the total dynamical susceptibility are determined by the condition $a_{ss}a_{\sigma\sigma}-a_{s\sigma}a_{\sigma s}=0$. The result are two complex roots: their real parts represent resonance frequencies, their imaginary parts represent the corresponding relaxation rates. We are interested in a solution close to the Kondo ion resonance frequency ω_s.

The coupling between the two systems is especially important if the relaxation rate of the conduction electrons toward the Kondo ions is much faster than to the lattice and the resonance frequencies are close to one another ("bottleneck" regime):

$$\Gamma_{\sigma\sigma}\gg\Gamma_{\sigma L},\ |\omega_s-\omega_\sigma|. \tag{29}$$

In the case of an isotropic system and equal Larmor frequencies ($g_\parallel=g_\perp=g_\sigma$) the solutions of (24) give the well known result [18, 20]: the ESR linewidth in the bottleneck regime is greatly narrowed due to conservation of the total magnetic moment (its operator commutes with the isotropic Kondo interaction and the latter disappears from the effective relaxation rate). In the opposite case of a strongly anisotropic Kondo interaction our results in the second order do not show any sufficient narrowing of the ESR linewidth in the bottleneck regime. The Kondo effect could change the situation but in this case the third order logarithmic terms in (28) become large and we need to improve the standard perturbation technique.

4. The renormalized relaxation rate

To develop a satisfactory theory in the low temperature regime one has to remove the logarithmic divergences by summing somehow the higher order terms in the perturbation expansion. In 1965 Abrikosov [21] carried out a summation of the leading logarithmic terms for the resistivity applying the Feynman diagrams technique to the s-d model. Later Anderson [22] proposed another method known as "Poor Man's Scaling" that allows one to extend the lowest order perturbation results, and effectively sums the leading order logarithmic terms. We use here the scaling approach because a considerable number of Kondo couplings makes it difficult to work with the diagrams technique.

The main idea of the "Poor Man's Scaling" approach is to take into account the effect of the high energy excitations on the low energy physics by a renormalization of coupling constants. We divide the conduction electrons band into the low and high energy states

$$0<|\varepsilon_{\mathbf{k}}|<W', \quad W'<|\varepsilon_{\mathbf{k}}|<W, \tag{30}$$

where W, W' are the initial and the running bandwidth, respectively. The projection of the original Kondo interaction $H_{s\sigma}$ (13) on to the low energy states yields a new Hamiltonian $H'_{s\sigma}$ with renormalized couplings $J'_{\alpha\beta}$. It is convenient to introduce the dimensionless parameters $U_{\alpha\beta}=(\rho J_{\alpha\beta})'$, with the symbol $(...)'$ marking all new quantities. The evolution of the renormalized parameters $U_{\alpha\beta}$ with W' is described with a set of equations derived to the second order in $U_{\alpha\beta}$:

$$\frac{dU_{\alpha\alpha'}}{dt}=-\frac{1}{2}\sum_{\beta\gamma\beta'\gamma'}\varepsilon_{\alpha\beta\gamma}\varepsilon_{\alpha'\beta'\gamma'}U_{\beta\beta'}U_{\gamma\gamma'}. \tag{31}$$

Here $\varepsilon_{\alpha\beta\gamma}$ denotes an antisymmetric third-rank tensor, $t=\ln(W'/W)$; α, β, γ run over x, y, z. The initial conditions for the system (31) are the "bare" parameters of the Kondo interaction,

$$U_{\alpha\beta}(0)=\rho J_{\alpha\beta} \tag{32}$$

with $J_{\alpha\beta}$ defined by (16).

The solution of equations (31) can be written in the form

$$\begin{aligned}
U_{xx}&=U_{\perp},\\
U_{yy}&=U_{\perp}\cos\theta\cos\psi+U_{\parallel}\sin\theta\sin\psi,\\
U_{zz}&=U_{\perp}\sin\theta\sin\psi+U_{\parallel}\cos\theta\cos\psi,\\
U_{yz}&=U_{\perp}\cos\theta\sin\psi-U_{\parallel}\sin\theta\cos\psi,\\
U_{zy}&=U_{\perp}\sin\theta\cos\psi-U_{\parallel}\sin\theta\cos\psi,\\
U_{xy}&=U_{yx}=U_{xz}=U_{zx}=0,
\end{aligned} \tag{33}$$

$$U_{\parallel}=\bar{J}\cot\varphi, \quad U_{\perp}=\bar{J}/\sin\varphi, \tag{34}$$

where ψ is defined by (18); $U_{\parallel}$, $U_{\perp}$ will represent the renormalized parameters of the Kondo interaction if the quantization axis is not rotated, $\bar{J}=\rho J\sqrt{g_{\perp}^2-g_{\parallel}^2}$, $\varphi=\bar{J}\ln(W'/T_{GK})$. The abbreviation "GK" indicates the ground Kramers state and T_{GK} is a characteristic temperature given as follows:

$$T_{GK}=W\exp\left[-\frac{1}{\bar{J}}\operatorname{arccot}\left(\frac{g_{\parallel}}{\sqrt{g_{\perp}^2-g_{\parallel}^2}}\right)\right]. \tag{35}$$

Journal of Physics: Conference Series **324** (2011) 012017 doi:10.1088/1742-6596/324/1/012017

However, the expressions just derived should be treated with care. One can see that a set of renormalized couplings (33) includes a new parameter U_{zy} which is absent among the parameters of the original Hamiltonian (16), as if the model would be nonrenormalizable. On the other hand we keep in mind the anisotropy of the Zeeman energy to be the same as that of the Kondo interaction, the property being likely independent of scaling procedure. It means that *g*-factors as well as the Kondo couplings are affected by the renormalization. The relation $J_{\|} / J_{\perp} = g_{\|} / g_{\perp}$ (see (5)) still holds to result, converting, after renormalization, to $U_{\|} / U_{\perp} = g'_{\|} / g'_{\perp} = \cos\varphi$. The latter immediately leads to the renormalizability of our model: $U_{zy}(t) = 0 = U_{zy}(0)$. Note, that the angle ψ is also renormalized following the equation $\cot\psi' = \cos\varphi \cot\theta$. With this correction, the expressions (33) transform to

$$
\begin{aligned}
&U_{xx} = \bar{J} / \sin\varphi, \\
&U_{yy} = U_{xx} \cos\varphi \Big/ \sqrt{\sin^2\theta + \cos^2\theta \cos^2\varphi}, \\
&U_{zz} = U_{xx} \sqrt{\sin^2\theta + \cos^2\theta \cos^2\varphi}, \\
&U_{yz} = U_{xx} \sin\theta \cos\theta \sin^2\varphi \Big/ \sqrt{\sin^2\theta + \cos^2\theta \cos^2\varphi}, \\
&U_{zy} = U_{xy} = U_{yx} = U_{xz} = U_{zx} = 0.
\end{aligned}
\tag{36}
$$

The characteristic temperature T_{GK} is obviously independent of initial and running values of the parameters, representing a universal energy scale to govern all low temperature physics:

$$T_{GK}(\rho J_{\alpha\beta}, W) = T_{GK}((\rho J_{\alpha\beta})', W'). \tag{37}$$

All physical quantities can be expressed in terms of the ratios T / T_{GK}, B / T_{GK}. By extending the scaling procedure down to the effective bandwidth $W' = T$ we obtain the temperature dependence of renormalized Kondo couplings:

$$\varphi(T) = \bar{J} \ln(T / T_{GK}). \tag{38}$$

Note, that in the isotropic case $(g_{\|} = g_{\perp} = g)$ equation (35) gives the standard result $T_{GK} = W e^{-\frac{1}{\rho J g}}$.

The temperature dependent parameters are readily seen to be logarithmically divergent at $T \to T_{GK}$ $(\varphi \ll 1)$: $U_{\alpha\beta}(T) \sim 1/\ln(T / T_{GK})$. It means that the conduction electron bandwidth can only be reduced to $W' \sim T_{GK}$, the value at which the perturbative scaling approach begins to break down. Consequently, all results derived by this method are only valid for temperatures well above T_{GK}.

Using the expressions (36) for the temperature dependent Kondo couplings we find the renormalized kinetic coefficients

$$
\begin{aligned}
&\Gamma_{ss}(\omega) = \pi \frac{\omega}{\omega_s} \bar{J}^2 T \left[\cot^2\varphi + \frac{1}{2} + \frac{\sin^2\theta}{4(\sin^2\theta + \cos^2\theta \cos^2\varphi)} \right], \\
&\Gamma_{\sigma\sigma}(\omega) = \pi \frac{\omega}{\omega_s} \bar{J}^2 \frac{g_s}{2 g_\sigma \rho (T + T_W)} T \left[\cot^2\varphi + \frac{1}{2} + \frac{\sin^2\theta}{4} \right], \\
&\Gamma_{\sigma s}(\omega) = \pi \frac{\omega}{\omega_s} \bar{J}^2 T \left[\sqrt{\sin^2\theta + \cos^2\theta \cos^2\varphi} + \cos\varphi \right] \Big/ 2\sin^2\varphi, \\
&\Gamma_{s\sigma}(\omega) = \Gamma_{\sigma s}(\omega) \frac{g_s}{2 g_\sigma \rho (T + T_W)}.
\end{aligned}
\tag{39}
$$

One can see that all kinetic coefficients logarithmically diverge at $T \to T_{GK}$ $(\varphi \ll 1)$: to the leading order in logarithmic terms they are of the form

$$\Gamma_{ss}(\omega) = \Gamma_{\sigma s}(\omega) = \pi \frac{\omega}{\omega_s} \frac{T}{\ln^2(T / T_{GK})},$$
$$\Gamma_{\sigma\sigma}(\omega) = \Gamma_{s\sigma}(\omega) = \Gamma_{ss}(\omega) \frac{g_s}{2 g_\sigma \rho (T + T_W)}. \tag{40}$$

At first glance, these results confirm the common belief that the ESR linewidth of Kondo ions (as well as conduction electrons) is expected to be too large for its detection. However, the coupling between two systems makes the situation quite different. We solve the coupled equations (24) under the condition of the bottleneck regime (29), taking into account the renormalization of kinetic coefficients. The relaxation rate of the collective spin mode with a frequency close to the Kondo-ion resonance now follows

$$\Gamma_{coll} = \Gamma_{sL} + \Gamma_{ss}^{eff} + \Gamma_{\sigma L}^{eff};$$
$$\Gamma_{ss}^{eff} = \Gamma_{ss} - \frac{\Gamma_{s\sigma}\Gamma_{\sigma s}}{\Gamma_{\sigma\sigma}}, \qquad \Gamma_{\sigma L}^{eff} = \Gamma_{\sigma L} \frac{\Gamma_{s\sigma}\Gamma_{\sigma s}}{\Gamma_{\sigma\sigma}^2}. \tag{41}$$

It is interesting to analyze the asymptotic behavior of an effective Korringa relaxation rate Γ_{ss}^{eff} and an effective relaxation rate of conduction electrons to the lattice $\Gamma_{\sigma L}^{eff}$ at $T \to T_{GK}$. In this case their expressions are essentially simplified to be written explicitly as functions of temperature and angle between the static magnetic field B and the crystal symmetry axis c:

$$\Gamma_{ss}^{eff} = \frac{\pi}{8} \bar{J}^2 (2 - \sin^4 \theta) T \ln^2(T / T_{GK}), \tag{42}$$

$$\Gamma_{\sigma L}^{eff} = \rho (T + T_W) \frac{2 g_\sigma}{g_s} \Gamma_{\sigma L}. \tag{43}$$

The result is rather unusual: instead of being divergent, the effective Korringa relaxation rate is greatly reduced and goes to zero at $T \to T_{GK}$. Although the Kondo interaction is strongly anisotropic, the Kondo effect leads to the common energy scale T_{GK} regulating the temperature dependence of different physical quantities. Indeed, all kinetic coefficients in (40) display a similar logarithmic temperature dependence leading to their complete mutual cancelation (to the leading order terms only!) in the collective spin mode. The effective relaxation rate of the conduction electrons to the lattice $\Gamma_{\sigma L}^{eff}$ is also greatly reduced, becoming proportional to temperature.

The angular dependence of the effective relaxation rate qualitatively agrees with the ESR linewidth as experimentally observed. Γ_{ss}^{eff} varies monotonically for $0 < \theta < \pi / 2$. The lowest and highest values of Γ_{ss}^{eff} correspond to the perpendicular and parallel orientation, respectively. The maximum of Γ_{ss}^{eff} is two times larger than minimum. This dependence is likely due to the partial breaking of the bottleneck regime condition (29) in the case of the parallel orientation, when the inequality $\omega_\sigma \gg \omega_s$ is fulfilled.

However, one must remember that all divergences and cancelations in relaxation rates should not be taken literally because the perturbational scaling approach is only valid for the temperatures well above T_{GK}. As a matter of fact, the singularities just indicate an increase of physical quantities with temperatures lowering to T_{GK}. Moreover, the considerations based on a single ion Kondo problem suggests that the Kondo anomalies are quenched, provided the static magnetic field is large compared with T_{GK}. At the temperature region $T_{GK} < T < B$ the thermodynamic functions become temperature independent and the ratio T / T_{GK} can be replaced with B / T_{GK}. In the previous work [4] the fitting to the experimental data revealed $T_{GK} = 0.36$ K, what is by two orders of magnitude smaller than the

Kondo temperature T_K derived thermodynamically [23] and by transport measurements [24]. The resonance field of the ESR experiments [1, 5] is of the order of T_{GK} : $B \equiv 0.46$ K at X-band and $B \equiv 1.62$ K at Q-band. For these reasons, the scaling approach for the ESR measurements seems to be valid for the temperature region $T_{GK} < B < T < T_K$.

For detailed analysis of the ESR linewidth angular dependence, as well as for a comparison with experimental data also contributions other than Kondo interaction need to be taken into account. Now we merely discuss the broadening of ESR linewidth which is represented by the kinetic coefficient Γ_{sL}. An obvious contribution comes from the usual magnetic dipole-dipole interactions of the Kondo ions. The corresponding local magnetic field can be easily evaluated and gives approximately $\Delta B_{loc} \sim 1000$ Oe [25]. This value is much larger than the observed ESR linewidth in $Yb_2Rh_2Si_2$ for X- and Q-bands ($\Delta B_{loc} \sim 200$ Oe at $T = 5$ K). The contribution from the RKKY interaction, which becomes highly anisotropic after projection onto the ground Kramers doublet, is expected to be much larger. So, it is evident that some narrowing mechanism for these type of contributions should exist. It is well established [18] that in the bottleneck regime the broadening of the ESR line by the distribution of local fields is sufficiently reduced due to fast reorientation of the Kondo ion moment caused by the Korringa relaxation. The corresponding contribution to the linewidth can be estimated as $\langle(\Delta\nu)^2\rangle / \Gamma_{ss}$ (but not Γ_{ss}^{eff} !), where $\langle(\Delta\nu)^2\rangle$ is the mean square distribution of the resonance frequencies due to the local fields. An additional broadening of the ESR line comes from the spin-phonon interaction of Kondo ions. It is enough to take into account the two-phonon Raman and Orbach processes at temperatures above a few K. This contribution can not be the subject for the narrowing described above.

5. Conclusion

We investigated the spin relaxation in the heavy fermion Kondo lattices $Yb_2Rh_2Si_2$ and $Yb_2Ir_2Si_2$ at temperatures well below thermodynamically measured Kondo temperature $T_K = 25$ K. The idea of a collective spin motion of Kondo ions with conduction electrons in presence of the Kondo effect successfully explains the ESR signal existence in these compounds. We conclude the article with a few comments.

The relaxation rate of the collective spin mode is found to be strongly dependent on the orientation of the static magnetic field to the crystal symmetry axis. Its angular dependence reflects the anisotropy of the Kondo interaction to confirm the idea of collective spin motion in the bottleneck regime. The broadening of the ESR linewidth with approaching to the parallel orientation can be attributed to the partial opening of the bottleneck due to significant difference between resonance frequencies of conduction electrons and Yb-ions. The treatment of the Kondo interaction alone provides for a qualitative agreement with experimental data on the angular dependence of the ESR linewidth. To obtain a quantitative agreement other contributions than the Kondo interaction needs be taken into account.

Another point to treat is the ratio of g-factors $g'_{\parallel} / g'_{\perp} = \cos\varphi$ used in derivation of the renormalized relaxation rate. One can see that at $T \to T_{GK}$ ($\varphi \ll 1$) the g-factors tend to be equal. This seems to contradict the experimentally observed strong anisotropy of the g-factor. However, the g-factors mentioned above are not the same as the experimentally observed ones. The resonance frequency of the collective spin mode is determined by the total dynamical susceptibility as a real part of its pole, whereas the g-factors in the Zeeman energy are just the matrix elements of the total electronic moment of an Yb-ion (see (4)). The equality $\left(g_{\parallel}^{\mathrm{Yb}}\right)' = \left(g_{\perp}^{\mathrm{Yb}}\right)'$ does not mean a similar relation in the case of resonance frequency of the collective mode, what is the quantity to be compared with experimental data.

Finally, we discuss the new characteristic temperature $T_{GK} = 0.36$ K revealed by the perturbational scaling approach and describing the g-data [4]. This parameter represents a common energy scale to regulate the behavior of the different kinetic coefficients. The scaling approach is valid at temperatures above T_{GK}. Hence, our theory is well suitable at the temperature region of the ESR measurements. At present the relation between T_{GK} and T_K remains an open question. We can only suggest them to be different energy scales, the former controlling the magnetic relaxation, and the latter is probably responsible for transport and thermal properties.

Acknowledgments
This work was supported by the Volkswagen Foundation (I/84689) and partially by the Ministry of Education and Science of the Russian Federation. The authors are grateful to J. Sichelschmidt for the fruitful discussions and for help in the preparation of the manuscript.

Appendix
Let us introduce a functional of two external random fields

$$V = \sum_1 \left(-\mathbf{v}_1\mathbf{S}_1 + \mathbf{w}_1\boldsymbol{\sigma}_1\right). \tag{A.1}$$

Here $1 \equiv (i_1, \tau_1)$, i_1 labels a lattice site, τ_1 is imaginary time at the interval $0 < \tau_1 < \beta$, $\beta = 1/k_B T$; $\mathbf{v}_1$, $\mathbf{w}_1$ are external random fields, $\sum_1 \equiv \sum_{i_1} \int_0^\beta$; $\mathbf{S}_1$, $\boldsymbol{\sigma}_1$ represent spin operators of Yb-ion and conduction electrons, respectively:

$$\mathbf{S}_1 = e^{\tau_1 H}\mathbf{S}_{i_1}e^{-\tau_1 H}, \quad \boldsymbol{\sigma}_1 = e^{\tau_1 H}\boldsymbol{\sigma}_{i_1}e^{-\tau_1 H}. \tag{A.2}$$

The spin operators $\mathbf{S}_1$, $\boldsymbol{\sigma}_1$ can be written in terms of variational derivatives of the functional (A.1) with respect to the external random fields:

$$\frac{\delta V}{\delta v_1^\alpha} = -S_1^\alpha, \quad \frac{\delta V}{\delta w_1^\alpha} = \sigma_1^\alpha, \quad \alpha = x, y, z. \tag{A.3}$$

For any set of time dependent operators we define the temperature Green function

$$\langle TA(\tau_1)B(\tau_2)...\rangle = \mathrm{Sp}\left(e^{-\beta H}TA(\tau_1)B(\tau_2)...e^{-V}\right)\big/\mathrm{Sp}\left(e^{-\beta H}Te^{-V}\right), \tag{A.4}$$

where T is the time ordering operator.

Functional derivation allows one to find composite averages, for example,

$$\begin{aligned}\frac{\delta\langle TA\rangle}{\delta v_1^\alpha} &= \langle TAS_1^\alpha\rangle - \langle TA\rangle\langle TS_1^\alpha\rangle,\\ \frac{\delta\langle TA\rangle}{\delta w_1^\alpha} &= -\langle TA\sigma_1^\alpha\rangle + \langle TA\rangle\langle T\sigma_1^\alpha\rangle.\end{aligned} \tag{A.5}$$

The Green functions (A.4) follow equation of motion

$$\frac{\partial\langle TA(\tau)\rangle}{\partial\tau} = \langle T\dot{A}(\tau)\rangle - \langle T\{A(\tau), V\}_+\rangle, \tag{A.6}$$

$$\frac{\partial\langle TA(\tau_1)B(\tau_2)\rangle}{\partial\tau_1} = \langle T\{A(\tau_1), B(\tau_2)\}_\eta\rangle + \langle T\dot{A}(\tau_1)B(\tau_2)\rangle - \langle T\{A(\tau_1), V\}_+ B(\tau_2)\rangle \tag{A.7}$$

with $\{A(\tau_1), B(\tau_2)\}_\eta = \delta(\tau_1 - \tau_2)\left(A(\tau_1)B(\tau_2) - \eta B(\tau_2)A(\tau_1)\right)$, $\eta = +1, -1$ in the cases of Bose- and Fermi operators, respectively.

To study the ESR response we need four two-spin Green functions

$$\begin{aligned} G_{12}^{\sigma_1\sigma_2} &= -\langle T c_{1\sigma_1} c_{2\sigma_2}^{+} \rangle, \\ D_{12}^{\alpha\beta} &= -\langle T S_1^{\alpha} S_2^{\beta} \rangle_c, \\ A_{12}^{\alpha\beta} &= -\langle T \sigma_1^{\alpha} \sigma_2^{\beta} \rangle_c, \\ B_{12}^{\alpha\beta} &= -\langle T \sigma_1^{\alpha} S_2^{\beta} \rangle_c. \end{aligned} \tag{A.8}$$

Here $\langle TAB\rangle_c = \langle TAB\rangle - \langle TA\rangle\langle TB\rangle$, σ_1, σ_2 denote the orientation of conduction electron spin. The Green functions (A.8) are coupled with obvious relations reducing the problem to calculation of two independent functions G, D:

$$A_{12}^{\alpha\beta} = \frac{\delta\langle T\sigma_1^{\alpha}\rangle}{\delta w_2^{\beta}} = \sum_{\lambda\mu} s_{\lambda\mu}^{\alpha} \frac{\delta G_{11}^{\mu\lambda}}{\delta w_2^{\beta}}, \qquad B_{12}^{\alpha\beta} = -\frac{\delta\langle T\sigma_1^{\alpha}\rangle}{\delta v_2^{\beta}} = -\sum_{\lambda\mu} s_{\lambda\mu}^{\alpha} \frac{\delta G_{11}^{\mu\lambda}}{\delta v_2^{\beta}}. \tag{A.9}$$

At the end of all calculations the external random fields $\mathbf{v}$, $\mathbf{w}$ are set to zero, and Green functions can be written in terms of their Fourier transforms

$$f_{12} = \frac{1}{N\beta} \sum_{\mathbf{k},n} e^{i\mathbf{k}(\mathbf{r}_1-\mathbf{r}_2)} e^{-i\omega_n(\tau_1-\tau_2)} f(\mathbf{k}, i\omega_n), \tag{A.10}$$

where $\omega_n = 2\pi n/\beta$, $\omega_n = \pi(2n+1)/\beta$ in the cases of Bose- and Fermi operators, respectively.

Using (A.7), (A.8) we obtain an equation of motion for the one-electron Green function $G_{12}^{\sigma_1\sigma_2}$

$$F_{11'}^{\sigma_1\sigma_1'} G_{1'2}^{\sigma_1'\sigma_2} = -\delta_{\sigma_1\sigma_2}\delta_{12} - \delta_{11'} J_{\alpha\beta} s_{\sigma_1\sigma_1'}^{\alpha} \left(\langle T S_1^{\beta} \rangle + \frac{\delta}{\delta v_1^{\beta}} \right) G_{1'2}^{\sigma_1'\sigma_2}, \tag{A.11}$$

where

$$F_{11'}^{\sigma_1\sigma_1'} = \delta_{\sigma_1\sigma_1'}\delta_{11'} \left(\frac{\partial}{\partial\tau_1} + \varepsilon^{\sigma_1} \right) + \delta_{\sigma_1\sigma_1'} t_{11'} + \delta_{11'} (\mathbf{w}_1 \mathbf{s}_{\sigma_1\sigma_1'}), \tag{A.12}$$

$\delta_{11'} = \delta_{i_1i_1'}\delta(\tau_1 - \tau_{1'})$, $t_{11'} = t_{i_1i_1'}\delta(\tau_1 - \tau_{1'})$; $s_{\sigma_1\sigma_1'}^{\alpha}$, ε^{λ}, $J_{\alpha\beta}$ are defined in (9), (15), (16), a sum over identical indices is implied.

The equation of motion for the localized spin Green function is derived in a similar way:

$$\begin{aligned} &\left(\lambda\frac{\partial}{\partial\tau_1} - g_s B + v_1^z \right) D_{12}^{\lambda\bar\lambda} = -\delta_{12}\langle T S_1^z\rangle + v_1^{\lambda} D_{12}^{z\bar\lambda} + \\ &+ J_{\alpha z}\left[D_{12}^{\lambda\bar\lambda} - \langle T S_1^{\lambda}\rangle \frac{\delta}{\delta v_2^{\lambda}} + \frac{\delta^2}{\delta v_1^{\bar\lambda}\delta v_2^{\lambda}} \right] s_{\lambda\mu}^{\alpha} G_{11}^{\mu\lambda} - J_{\alpha\lambda}\left[D_{12}^{z\bar\lambda} - \langle T S_1^{z}\rangle \frac{\delta}{\delta v_2^{\lambda}} + \frac{\delta^2}{\delta v_1^{z}\delta v_2^{\lambda}} \right] s_{\lambda\mu}^{\alpha} G_{11}^{\mu\lambda} + \\ &+ I_{11'}^{z\alpha}\left[\langle T S_1^{\lambda}\rangle D_{1'2}^{\alpha\bar\lambda} + \langle T S_{1'}^{\alpha}\rangle D_{12}^{\lambda\bar\lambda} - \frac{\delta}{\delta v_{1'}^{\alpha}} D_{12}^{\lambda\bar\lambda} \right] - I_{11'}^{z\alpha}\left[\langle T S_1^{z}\rangle D_{1'2}^{\alpha\bar\lambda} + \langle T S_{1'}^{\alpha}\rangle D_{12}^{z\bar\lambda} - \frac{\delta}{\delta v_{1'}^{\alpha}} D_{12}^{z\bar\lambda} \right]. \end{aligned} \tag{A.13}$$

Here $I_{11'} = I_{i_1i_1'}\delta(\tau_1 - \tau_{1'})$, $J_{\alpha\lambda} = (J_{\alpha x} + i\lambda J_{\alpha y})/\sqrt{2}$, $I^{\lambda\alpha} = (I^{x\alpha} + i\lambda I^{y\alpha})/\sqrt{2}$, $\lambda = +1, -1$, $\bar\lambda = -\lambda$.
Equations of motion are solved by a perturbational expansion on Kondo couplings $J_{\alpha\beta}$. The treatment of both equations (A.11) and (A.13) are basically similar to each other, but the latter is much more complicated, so we shall not write it out explicitly.

It is convenient to introduce a self energy for the one electron Green function

$$G^{-1} = G_0^{-1} - \Sigma, \tag{A.14}$$

G_0 is the Green function in the absence of the Kondo interaction, indices 1, 2, σ_1, σ_2 are dropped. Combining (A.12) with (A.14) we immediately obtain

$$G_0^{-1} = -F\,, \tag{A.15}$$

$$\Sigma_{12}^{\sigma_1\sigma_2} = \delta_{12} J_{\alpha\beta} s^{\alpha}_{\sigma_1\sigma_2} \langle TS_1^{\beta}\rangle + J_{\alpha\beta} s^{\alpha}_{\sigma_1\lambda} G_{11'}^{\lambda\lambda'} \frac{\delta\Sigma_{1'2}^{\lambda\sigma_2}}{\delta v_1^{\beta}}\,. \tag{A.16}$$

An iterative process with equation (A.16) yields $\Sigma = \Sigma_1 + \Sigma_2 + \Sigma_3 + O(J^4)$, where

$$\begin{aligned}
(\Sigma_1)_{12}^{\sigma_1\sigma_2} &= \delta_{12} J_{\alpha\alpha'} s^{\alpha}_{\sigma_1\sigma_2} \langle TS_1^{\alpha'}\rangle,\\
(\Sigma_2)_{12}^{\sigma_1\sigma_2} &= -J_{\alpha\alpha'} J_{\beta\beta'} s^{\alpha}_{\sigma_1\lambda} s^{\beta}_{\lambda'\sigma_2} G_{12}^{\lambda\lambda'} D_{21}^{\beta'\alpha'},\\
(\Sigma_3)_{12}^{\sigma_1\sigma_2} &= -J_{\alpha\alpha'} J_{\beta\beta'} J_{\gamma\gamma'} s^{\alpha}_{\sigma_1\lambda} s^{\beta}_{\lambda'\mu'} s^{\gamma}_{\mu\sigma_2} G_{11'}^{\lambda\lambda'} G_{1'2}^{\mu'\mu} \frac{\delta D_{1'1}^{\beta'\alpha'}}{\delta v_{2'}^{\gamma'}},
\end{aligned} \tag{A.17}$$

the variational derivative $\dfrac{\delta D}{\delta v}$ is a three spin Green function:

$$-\frac{\delta D_{12}^{\alpha\beta}}{\delta v_3^{\gamma}} = \langle TS_1^{\alpha} S_2^{\beta} S_3^{\gamma}\rangle_c = \langle TS_1^{\alpha} S_2^{\beta} S_3^{\gamma}\rangle + \langle TS_1^{\alpha}\rangle D_{23}^{\beta\gamma} + \langle TS_2^{\beta}\rangle D_{13}^{\alpha\gamma} + \langle TS_3^{\gamma}\rangle D_{12}^{\alpha\beta} + \langle TS_1^{\alpha}\rangle\langle TS_2^{\beta}\rangle\langle TS_3^{\gamma}\rangle\,.$$

Putting $V = 0$ we obtain the first order self energy

$$(\Sigma_1)_{12}^{\lambda\lambda} = \delta_{12} \lambda J_{zz} \langle S^z\rangle / 2, \qquad (\Sigma_1)_{12}^{\lambda\bar\lambda} = \delta_{12} J_{\bar\lambda z} \langle S^z\rangle / \sqrt{2}, \qquad \langle S^{\alpha}\rangle \equiv \langle TS_1^{\alpha}\rangle_{V=0}\,. \tag{A.18}$$

Diagonal part of Σ_1 gives the shift of the conduction electrons resonance frequency, which, after Fourier transform, can be included into the zero order Green function

$$G_0^{\lambda\lambda}(\mathbf{k}, i\omega_n) = \frac{1}{i\omega_n - \varepsilon_{\mathbf{k}}^{\lambda}}\,, \tag{A.19}$$

where $\omega_n = \pi(2n+1)/\beta$, $\varepsilon_{\mathbf{k}}^{\lambda} = t(\mathbf{k}) - \mu + \lambda\omega_{\sigma}/2$, $\omega_{\sigma} = g_{\sigma}B + J_{zz}\langle S^z\rangle$. Exact expressions for $G^{\sigma_1\sigma_2}(\mathbf{k}, i\omega_n)$ are of the form

$$G^{\lambda\lambda} = \left[(G_0^{-1})^{\lambda\lambda} - \Sigma^{\lambda\lambda} - \frac{\Sigma^{\lambda\bar\lambda}\Sigma^{\bar\lambda\lambda}}{(G_0^{-1})^{\bar\lambda\bar\lambda} - \Sigma^{\bar\lambda\bar\lambda}} \right]^{-1}, \qquad G^{\lambda\bar\lambda} = \frac{\Sigma^{\lambda\bar\lambda}}{(G_0^{-1})^{\bar\lambda\bar\lambda} - \Sigma^{\bar\lambda\bar\lambda}} G^{\lambda\lambda}\,. \tag{A.20}$$

Substitution of (A.17) into (A.20) gives the non-diagonal Green function as follows

$$G_{12}^{\lambda\bar\lambda} = \frac{J_{\bar\lambda z}}{\sqrt{2}} \frac{\langle S^z\rangle}{\omega_{\sigma}} \sum_{\mu} \mu G_{12}^{\mu\mu} + O(J^2)\,. \tag{A.21}$$

A similar treatment of equation (A.13) yields the localized spin Green function to the first order in J, the RKKY interaction taken into account in molecular field approximation only. The result is

$$D^{\lambda\bar\lambda}(\mathbf{k}, i\omega_n) = \frac{\lambda\langle S^z\rangle}{i\omega_n - \lambda\omega_s} + O(J^2)\,, \qquad D^{zz}(\mathbf{k}, i\omega_n) = -\frac{\beta}{4}\delta_{n0} + O(J^2)\,, \tag{A.22}$$

where $\omega_n = 2\pi n/\beta$, ω_s is given by (26).

Two-spin Green functions A, B are derived from (A.9) via the obvious relations

$$\begin{aligned}
&\delta G = -G\delta G_0^{-1} G + G\delta\Sigma G,\\
&\frac{\delta (G_0^{-1})_{12}^{\lambda\mu}}{\delta w_3^{\alpha}} = -\delta_{12}\delta_{23} s^{\alpha}_{\lambda\mu}, \quad \frac{\delta G_0^{-1}}{\delta v} = 0.
\end{aligned} \tag{A.23}$$

The calculation of all Green functions up to the third order in J results in the system of coupled equations (24). Here we set out explicit expressions for A, B to the first order in J only, since their higher order terms are too lengthy.

$$A_{12}^{zz} = \frac{1}{4}\sum_{\lambda} G_{12}^{\lambda\lambda} G_{21}^{\lambda\lambda} + O(J^2), \qquad A_{12}^{\lambda\bar{\lambda}} = \frac{1}{2} G_{12}^{\bar{\lambda}\bar{\lambda}} G_{21}^{\lambda\lambda} + O(J^2),$$
$$A_{12}^{\lambda z} = A_{21}^{z\lambda} = J_{\lambda z} \frac{\langle S^z \rangle}{\omega_\sigma} \left(A_{12}^{zz} - A_{12}^{\lambda\bar{\lambda}} \right) + O(J^2), \qquad A_{12}^{\lambda\lambda} = O(J^2). \tag{A.24}$$

$$B_{12}^{\alpha\beta} = J_{\alpha'\beta'} A_{11'}^{\alpha\alpha'} D_{1'2}^{\beta'\beta} + O(J^2). \tag{A.25}$$

References

[1] Sichelschmidt J, Ivanshin V A, Ferstl J, Geibel C and Steglich F 2003 *Phys. Rev. Lett.* **91** 156401
[2] Abrahams E and Wölfle P 2008 *Phys. Rev.* B **78** 104423
[3] Schlottmann P 2009 *Phys. Rev.* B **79** 045104
[4] Kochelaev B I, Belov S I, Skvortsova A M, Kutuzov A S, Sichelschmidt J, Wykhoff J, Geibel C and Steglich F 2009 *Eur. Phys. J.* B **72** 485–489
[5] Sichelschmidt J, Wykhoff J, Krug von Nidda H-A, Fazlishanov I I, Hossain Z, Krellner C, Geibel C and Steglich F 2007 *J. Phys.: Condens. Matter* **19** 016211
[6] Kutuzov A S, Skvortsova A M, Belov S I, Sichelschmidt J, Wykhoff J, Eremin I, Krellner C, Geibel C and Kochelaev B I 2008 *J. Phys.: Condens. Matter* **20** 455208
[7] Kadanoff L P and Baym G 1962 *Quantum Statistical Mechanics. Green's Function Methods in Equilibrium and Nonequilibrium Problems.* (New York: W. A. Benjamin, Inc.)
[8] Taylor K N R and Darby M I 1972 *Physics of Rare Earth Solids* (London: Chapman and Hall)
[9] Liu S H 1961 *Phys. Rev.* **121** 451–455
[10] Liu S H 1961 *Phys. Rev.* **123** 470–474
[11] Kaplan T A and Lyons D H 1963 *Phys. Rev.* **129** 2072–86
[12] Specht F 1967 *Phys. Rev.* **162** 389–398
[13] Druzhinin V V and Irkhin Yu P 1966 *Zh. Eksp. Teor. Fiz.* **51** 1856–62
[14] Kutuzov A S and Skvortsova A M, these proceedings
[15] Kutuzov A S and Skvortsova A M 2009 *Magn. Reson. Solids* **11** 7–13
[16] Stockert O, Koza M M, Ferstl J, Murani A P, Geibel C and Steglich F 2006 *Physica* B **378–380** 157–158
[17] Hiess A, Stockert O, Koza M M, Hossain Z and Geibel C 2006 *Physica* B **378–380** 748–749
[18] Barns S E 1981 *Adv.Phys.* **30** 801–938
[19] Kochelaev B I and Safina A M 2004 *Phys. Solid State* **46** 226–230
[20] Fazleev N G, Mironov G I and Fry J L 1992 *J. Magn. Mag. Mat.* **108** 123–124
[21] Abrikosov A A 1965 *Physics* **2** 5–20
[22] Anderson P W 1970 *J. Phys.* C **3** 2436–42
[23] Trovarelli O, Geibel C, Mederle S, Langhammer C, Grosche F M, Gegenwart P, Lang M, Sparn G and Steglich F 2000 *Phys. Rev. Lett.* **85** 626–629
[24] Köhler U, Oeschler, Steglich F, Maquilon S and Fisk Z 2008 *Phys. Rev.* B **77** 104412
[25] Vishina A A, these proceedings

Journal of Physics: Conference Series **324** (2011) 012018 doi:10.1088/1742-6596/324/1/012018

Motional narrowing of EPR line in Kondo-lattice with heavy fermions

Alena Vishina
KFU, 420008, 18 Kremlyovskaya St., Kazan, Russian Federation.
E-mail: alenavishina@gmail.com

Abstract. An additional contributions to the EPR linewidth besides the Kondo interaction are calculated for the Kondo lattice with heavy fermions $YbRh_2Si_2$. It is found that the broadening of the Electron Paramagnetic Resonance (EPR) line due to the usual dipole-dipole and anisotropic RKKY interactions should be much larger than the experimentally observed total EPR linewidth. It is shown that these contributions can be considerably reduced due to motional narrowing of the EPR line caused by the translational diffusion of quasi-localized $4f$-electrons.

1. Introduction

According to the common belief based on the single ion Kondo effect the EPR signal could not be observed in compound $YbRh_2Si_2$, since magnetic moments of the Kondo ions should be screened by conduction electrons at temperatures below the Kondo temperature $T_K = 25$ K, and the EPR linewidth is expected to be of the order of $\Delta\nu = k_B T_K / 2\pi\hbar \sim 500$ GHz. However an intensive EPR signal having a rather narrow linewidth has been found in this compound [1]. The EPR signal is highly anisotropic; both resonance frequency and the linewidth strongly depend on the magnetic field orientation with respect to the crystallographic axes. The angular dependence of the EPR g-factor and linewidth reflects the local properties of Yb^{3+} ion in a crystal electric field [2].

An existence of the EPR signal in a heavy fermion Kondo lattice was explained recently by the formation of a collective spin mode of quasi-localized f-electrons and conduction electrons due to exchange coupling between them [3].

At the same time one should expect additional contributions to EPR linewidth due to the other reasons: anisotropic RKKY interactions, usual magnetic dipole-dipole interactions, crystal imperfections and so on. In this paper we calculate these contributions to the EPR linewidth and show that this type of the EPR broadening can be a subject of the well known motional narrowing due to translational mobility of the $4f$-electrons.

2. Analysis

2.1. The EPR linewidth due to spin-spin interactions and imperfections of a crystal

At first we consider the EPR line broadening due to the dipole-dipole interaction. The Hamiltonian for a system of N ions having magnetic moments in the external magnetic field H_0 contains Zeeman and magnetic dipole-dipole interaction:

$$H = H_z + H_{dd}, \tag{1}$$

$$H_z = \mu_B H_0(lg_x S_x + mg_y S_y + ng_z S_z), \tag{2}$$

for i-th and j-th ions dipole-dipole interaction Hamiltonian takes the form

$$H_{dd}^{ij} = \frac{\mu_B^2}{r_{ij}^3}\{S_{ix}S_{jx}(1-3l^2)g_{ix}g_{jx} + S_{iy}S_{jy}(1-3m^2)g_{iy}g_{jy}S_{iz}S_{jz}(1-3n^2)g_{iz}g_{jz} -$$
$$(g_{ix}g_{jy}S_{ix}S_{jy} + g_{iy}g_{jx}S_{iy}S_{jx})3lm - (g_{iy}g_{jz}S_{iy}S_{jz} + g_{iz}g_{jy}S_{iz}S_{jy})3mn - \tag{3}$$
$$(g_{ix}g_{jz}S_{ix}S_{jz} + g_{iz}g_{jx}S_{iz}S_{jx})3nl\},$$

here μ_B is the Bohr magneton; r_{ij} is the distance between i-th and j-th ions; l, m and n are magnetic field direction cosines with respect to the main axes of g-factor.

For magnetic field directed at the angle θ with respect to the crystal axis it is convenient to perform an axes rotation to diagonalize the Zeeman interaction (for the definiteness we consider the field lying in $x = 0$ plane). After the ψ-angle rotation around the x-axis, where

$$\tan\psi = \frac{g_\perp}{g_\parallel}\tan\theta, \tag{4}$$

Zeeman Hamiltonian takes the form

$$H_z = \mu_B g H_0 S_z'; \tag{5}$$

here $g = \sqrt{g_\parallel^2\cos^2\theta + g_\perp^2\sin^2\theta}$, where $g_\parallel$, $g_\perp$ are their g-factors for parallel and perpendicular field orientation relative to crystal symmetry axis. We also take into account the tetragonal symmetry and put $g_x = g_y = g_\perp$ and $g_z = g_\parallel$; S_z' is spin operator z-component for a new direction of axes.

Since we consider the main resonance line only we should drop all the non-diagonal terms in the Hamiltonian. We keep the terms of the form $S_+'S_-'$ ($S_+' = S_x' + iS_y'$, $S_-' = S_x' - iS_y'$) and S_zS_z only, since they don't change the magnetic quantum number.

After the rotation to the new quantization axes and a semi-diagonal part extraction dipole-dipole Hamiltonian takes the form (we drop the accents at S for the sake of brevity)

$$H_{dd} = \sum_{ij}\{A_{ij}\mathbf{S}_i\mathbf{S}_j + B_{ij}S_{zi}S_{zj}\} \tag{6}$$

with

$$A_{ij} = \frac{\mu_B^2}{2r_{ij}^3}\left\{g_\perp^2(3n_{ij}^2 - 1) + g_\parallel^2(1-3n_{ij}^2)\sin^2\psi - (1-3m_{ij}^2)g_\perp^2\sin^2\psi + \right. \tag{7}$$
$$\left. 3g_\perp g_\parallel m_{ij}n_{ij}\sin 2\psi\right\},$$

$$B_{ij} = \frac{3}{2}\frac{\mu_B^2}{r_{ij}^3}\left\{\frac{2g_\parallel^2 + g_\perp^2}{3}(1-3n_{ij}^2) - (1-3n_{ij}^2)g_\parallel^2\sin^2\psi - \right. \tag{8}$$
$$\left.(1-3m_{ij}^2)g_\perp^2\sin^2\psi - 3g_\perp g_\parallel m_{ij}n_{ij}\sin 2\psi\right\},$$

where l_{ij}, m_{ij}, n_{ij} are direction cosines for the line, connecting i-th and j-th ions with respect to crystallographic axis c.

Here for the linewidth calculation we use the Moments method [4]. Formulas for the second and the forth moments of the line are

$$M_2 = -\frac{\mathrm{Sp}([S^+, H_{int}][S^-, H_{int}])}{\mathrm{Sp}(S^+S^-)}. \tag{9}$$

$$M_4 = \frac{\mathrm{Sp}\left([[S^+, H_{int}], H_{int}]\,[[S^-, H_{int}], H_{int}]\right)}{\mathrm{Sp}(S^+S^-)}. \tag{10}$$

For dipole-dipole interaction we calculate the second moment only since the form of the line for it in a dense lattice is known to be almost Gaussian. Making calculations for the nearest neighbors inside $YbRh_2Si_2$ we obtain the following expression for an arbitrary direction of external magnetic field

$$M_2 = \frac{9}{16}\frac{g_\perp^4 \mu_B^4}{a^6}(0.672 - 2.015\sin^2\psi + 10.511\sin^4\psi)\frac{1}{a^6}, \tag{11}$$

where a is the distance between the closest Yb-ions.

Considering the line form to be Gaussian we can estimate the linewidth by the formula $\Delta H = \sqrt{M_2}/g_\perp \mu_B$. For $YbRh_2Si_2$ $g_\parallel = 0.17$, $g_\perp = 3.56$ [1] and the distance between the closest Yb-ions $a = 4.007$ Å. For example we have $\Delta H_{\pi/2} \approx 1160$ G for $\psi = 90°$, while experimental value is $\Delta H_{\pi/2,\ exp} \approx 100$ G [1] (this value is obtained by an extrapolation of the experimental data for $T \to 0$, since we estimate the dipole-dipole interaction and its contribution does not depend on temperature). So the linewidth due to the dipole-dipole interaction occurs to be an order of magnitude larger than experimentally observed one. It means that there must be some mechanism for the EPR line narrowing.

Next we consider the exchange interaction between Yb ions via conduction electrons (RKKY interaction). RKKY interaction after projection onto the Kramers ground state [3] becomes anisotropic. And anisotropic RKKY interaction unlike the isotropic one leads to additional line widening. RKKY interaction Hamiltonian has the form (in the coordinate system, where magnetic field is directed along the crystallographic axis c)

$$H_{RKKY} = \sum_{ij}\{J_\parallel^{ij} S_i^z S_j^z + J_\perp^{ij} S_i^+ S_j^-\} \tag{12}$$

and after the axes rotation (described above for dipole-dipole interaction) it takes the form similar to the form of dipole-dipole interaction (6), but with the coefficients

$$A_{ij} = \frac{1}{2}J_\parallel^{ij}\sin^2\psi + J_\perp^{ij}\left(\frac{1}{2} + \frac{1}{2}\cos^2\psi\right), \tag{13}$$

$$B_{ij} = J_\parallel^{ij}\left(1 - \frac{3}{2}\sin^2\psi\right) + J_\perp^{ij}\left(-\frac{1}{2} + (1 - \frac{3}{2}\cos^2\psi)\right), \tag{14}$$

where $J_{\parallel,\perp}$ are the RKKY coupling constants and S_j is the spin-1/2 operator of j-th Yb-ion.

Applying the formulas for the second and forth moments [5] to the given Hamiltonian and taking into account the fact, that $g_\parallel \ll g_\perp$, $|A_{ij}| \approx |B_{ij}|$, and $S = 1/2$, for the field directed along the crystal symmetry axis we obtain

$$M_2 = \frac{1}{4\hbar^2}\sum_j B_{ij}^2, \tag{15}$$

$$M_4 = \frac{1}{16\hbar^4}\sum_{j\neq k}[3B_{jk}^2 B_{ji}^2 + 2B_{jk}^2(B_{ji} - B_{ki})^2 + \tag{16}$$

$$2B_{jk}B_{ki}(B_{ji} - B_{jk})(B_{ji} - B_{ki}) + 2B_{jk}B_{jk}(B_{ji} - B_{ki})^2] + \frac{1}{8\hbar^4}\sum_j B_{ji}^4.$$

Computations for the nearest neighbors give $M_4/M_2^2 = 8.5$. An isotropic exchange interaction usually results in a line narrowing, but here due to the high anisotropy, RKKY interaction itself contributes the EPR linewidth. As it is seen from the moments ratio (that is larger than the one for the Gaussian lineform) RKKY lineform is closer to Lorentzian than to Gaussian. Still it contributes the linewidth so there must be some narrowing mechanism.

We also consider the impact of crystal imperfections since EPR spectra are sensitive to the value and symmetry of the inner electric field produced by crystal ions. We analyze the special case of helical dislocation. In this case the Hamiltonian of an interaction between the Yb ions and lattice elastic deformations takes the form [6]

$$H_{int} = \sum_{\alpha\beta\gamma\delta i} G_{\alpha\beta\gamma\delta}\varepsilon_{\alpha\beta}(\mathbf{r})S_\gamma^i H_\delta, \tag{17}$$

where $\varepsilon_{\alpha\beta}$ are the components of a deformation tensor and $G_{\alpha\beta\gamma\delta}$ is a tensor characterizing the spin-elastic interaction. Let us once again find the second moment of the line shape using the formula (9). For a helical dislocation and magnetic field both directed along the z-axis

$$M_2 = \frac{bc^2}{32}\left\{G_{zxzz}^2 + G_{zyzz}^2\right\}\ln\left(\frac{1}{\pi c r_0^2}\right). \tag{18}$$

where c is a dislocation concentration, r_0 – the distance between spins and b is a slip (Burgers) vector.

To estimate the crystal imperfections input to the EPR linewidth we need to know G-tenzor components. Here we consider the lowest level which is a Kramers doublet only (it is well-separated from the higher levels [3]). It splits due to addition of the excited states by magnetic field with the weight $g\mu_B H/\Delta$ [7], where Δ is the crystal field splitting. Hence additional electric field potential due to a lattice deformation can be replaced by [7] $g\mu_B HV/\Delta \sim g\mu_B H$, here V is the matrix element of interaction between the ion and crystal field modulation, which appears due to the lattice deformation. So we can estimate the G-components to be $\sim g\mu_B$. If we put $g \approx 3.5$; b, $r_0 \sim 4\mathring{A}$ and $c \sim 10^9$ we obtain the linewidth $\Delta H \approx 8$ Oe. So the contribution of dislocations into the line shape is negligible.

2.2. Line narrowing

Now we consider the mechanism that can lead to the EPR line narrowing – a motional narrowing due to quasi-localized f-electrons translational diffusion.

To analyze the influence of electron motion over the vacant lattice sites onto the width of the EPR line, we use the Hamiltonian that contains a kinetic part, describing electron motion over the vacant lattice sites H_{kin}, an spin-spin interaction of magnetic moments H_{ss} (here we consider the specific case of anisotropic exchange) and the energy of moments in magnetic field H_z.

It is convenient to write the kinetic energy using the Hubbard X-operators. The operator X_i^{pq} is defined on each site i and describes the transitions between the states p and q, that take three possible values $|\sigma\rangle$, $|0\rangle$ ($\sigma = \uparrow, \downarrow$): with (spin up and spin down) and without electron on the site; so the operator $X_i^{\sigma 0}$, for example, creates an electron with spin σ on the site i.

So, the Hamiltonian here takes the form

$$H = H_z + H_{kin} + H_{ss}, \tag{19}$$

where

$$H_{kin} = \sum_{\sigma ij} t_{ij} X_i^{\sigma 0} X_j^{0\sigma}, \tag{20}$$

$$H_{ss} = \sum_{\alpha\sigma ij} (A_{ij} + B_{ij}) X_i^{\alpha\alpha} X_j^{\sigma\sigma} \alpha\sigma + \sum_{ij} \frac{1}{2} A_{ij} \left(X_i^{\uparrow\downarrow} X_j^{\downarrow\uparrow} + X_i^{\downarrow\uparrow} X_j^{\uparrow\downarrow} \right) \tag{21}$$

and

$$H_z = g\mu_B H_0 \sum_{\alpha i} \alpha X_i^{\alpha\alpha}. \tag{22}$$

Here t_{ij} is hopping energy. We should note that in H_{ss} and H_z the summations are performed over the occupied lattice sites only.

In order to determine the second and fourth moments of the line form we use the relations (9) and (10) with $S^+ = \sum_i X_i^{\uparrow\downarrow}$, $S^- = \sum_i X_i^{\downarrow\uparrow}$.

Using the property of the Hubbard operators, $X_i^{pq} X_i^{rs} = \delta_{qr} X_i^{ps}$, we obtain formulas for the second and fourth moments of the line.

$$M_2 = (1 - n_0) \sum_j B_{ij}^2 \tag{23}$$

$$M_4 = (1 - n_0) \sum_j B_{ij}^4 + 2(1 - n_0)^2 \sum_{jk} B_{ij}^2 B_{ik}^2 + 2n_0(1 - n_0) \sum_{jk} (B_{kj} - B_{ij})^2 t_{ki}^2, \tag{24}$$

where n_0 is a holes concentration. Since it is convenient to sum over all the lattice sites we replace the confined sums $\tilde{\sum}_{ij}$ (over the occupied sites only) by the sums $(1 - n_0) \sum_{ij}$, taken over all the sites. Therefore the factor $(1 - n_0)$ appears.

To estimate how much does electron motion affect the linewidth we calculate the moments for the 'nearest neighbors' in a flat lattice (since calculation for the next neighbors is more complicated):

$$M_2 = 2(1 - n_0) \left[B_0^2 + B_1^2 \right], \tag{25}$$

$$M_4 = 2(1 - n_0) \left[B_0^4 + B_1^4 \right] + 8(1 - n_0)^2 \left[B_0^4 + 2B_0^2 B_1^2 \right] + 16n_0(1 - n_0) \left[B_0 - B_1 \right]^2 t_0^2, \tag{26}$$

where "0" denotes the nearest neighbors and "1" - the next ones.

Now we can estimate the linewidth by the formula $\widetilde{\Delta H} = M_2^{3/2} / M_4^{1/2}$:

$$\widetilde{\Delta H} = \frac{2(1 - n_0) B_0^3}{\left[B_0^4 + 4(1 - n_0) B_0^4 + 8n_0 B_0^2 t_0^2 \right]^{1/2}}. \tag{27}$$

Now we can estimate the line narrowing due to electron motion for the case of dipole-dipole interaction. Above we have already calculated the second moment and the linewidth $\Delta H_{dd} \approx 1200$ G ≈ 0.081 K for dipole-dipole interaction. Since we can see the expression, almost similar to M_2 in front of t_0^2 in M_4 (26), we can estimate the linewidth for dipole-dipole interaction with electron motion taken into account.

Hopping energy t_0 multiplied by the number of nearest neighbors is expected to be of the order of a bandwidth. It is possible to estimate it using the data of angle-resolved photoemission experiments [8]. It has been shown in [8] that Kramers doublet energy depends on the wave vector k which allows to roughly estimate the lowest doublet bandwidth to be ~ 100 K. We take $t_0 = 20$ K (since t_0 is for the one neighbor only) and $n_0 = 0.01$, then $\widetilde{\Delta H}_{dd} \approx 17$ G, which is ~ 70 times narrower than ΔH_{dd}, and much narrower than experimentally observed value, so even $t_0 = 20$ K is enough to compensate the dipole-dipole linewidth.

We should expect exchange interaction to be larger than dipole-dipole one. It's difficult to estimate its value since J is unknown, but by the formulas we can see that it could also be narrowed by electron motion.

Journal of Physics: Conference Series **324** (2011) 012018 doi:10.1088/1742-6596/324/1/012018

3. Discussion

We considered several mechanisms affecting the EPR linewidth in $YbRh_2Si_2$ compound. It is shown that spin-spin interactions (Yb-ions dipole-dipole interaction and anisotropic RKKY-interaction) result in the linewidth much larger than experimentally observed one. Therefore, it is obvious that some narrowing mechanism must exist.

Electron motion is only one of the possible narrowing mechanisms. We believe it might be the case for $YbRh_2Si_2$ since due to the f-electrons and conduction electrons hybridization the jumping of f-electrons between the nearest Yb sites appears. There are several possible line narrowing mechanisms, for example a fast reorientation of Kondo ion moments in the bottleneck regime, and it is not yet possible to determine, which one is more effective.

Acknowledgments

This work was supported by the Volkswagen Foundation (I/84689). The author thanks Boris Kochelaev for the suggested topic of this research and many helpful discussions.

References

[1] Sichelschmidt J, Ivanshin V, Ferstl J, Geibel C and Steglich F 2003 *Phys. Rev. Lett.* **91** 156401
[2] Kutuzov A S, Skvortsova A M, Belov S I, Sichelschmidt J, Wykhoff J, Eremin I, Krellner C, Geibel C and Kochelaev B I 2008 *J. Phys.: Condens. Matter.* **20** 455208
[3] Kochelaev B I, Belov S I, Skvortsova A M, Kutuzov A S, Sichelschmidt J, Wykhoff J, Geibel C and Steglich F 2009 *Eur. Phys. J.* **72** 485-489
[4] Abragam A 1961 *Nuclear magnetism* (Oxford: Clarendon Press)
[5] Lösche A 1957 *Nuclear induction* (Berlin)
[6] Koloskova N G, *Paramagnetic resonance (collected book)* (Kazan: KSU) 1964 p.115
[7] Abragam A and Bleaney B 1970 *Electron paramagnetic resonance of transition ions* (Oxford: Clarendon Press)
[8] Vyalikh D V, Danzenbächer S, Kucherenko K, Krellner C, Geibel C, Holder M G, Kim T K, Laubschat C, Shi M, Patthey L, Follath R and Molodtsov S L 2010 *Phys. Rev. Lett.* **105** 237601

Journal of Physics: Conference Series **324** (2011) 012019 doi:10.1088/1742-6596/324/1/012019

ESR study of spin dynamics in the ternary phosphide $YbRh_6P_4$

V A Ivanshin[1], E M Gataullin[1] and A A Sukhanov[2]

[1]Kazan (Volga region) Federal University, Kremlevskaya str. 18, 420008 Kazan, Russia
[2]Zavoisky Physical-Technical Institute, Sibirsky trakt 10/7, 420029 Kazan, Russia

E-mail: Vladimir.Ivanshin@ksu.ru

Abstract. We have used electron spin resonance (ESR) to investigate the $YbRh_6P_4$ powder samples. . It is shown that the exotic ESR absorption is caused here by hybridization effects between the Yb *4f*-orbitals and the wave functions of the conduction electrons (CE) of the outer electronic shells. A broadening and disappearance of the ESR line at temperatures above 10 K can be explained by the processes of the spin-lattice relaxation of the Yb^{3+} ions through the first excited Stark sublevel with an activation energy $\Delta \approx 82.1$ K.

1. Introduction

The explanation of transformation of localized electrons at high temperatures into itinerant quasi-particles with an enhanced mass at lower temperatures belongs to the most important challenges in the physics of the solid state. The mechanism of this evolution depends on the Kondo interactions between the CE and the localized *d* or *f* electrons. ESR probes microscopically both the local moment (LM) spins and CE in different strongly correlated electron systems (SCES) such as high-temperature superconductors, pnictides, heavy fermion systems. As a rule, LM ESR is studied in compounds with doped paramagnetic ions and localized electrons. ESR from CE can be detected in metallic systems with an enhanced Pauli susceptibility. Surprisingly, ESR could be also observed in some dense SCES without any kind of paramagnetic doping (s. [1] and references therein). According to several recent theories [2-5], a measurable ESR signals in these experiments are caused by an effective hybridization between localized *f* electrons and conduction bands in conjunction with ferromagnetic (FM) correlations which reduce usually a very fast spin-lattice relaxation. In the present contribution, we report the temperature-dependent ESR spectra in the recently synthesized phosphide $YbRh_6P_4$.

2. Experimental

$YbRh_6P_4$ belongs to the large number of metal-rich compounds with a metal-to-phosphorus ratio near 2:1. The twinned crystal $YbRh_6P_4$ samples were prepared by the bismuth flux technique and were refined using X-ray diffractometer [6]. ESR spectra were recorded on an X-band EMX/plus spectrometer (Bruker BioSpin GmbH, Germany) in the temperature range between 4.2 and 20 K. Figure 1 presents the temperature evolution of the ESR spectrum with $g_{eff} \approx 3.34$ (at T = 6 K) in the $YbRh_6P_4$ powder samples which shows typical properties of the Yb^{3+}-ESR similar to $YbRh_2Si_2$ [7], $YbIr_2Si_2$ [8], and $YbRh_2Pb$ [9].

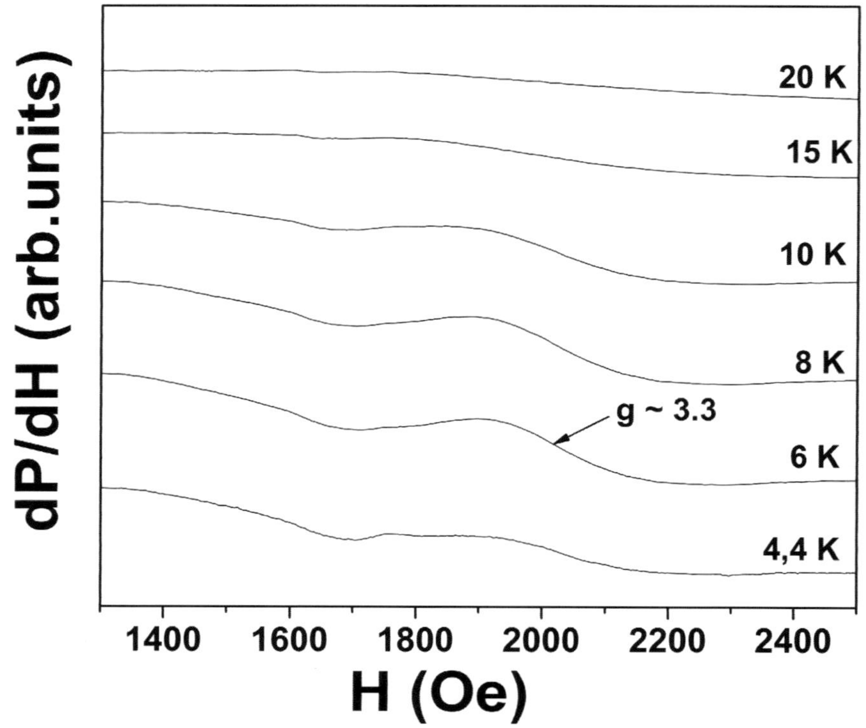

Figure 1. ESR spectra of $YbRh_6P_4$ at 9.4 GHz at selected temperatures.

3. Discussion

The twinned sample structure has prevented the detailed investigation of the anisotropy of the LM centers. The nature of the relaxation mechanism of the ESR-active spins has been analyzed using the temperature dependence of the ESR peak-to-peak linewidth ΔH which is shown in Figure 2. These data can be well described by the dashed line which displays the behaviour for the LM relaxation in metallic environment [7-9]:

$$\Delta H = A + BT + C\exp(-\Delta/T), \tag{1}$$

where $A \approx 348.7$ Oe, $B \approx 5$ Oe/K, $C \approx 5.6$ kOe, and $\Delta \approx 82.1$ K. The parameter A is caused by spin–spin interactions and inhomogeneous ESR line broadening. The term BT belongs to the contribution to the ESR linewidth from Korringa relaxation. The third term in Eq. (1) is the result of thermal fluctuations of the electronic states of the Yb^{3+} ions proposed in [10] for $YbRh_2Si_2$.

This spin-lattice relaxation mechanism which stimulates random transitions of the Yb^{3+} ions from the ground state to the first excited Kramers doublet with an activation energy $\Delta \approx 82.1$ K can be also responsible for the temperature dependence of the effective ESR g-factor in $YbRh_6P_4$ at T > 10 K (figure 3):

$$g(T) = g_0 + \Delta g_0 \exp(-\Delta/T),$$
$$\Delta g_0 = g_{exc} - g_0, \tag{2}$$

where $g_0 \approx 3.34$ and $g_{exc} \approx -25.66$ are the effective g factors of the ground and the first excited Stark sublevels of the Yb^{3+} ion, respectively.

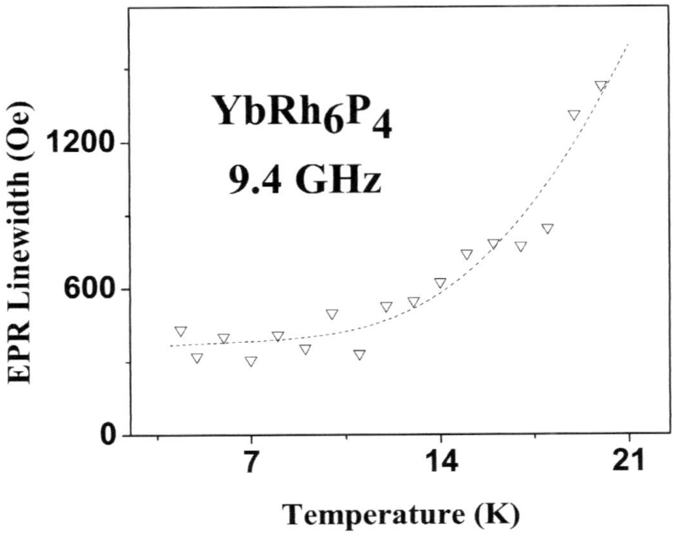

Figure 2. Temperature dependence of the ESR linewidth in $YbRh_6P_4$. The dashed line presents the equation (1) with the parameters $A \approx 348.7$ Oe, $B \approx 5$ Oe/K, $C \approx 5.6$ kOe, and $\Delta \approx 82.1$ K.

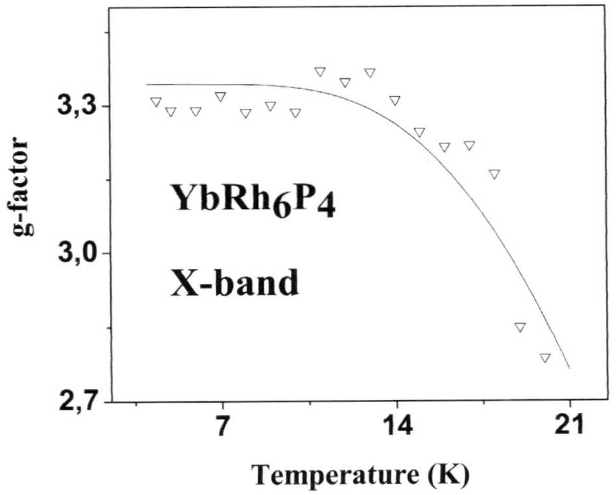

Figure 3. Temperature dependence of the ESR g-factor measured at a frequency 9.4 GHz. The solid line is the theoretical fit using equation (2).

Functional dependence (2) which is shown as a solid line in figure 3 illustrates a transition from the mixed correlated state of localized 4f states of ytterbium and collectivized electrons from the unfilled outer d and p shells of rhodium and phosphorus below 10 K to a paramagnet with the magnetic moments of the 4f shells close to the magnetic moments of free Yb^{3+} ions at higher temperatures.. Although a very similar behaviour has been observed in $YbRh_2Si_2$ and $YbIr_2Si_2$, the ESR signal intensity in $YbRh_6P_4$ was significantly smaller demonstrating a weaker hybridization effects in the phosphide in contrast to the both these silicides. Most likely, a weak extent of hybridization results also in the small value of an indirect interspin Rudermann-Kittel-Kasuya-Yosida (RKKY) interaction and the corresponding FM correlations [11] as it also has been detected in the EPR experiments on the hybridized electronic states in the Heusler alloy $YbRh_2Pb$ [9]. It means that the replacement of Si ([Ne]$3s^23p^2$) by Pb ([Hg]$6p^2$) or by P ([Ne]$3s^23p^3$) does not lead to a significant qualitative change of the ESR behaviour but dramatically reduces the *f-p* hybridization in $YbRh_6P_4$ and $YbRh_2Pb$ as compared to $YbRh_2Si_2$. Thus, a comparison of spin dynamics in all Yb-Rh-based systems investigated by ESR method [7-10] allows us to suggest that the *f-d* hybridization is mainly responsible for the contribution of CE in the collective spin mode of LM and CE. The strength of the *f-p* hybridization which determines the intensity of ESR absorption is essentially more in the both Si-containing intermetallics ($YbRh_2Si_2$ and $YbIr_2Si_2$) than in $YbRh_6P_4$ or $YbRh_2Pb$.

In conclusion, the low temperature ESR absorption has been found out in the new phosphide $YbRh_6P_4$. Electronic spin-lattice relaxation processes with an involvement of the first excited Stark sublevel of the Yb^{3+} ion with an energy $\Delta \approx 82.1$ K describe quite well a drastic broadening and vanishing of the ESR signal above 10 K. The observed ESR behavior provides evidence for the relatively weak *f-p* hybridization effects in $YbRh_6P_4$ in contrast to silicides $YbRh_2Si_2$ and $YbIr_2Si_2$.

We would like to thank U. Pfannenschmidt and R. Pöttgen for sample preparation.

References

[1] Ivanshin V A, Litvinova T O, Sukhanov A A, Ivanshin N A and Enderward B 2011 *Solid State Phenomena* **170** 170
[2] Abrahams E and Wölfle P 2008 *Phys. Rev.* B **78** 104423
[3] Schlottmann P 2009 *Phys. Rev.* B **79** 045104
[4] Huber D L 2009 *J. Phys.: Condens. Matter* **21** 322203
[5] Wölfle P and Abrahams E 2009 Phys. Rev. B **80** 235112
[6] Pfannenschmidt U, Rodewald U C and Pöttgen R 2011 *Monatsh. Chem.* **142** 219
[7] Sichelschmidt J, Ivanshin V A, Ferstl J, Geibel C and Steglich F 2003 *Phys. Rev. Lett.* **91** 156401
[8] Sichelschmidt J, Wykhoff J, Krug von Nidda H-A, Fazlishanov I I, Hossain Z, Krellner C, Geibel C and Steglich F 2007 *J. Phys.: Condens. Matter* **19** 016211
[9] Ivanshin V A, Litvinova T O, Sukhanov A A, Sokolov D A and Aronson M C 2009 *JETP Lett.* **90** 116
[10] Ivanshin V A, Aminov L K, Kurkin I N, Sichelschmidt J, Stockert O, Ferstl J and Geibel C 2003 *JETP Lett.* **77** 526.
[11] Krellner C, Förster T, Jeevan H, Geibel C and Sichelschmidt J 2008 *Phys. Rev. Lett.* **100** 066401

Investigation of Eu Ternary Pnictide by ESR

T S Altshuler [1], Yu V Goryunov [1] and A V Nateprov [2]

[1] Kazan Physical-Technical Institute of the RAS, 420029, Kazan, Russia
[2] Institute of Applied Physics of the ASM, 2028, Kishinev, Moldova

tatiana@kfti.knc.ru

Abstract. The investigation of the ESR in EuZ_2nAs_2 was performed at the frequency 9.3 GHz in the temperature range 4.2 - 300 K. The observed resonance line with the g-factors 1.96 of the Eu^{2+} ions was symmetric and it was described very good by a Lorentzian lineshape at temperatures above 120 K. At the temperature decreasing well before the magnetic ordering temperature ($\mathbf{T_N}$) we was observed an exponential-like increasing of linewidth and a decreasing of the resonance field, indicating the magnification of antiferromagnetic fluctuations and the effective molecular Weis's field. The shape of the resonance line changes from Lorentzian to Gaussian at the temperature close to the $\mathbf{T_N}$. Below T_N we observed a decrease of the lineswidth and the changing of the lineshape to the shape close Dysonian.

1. Introduction

Electron spin resonance (ESR) has proved its effectiveness in the investigation of magnetic phenomena in rare-earth compounds. F-electron compounds showed us an amazingly wide range of different ground states, including complex forms of magnetic order, heavy-fermion and Kondo-lattice. The large variety of ground states is generally attributed to the presence of strong electronic correlations. However, the nature of the electronic correlations in f-electron materials is as diverse as the row of different ground states which they assume. Europium has a half-filled 4f - shell. The electronic configuration of the f-shell in Eu - compounds is $4f^6$ (nonmagnetic) or $4f^7$ (magnetic) depending on whether the valence of the europium is either 3+ or 2+. The divalent $4f^7$ state is a purely spin state S= 7/2. The energetic proximity of these configurations may lead to valence fluctuations. This is the reason why the properties of the intermediate-valent systems have attract great scientific interest [1-3]. Relevance of researches of an electronic interactions in the "122" superconducting iron pnictides attaches great interest in studying similar pnictides of rare-earth metal.

2. Results and discussion

2.1. Experiment

Polycrystalline $EuZn_2As_2$ was prepared by direct reaction of the stoichiometric amounts of Zn, As (99,999 %) and Eu (99,99 %) in an Al_2O_3 crucible enclosed in an evacuated quartz ampoule. The trigonal $CaAl_2Si_2$-type crystal structure (space group $\boldsymbol{P3^{-}m1}$, Int. tables No. 164) was confirmed by X-ray powder difraction. The ESR (electron spin resonance) measurements of the $EuZn_2As_2$ powder samples (size of particles ~ 4 μm) were performed on frequency 9.3 GHz in TE_{102} rectangular cavity of EPR-spectrometer "Bruker" in the temperature range from 4.2 to 300 K. The dc susceptibility M/B was measured at 2 - 400 K with a commercial Vibrating Sample Magnetometer (Oxford Instruments).

2.2. Electron spin resonance in $EuZn_2As_2$

2.2.1. Resonance fields, linewidth and g- factor

The observed single resonance lines of the Eu^{2+} ions were symmetric and they are described very good by a Lorentzian lineshape with linewidths 530 Oe at temperatures above 120 K (see fig. 1). Position and width of the resonance lines were not changed and the g-factors was 1.96(3) in this temperature range. Bivalent europium has the electron spin 7/2 and two stable isotopes with a nuclear spin of 5/2. Therefore 84 resonance lines should be observed. The absence of the fine and hyperfine structures is associated with the averaging due to strong exchange interactions. At the temperature decreasing well before the magnetic ordering temperature we was observed an exponential-like increasing of linewidth and a decreasing of the resonance field, indicating the magnification of antiferromagnetic fluctuations and the effective molecular Weis's field. Also this is confirmed by changing the shape of the resonance line from Lorentzian to Gaussian. Large deviation of the g-factor ($\Delta g \sim 0.03$) from 2.0 (g-factor of free electron, see also [4]) indicates on the strong hybridization of the europium f-electron states with the p- s- states of the band electrons.

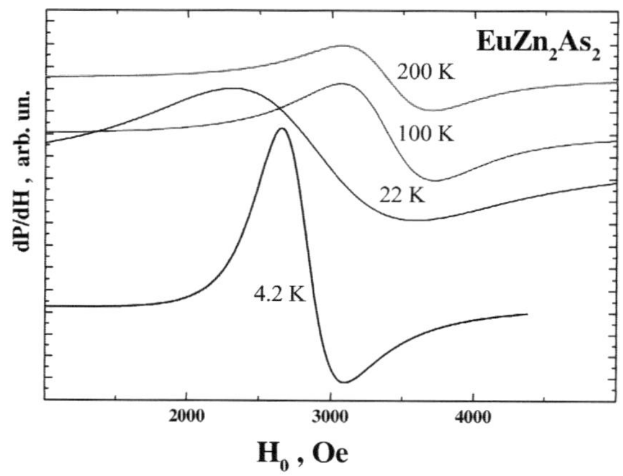

Figure 1. Types of the ESR lines under changes of temperature for $EuZn_2As_2$ powder sample. Parameter of line asymmetry A/B for T=4.2 K is 2.31. It is equal to 1.2 for high temperature.

Since $4f^7$ configuration of paramagnetic ions Eu^{2+} is pure spin state ${}^8S_{7/2}$, the resultant orbital angular momentum equal to zero. The crystal field may split its main levels (the ground state). In a crystalline field of cubic symmetry, the averaged energy of the spin-spin interaction between electrons of a single paramagnetic ion does not depend on their orientation with respect to each other and the ground state ion is completely degenerate in spin. Under the influence of crystal field of tetragonal or trigonal symmetry of the ion electronic cloud deforms slightly and averaged over the electron cloud energy of the spin-spin interaction depends on the relative orientation of electron spins of magnetic ions - the spin degeneracy is lifted. However, these interactions leading to some change in g - factor leave its isotropic. Given that the measurements were made on the powder sample, the isotropy of g - factor allows us to conclude that we are dealing with a real resonance line, but not with the convolution of the line with an anisotropic g - factor in the directions.

2.2.2. *Shape of ESR line*

At the temperature decreasing from room temperature to about 120 K, we observed a nearly symmetric resonance line of Lorentzian shape. A small asymmetry is described by a small (few percent) admixture of the dispersion associated with the Dyson's distortions caused by the presence of the skin layer. At temperatures around 100 K, the line asymmetry parameter A /B was 1.2. (A and B are values of dP/dH, respectively, for the high field peak and the low field peak in the first derivative of the field absorption line ESR). At further decrease in temperature occurs the deviation from the Lorentzian lineshape which increased along with broadening of the resonance line. The shape of the ESR line is getting shape closer to a Gaussian. This is seen most dramatically in the temperature range

10 – 23 K, where it was required of additions of the 40% of the impurity of Gaussian shape line for an acceptable description of the shape (see fig. 2). Also in the spectra measured at temperatures below 21 K, appeared and became a prominent feature of the low field, similar in form to the high field wing of the resonance line width of about 300 Oe and a zero resonant field. Spectra of this band is best described by three resonance lines of Gaussian shape: a line called the zero-field, the main line with a low field peak near 2500 Oe and the line with a resonant field near 1800 Oe. The width of the last line was close to the width of the main line, and the intensity was about 10% of the intensity of the main line. How such a description reflects the actual spectra can be judged from the figure 2, where are shown two calculated spectra for the temperatures 17.6 K and 12.5 K. The measurements of ESR at 4.2 K showed that markedly decreased the intensity and the width of the ESR, and line was shifted even more to lower field. The lineshape at 4.2 K was once again the ideal Lorentzian shape. But the Dyson's distortion increased so much that it became consistent with the ESR lineshape of the bulk metallic sample: the asymmetry parameter A / B line ESR was 2.31 against 2.56 of its value for a bulk sample (see figure 1).

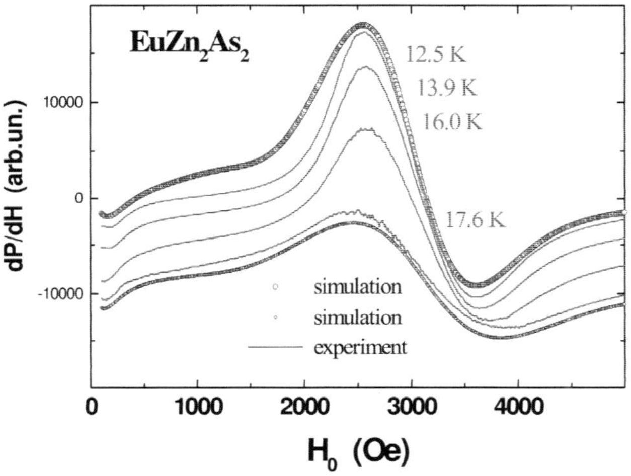

Figure 2. Types of the ESR spectrum at low temperatures. The upper and lower spectra are the result of calculation. The initial part of the spectrum at temperatures near to T_N is interpreted as due to the spin-flop transition.

Another possible explanation for the distortion of the line shape of ESR, like Dyson's distortion, may be anisotropy of the antiferromagnetic fluctuations. But this explanation does not currently seem likely. If we assume the resonant field that substantially exceeds the value of the field of spin-flop transition (as is evident from the behavior of the baseline in low fields on fig.2), the magnetic sub-lattices in such magnetic field are known to be able to slant, regardless of the orientation of the antiferromagnetic axis with respect to the external magnetic field. Therefore to believe there is a specific anisotropy of the resonance fields leading to such "Dyson's distortions" was not warranted. Need for additional line of small intensity in the field around 1800 Oe is consistent with the possibility of formation of bound states of magnetic ions with electrons and valence band and conduction band (magnetic polarons). If we assume that this additional line due to the interaction with conduction electrons, and its position affect the fluctuations of the exchange field in the same manner as they impact on the main line, the difference of resonance fields of this line and main line is ~ 1250 Oe. It gives origin g – factor of the small intensity line ~ 3, 0.

2.2.3. *Intensity of the ESR line.*

Intensity of the ESR spectrum is proportional to magnetic susceptibility of a spin systems. If there are not effects which influense on quantity of spins, we have selective susceptibility of spin sub-system. Peculiarities in the behavior of the spectrum baseline ESR is usually associated features of the dependence of microwave surface impedance of the sample on the magnetic field. These relationships can be caused by mutations occurring in the spin system under the influence of an external magnetic field, magnetoresistance, superconductivity. Features, associated with the magnetoresistance, have the following character with decreasing temperature. Originating in the high fields they move to low field. Features, associated with superconductivity, have the opposite character. Nucleating at low fields, they quickly moves to the high fields. In addition, this features are significantly stronger than the intensity of the resonance lines. The most suitable explanation for our case of the low field ESR features of the spectrum is the restructuring of the spin system $EuZn_2As_2$ in relatively small fields. If we turn to the previously investigated similar compounds $EuZn_2Sb_2$ [2], we see that in fields less than 400 Oe is the spin-flop transition in the magnetic sub-lattices of the antiferromagnet. In our case, the region of features of the baseline, which we associate with the spin-flop transition, begins with a temperature slightly higher than the Neel temperature, determined by the maximum intensity of the ESR. This may be due to the induced external magnetic field induced ferromagnetic order on the background of strong antiferromagnetic fluctuations. Low value of the Neel temperature obtained from measurements of the intensity of the ESR could be related to the peculiarities of these measurements are associated with the use of microwave radiation. As the figure shows the changing of intensity of the ESR line with temperature corresponds to a typical temperature dependence of the susceptibility of an antiferromagnet. Also the temperature dependence of inverse intensity is typical. However, it should be emphasized that, unlike the typical for antiferromagnet negative paramagnetic Curie-Weiss temperature θ_{CW} we obtained positive θ_{ESR} and significantly greater than the Neel's temperature T_N. This fact of positiveness paramagnetic Curie-Weiss tempereature for the antiferromagnet is confirmed by direct measurements of the dc susceptibility. The obtained values of the paramagnetic temperatures were quite close: $\theta_{ESR} = +20.4$ K and $\theta_{CW} = +21.2$ K.

2.3. DC – susceptibility

The magnetic properties of $EuZn_2As_2$ were studied on polycrystalline samples. In the temperature range 10 - 300 K the susceptibility M/B follows a Curie-Weiss law (see Fig. 4). The effective paramagnetic moment $\mu = 7.1 \mu_B$/Eu and antiferromagnetic order with $T_N = 16.5$ K (Fig. 4). As the ESR date the inverse susceptibility shows a positive paramagnetic Curie-Weiss temperature $\theta_{CW} = 21.2$ K. The saturation moment of the magnetization is 5.7 μ_B/Eu. The reduced values point out the presence of some Eu^{3+} ions in $EuZn_2As_2$.

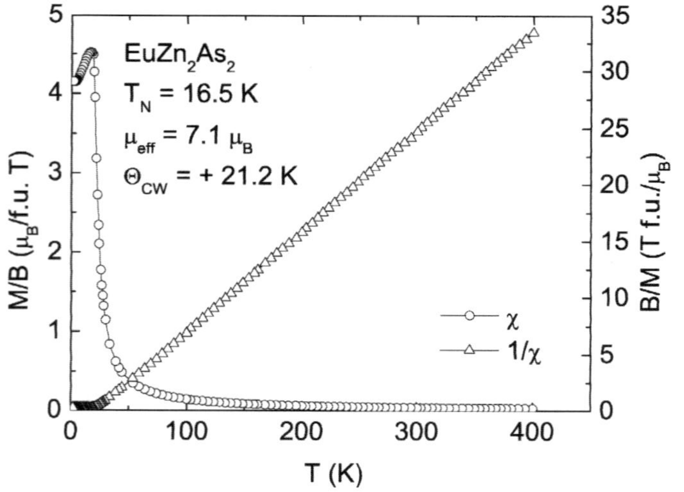

Figure 4. DC Susceptibility of the polycrystalline $EuZn_2As_2$.

Journal of Physics: Conference Series **324** (2011) 012020 doi:10.1088/1742-6596/324/1/012020

We found no evidence for ferromagnetic order], however the positive Weiss temperature points to the existence of ferromagnetic correlations in this material.

It is clear that the antiferromagnetic fluctuations are caused by fluctuations in the exchange field, which in turn is related to density fluctuations of the conduction band electrons and the valence band electrons. Fluctuations in the density of valence band electrons may well cause fluctuations in the valence state of europium between the magnetic state of Eu^{2+} and nonmagnetic Eu^{3+}. This explains small value of the saturation magnetization obtained in the measurements of dc susceptibility.

3. Conclusion

Thus, we can conclude on the basis of obtained data that $EuZn_2As_2$ is an antiferromagnet in which there are strong ferromagnetic correlations in the fields above 1000 Oe. This is reflected in the positive sign of the paramagnetic Curie-Weiss temperature. Strong Dyson's distortions of the ESR lineshape for powder sample are associated with a sharp decrease in 5-10 times of the skin layer and with a corresponding increase of the conductivity of the sample, i.e. its transition from a semiconducting to metallic character and the disappearance of the narrow gap between the valence and conduction bands. The observed strong decrease of the g-factor we can explain by the formation of bound states of the magnetic moment of europium with free charge carriers. Taking into account that the density of electrons in the conduction band is very low, we believe that this shift is due to hybridization and the formation of bound states of 4f-electrons Eu^{2+} ion with the electrons of the valence band. As a consequence the indirect exchange interaction between the divalent ions of europium through the electrons of the valence band, the so-called modified-RKKY Bloembergen-Rowland's [5, 6], is responsibly behind anti-ferromagnetism of $EuZn_2As_2$.

Acknowledgments

This work was supported by grants from RFBR and Presidium of RAS.

References

[1] Yu C, Zhu T J, Zhang S N, Zhao X B, He J, Su Z, Tritt T M, 2008 *J. Appl. Phys.* **104** 013705
[2] Weber F, Cosceev A, Drobnik S, Faisst A, Grube K, Nateprov A, Pfeiderer C, Uhlarz M and Löhneysen H 2006 *Phys. Rev.* **B 73** 014427
[3] Nateprov A et al., 2009 *Abstracts book of ICM*-2009 (26-31.07.09, Karlsruhe) p 342
[4] Altshuler S A and Kozyrev B M 1972 *Electron paramagnetic resonance of the intermediate group compounds.*(Moscow: Science)
[5] Ruderman M A and Kittel C 1954 *Phys. Rev.* **96** 99
[6] Bloembergen N and Rowland T J 1955 *Phys. Rev.* **97** 1679

Paramagnetic Curie Temperature of $EuB_{6-x}C_x$ from ESR data

T S Altshuler[1], Yu V Goryunov[1] and A V Levchenko [2]

[1] Kazan Physical-Technical Institute of the RAS, 420029, Kazan, Russia
[2] Institute of Material Science Problems of the NASU, 03680, Kiev, Ukraine

E-mail: tatiana@kfti.knc.ru

Abstract. The ESR measurements of the $EuB_{6-x}C_x$ (x<0.1) single crystal samples were performed on frequency 9.25 GHz in TE_{102} rectangular cavity in the temperature range from 10 to 300 K. We found that the single-crystal europium hexaboride with substituting 2% of boron by carbon behaves in fields around 0.4 T along the [111] - direction as a ferromagnetic, while for field along the [100] - direction its behavior resembles the behavior of an anti-ferromagnetic.

1. Introduction

It is known, that complex forms of magnetic order are realized in rare-earth compounds. However, unusual behavior of these compounds may be also in paramagnetic region. Earlier, we have observed [1] in single crystal of europium hexaboride the anisotropy of the induced magnetization $\boldsymbol{M}$, which associated with anisotropy magnetic susceptibility. Unusual behavior of magnetic susceptibility in paramagnetic state was observed by other researchers [2, 3]. A practice shows that the method of electron magnetic resonance (ESR) may be useful for the selective measurement of magnetic susceptibility. It is well known that the intensity of the ESR spectrum is proportional to the magnetic susceptibility of the sample. In general, the magnetic susceptibility χ is a tensor and components of vector of magnetization are associated with components of magnetic field vector as follows:

$$\boldsymbol{M}_i = \Sigma\, \chi_{ij} \cdot \boldsymbol{H}_i \tag{1}$$

In this approximation the magnetic susceptibility dependence on magnetic field and temperature can be presented as follows:

$$\chi = \frac{\mu_0 \mu_B g_e}{2\boldsymbol{H}_0}\left[(2S+1)\coth\left(\frac{(2S+1)\mu_B g_e \boldsymbol{H}_0}{2kT^*}\right) - \coth\left(\frac{\mu_B g_e \boldsymbol{H}_0}{2kT^*}\right)\right] \tag{2}$$

$T^* = T + \theta_p$; $\theta_p = 1/3k \cdot S \cdot (S+1) \cdot \Sigma\, J_{ij}$,

where $\boldsymbol{H}_0$ – magnetic field, $\boldsymbol{M}$ – magnetization , T – temperature, θ_p - paramagnetic Curie temperature J_{ij} – exchange integral of the i- -j- atom interactions.

In this regard, the components of magnetization can have not a proportional relation with the components of the external magnetic field and the Curie-Weiss's law can be different for different directions. Thus, paramagnetic Curie temperature can be anisotropic in case of anisotropic exchange interaction. Present work is devoted to experimental study of this question.

2. Results and Discussion

2.1. Experimental details

The ESR measurements of the EuB6 single crystal samples were performed on frequency 9.25 GHz in TE_{102} rectangular cavity in the temperature range from 10 to 300 K. We used samples of identical form and size but different orientation of crystal axis, to estimate the temperature dependences of the ESR spectra parameters, namely: resonance field, linewidth, intensity, lineshape. Paramagnetic Curie temperature was estimated by measuring the temperature dependences of inverse intensity of ESR signal. The induced by magnetic field magnetization of the sample was studied by measuring the shift of the resonance line under the influence of the demagnetizing field at perpendicular orientation of the sample plane to external magnetic field as [1, 4].

2.2. Influence of Carbon concentration on Magnetic order and Paramagnetic Curie temperature

As a rule, in temperature range 150 -300 K lineshape was lorentzian shape with dysonian distortion. Also as in [4], at T = 10 – 30 K was observed the magnetic phase separation in samples with small impurities of carbon. For sample with more content of carbon splitting of single ESR line was observed at higher temperature and the dependences of resonance field and linewidth are quite different. Figure 1 shows the temperature dependence of the inverse ESR intensity of the spectrum for the orientation of the plates [100] pure europium hexaboride and europium hexaboride, in which 7% of boron substituted carbon. Dependence in figure 1a is a classic example of the temperature dependence of the susceptibility of the ferromagnet. Dependence in figure 1b is a typical temperature dependence of susceptibility for not compensated antiferromagnet - ferrimagnet. These curves show quite high in comparison with the magnetic ordering temperature paramagnetic Curie temperature with opposite signs. A sample of europium hexaboride single crystal plate in the orientation [100] with a smaller (2%) admixture of carbon also showed negative a paramagnetic temperature. Thus, the substitution of boron with carbon leads to a monotonic amplification of antiferromagnetic properties of europium hexaboride.

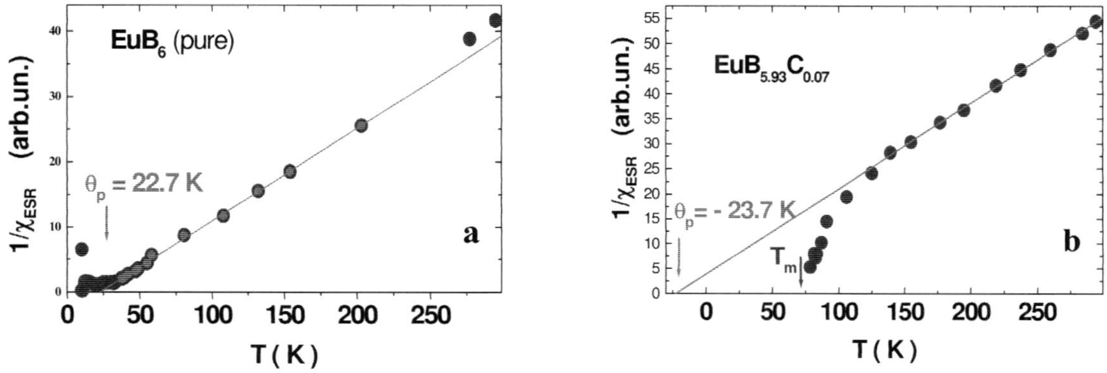

Figure 1. The temperature dependence on inverse intensity of the ESR spectrum of the $EuB_{6-x}C_x$ single crystal.

2.3. Influence of Magnetic field direction on Paramagnetic Curie temperature

Earlier in [1], we found a strong anisotropy of the induced magnetization of pure europium hexaboride (see Fig. 2). For pure hexaboride this anisotropy was in considerable amplification of the magnetization along [111] direction. Also, there is an amplification of the ferromagnetic properties along this direction and to antiferromagnetic samples with impurity carbon. This is most clearly evident for the samples of the $EuB_{6-x}C_x$. We found that the single-crystal europium hexaboride with substituting 2% of boron by carbon behaves in fields around 0.4 T along the [111] - direction as a ferromagnetic, while for field along the [100] - direction its behavior resembles the behavior of an anti-ferromagnetic

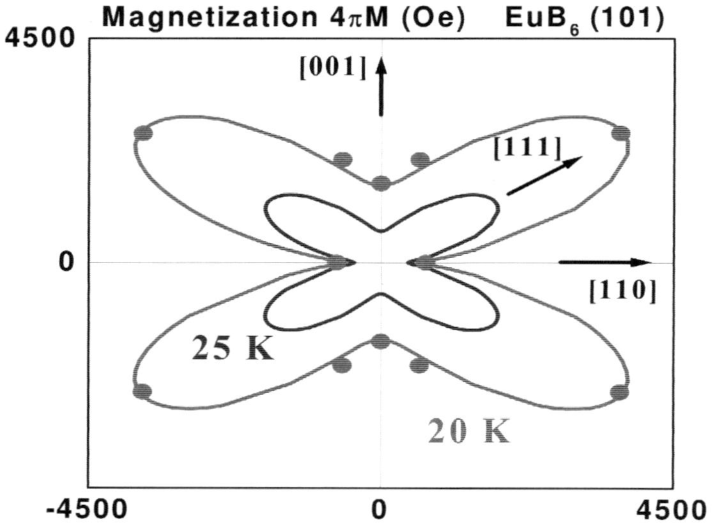

Figure 2. Effective magnetization induced by resonance magnetic field along different crystal axes of the EuB_6 sample for various temperatures.

Paramagnetic Curie's temperature was about +8 K in case an external magnetic field parallel to [111] axis and it is negative temperature about -7 K for the magnetic field along the [100] axis (see fig. 3).

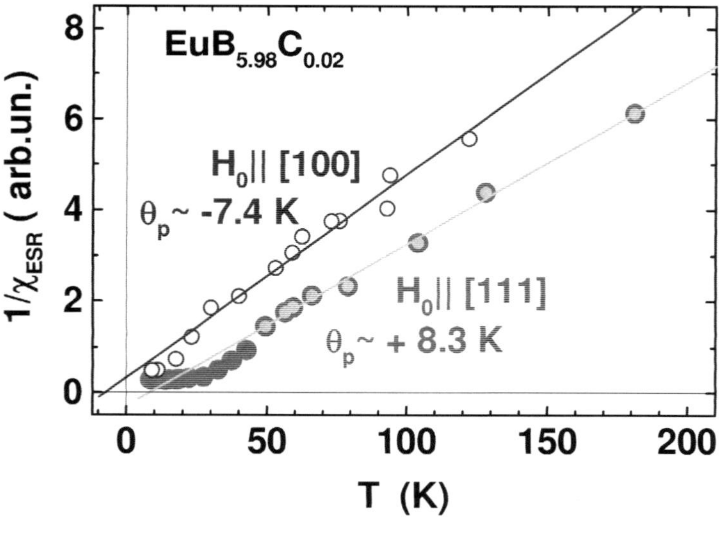

Figure 3. Temperature dependences of the inverse intensity of the ESR spectra for two orientations of the EuB_6 single crystal in sweep magnetic field.

It should be noted that for the orientation [100] the deviation of inverse susceptibility from the linear dependence is similar to figure 2, but point T_m finds at lower temperatures

2.4. Structure of EuB_6 and the role of boron in the indirect exchange interaction between Eu ions

From the viewpoint of possibility of implementing those or other interactions between ions europium it is useful to consider the structure of europium hexaboride more detail than is usually done. As is known, the structure of europium hexaboride represents two inserted into each other the simple cubic lattices of europium ions and B_6 octahedra with size of the cube 4.18 Å. In general, the boron octahedra are also charged, negatively, and they have a deficiency of valence electrons.

Assuming equidistant (1.73 Å) of the boron-boron bonds we obtain that europium ion is surrounded by 24 boron atoms, placed on a sphere of radius 3.08 Å. The 24 atoms are organized in a triples of boron atom from 8 different octahedra B6. The spheres have a common boron atoms in the directions {100} and {110}, and in the direction of {111} there were none. In other words, the planes (111) of an europium ions are separated by two planes of boron atoms, the planes (100) of an europium ions are separated by a single layer of boron atoms and the planes (110) contain a boron atoms.

Direct exchange interaction between europium magnetic moments is practically excluded in such an environment of 24 boron atoms. Conventional RKKY interaction can not be effective due to the small number of electrons in the conduction band. From effective mechanisms of the indirect exchange interaction remain the mechanisms of indirect exchange via the valence electrons, namely superexchange Kramers-Anderson [8-10] and the modified RKKY-interaction [11] Bloembergen-Rowland [12]. These mechanisms are practically indistinguishable for such alternation of layers.

Indirect exchange interaction (superexchange Kramers-Anderson) via p-orbitals has a anisotropic character, it can be both antiferromagnetic and ferromagnetic, thus leading to the formation of helical magnetic structures [6, 7]. In our case, assuming a character of indirect exchange interaction via sp-orbitals of boron is antiferromagnetic, we automatically obtain the same for directions {100} and {110}, and the ferromagnetic exchange for the directions {111}. It should be also noted that the boron-boron bonds have the directions only {100} and {110}.

Because the sp-orbitals boron of the octahedra impede to the direct moving of the valence 4f7 electron from one europium ion to other ion, the localized 4f- electrons impede the free movement of sp - electrons, forming with them the bound states.

The appearance of free electrons leads to filling of sp- hybrid orbitals of the octahedra B_6, which should lead to their local compression, as by strengthening the real covalent bonds, and due to their compression by positive ions europium which are attracted to the excess negative charge of the B_6 octahedra. In fact, this is the formation of lattice polarons with the ferromagnetic axis along the {111} direction and symmetry breaking. In this case, the behavior of the B_6 octahedron is somewhat similar the behavior of the Jahn-Teller ion.

3. Conclusion

Thus, summing up the discussion, we conclude the following. The external magnetic field changes the distribution of spin density in electron shells by means polarization of the localized spins and these shells, thereby changing the character of the exchange interaction depending on the direction of the external magnetic field. Changing the exchange integrals leads to a change in the average energy of interaction between the localized moments, to the change of the magnetic structure in the ground state and, consequently, a change in the paramagnetic Curie temperature.

Acknowledgments

This work was supported by grants from RFBR and Presidium of the RAS.

References

[1] Altshuler T S, Goryunov Y V, Shitsevalova N Y et al. 2010 *J. Phys: Conf. Series* **200** 032019
[2] Nateprov A et al. 2009 *Abstracts book of ICM*-2009 (26-31.07.09, Karlsruhe) 342
[3] Weber F, Cosceev A, Drobnik S, Faisst A, Grube K et al. 2006 *Phys. Rev.* B **73** 014427
[4] Altshuler T S, Goryunov Y V, Shitsevalova N Y and Dukhnenko A 2008 *JETP Lett.* **88** 779
[5] Altshuler S A and Kozyrev B M 1972 *Electron paramagnetic resonance of the intermediate group compounds* (Moscow: Science) 478
[6] Vonsovskii S V 1971 *Magnetism: magnetic propertis of dia-, para-, ferro-, antiferro- and*

ferrimagnets (Moscow: Science) 102-410
[7] Nagaev E L 1979 *Physics of magnetic semiconductors* (Moscow: Science) 252-344
[8] Kramers H A 1934 *Physica* **1** 182
[9] Anderson P W 1950 *Phys. Rev.* **79** 350
[10] Anderson P W 1963 Exchange in insulators. Super-exchange, direct exchange and double exchange, in *Magnetism* (New York.-L.) vol. 1
[11] Ruderman M A and Kittel C 1954 *Phys. Rev.* **96** 99
[12] Bloembergen N and Rowland T J 1955 *Phys. Rev.* **97** 1679

Journal of Physics: Conference Series **324** (2011) 012022 doi:10.1088/1742-6596/324/1/012022

Possible mechanisms of magnon sidebands formation in transition metal compounds

M V Eremin and M A Fayzullin

Institute for Physics, Kazan Federal University, 19 Kremlyovskaya Street, Kazan, 420008, Russia

E-mail: Marsel.Fayzullin@ksu.ru

Abstract. In this short report possible mechanisms of magnon sidebands formation in transition metal compounds are shortly discussed. Besides the well known exchange electric dipole mechanism, suggested by Tanabe, Moriya and Sugano, recently a new the exciton-magnon absorption mechanisms assisted by phonons become actual. Special attention is paid to possible contribution of phonon-assisted electric dipole exchange transition in quasi-one-dimensional antiferromagnet $KCuF_3$.

1. Introduction

The presence of the magnetic excitations in infrared (IR) absorption spectra of antiferromagnetic (AFM) insulators have been the subject of intense experimental and theoretical studies that provided information on the spin dynamics in these materials. Often, a new experimental results in such compounds required a new theoretical approach which could more precisely describe the coupling of a photon with electron spins.

Optically active spin-wave excitations had been observed first in MnF_2 AFM [1]. The peculiarities of the sideband structure were understood on the basis exchange-induced electric dipole (EIED) mechanism suggested by Tanabe, Moriya and Sugano [2]. It is supposed weakly allowed electric dipole (ED) transition accompanied by spin-wave excitation due to effective ED moment coming from the second order of perturbation theory with off-diagonal exchange interaction giving the correct polarization [3] as well as integrated intensity [4] dependences of the sideband peaks in MnF_2. Moreover, they predicted direct two-magnon emission for the pairs of magnetic ions without inversion symmetry. The theory of two-magnon absorption was successfully applied to the rutile type iron group fluorides [5].

Perkins *et. all* [6] reported optical measurements for a series of undoped single-crystal copper oxides: La_2CuO_4, Nd_2CuO_4, Pr_2CuO_4, and $Sr_2CuO_2Cl_2$. For all four compounds in the case of $\mathbf{E} \parallel CuO_2$ layers the absorption spectra have common feature - the low-energy peak centered near 0.4 eV, which was subsequently attributed to the simultaneous excitation of two magnons (bimagnon) and an optical phonon. These features can be explained on the basis phonon-assisted multimagnon mechanism suggested by Lorenzana and Sawatzky [7]. Optical absorption in La_2NiO_4 [8] confirmed phonon-assisted bimagnon creation scenario but fails multimagnon one concerning a broad bands on the energy range 0.4 -1 eV in La_2CuO_4. The similar bands were observed at 1.5 eV for both La_2CuO_4 and $Sr_2CuO_2Cl_2$. Perkins *et. all* [9] have concluded that the origin of these bands can be treated as result of the exciton-phonon (magnon) sideband transition.

Experiments on near-IR optical absorption in $KCuF_3$ at 10 K [10] revealed two sharp spectral lines MD_1=8508 cm^{-1} and MD_2=9774 cm^{-1} accompanied by broad multiphonon band for both π and σ polarizations of incoming light. Each of these lines is followed by own good resolved magnon sideband structure which disappear as temperature increase towards T_N. Unlike the copper oxides gives clear evidence for the MD_1 and MD_2 transitions as well as neutron scattering study [11] allowed to assign sideband absorption as simultaneous creation a bare MD_1 or MD_2 exciton and magnons.

The conventional EIED mechanism, as it turned out, can not explain the observed low-energy sideband features but can be responsible to high-energy shoulders in absorption spectrum. The suggested exchange-induced magnetic dipole (EIMD) mechanism [12] explains the basic experimental observation concerning fine structure of phonon related absorption. However, an additional possible contribution the phonon-assisted mechanism has to be discussed.

2. Exchange-induced ED and MD transitions in $KCuF_3$

The perovskite structured $KCuF_3$ crystal is a quasi 1D AFM with the Neel temperature 39 K: Cu ions along the chain (parallel to c axis) suffer strong AFM exchange interaction, while the interchain exchange coupling is a weak ferromagnetic $J^{\|}/J^{\perp} \simeq -0.01$. Each copper ion Cu^{2+} (electronic configuration $3d^9$, S=1/2) is surrounded by a distorted octahedron. Under local crystal field and spin-orbit interaction the ground term ^2D splits into five doublets (see Fig. 1).

It is known that ED transitions between d levels are parity forbidden since the copper ion in the center of inversion, whereas MD transitions A_1-A_4 (in the notation of [10]) are allowed. The orbital ordering is such that two unlike copper ion positions with the ground $d_{x^2-z^2}$ and $d_{y^2-z^2}$ orbital states of a hole persist. The last one is a result of rotation the former by $\pi/2$ about the $c(z)$ axis. Using a modern conceptions of the crystal field theory we have calculated the relative probabilities of the A_1-A_4 optical transitions per two copper ion positions [13]. The calculations show that probabilities of the $A_2(MD_1)$ and $A_3(MD_2)$ transitions are large with respect to A_1, A_4 and, what is more, the observed relative intensities for the MD_1 and MD_2 transitions in π and σ polarizations coincides qualitatively with calculated one.

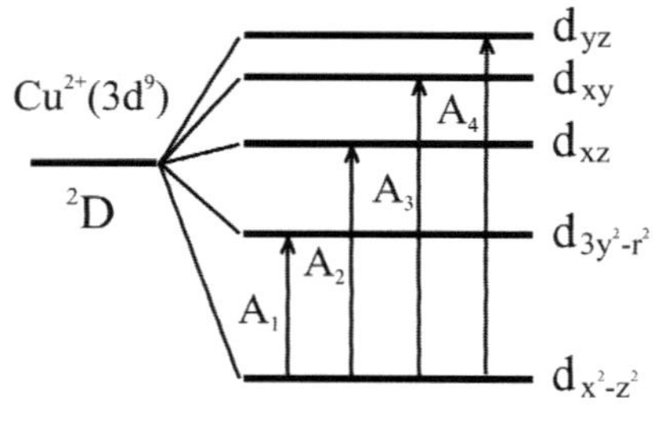

Figure 1. The energy level scheme of Cu^{2+} ion in $KCuF_3$. The calculated energies of the A_2, A_3 transitions are 8897 and 10097 cm^{-1}, respectively [13].

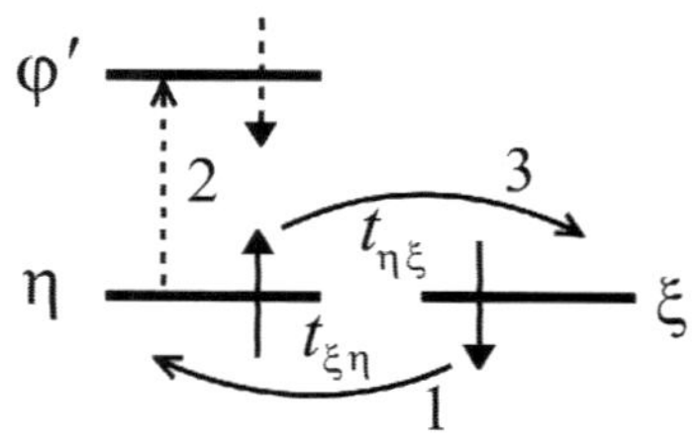

Figure 2. Virtual process scheme of the EIMD transition [12]. Solid arrows 1, 3 correspond to a hole hopping and dashed arrow 2 is referred to MD transition.

The first two magnon sidebands appear at 88 and 114 cm^{-1} that correspond to X[π/a,0,0] and M[π/a,π/a,0] low-energy singularities in the magnon density of states (DOS), respectively [12]. The other sidebands is due to multimagnon processes of the X and M magnons. Besides X,

M singularities the DOS showed a huge high-energy peaks for the Z$[0,0,\pi/2c]$ and R$[\pi/a,0,\pi/2c]$ boundary points which were not observed experimentally.

Like $KCuF_3$ in the MnF_2 an exciton transition is MD allowed but sidebands have ED origin. For $KCuF_3$ the Hamiltonian corresponding to absorption on the a site copper ion according to EIED scenario can be written as follows

$$H_{eff}^{ED} = \sum_b \frac{J_{ab}^{\|}}{t_{ab}} < \xi|\mathbf{Ed}|\varphi\prime > [\mathbf{S}_a^*\mathbf{S}_b - 1/4], \quad (1)$$

where $J_{ab}^{\|}$ is a parameter of conventional superexchange coupling between the nearest neighbors copper ions in the ground η and ξ orbital states, t_{ab} is an effective hopping integral, $\mathbf{E}$ - an electric field and, $\mathbf{d}$ is dipole moment required to a hole transfer from the b site to the excited $\varphi\prime$ orbital state of the a ion. The asterisk in (1) denotes that the spin operator for the a site is not a true spin operator because it has matrix elements between different orbital states.

We found that the matrix element in (1) is nonzero when the $\mathbf{E}$ is perpendicular to the c axis for the MD_1 exciton and hence can contributes only in σ polarization, whereas in the case of MD_2 exciton it is nothing for any direction of the $\mathbf{E}$. Performing the sum over copper spins and doing the Fourier transform like it explained in papers [3, 4] one finds that intensity of magnon sideband transitions is proportional to $\sin^2(k_z c)$. This factor vanishes at X and M points of Brillouin zone and therefore becomes clear way this mechanism is not effective at 88 and 114 cm^{-1} in $KCuF_3$. These facts gave us the idea about MD nature of the magnon sidebands, i.e. EIMD mechanism.

The effective Hamiltonian had been derived in [12]

$$H_{eff}^{MD} = \frac{\mu_B}{2} \sum_b \frac{J_{ab}^{\|}}{|\Delta_{ab}|} < \eta|\mathbf{Hl}_a|\varphi\prime > [\mathbf{S}_a^*\mathbf{S}_b - 1/4], \quad (2)$$

where μ_B is Bohr's magneton, Δ_{ab} is the charge-transfer energy, $\mathbf{l}_a$ - a hole orbital momentum on the a site and, $\mathbf{H}$ is a magnetic field. One of the virtual process scheme is shown in Fig. 2. The derived Hamiltonian gives a correct description of relative intensities at least for the first two sidebands in π and σ polarizations. The calculation of the absorption coefficient showed that not all singularities in DOS in magnon spectrum are responsible for observed sideband. After summation over nearest neighbors copper ions an important form-factor is appeared. In particular for $KCuF_3$, the singularities corresponding to X,M points in magnetic Brillouin zone remain active. Whereas the singularities related to Z and R points are filtered out by $\cos^2(k_z c)$ factor.

3. Phonon-assisted EIED mechanism

Like in paper [12] let us consider two exchange interacting Cu^{2+} ($3d^9$) ions at the a and b sites with the ground η and ξ orbital states, respectively. One of the virtual process scheme on which we focus here is illustrated on Fig. 3. Firstly, the spin up of the hole from the a center jumps to the b center (arrow 1), while the spin down on the b center via phonon induced effective $t_{\xi\lambda}$ hopping integral, excites the a center to the electron-phonon product $|\lambda >= |\varphi\prime > |ph >$ state (arrow 2), then due to coupling with electric field the phonon disappear and the spin down transfer to the $\varphi\prime$ state (arrow 3). The effective Hamiltonian for all possible spin configurations of the two centers can be obtained in third order of perturbation theory.

The perturbation here is a phonon coupling to the electric field $\mathbf{E}$

$$F_{ph} = |e| \sum_{\lambda,\varphi\prime} a_\lambda^+ < \lambda|\mathbf{Eu}_F|\varphi\prime > a_{\varphi\prime} + h.c., \quad (3)$$

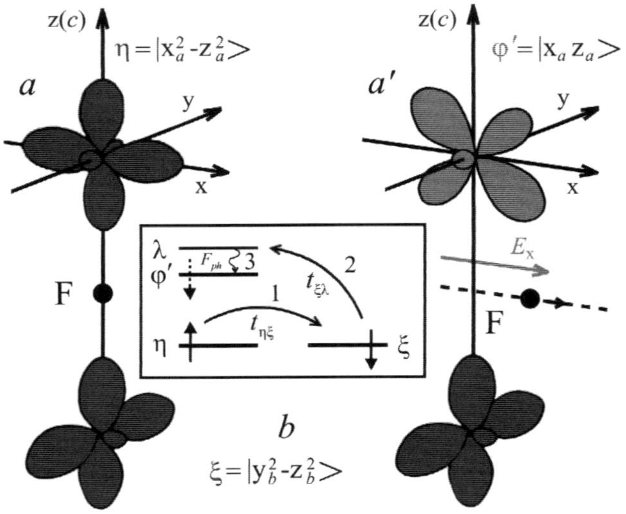

Figure 3. Spatial distribution of the ground η, ξ (blue) orbital states and excited $\varphi\prime$ (red) orbital state in the $x = 0, y = 0, z = 0$ position [13] of copper ions along the chain. Inset: virtual process scheme of the phonon-assisted EIED transition.

where e is an electron charge and $\mathbf{u}_F$ is vibration coordinate of fluorine bridge ion. Like in paper [12] using the canonical transformation method for S operator one has

$$S = \sum_{\xi,\rho\prime} \frac{t_{\xi\rho\prime}}{|\Delta_{\xi\rho\prime}|} b_\xi^+ a_{\rho\prime} - \sum_{\rho,\xi\prime} \frac{t_{\rho\xi\prime}}{|\Delta_{\rho\xi\prime}|} a_\rho^+ b_{\xi\prime}. \tag{4}$$

Here, summation is running over $\rho = (\eta, \lambda)$. The case of $\rho = \eta$ corresponds to conventional hopping between the ground states of copper ions (left hand side on Fig. 3), whereas the case of $\rho = \lambda$ is responsible to excitation the a ion due to an effective $t_{\xi\lambda} = \frac{\partial t}{\partial \mathbf{u}} \mathbf{u}$ hopping integral as a result of displacement of the fluorine ion away from c axis (right hand side on Fig. 3).

Using (3) and (4) for the effective Hamiltonian corresponding to absorption on the a center we get

$$H_{eff}^{ph-ED} = |e| \sum_b \frac{t_{ab}}{|\Delta_{ab}|} < \lambda|\mathbf{E}\mathbf{u}_F|\varphi\prime > \frac{t_{ba(\lambda)}}{|\Delta_{ba(\lambda)}|} [\mathbf{S}_a^*\mathbf{S}_b - 1/4]. \tag{5}$$

Performing sum over neighbors copper ions and doing Fourier transform one arrives to form-factor $\cos(k_z c)$. Therefore, the singularities in DOS of magnon excitations, corresponding to X,M points in magnetic Brillouin zone are also active in this mechanism. Besides, from Fig. 3 it is clear that it contributes to total integral intensity when the external electric field perpendicular to the c axis, i. e. in the σ polarization.

4. Conclusion

We shortly have discussed the microscopic models of magnon sidebands transitions in low dimensional transition metal compounds. Two important points have been emphasized. Important role plays the k-dependent form-factor, which can filter out the contribution of exchange electric dipole mechanism for low frequency peaks in magnon density of state. Instead it the magnetic exchange dipole mechanism becomes relevant. Second - for correct description of the origin of the magnon sideband transition the relative intensity analysis in different polarizations is very useful. The effective transition operator of phonon-assisted electric dipole mechanism has been derived. We found that it makes an additional contribution to low frequency magnon sideband in $KCuF_3$ in σ polarization only.

References

[1] Greene R L, Sell D D, Yen W M, Schawlow A L and White R M 1965 *Phys. Rev. Lett.* **15** 656
[2] Tanabe Y, Moriya T and Sugano S 1965 *Phys. Rev. Lett.* **15** 1023
[3] Tanabe Y and Gondaira K.-I 1967 *J. Phys. Soc. Japan* **22** 573
[4] Shinagawa K and Tanabe Y 1967 *J. Phys. Soc. Japan* **30** 1280
[5] Moriya T 1966 *J. Phys. Soc. Japan* **21** 926
[6] Perkins J D, Graybeal J M, Kastner M A, Birgeneau R J, Flack J P and Greven M 1993 *Phys. Rev. Lett.* **71** 1621
[7] Lorenzana J and Sawatzky G A 1995 *Phys. Rev.* B **52** 9576
[8] Perkins J D, Kleinberg D S, Kastner M A and Birgeneau R J 1994 *Phys. Rev.* B **52** R9863
[9] Perkins J D, Birgeneau R J and Graybeal J M 1998 *Phys. Rev.* B **58** 9390
[10] Deisenhofer J, Leonov I, Eremin M V, Kant Ch, Chigna P, Mayr F, Iglamov V V, Anisimov V I and van der Marel D 2008 *Phys. Rev. Lett.* **101** 157406
[11] Satija S K, Axe J D, Shirane G, Yoshizawa H and Hirakawa K 1980 *Phys. Rev.* B **21** 2001
[12] Eremin M V, Deisenhofer J, Fayzullin M A, Kant Ch and Loidl A 2011 *Solid State Phenomena* **168-169** 113
[13] Eremin M V, Iglamov V V and Fayzullin M A 2011 *Optika i Spektroskopiya* **111** 414

Narrowing of the exciton lines using WAHUHA method of solid state NMR spectroscopy

I I Geru

Institute of Chemistry of the Academy of Sciences of Moldova, 3 Academiei str., MD-2028, Chisinau, Moldova

E-mail: iongeru@yahoo.com

Abstract. A method for allocating of two thermodynamic reservoirs which are not related to each other in high density exciton gas in semiconductors using the Schwinger representation for the projection operators of the effective spin (pseudospin) of the excitons is proposed. It is shown that the method of narrowing the NMR resonance lines in solids under the influence of a sequence of four electromagnetic pulses of the WAHUHA type can be applied for the line narrowing of the recombination emission and absorption of light by excitons in optical spectroscopy. We found out that, in contrast to the averaging to zero of the magnetic dipole-dipole interaction between nuclei, the exciton-exciton interaction is averaged only partially. This analysis for the excitons can be extended to other quasiparticles of boson type, i.e. biexcitons, polaritons and magnons, in solids.

1. Introduction

Methods for the high-resolution NMR in solids, leading back to the pioneering work of Waugh, Huber and Haeberlen [1] (method WAHUHA), allow us to detect and investigate the NMR spectra, in spite of the large broadening of resonance lines (exceeding the spectral bandwidth of recording systems), caused by magnetic dipole-dipole interactions between nuclei [1-5]. The width of the optical resonance lines corresponding to the electron and exciton transitions (in the visible region) or to the electron-vibrational transitions (in the infrared region) are due to the interaction of electrons, holes and excitons with phonons and with each other. In particular, at high levels of optical excitation of semiconductor crystals, followed by the generation of the excitons of high density, the main mechanism of broadening of the lines for recombination emission and absorption of light by excitons at low temperatures is the mechanism of the pair interaction between excitons during their intra-band scattering on each other.

At low temperatures, with increasing of the concentration n_{ex} of Wannier-Mott excitons in semiconductors, when the average distance between them is comparable to the exciton radius a_{ex} $\left(n_{ex}a_{ex}^3 \leq 1\right)$ depending on the parameters of the crystal, the exciton system evolves by one of the following ways: the formation of excitonic molecules (biexcitons) [6, 7] , the formation of electron-hole drops and electron-hole liquid (condensation in coordinate space) [8-10], Bose-Einstein condensation of excitons and biexcitons (condensation in quasi-momentum space) [11-16], Bose condensation of excitons in a strong electromagnetic field [16]. In all these cases, the exciton-exciton interaction is dominant and makes a major contribution to the widths of the lines of recombination radiation. Except the case for the δ-shaped peak emission due to the exciton Bose condensate and associated with the behavior patterns of macro-doped exciton mode, the exciton-exciton interactions lead to a broadening of the exciton lines of recombination emission and absorption of light.

In this paper we discuss the possibility for narrowing of the exciton lines of absorption of light, whose widths are due to the intra-band exciton-exciton collisions, under the influence of a sequence of four terasound pulses, similar to a WAHUHA sequence of four electromagnetic pulses used in multi-pulse solid state NMR spectroscopy. To this end, the Hamiltonian of exciton-exciton interaction, written in the occupation number representation, is transformed into a convenient form to perform coherent averaging using average Hamiltonian theory. The transformed Hamiltonian of exciton-exciton interactions are reminiscent (up to a factor 3) of the form for the secular part of the spin-spin interactions operator in the theory of NMR. It is shown that due to no exact match in the form of the operator of exciton-exciton interactions with the secular part of spin-spin interactions, application of the WAHUHA method leads to a partial coherent averaging of the exciton-exciton interactions, in contrast to the complete averaging (up to zero) of the magnetic dipole-dipole interactions in a spin system.

2. Hamiltonian of the exciton system in Schwinger representation

Let's consider a crystal containing high density excitons, taking into account only one (the lowest) of the exciton band. It is assumed that the exciton ground state is non-degenerate in the orbital and spin quantum numbers. The interaction between excitons is caused by the elastic exciton-exciton scattering within a non-degenerate exciton band. In this case, the width of optical absorption line related to the quantum transitions from the ground state of the crystal in a non-degenerate exciton band will be determined by the processes of intra-band exciton-exciton collisions. Hamiltonian of the exciton system in the occupation number representation is given by [15]

$$\mathcal{H} = \sum_{\mathbf{k}} [E_{ex}(\mathbf{k}) - \mu] A_{\mathbf{k}}^{+} A_{\mathbf{k}} + \frac{1}{2\mathrm{V}} \sum_{\mathbf{k}_1, \mathbf{k}_2, \mathbf{q}} V_{\mathbf{q}} A_{\mathbf{k}_1}^{+} A_{\mathbf{k}_2}^{+} A_{\mathbf{k}_2 - \mathbf{q}} A_{\mathbf{k}_1 + \mathbf{q}}, \tag{1}$$

where $A_{\mathbf{k}}^{+}$ and $A_{\mathbf{k}}$ are the exciton operators of creation and annihilation, $E_{ex}(\mathbf{k})$ is the energy of exciton formation with wave vector $\mathbf{k}$ and quadratic dispersion law, $V_{\mathbf{q}}$ is the Fourier transform of the energy of interaction between two excitons, μ is the chemical potential of the system, and V is the volume of the crystal.

The form of Hamiltonian (1) differs from the Hamiltonian which describes the Zeeman interaction of nuclear spins with a constant magnetic field, and the secular part of the operator of magnetic dipole-dipole interactions between nuclear spins [3]

$$\mathcal{H}_{spin} = \mathcal{H}_Z + \mathcal{H}_d^0, \tag{2}$$

$$\mathcal{H}_Z = \gamma H_0 \sum_j I_{Zj}, \tag{3}$$

$$\mathcal{H}_d^0 = \gamma^2 \hbar \sum_i \sum_{\langle j} r_{ij}^{-3} P_2(\cos\theta_{ij})(\boldsymbol{I}_i \boldsymbol{I}_j - 3 I_{Zi} I_{Zj}), \tag{4}$$

where γ is the gyromagnetic ratio of studied nuclei, I_{Zi} and I_{Zj} are operators of the projections of nuclear spin on Z-axis (the direction of the static magnetic field H_0), θ_{ij} is the angle between this direction and the direction of a vector between spins i and j, and $P_2(X) = (1/2)(3X^2-1)$ is the Legendre polynomial.

For the implementation of narrowing of NMR lines in solids under the influence of a sequence of WAHUHA electromagnetic pulses, it is well known that it is essential to depict the main contribution to the width of the NMR lines by the secular part $\mathcal{H}_d^0$ of the operator of magnetic dipole-dipole interactions at high magnetic fields, which commutes with the operator $\mathcal{H}_Z$

$$[\mathcal{H}_Z, \mathcal{H}_d^0] = 0. \tag{5}$$

This leads to a separation of two non-related to each other thermodynamic reservoirs: Zeeman, due to the interaction with a constant magnetic field H_0, and dipole-dipole, due to the secular part of dipole-dipole Hamiltonian $\mathcal{H}_d^0$ [3]. In this case, the energy of each system separately is an integral of motion and one might obtain a situation when the temperatures of each reservoir will be different.

A comparison of Hamiltonians $\mathcal{H}$ and $\mathcal{H}_{spin}$, given by equations (1) and (2), has not determined if the exciton system can be divided into two independent thermodynamic reservoirs similar to the case of interacting nuclear spins in a strong magnetic field. However, the analogy between $\mathcal{H}$ and $\mathcal{H}_{spin}$, and, accordingly, the possibility of separation of two independent thermodynamic reservoirs in the exciton system can be brought out, if the Schwinger representation for angular momentum is used in Eq. (1) [17].

We introduce now the effective spin operator $\boldsymbol{I}_{\mathbf{k}_1\mathbf{k}_2}$, the projection of which on the directions *X, Y, Z* can be presented in the form

$$\begin{aligned} I^X_{\mathbf{k}_1\mathbf{k}_2} &= \frac{1}{2}\left(A^+_{\mathbf{k}_1}A_{\mathbf{k}_2} + A_{\mathbf{k}_1}A^+_{\mathbf{k}_2}\right), \\ I^Y_{\mathbf{k}_1\mathbf{k}_2} &= \frac{1}{2i}\left(A^+_{\mathbf{k}_1}A_{\mathbf{k}_2} - A_{\mathbf{k}_1}A^+_{\mathbf{k}_2}\right), \\ I^Z_{\mathbf{k}_1\mathbf{k}_2} &= \frac{1}{2}\left(A^+_{\mathbf{k}_1}A_{\mathbf{k}_1} - A^+_{\mathbf{k}_2}A_{\mathbf{k}_2}\right). \end{aligned} \tag{6}$$

After some transformations using Eqs. (6), the Hamiltonian (1) can then be presented as follows:

$$\mathcal{H} = \mathcal{H}_0 + \mathcal{H}^Z_{ex}, \tag{7}$$

where

$$\mathcal{H}_0 = \sum_{\mathbf{k}_1}[E_{ex}(\mathbf{k}_1) - \mu]S^Z_{\mathbf{k}_1,\mathbf{k}_1+\mathbf{q}} - \sum_{\mathbf{k}_2}[E_{ex}(\mathbf{k}_2) - \mu]S^Z_{\mathbf{k}_2-\mathbf{q},\mathbf{k}_2} + \frac{1}{2}\sum_{\mathbf{k}}[E_{ex}(\mathbf{k}_1) - \mu], \tag{8}$$

$$\mathcal{H}^Z_{ex} = \frac{1}{2\mathrm{V}}\sum_{\mathbf{k}_1\mathbf{k}_2\mathbf{q}} V_q\left(\mathbf{S}_{\mathbf{k}_1,\mathbf{k}_1+\mathbf{q}}\mathbf{S}_{\mathbf{k}_2-\mathbf{q},\mathbf{k}_2} - S^Z_{\mathbf{k}_1,\mathbf{k}_1+\mathbf{q}}S^Z_{\mathbf{k}_2-\mathbf{q},\mathbf{k}_2}\right). \tag{9}$$

It was taking into account when deriving the Hamiltonian $\mathcal{H}_0$ from (8) that the operators $A^+_{\mathbf{k}_1+\mathbf{q}}A_{\mathbf{k}_1+\mathbf{q}}$ and $A^+_{\mathbf{k}_2-\mathbf{q}}A_{\mathbf{k}_2-\mathbf{q}}$ which realize the two-boson representation of the pseudospin *z*-component for $\mathbf{S}_{\mathbf{k}_1+\mathbf{q},\mathbf{k}_2-\mathbf{q}}$ operator, i.e.

$$S^Z_{\mathbf{k}_1+\mathbf{q},\mathbf{k}_2-\mathbf{q}} = \frac{1}{2}\left(A^+_{\mathbf{k}_1+\mathbf{q}}A_{\mathbf{k}_1+\mathbf{q}} - A^+_{\mathbf{k}_2-\mathbf{q}}A_{\mathbf{k}_2-\mathbf{q}}\right), \tag{10}$$

satisfy the condition

$$A^+_{\mathbf{k}_1+\mathbf{q}}A_{\mathbf{k}_1+\mathbf{q}} + A^+_{\mathbf{k}_2-\mathbf{q}}A_{\mathbf{k}_2-\mathbf{q}} = 1. \tag{11}$$

The Hamiltonian $\mathcal{H}^Z_{ex}$ from (9) was obtained neglecting the rare processes of pair collisions during which two excitons, having before collision impulses $\mathbf{k}_1$ and $\mathbf{k}_2$ (in $\hbar$ units), get after collision the impulses $\mathbf{k}'_1 = \mathbf{k}_2$ and $\mathbf{k}'_2 = \mathbf{k}_1$, respectively. Taking into account these processes, this leads that the averaged energy (on the exciton occupation numbers) of interaction between excitons decreases by the value of

$$\delta E = \frac{1}{2}n_{ex}\sum_{\mathbf{q}} V_q\,. \tag{12}$$

Comparing equations (8) and (9) with equations (3) and (4), respectively, we can see that there is a deep analogy between the problem of elastic intra-band exciton-exciton collisions in semiconductors and the problem of the magnetic dipole-dipole interactions between nuclear spins in a strong magnetic field. In a system of excitons of high density, as well as in the case of interacting nuclear spins in a strong magnetic field, there are two unrelated with each other thermodynamic reservoirs, similar to Zeeman and dipole-dipole reservoirs in the case of nuclear spins systems. However, there is a significant difference between exciton-exciton and dipole-dipole reservoirs consisting in the absence of factor 3 in the operator $S^Z_{\mathbf{k}_1,\mathbf{k}_1+\mathbf{q}} S^Z_{\mathbf{k}_2-\mathbf{q},\mathbf{k}_2}$ found in (9), and in the presence of such a factor in the operator $I_{Zi} I_{Zj}$ (see equation (4)). As it will be shown in the next section, this leads to the fact that, in contrast to spin systems for which the magnetic dipole-dipole interactions are averaged to zero under the influence of a sequence of four WAHUHA electromagnetic pulses, the impact of a sequence of four WAHUHA terasound pulses leads only to a partial averaging of the exciton-exciton interactions in the case of exciton systems.

3. Coherent averaging of the exciton-exciton interactions

In contrast to the system of interacting nuclear spins for which the dipole-dipole reservoir can be identified only in the presence of a strong magnetic field, the "exciton-exciton" reservoir with the corresponding Hamiltonian $\mathcal{H}^Z_{ex}$ always exists due to the fact that the energy of exciton-exciton interaction is much smaller than the eigenvalues of the operator $\mathcal{H}_0$, which is an analog of the Hamiltonian for the interactions of nuclei with magnetic field. Due to a quasi-discrete energy spectrum of excitons and a finite reduced Brillouin zone, the projection operators of the total effective spins $S^Z_1 = \sum_{\boldsymbol{k}_1} \sum_{\boldsymbol{q}} S^Z_{\boldsymbol{k}_1,\boldsymbol{k}_1+\boldsymbol{q}}$ and $S^Z_2 = \sum_{\boldsymbol{k}_2} \sum_{\boldsymbol{q}} S^Z_{\boldsymbol{k}_2-\boldsymbol{q},\boldsymbol{k}_2}$ have though a large, but finite number of eigenvalues. Therefore, the formalism developed for systems with a finite number of degrees of freedom can be applied to them.

Let's consider now the impact of a sequence of four WAHUHA 90-degree terasound pulses on the system of effective spins periodically repeated with a period $t_c < T_2$ (T_2 is the transverse relaxation time of excitons). For pure direct-band semiconductors $T_2 \sim 10^{-9}$-10^{-7} s [8] and, therefore, in order to perform such experiments as WAHUHA in the case of high-density excitons in direct-band semiconductor, when defining processes are intra-band scattering of excitons on excitons, the terasound impulses are needed. In the case of indirect excitons, for which the value of T_2 is three orders of magnitude higher, one can also use a sequence of hypersonic pulses.

In accordance with the criterion of applicability of the method of coherent averaging [3, 4], we assume that the energy of the exciton-exciton interaction is a small perturbation in comparison with the energy of interaction of coherent terasound with excitons. The interaction Hamiltonian of terasound with excitons can be presented in the form

$$\mathcal{H}_I = \sqrt{N_{\mathbf{q}\sigma}} \sum_{\mathbf{k}} [g_\sigma(\mathbf{q}) S^+_{\mathbf{k},\mathbf{k}+\mathbf{q}} \, exp(\text{-}i\omega_{\mathbf{q},\sigma} t) + H.c.], \tag{13}$$

where $S^+_{\mathbf{k},\mathbf{k}+\mathbf{q}} = S^X_{\mathbf{k},\mathbf{k}+\mathbf{q}} + iS^Y_{\mathbf{k},\mathbf{k}+\mathbf{q}}$, $g_\sigma(\mathbf{q})$ is the imaginary exciton-phonon coupling constant, and $N_{\mathbf{q}\sigma}$ is number of terasound phonons with frequency $\omega_{\mathbf{q}\sigma}$, wave vector $\mathbf{q}$ and polarization σ. The equation (13) corresponds to the interaction of excitons with monochromatic terasound wave of high amplitude (macro-completed mode) when the corpuscular structure of terasound field is neglected ($\langle b_{\mathbf{q},\sigma} \rangle = \langle b^+_{\mathbf{q},\sigma} \rangle \approx \sqrt{N_{\mathbf{q}\sigma}}$).

Evolution operator of the system (propagator) at the time moment corresponding to the end of the cycle ($t_c = 6\tau$) is given by

$$U_{int}(t_c) = exp\left(-\frac{i}{\hbar}\mathcal{H}^Z_{ex}\tau\right) P_X \, exp\left(-\frac{i}{\hbar}\mathcal{H}^Z_{ex}\tau\right) P_{-Y} \, exp\left(-2\frac{i}{\hbar}\mathcal{H}^Z_{ex}\tau\right) P_Y \, exp\left(-\frac{i}{\hbar}\mathcal{H}^Z_{ex}\tau\right) P_{-X} \, exp\left(-\frac{i}{\hbar}\mathcal{H}^Z_{ex}\tau\right), \tag{14}$$

which, taken into account the result of action of the operators

$$P_{\pm\xi} = exp\left(\mp i\frac{\pi}{2}\sum_{\mathbf{k}_1\mathbf{k}_2} S^{\xi}_{\mathbf{k}_1\mathbf{k}_2}\right), \quad \xi = \mathrm{X}, \mathrm{Y} \tag{15}$$

on $exp\left(-\frac{i}{\hbar}\mathcal{H}^{Z}_{ex}\tau\right)$, takes the form

$$U_{int}(t_c) = exp\left(-\frac{i}{\hbar}\mathcal{H}^{Z}_{ex}\tau\right)exp\left(-\frac{i}{\hbar}\mathcal{H}^{Y}_{ex}\tau\right)exp\left(-2\frac{i}{\hbar}\mathcal{H}^{X}_{ex}\tau\right)exp\left(-\frac{i}{\hbar}\mathcal{H}^{Y}_{ex}\tau\right)exp\left(-\frac{i}{\hbar}\mathcal{H}^{Z}_{ex}\tau\right), \tag{16}$$

where $\mathcal{H}^{X}_{ex}$ and $\mathcal{H}^{Y}_{ex}$ are obtained from $\mathcal{H}^{Z}_{ex}$ by replacing Z by X and Y, respectively. Here, τ is the time during which the system develops under the influence of the effective "exciton-exciton" Hamiltonian (free development), and after then the first terasound impulse is applied to it.

After using the Magnus expansion to the operator $U_{int}(t_c)$ in (16) (see, for example, [4]), the following expression for the average Hamiltonian is derived in zero approximation

$$\overline{\mathcal{H}^{Z(0)}_{ex}} = \frac{1}{3\mathrm{V}}\sum_{\mathbf{k}_1,\mathbf{k}_2,\mathbf{q}} V_q S_{\mathbf{k}_1,\mathbf{k}_1-\mathbf{q}} S_{\mathbf{k}_2+\mathbf{q},\mathbf{k}_2}\,. \tag{17}$$

Equation (17) shows that, in contrast to the spin systems with magnetic dipole-dipole interaction, the exciton-exciton interaction is not averaged to zero. A scalar part of the interaction diminished by one and half times remains. This is due to the different forms for the effective Hamiltonian $\mathcal{H}^{Z}_{ex}$ and the truncated Hamiltonian of the magnetic dipole-dipole interactions.

4. Discussion and conclusions

In this paper we show that, by introducing the projection operators of the effective spin (pseudospin), the Hamiltonian of intra-band exciton-exciton scattering can be reduced to the form which is almost similar to the secular part of the operator of magnetic dipole-dipole interactions in spin systems. In this pseudospin formalism two independent thermodynamic reservoirs can be separated in the exciton system, by analogy with the Zeeman and dipole-dipole reservoirs in spin systems, with energies as integrals of motion. In contrast to the magnetic dipole-dipole reservoir, which can be separated only in the presence of a strong static magnetic field, the “exciton-exciton” reservoir exists even in the absence of interaction with any external constant fields.

For all elementary excitations in solids obeying Bose-Einstein statistics, the paired processes of elastic collisions, in which the quasi-particles remain within the same energy band, can be described by similar Hamiltonians (in the form of second term in (1)). Our analysis for excitons can then be extended to other elementary excitations of bosonic type (biexcitons, polaritons and magnons). In each of these cases it is possible to introduce a pseudospin, the selection of two non-connected with each other thermodynamic reservoirs, and, in the framework of coherent averaging techniques, one can use multi-pulse methods of NMR spectroscopy of solids for narrowing of the corresponding spectral lines of emission or absorption. A similar analysis can be also done in regard to the processes of inelastic scattering of quasi-particles with a turn to other branches of the elementary excitations spectrum.

According to the averaging Krylov-Bogolyubov-Mitropolsky method, it was shown in the article [18] that the expression for the average Hamiltonian, based on the Magnus expansion, is non-secular in higher orders of the perturbation theory. Authors of the article [19] have developed a method of canonical transformations which allows to study the dynamics of spin systems in a variety of multi-pulse experiments, and the thermodynamic effects in multi-pulse NMR spectroscopy of solids have been studied based on this method. These generalizations of the average Hamiltonian method should also be taken into account when the effects of multi-pulse sequences on excitons and other quasi-particles of bosonic type are considered. In this framework of the average Hamiltonian method, we have shown only that the multi-pulse techniques developed in the solid state NMR spectroscopy can be applied, in particular, to the optical spectroscopy of excitons.

Acknowledgements

The author is thankful to the Supreme Council for Science and Technological Development of the Academy of Sciences of Moldova for financial support, project number 11.832.08.06 A.

References

[1] Waugh I S, Huber L M and Haeberlen U 1968 *Phys. Rev. Lett.* **20** 180
[2] Haeberlen U 1976 *High Resolution NMR in Solids. Selective Averaging* (New York: Academic Press)
[3] Waugh I S 1978 *New NMR methods in solid state* (Moscow: Mir, in Russian)
[4] Mehring M 1976 *High Resolution NMR Spectroscopy in Solids* in *NMR 11, Basic Principles and Progress* (Berlin: Springer-Verlag)
[5] Slichter C P 1980 *Principles of Magnetic Resonance* (New York: Springer-Verlag)
[6] Moskalenko S A 1958 *Optika i spectroscopia* **5** 147
[7] Lampert M A 1958 *Phys. Rev. Lett.* **1** 450
[8] Keldysh L V 1971 *Kollektivnye svoistva eksitonov v poluprovodnikah,* in *Eksitony v poluprovodnikah* (Moskva: Nauka) p 5
[9] Keldysh L V 1970 Uspehi Fizicheskih Nauk **100** 514
[10] Rice I M 1976 *Proc. XII Intern. conf. phys. semicond* (Stuttgart) (Amsterdam: Elsevier) p 23
[11] Moskalenko S A 1962 *Solid State Physics* (Leningrad) **4** 276
[12] Blatt I M, Böer K W and Brandt W 1962 *Phys. Rev.* **126** 1691
[13] Cassela R C 1963 *J. Appl. Phys.* **34** 1703
[14] Keldysh L V and Kozlov A N 1968 *JETF* **54** 978
[15] Moskalenko S A and Snoke D V 2000 *Bose-Einstein condensation of excitons and biexcitons: and coherent nonlinear optics with excitons* (Cambridge University Press)
[16] Elesin V F and Kopaev Yu V 1972 *JETF* **63** 1477
[17] Schwinger J 1965 *Quantum theory of Angular Momentum*, ed. L Beidenharn and H van Dam (New York: Academic Press) p 229
[18] Buishvili L L and Menadbe M G 1979 *JETF* **77** 2435
[19] Provotorov B P and Feldman E B 1980 *JETF* **79** 2206

Manifestation of non-centrality effect in the EPR spectrum of Fe^{3+} ion in the polycrystalline materials

V N Vasyukov

JSC “Gazprom promgaz”, Ulitsa Nametkina, 6, Moscow, Russian Federation

V.Vasyukov@promgaz.gasprom.ru

Abstract. In a number of works [1-6], the study of the electron paramagnetic resonance (EPR) spectrum of polycrystalline non-organic and organic matters revealed presence of EPR spectrum of impurity Fe^{3+} ion which has essentially different forms at helium and room temperatures. At the helium temperature intensity of the resonance line with g-value close to $g_1 \cong 4.3$ prevails (Line 1 of the low temperature spectrum). The resonance line with $g \cong 4.3$ corresponds to magnetic center with spin S = 5/2 by the spin Hamiltonian parameter $D >> h\nu$.The resonance line with g-value close to 2.0 (Line 2 of the high temperature spectrum) prevails at the room temperatures. The resonance line with $g_2 \cong 2.0$ corresponds to magnetic center with spin S = 5/2 by the spin Hamiltonian parameter $D << h\nu$. The transition from the low- to high-temperature spectrum is gradual and it is accompanied by a redistribution of the absorption intensity. The observed properties of the temperature dependence of the EPR spectrum are characteristic of systems with a multiminimum potential. The described effect of change of integral intensity of low-temperature and high-temperature EPR spectrum of Fe^{3+} ion is investigated in substances $Na[FeO_6(C_{10}H_8N)_3]$, $C_{21}H_{14}Br_4O_5S$, $C_{25}H_{30}ClN_3$, $Al_2[Si_2O_5](OH)_4$, polyaniline and polyparaphenylene.

1. Introduction

This work represents the review of results of researches of EPR spectrum of Fe^{3+} ion in a number polycrystalline organic substances and kaolinite [1-6]. Feature of the given researches is the experimental detection of unusual structure of EPR spectrum of Fe^{3+} ion in polycrystalline substances and unusual temperature dependence of integral intensity of resonance lines of EPR spectrum.

In works [1-6] EPR spectrum of Fe^{3+} ion is investigated in an interval of temperatures from temperature liquid Helium up to room temperature in substances $Na[FeO_6(C_{10}H_8N)_3]$, $C_{21}H_{14}Br_4O_5S$, $C_{25}H_{30}ClN_3$, $Al_2[Si_2O_5](OH)_4$, polyaniline and polyparaphenylene.

In all investigated substances EPR spectrum represents a superposition of two resonance lines, intensity of which behave opposite with change of temperature. At temperature decreases the intensity of one line increases, and intensity of the other decreases. One of the resonance lines dominates at low (helium) temperatures, and the other at room temperatures. The first line was determined as low-temperature (LT) component of spectrum, and the second one as high-temperature (HT) component. The change of temperature results in redistribution of intensity of absorption between LT and HT components of EPR spectrum. The similar behavior of EPR spectrum testifies unusual dynamics of molecular environment of Fe^{3+} ion.

The research of these dynamic transitions has independent significance, because systems which showing such properties are systems with multiminimum potential. The systems with multiminimum potential can have various physical nature, but the most investigated system with the multiminimum potential is the Jahn-Teller ion (divalent copper ion in an octahedral environment) [7].

Fe^{3+} ion is S-ion and consequently the display of Jahn-Teller effect is impossible. It is more probable that properties of Fe^{3+} ion revealed in works [1-6] are associated with display of an off centre effect [8].

The off centre features appear in EPR spectrum in a number of characteristics: 1) presence LT and HT components of EPR spectrum, 2) redistribution of intensity between LT and HT components at temperature change.

2. Experimental results

The research of EPR spectrum was performed with microwave frequency close to $\nu \approx 9.3$ GHz in temperatures interval $T = 4.2 \div 300$ K.

In the investigated substances the resonance lines of EPR spectrum of Fe^{3+} ion have significant width. In this work for efficient analysis of the contributions EPR spectra are submitted in the form of absorption line, instead of common form of a derivative of an absorption line.

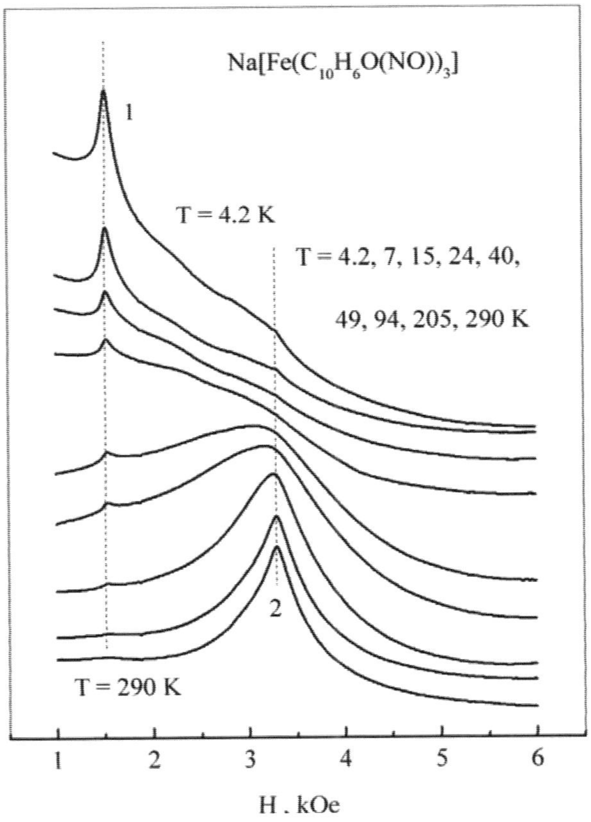

Figure 1. Temperature change of the form of the EPR spectrum in 2-hydroxy-1-nitrozonaphathalene $Na[Fe(C_{10}H_6(NO))_3]$

Figures 1-4 show absorption lines of EPR spectrum of Fe^{3+} ion for organic substances $Na[Fe(C_{10}H_6(NO))_3]$, $C_{21}H_{14}Br_4O_5S$, $C_{25}H_{30}ClN_3$ and polyaniline for several temperatures. The absorption lines are obtained by numerical integration of experimental EPR spectra. Figure 5 shows

EPR spectrum of Fe^{3+} ion in $Na[Fe(C_{10}H_6(NO))_3]$ at temperature T = 4.2 K given in the form of a absorption line and in the form derivative simultaneously.

The analysis of EPR spectra (figures 1-4) shows that EPR spectra of the investigated substances contain two resonance lines, which are named resonance lines 1 and 2. On the ground of the g-values and character of temperature change of intensity we can draw a conclusion that these resonance lines have an essentially different physical nature. For investigated substances the g-values of resonance lines 1 and 2are given in the table 1.

As shown in figures 1-4, the value of resonance magnetic field of lines 1 and 2 practically does not vary at change of temperature.

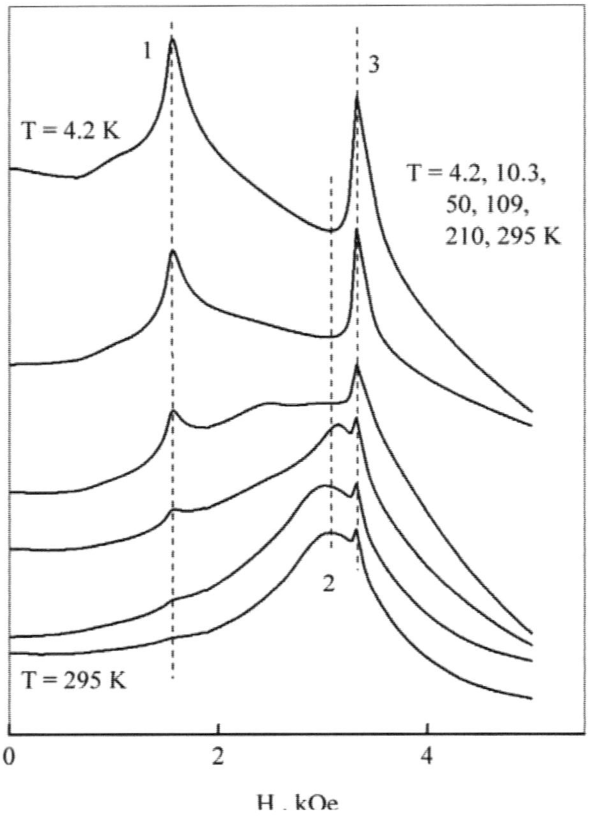

Figure 2. Temperature change of the form of the EPR spectrum in polyaniline.The resonance lines 1 and 2 belong to Fe^{3+} ion, and the resonance line 3 belongs to a radical.

Table 1. The g-values of resonance lines of EPR spectrum of Fe^{3+} ion

Substance	Line 1		Line 2	
	g_1	T, K	g_2	T, K
$Na[Fe(C_{10}H_6(NO))_3]$	4.32 ± 0.02	4.2	2.00 ± 0.01	290
Polyaniline	4.22 ± 0.03	4.2	2.13 ± 0.05	295
$(C_{21}H_{14}Br_4O_5S)$	4.39 ± 0.03	4.2	2.03 ± 0.02	295
$(C_{25}H_{30}ClN_3)$	4.5 ± 0.1	4.2	2.15 ± 0.15	185
Polyparaphenylene	4.21 ± 0.05	4.2	2.00 ± 0.09	300
$Al_2[Si_2O_5](OH)_4$	4.13 ± 0.16	4.2	2.15 ± 0.1	288

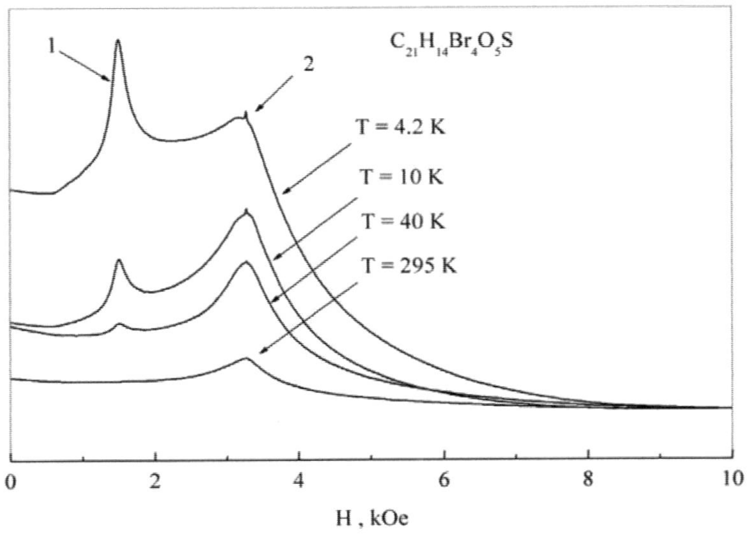

Figure 3. Temperature change of the form of EPR spectrum in $C_{21}H_{14}Br_4O_5S$

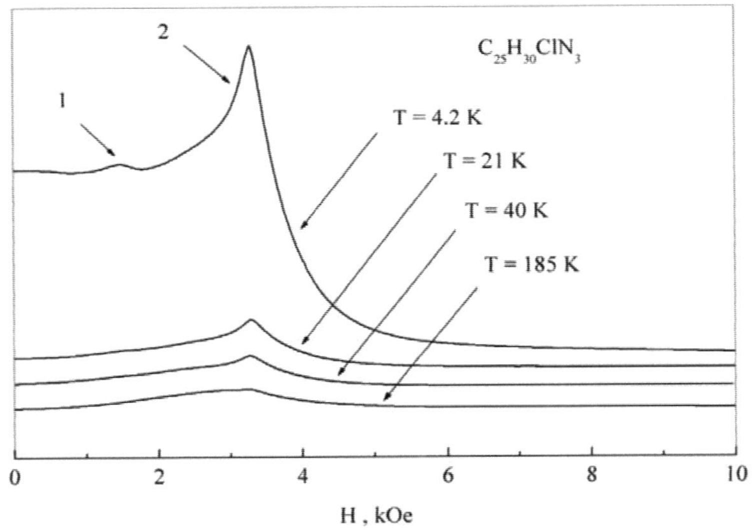

Figure 4. Temperature change of the form of EPR spectrum in $C_{25}H_{30}ClN_3$

For more detailed analysis of temperature dependence of integral intensities and resonance line widths modeling of forms of resonance lines and numerical decomposition of absorption lines is carried out for all investigated substances and temperatures.

For example, figures 6 and 7 show the decomposition of lines of absorption for $Na[Fe(C_{10}H_6(NO))_3]$ и $Al_2[Si_2O_5](OH)_4$ for two temperatures. At temperature 4.2 K intensities of a lines 1 for $Na[Fe(C_{10}H_6(NO))_3]$ и $Al_2[Si_2O_5](OH)_4$ are maximal, and the intensities of a lines 2 are minimal. On the contrary, intensity of the resonance line 1 for $Na[Fe(C_{10}H_6(NO))_3]$ is minimal at temperature 277 K (line 1 is not visible almost), and the intensity of a line 2 is maximal. Similarly to $Al_2[Si_2O_5](OH)_4$, intensity of the resonance line 1 is minimal, and the intensity of the line 2 is maximal at temperature 288. The intensity and form of lines 1 and 2 were chosen on condition that there was a minimal deviation between the model line, being a superposition of lines 1 and 2, and the experimental form of the absorption line.

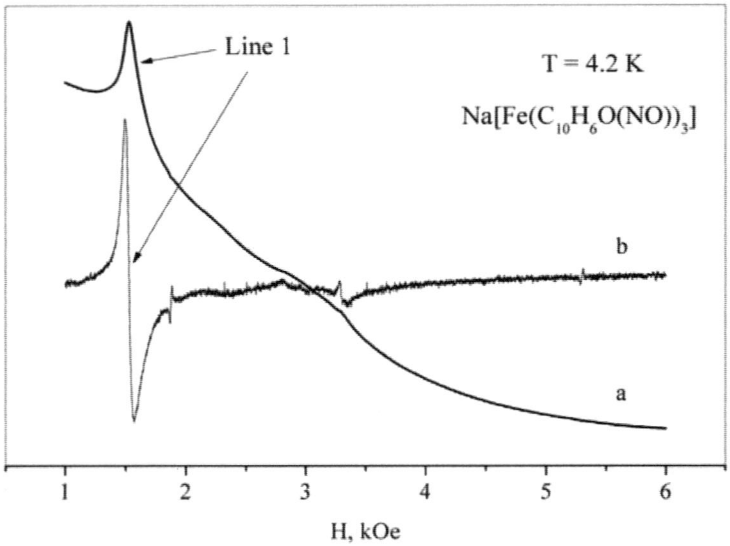

Figure 5. EPR spectrum in the form of the absorption curve and in the form derivative obtained experimentally.

It is necessary to pay attention that modeling of the form of resonance lines and decomposition of absorption lines show that absorption lines (except for resonance lines 1 and 2) contain addition non-resonance contribution as it shown in figure 6.

Figure 8 shows the temperature change of the absorption line of EPR spectrum of Fe^{3+} ion in inorganic substance kaolinite ($Al_2[Si_2O_5](OH)_4$). On top of resonance line 2 narrow resonance line of small intensity belonging to a radical is observed. This radical is used for magnetic field calibration. It is necessary to note that in figure 8 non-resonance contribution previously was removed during the analysis of an absorption line for $Al_2[Si_2O_5](OH)_4$.

3. Temperature change of EPR spectra of Fe^{3+} ion

The qualitative analysis of the results given in figures 1 - 4 and 8 shows, that temperature change of intensity of resonance lines 1 and 2 is different.

To implement the quantitative analysis of temperature changes of a spectrum for all investigated substances and temperatures modeling resonance lines and numerical decomposition of absorption lines has been carried out as shown in figures 6 and 7.

The analysis of temperature dependence of integral intensity of resonance lines shows that temperature dependence of a spectrum has two mechanisms. It is possible to name the first mechanism usual. It results in reduction of integral intensity of all contributions of a EPR spectrum. This reduction of intensity is described by ratio $J(T) = J_0*\tanh(h\nu/2kT)$. For example, figure 9 shows the temperature dependence of integral intensity of resonance line 1 for $Al_2[Si_2O_5](OH)_4$.

The second mechanism of temperature dependence is shown in redistribution of intensity between the various contributions to EPR spectrum and is defined the multiminimum potential of the magnetic centre. Figures 10 - 13 show the temperature dependences relative integral intensities of resonance lines 1 and 2, determined by ratio $J_{1R} = J_1/(J_1 + J_2)$, $J_{2R} = J_2/(J_1 + J_2) - J_0$, where J_1, J_2 are integral intensity of resonance lines 1 and 2. Parameter J_0 is connected to height of a potential barrier between minima. At plotting of figures 10 – 13, parameter J_0 is used for mutual scaling of intensities of lines 1 and 2. These dependences describe the second mechanism of temperature change in the EPR spectrum which describes the process of the integral intensity redistribution between lines 1 and 2.

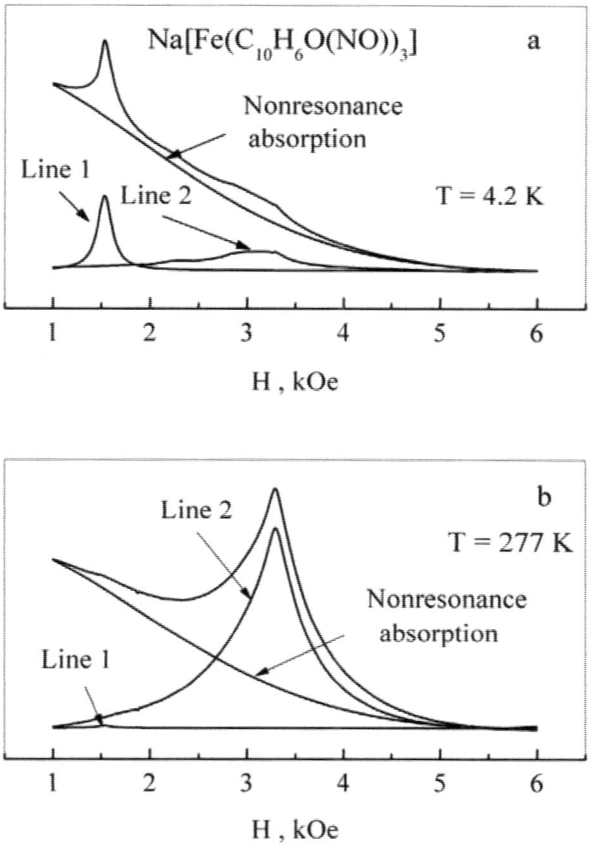

Figure 6. Decomposition of the resulting absorption line in Fe^{3+} EPR spectrum in $Na[Fe(C_{10}H_6ONO)_3]$ into component resonance lines 1, 2 and non-resonance absorption.
a) T = 4.2 K, b) T = 277 K.

As a result of modeling the form of resonance lines and numerical decomposition of a absorption lines on contributions not only temperature dependences integral intensities of resonance lines have been obtained, but also temperature dependences of width of resonance lines. Figure 13 shows the dependences line width of EPR spectrum of Fe^{3+} ion in $Al_2[Si_2O_5](OH)_4$ and $Na[Fe(C_{10}H_6(NO))_3]$. As shown in figure 13 width of a resonance line 1 in $Al_2[Si_2O_5](OH)_4$ in interval of temperatures T = 4.2 - 300 K changes not considerably. Width of a resonance line 2 increases with temperature decrease as for $Al_2[Si_2O_5](OH)_4$ and for $Na[Fe(C_{10}H_6(NO))_3]$. When the temperature decrease, the width of line 2 grows exponentially. Such change of line width is not caused by relaxation mechanisms of line-broadening, and is defined by specificity of formation of a resonance line of EPR spectrum of Fe^{3+} ion in polycrystalline substance in a case multiminimum potential of the magnetic centre.

4. Discussion of the results

EPR spectra, investigated in [1-6], contain two resonance lines (figures 1-4, 8). One line with the effective g-value $g_1 \approx 4.3$, and another with the effective g-value $g_2 \approx 2$. The lines of EPR spectrum with such g-values are characteristic for Fe^{3+} ions in polycrystalline substances. In [9, 10] the different reasons for simultaneous observation of resonance lines with the g-values $g \cong 2$ and $g \cong 4.3$ are considered. In [10] the analysis of results of research of EPR spectra of ions of iron in silicate glasses is carried out, and in article [9] the temperature change of EPR spectrum of Fe^{3+} ions in polycrystalline substances in the presence of the off centre effect is theoretically investigated.

In [10] it is assumed that the spectrum consisting of two resonance lines with g-values $g \cong 2$ and $g \cong 4.3$ belongs to different nonequivalent magnetic centers of the Fe^{3+} ion.

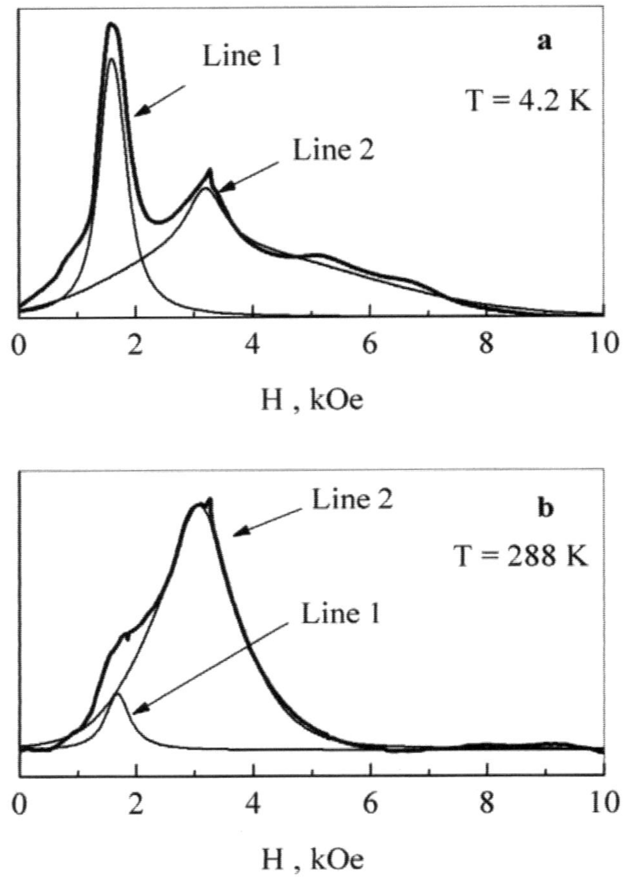

Figure 7. Computer resolution of the absorption line for $Al_2[Si_2O_5](OH)_4$. a) T = 4.2 K, b) T = 288 K.

The distinction of the centres consists in distinction of the low-symmetry components of a crystalline field acting on a magnetic ion. The line with $g \cong 2$ corresponds to the centre, low-symmetry component of crystal field of which there is less Zeeman energy. Under this condition, zero splitting parameter D satisfies to a condition $D \ll h\nu$, where $h\nu$ quantum of a microwave field. The line with $g \cong 4.3$ corresponds to the centre, low-symmetry component of crystal field of which is much greater of Zeeman energy. Under this condition, zero splitting parameter D satisfies to a condition $D \gg h\nu$.

In [1-6] it is supposed that the resonance lines with the g-values $g \cong 2$ and $g \cong 4.3$ belong to various vibronic states of the same magnetic centre of Fe^{3+} ion. This assumption is made on the basis of the analysis of temperature dependence of EPR spectra of substances containing noncentral Fe^{3+} ions [1-6] and model off centres [9], which takes into account whole set vibronic states. This model is similar to description of the temperature change of EPR spectrum of Jahn-Teller ion having ground orbital ^2E-state [7, 11].

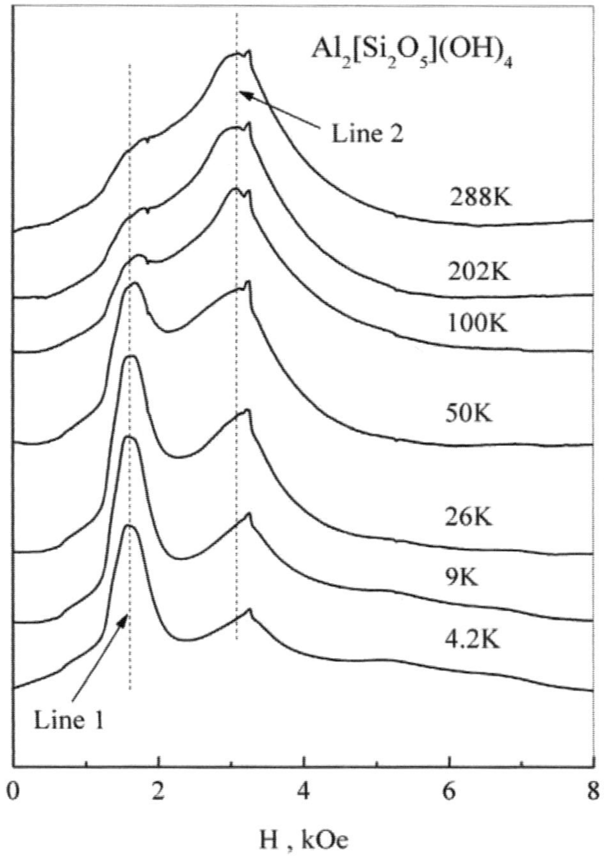

Figure 8. Temperature change of the form of the EPR spectrum in kaolinite $Al_2[Si_2O_5](OH)_4$ in the form of the absorption curve for the temperatures T = 4.2, 9, 26, 50, 100, 202 and 288 K.

According to the temperature dependence of the EPR spectrum obtained in [1-6] the Fe^{3+} ions are distributed over states having different values of the low-symmetry component of the crystalline field. At low temperatures, the states with the low-symmetry component possessing much more than the Zeeman energy are mainly occupied. When the temperature increases, the excited states with the low-symmetry component much less than the Zeeman energy are occupied. Since the total number of magnetic ions is constant, the transition of ions to the excited states results in the decrease of the intensity of line 1 as the temperature increased. Line 2 becomes more intense in this case (figures 1-4, 8).

We thus pay attention to the following three characteristics. First, the low-temperature states are characterized by the low-symmetry properties; while the excited states are typical of the properties of much higher symmetry. Secondly, a region exists in which the low- and high-symmetry spectra coexist. Thirdly, a decrease of the intensity of the low-symmetry spectrum when the temperature increases, and a corresponding increase of the intensity of the high-symmetry spectrum is observed. Such characteristics are inherent in the systems with the multiminimum potential.

The best studied system with the multiminimum potential is the Jahn-Teller ion of a divalent copper ion in an octahedral environment. In such systems, the temperature-induced changes in the intensity of spectra which correspond to the low- and high-symmetric states are due to changes in the population of vibronic states [7, 11]. It should be noted that Fe^{3+} is an S-ion, and the appearance of the Jahn-Teller effect for this ion is hardly probable. In the present case, the ion plays the role of a paramagnetic probe enabling one to observe the unusual dynamics of the ligand environment. The coincidence of the temperature changes of the EPR spectrum of Fe^{3+} in investigated substances with similar temperature changes in systems with the multiminimum potential provides grounds to propose that the magnetic center under consideration possesses similar properties.

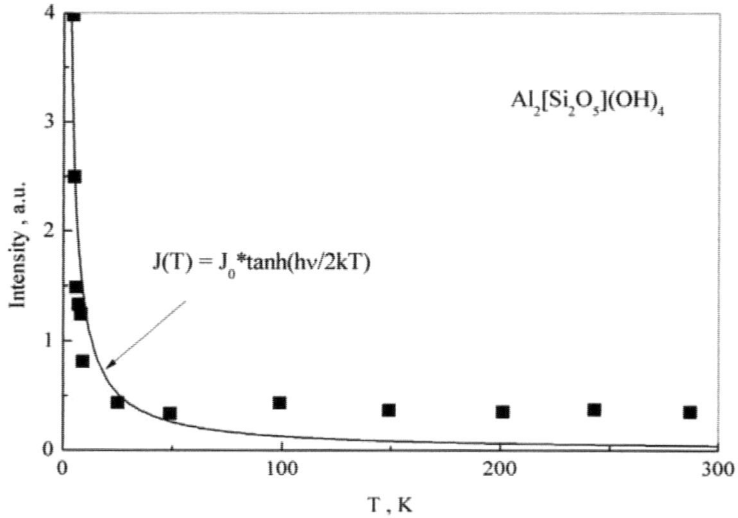

Figure 9. Temperature dependence of the integral intensities of lines 1 in the EPR spectrum of Fe^{3+} in kaolinite $Al_2[Si_2O_5](OH)_4$. The solid line describes dependence J(T) = J_0*tanh(hν/2kT).

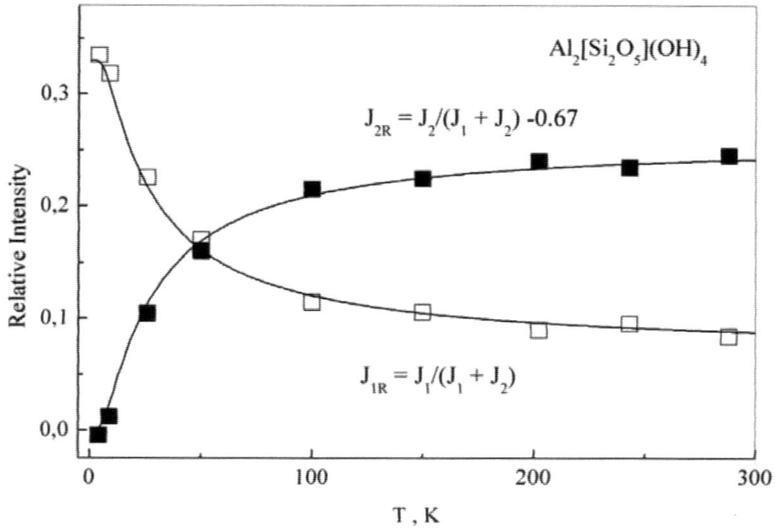

Figure 10. Temperature dependence of the relative integral intensities of lines 1 and 2 in kaolinite $Al_2[Si_2O_5](OH)_4$

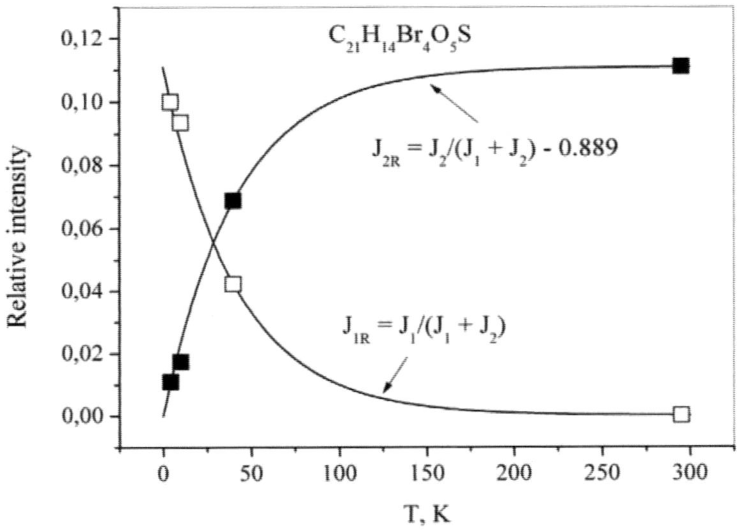

Figure 11. Temperature dependence of the relative integral intensities of lines 1 and 2 in bromide cresol green $C_{21}H_{14}Br_4O_5S$.

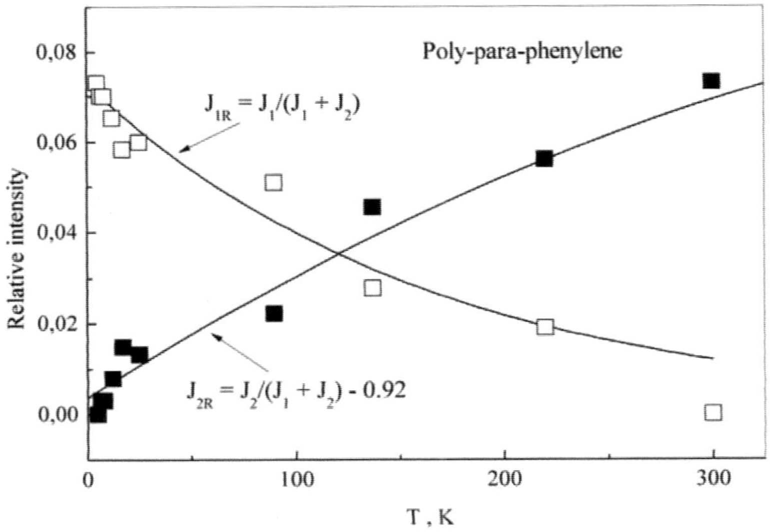

Figure 12. Temperature dependence of the relative integral intensities of lines 1 and 2 in polyparaphenylene.

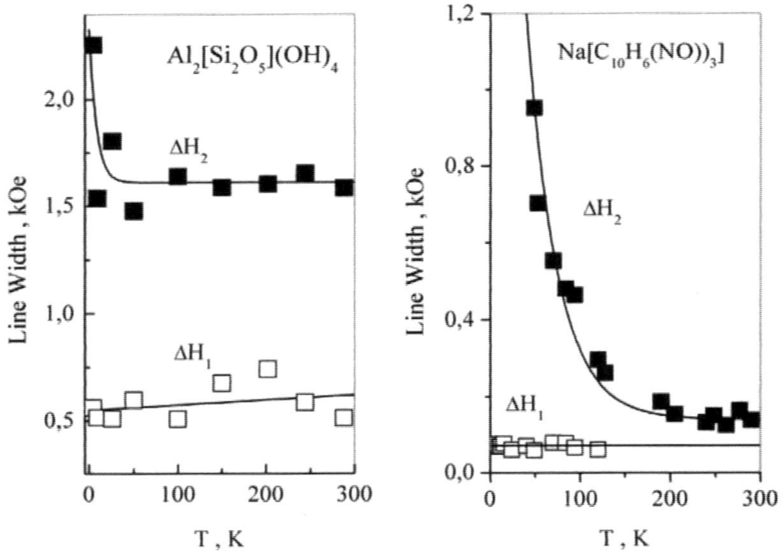

Figure 13. Temperature dependence of the width of resonance lines in kaolinite $Al_2[Si_2O_5](OH)_4$ and $Na[Fe(C_{10}H_6(NO))_3]$

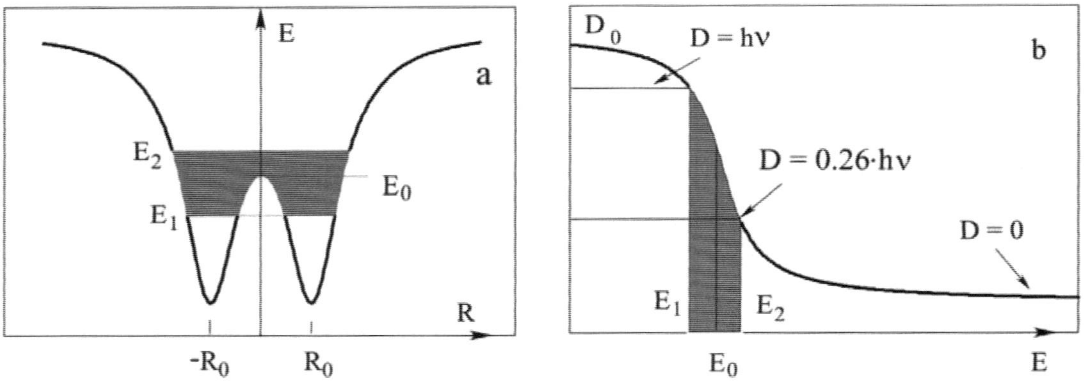

Figure 14. a) Potential surface form of the magnetic center and b) dependence of the zero splitting parameter on vibronic states energy

In figure 14a for visibility representation of structure and vibronic states properties, the form of a potential surface with two potential holes is schematically given. The quantity of holes in different substances can be different. For example, as shown in article [3] in polyparaphenylene the magnetic Fe^{3+} ion has two potential holes, and as theoretically shown in article [8] noncentral ions in cubic crystals can have six or eight potential holes. In figure 14a the parameters E_0 and R_0 designate height of a barrier dividing potential holes and spontaneous shift of Fe^{3+} ion from a point of symmetry of a molecular environment.

It is convenient to divide whole set vibronic states into three parts according to their properties. By analogy with [7] these groups of states it is possible to name under barrier, about barrier and above barrier states. In figure 14a the area about barrier states is allocated with a grey background. The magnetic ion in under barrier state is actually located in one of potential holes and for this reason its molecular environment is deformed much more strongly, than the environment of an ion, which is in any above barrier state. About a barrier states have intermediate properties. The properties about barrier states strongly depend on energy of excitation and vary from isotropic properties of above barrier states up to strongly anisotropic properties of under barrier states.

The distortion of an environment results in occurrence of low-symmetry component of the crystalline field working on a magnetic ion. This low-symmetry component of the crystalline field determines zero splitting parameter D, which corresponds with given vibronic state. Different vibronic states have the different values of zero splitting parameter D. Figure 14b schematically shows the dependence of parameter D on activation energy of vibronic states. The maximal value of parameter D_0 corresponds with the ground vibronic state. For above barrier vibronic states parameter D is equal to zero.

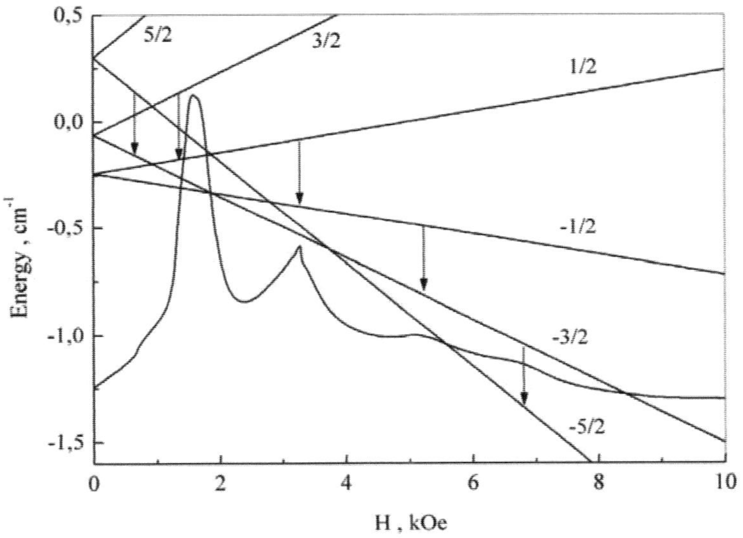

Figure 15. Structure of the energy states of the spin multiplet S=5/2 and a form of the EPR spectrum at T=4.2K in kaolinite.

In article [9] is shown, that at the given value of microwave field quantum hν the line with the g-value $g \cong 4.3$ is observed in a EPR spectrum of a polycrystalline sample under condition, if zero splitting parameter of ground vibronic state fulfils condition $D_0 \geq h\nu$. This condition less rigid than condition $D_0 >> \cdot h\nu$, given in [10]. For an Fe^{3+} ion the condition $D_0 = h\nu$ defines border E_1 between under barrier and about barrier vibronic states.

In article [9] is shown that a resonance line with the g-value $g \cong 2$ in EPR spectrum of a polycrystalline sample is observed under condition $D_0 \leq 0.26 \cdot h\nu$. For the Fe^{3+} ion the condition $D_0 = 0.26 \cdot h\nu$ defines border E_2 between about barrier and above barrier vibronic states.

It should be noted that the three different states of the Fe^{3+} magnetic ion are observed experimentally. In the first low-symmetry state, the parameter of spin Hamiltonian $D >> h\nu = 0.3$ см$^{-1}$. The presence of line 1 testifies to the existence of such a state.

The third state is highly symmetric and it is defined by the spin Hamiltonian parameter $D << D_0 < h\nu$. The form of line 2 at $T \approx 300$ K corresponds to this state.

The second state is intermediate on degree of molecular environment distortion. In kaolinite at T = 4.2 K analyses of form of resonance line 2 allow to estimate the spin Hamiltonian parameter $D \approx 0.09$ см$^{-1}$ ($D < h\nu$) (Fig.15).

In the high-field part of resonance line 2 of the EPR spectrum of figure 8 (T = 4.2, 9, 26 K), one can see the maxima rather well. The maxima are typical of the EPR spectrum which is averaged over directions of the magnetic field. Figure 15 shows the structure of energy levels of the multiplet S=5/2 and the EPR spectrum for T = 4.2 K in kaolinite. The resulting form of line 2 averaged over magnetic field directions has 5 maxima. The central maximum corresponds to transition 1/2 ⟷ -1/2. The other maxima correspond to another four transitions and appear as a background of the averaged line 2. The two high-field maxima correspond to transitions -1/2 ⟷ -3/2, -3/2 ⟷ -5/2. The two close maxima corresponding to transitions -5/2 ⟷ -3/2 and 3/2 ⟷ 1/2 are near the field $H \approx 1$ kOe.

The maxima on the high-field part of curve 2 permit one to estimate the parameter of the zero splitting $D = 0.09$ см$^{-1}$ ($D < h\nu$).

Modeling the form of a absorption line of EPR spectrum that was presented in [9], shows that magnetic dipole transitions in polycrystalline substances result in occurrence only resonance lines 1 and 2. Non-resonance contribution experimentally observed and shown in figure 6 is not connected with magnetic dipole transitions.

The possible reason for this absorption is electric dipole transitions between vibronic states. Noncentral ions in under barrier and about barrier vibronic states have the electrical dipole moment d not equal to zero. For this reason these ions can interact with a microwave electrical field. In antinode of microwave electrical field the interaction $-(\boldsymbol{d}\cdot\boldsymbol{E})$ is several orders more than interaction $-g\cdot\mu_B\cdot(\boldsymbol{S}\cdot\boldsymbol{H})$ in antinode of microwave magnetic field. For this reason even the small electrical field, which affects a part of a sample near antinode of microwave magnetic field, can result in microwave field absorption greater than absorption caused by a microwave magnetic field.

5. Conclusions

Within the framework of the proposed model of the excited-state population, it is possible to explain the character of the temperature dependence of the integral intensity of the EPR spectrum resonance lines, the width of resonance for line 2.

At helium temperature, magnetic centers are "frozen" at the bottom of the potential well (or wells) which correspond to the largest value of the low-symmetry component of the crystalline field. Simultaneously, the anisotropy of the crystalline electrical field is maximum, and therefore line 1 is the most intense. As the temperature is increased, a number of magnetic centers pass to the excited vibronic states. The excited states of magnetic centers are less anisotropic than the ground states. The resonance line 2 of the EPR spectrum results from the population of weakly anisotropic excited states (figures 1-4, 8). Increasing the number of magnetic centers in the excited states leads to a decrease of the number of magnetic centers in the ground state. It is the cause of the increase of the intensity of resonance line 1 when the temperature is raised.

The unusual temperature dependence of the width of resonance line 2 (figure 13) is due to the change in distribution of magnetic centers over the excited states.

At high temperatures, the greater part of the magnetic centers is found in the excited states for which $D \ll h\nu$. The EPR spectrum of these centers is characterized by the g value of the transition 1/2 ⟷ -1/2. The orientational averaging of the rest of transitions does not broaden the resonance line.

At low temperatures, a major portion of the magnetic centers is in states close to the ground state. The lower the excitation energy, the higher the value of the low-symmetry component of the crystalline field and the farther the resonance lines of transitions ±5/2 ⟷ ±3/2 are located from the line of the transition 1/2 ⟷ -1/2. As a result of the orientational averaging over directions of the axes of symmetry, the contribution of the transitions ±5/2 ⟷ ±3/2 and ±3/2 ⟷ ±1/2 at low temperatures will result in the broadening of resonance line 2 (figure 15).

References

[1] Vasyukov V N, Dyakonov V P, Shapovalov V A, Aksimentyeva E I, Szymczak H and Piechota S 2000 *Low Temp. Phys.* **26** 265
[2] Chabanenko V V, Vasyukov V N, Kochkanjan R O, Nechitailov M M, Szymchak H, Piechota S and Nabialek A 2002 *Low Temp. Phys.* **28** 49
[3] Vasyukov V N, Shapovalov V A, Dyakonov V P, Dmitruk A F, Aksimentjeva E I, Szymczak H and Piechota S 2002 *Intern. J. Quantum Chem.* **88** 525
[4] Vasyukov V N, Shapovalov V V, Schwarz S A, Rafailovich M H, Sokolov J C, Shapovalov V A 2002 *J. Magn. Resonance* **154** 15
[5] Vasyukov V N, Shapovalov V A, Dyakonov V P, Shapovalov V V, Aksimentyeva E I, Szymczak H and Piechota S 2002 *Molecular Crystals and Liquid Crystals* **384** 13
[6] Vasyukov V N, Chabanenko V V, Kochkanjan R O, Nechitailov M M, Nabialek A and Piechota S 2004 *Low Temp. Phys.* **30** 1272
[7] Vasyukov V N and Sukharevskii B Ya 1994 *Low Temp. Phys.* **20** 644
[8] Deigen M F and Glinchuk M D 1975 Paraelectric resonance of noncentral ions *Sov. Phys. Usp.* **17** 691
[9] Vasyukov V N 2002 *Low Temp. Phys.* **28** 199
[10] Klyava Ya G 1988 *EPR spectroscopy of disordered solids* (Riga: "Znanie")
[11] Vasyukov V N 1986 *Phys.Status Sol.* (b) **137** 623

Journal of Physics: Conference Series **324** (2011) 012025 doi:10.1088/1742-6596/324/1/012025

Hexamer clusters in MeF_2:Yb^{3+} (Me=Ca,Sr,Ba)

V A Chernyshev, A E Nikiforov and A D Nazemnikh

Ural State University, 51 Lenin av., Ekaterinburg 620000, Russia

e-mail: anatoliy.nikiforov@usu.ru

Abstract. The electronic structure and g factors of simple impurity centers and hexamer ytterbium clusters in fluorites have been calculated in terms of the exchange charge model. Preliminarily, their local crystal structures have been calculated in the framework of the shell model with the inclusion of lattice distortions near the impurity. Crystal structure and g factors have been calculated of linear cluster Yb–F–Yb–F oriented along a trigonal axis in BaF_2.

1. Introduction

Fluorites MeF_2 (Me = Ca, Sr, Ba) doped with ytterbium Yb^{3+} have attracted attention of researchers for several decades due to their practical application as laser media, scintillation materials, and phosphors. For example, phosphors that emit white light under excitation by a diode laser owing to the up-conversion mechanism have been developed in recent years [1]. Moreover, fluorites have a high-symmetry cubic structure that is retained even after significant doping [1], which makes them convenient for investigations.

At low impurity concentrations (of the order of 0.1 at %), single centers in which Yb^{3+} substitutes for a Me^{2+} cation are formed in MeF_2. The compensation of the excess charge can be nonlocal (cubic center) or local, e.g., by an interstitial fluorine ion F^- located at the interstitial site of the fluorine sublattice on the C_3 or C_4 axis (trigonal or tetragonal centers). For impurity concentrations of the order of 1 at % and higher, clusters are formed from Yb^{3+} and F^- ions.

Spectra of single centers and clusters have been investigated by different methods, such as electron paramagnetic resonance, electron–nuclear double resonance [2–6], selective laser excitation [7], and optical detection of electron paramagnetic resonance [8, 9]; however, the Stark structure of even single Yb^{3+} impurity centers remains unknown in many respects. Only recently, the reliable interpretation of the spectra of the simplest (cubic) Yb^{3+} centers in CaF_2 and CdF_2 was proposed by Gerasimov and Falin [8]. Up to now, there are disagreements in the interpretation of spectra of tetragonal centers, e.g., in the classification of levels according to representations [7, 10].

In recent years, the optical spectra of CaF_2:Yb^{3+} have been studied experimentally for impurity concentrations of the order of 0.1 at % and higher (to a few percent) [7, 11, 12] when nanoclusters containing several Yb^{3+} ions can be formed. Only the main types of nanoclusters formed according to the fluorite motif, their symmetry, and the maximum number of the rare-earth ions forming a cluster were determined experimentally [13]. Therefore, in order to interpret the spectra at high impurity concentrations, it is necessary to perform consistent calculations of the crystal structure and the electronic spectrum of the cluster. This simulation and interpretation of the spectrum were carried out for hexamer clusters M_6F_{36} (M is a trivalent rare-earth ion) in fluorites by Kazanskii, Nikiforov et al.

[14], who calculated the crystal structure in the framework of the shell model in the pair potential approximation. In this case, the relaxation of the crystal lattice near the cluster was taken into account. The parameters for Gd^{3+} located in the middle of the rare_earth series were taken as parameters of the pair interactions between rare_earth and crystal ions. These parameters were determined from the condition for the reproduction of distortions of the fluorine sublattice in the $MeF_2:Gd^{3+}$ impurity centers investigated by electron–nuclear double resonance [15]. This approach allows one to adequately reproduce the crystal lattice distortions near the impurity that makes the main contribution to the parameters of the crystal field. In [14], the electronic spectrum was calculated in the framework of the exchange charge model developed by Malkin [16–18]. In the calculation of the crystal field parameters, the model accounts for the effects of covalence, overlapping, and the electron density transfer between the rare-earth ion and ligands. This model has been successfully used to describe the spectra of almost all single rare-earth ion centers in fluorites. In [14], the authors calculated the g factors and demonstrated that each rare-earth ion in the M_6F_{36} hexacluster is in the tetragonal crystal field with a large axial component induced by the antiprism of eight fluorine ions (Fig. 1).

All levels of the cubic centers $CaF_2:Yb^{3+}$ and $CdF_2:Yb^{3+}$ were classified according to their representations in [8]. This allows one to determine the parameters of the exchange charge model for Yb^{3+} and to use them in the calculations of the spectra of single centers and nanoclusters. In our earlier work [19], we calculated the crystal structure of $MeF_2:Yb^{3+}$ (Me = Ca, Sr, Ba) impurity centers. In this case, the parameters of the Yb^{3+}–F pair interactions were determined from the condition for the reproduction of the distortions in the fluorine sublattice of these centers investigated by the electron–nuclear double resonance method.

Thus, we can perform consistent simulation of the Stark structure of Yb^{3+} in different centers: in the shell model, with the parameters of interionic interactions Yb^{3+}–F unique for all the fluorites $MeF_2:Yb^{3+}$, we can calculate the crystal structure of the single impurity centers and also ytterbium nanoclusters and take into account the lattice relaxation near the impurity. Then, with the calculated structure, we can calculate the parameters of the exchange charge model from the interpreted cubic center spectra [8] and use them to calculate the electronic structure of the tetragonal impurity centers and nanoclusters.

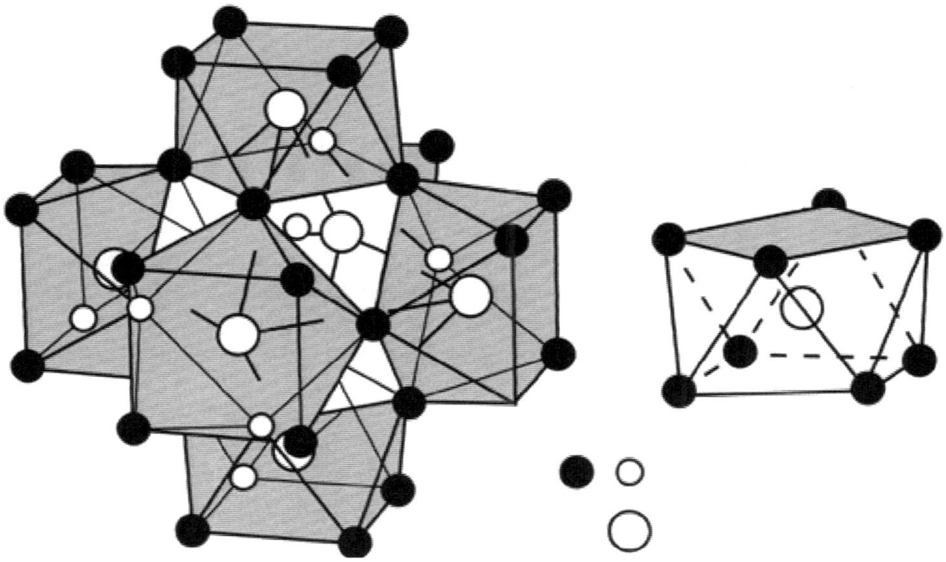

Figure 1. (a) Hexamer cluster and (b) local environment of a rare-earth ion.

2. Calculation of the crystal structure

The expression for the crystal lattice energy, the parameters of the pair interionic potentials and methods of their calculations are given in [15, 19]. The lattice parameters of MeF_2 calculated with these parameters agree well with the experiment (table 1).

Table 1. Lattice parameters of the crystals (pm)

Crystal	Experiment [20]	Calculation
CaF_2	544.3	544.7
SrF_2	578.0	579.3
BaF_2	617.8	623.2

The calculations were carried out using the GULP program [21]. In this program, the crystal lattice structure near the impurity is calculated by the Mott–Littleton method, according to which a spherical region containing a defect and an adjacent spherical layer are considered. Ions in the region containing the defect can be shifted within the limits of its symmetry, and a polarization is allowed in the adjacent layer. The sizes of the region and the layer were chosen in such a way as to adequately describe the lattice distortion, i.e., the sizes were increased until the ion coordinate near the defect determined from the energy minimum become unchanged. For the purposes of verifying, we calculated the crystal field parameters. In the noncubic impurity centers and clusters, it is convenient to use the parameter B_{20}, to which only the distorted region of the lattice near the defect contributes. For example, in the tetragonal impurity centers CaF_2:Yb^{3+}, the lattice relaxation (including the polarization layer) should be taken into account in the radius of 17 Å, which corresponds to 1500 ions. In the calculations of the structure of the hexamer cluster M_6F_{36} (figure 1) which fits into the unit cell, the lattice relaxation was taken into account in the radius 25Å, which corresponds to approximately 10^4 ions. The relaxation in the same region was taken into account when calculating the structure of the M_6F_{37} cluster that differs from the above M_6F_{36} cluster only by the presence of the additional interstitial fluorine ion at the cluster center. We verified to what extent the inclusion of the lattice distortion near the cluster is substantial when calculating the crystal field parameters. For comparison, we calculated the crystal field parameters in the M_6F_{36} cluster without inclusion of the relaxation, i.e., the ion coordinates near the cluster were taken equal to the coordinates of the pure lattice. The crystal field parameters calculated without inclusion of the relaxation are significantly different. For example, the crystal field parameters B_{20}, B_{60}, and B_{64} have other sign and they are different in magnitude by a factor of three or four. Note here that the lattice distortions near the cluster decay very fast. In the Mott–Littleton method that allows ion displacements, the increase in the first region radius from 10 to 18Å, i.e., from 328 to 1822 ions, respectively, does not practically influence the values of the crystal field parameters, e.g., as a result, the Stark levels are shifted by 5–7 cm^{-1}.

The ion coordinates in the M_6F_{36} cluster are listed in tables 2 and 3; the origin of coordinates is in the cluster center. The cluster can be conveniently represented as three coordination spheres F(12), Yb(6), and F(24). Each of the spheres can be characterized by one ion (tables 2 and 3). First two spheres F(12) and Yb(6) are strongly displaced during the relaxation. After the relaxation, the F(12) atoms have practically such a radial distance as the Yb(6) atoms have; i.e., they lie at the same spherical surface (table 3). Twenty four fluorine atoms F(24) forming the outer part of the cluster are slightly displaced with respect to their positions in the pure crystal (table 3).

Table 2. Ion coordinates in the M_6F_{36} cluster (Å)

Ion	Before relaxation			After relaxation		
	x	y	z	x	y	z
F(12)	0.0000	−1.3619	−1.3619	0.0000	−2.0909	−2.0909
Yb(6)	2.7238	0.0000	0.0000	3.0673	0.0000	0.0000
F(24)	−1.3619	−1.3619	4.0858	−1.4038	−1.4038	4.2843

Table 3. Radial and angular coordinates of ions in the M_6F_{36} cluster

Ion	Before relaxation		After relaxation	
	R, Å	Θ, deg	R, Å	Θ, deg
F(12)	1.926	45	2.957	45
Yb(6)	2.724	0	3.067	0
F(24)	4.517	25.24	4.722	24.86

3. Exchange charge model

In the exchange charge model [16, 17], the crystal field parameters are written as the sum of the electrostatic and exchange contributions

$$B_p^q = B_p^q + B_{pS}^q \tag{1}$$

The electrostatic contribution is

$$B_p^q = K_p^q \cdot (1-\sigma_p) \cdot e^2 \cdot \langle r^p \rangle \cdot \sum_{i=1}^{N} (-Z_i) P_p^q(x_i, y_i, z_i, r_i) \cdot r_i^{-(2\cdot p+1)} \tag{2}$$

where (x_i, y_i, z_i, r_i) are polynomials [22], are numerical factors [23], $<r^p>$ is the average value of r^p of $4f$ electron [24], σ_p is the screening parameter of the $4f$ shell with external 5s and 5p shells [25], x_i, y_i, z_i are the ion coordinates; r_i is the distance from the origin of the coordinates to an ion. Relationship (2) is written in the Hartree unit system. When summing, the index i runs over all the lattice ions and, since an ion is represented in the shell model as the sum of two point charges, namely, the core and the shell, the summation is performed over both the cores and the shells; thus, the contribution of the lattice polarization is taken into account. The exchange contribution is

$$B_{pS}^q = K_p^q \cdot \frac{2\cdot(2p+1)}{(2\cdot l+1)} \cdot e^2 \cdot \sum_{i=1}^{M} S_p(r_i) \cdot r_i^{-(p+1)} P_p^q(x_i, y_i, z_i, r_i) \tag{3}$$

Here, when summing, the index i runs only over the ligand cores, l is the orbital moment, S_p is a combination of the squared overlapping integrals of Yb $4f$ functions with the $2s$ and $2p$ functions of fluorine

$$S_p = G_s \cdot S_s^2 + G_\sigma \cdot S_\sigma^2 + G_\pi \cdot k_p \cdot S_\pi^2 \tag{4}$$

The numerical coefficients k_p were obtained when general relationship (4) was deduced in [16]

$$k_2 = -k_6 = \frac{3}{2}, \quad k_4 = \frac{1}{3}.$$

The overlapping integrals S_s, S_σ, and S_π were calculated by formulas presented in [26] on the wave functions [26, 27].The dependence of the overlapping integrals on the rare-earth ion–ligand distance was approximated by an exponential function

$$S_s(r) = 0.2373 \cdot \exp(-0.9286 \cdot r^{1.6183}),$$

$$S_\sigma(r) = 0.1317 \cdot \exp(-0.6699 \cdot r^{1.593}),$$

$$S_\pi(r) = 2.0152 \cdot \exp(-2.7567 \cdot r^{0.8927}).$$

Here, r is expressed in angstroms, the approximation range 1.8–3.2 Å corresponds to the Yb–F distance in the impurity centers and the clusters. In relationship (4), G_s, G_σ, and G_π are dimensionless parameters of the model, i.e., exchange charges. In this case, they are determined from the condition of the best reproduction of the spectrum of the cubic centers CaF_2:Yb^{3+} and CdF_2:Yb^{3+} [8]. The electronic structure of Yb^{3+} for the known crystal field parameters was calculated using the Reid program obtained from its author (Dr M.F. Reid, University of Canterbury). This program takes into account the *J–J* mixing and uses the effective Hamiltonian for 4*f* electrons of rare earths [28]. The parameters for Yb^{3+} were taken from [29].

4. Electronic structure of Yb^{3+} in single impurity centers and nanoclusters

The parameters of the exchange charges ($G_s = 11$, $G_\sigma = 30$, and $G_\pi = 38$) were determined from the condition of the best reproduction of the levels and the g factors of the cubic Yb^{3+} centers in CaF_2 and CdF_2.

Table 4. Energy levels (cm^{-1}) and *g* factors of the cubic Yb^{3+} centers in CaF_2 and CdF_2

J	Irrep	CaF_2		CdF_2	
		calculation	experiment [8]	calculation	experiment [8]
5/2	$^2\Gamma_8$	10867	10840	10880	10867
	$^2\Gamma_7$	10384	10381	10389	10392
7/2	$^1\Gamma_6$	717	778	731	755
	$^1\Gamma_8$	660	644	671	676
	$^1\Gamma_7$	0	0	0	0
	$^2\Gamma_7$	g=-1.41	g=-1.414	g=-1.41	g=-1.411
	$^1\Gamma_7$	g=3.37	g=3.438	g=3.37	g=3.429

Table 5. Energy levels (cm^{-1}) and *g* factors of the cubic Yb^{3+} centers in SrF_2 and BaF_2

J	Irrep	SrF_2		BaF_2	
		calculation	experiment [30]	calculation	experiment [30]
5/2	$^2\Gamma_8$	10785	10783	10715	10739
	$^2\Gamma_7$	10356	10358	10333	10339
7/2	$^1\Gamma_6$	629	600	556	542
	$^1\Gamma_8$	581	586	514	535
	$^1\Gamma_7$	0	0	0	0
	$^2\Gamma_7$	g=-1.41	-	g=-1.41	-
	$^1\Gamma_7$	g=3.47	g=3.441	g=3.47	g=3.422

The agreement with the experiment is good (table 4). Thus, to determine the three empiric parameters, more than ten experimental quantities were used. For purposes of additional verification of the G_i parameters, they were used in calculations of the cubic Yb^{3+} centers in SrF_2 and BaF_2. We also obtained good agreement with the experiment [30] (table 5) in the values of the energy levels and in their representation classification. The spectrum of the tetragonal Yb^{3+} center in CaF_2 was calculated as well (table 6). In the representation classification of the levels, the calculated results agree with the data of [7]. Thus, using the three parameters G_s, G_σ, and G_π, we succeeded in describing the electronic levels and the *g* factors of the cubic and tetragonal centers, in the aggregate about 30 experimental quantities. This allows us to use them in what follows in calculations of the spectra of the nanoclusters.

Table 6. Energy levels (cm^{-1}) and *g* factors of the tetragonal Yb^{3+} centers in CaF_2

J	Irrep	CaF_2:Yb^{3+}	
		calculation	experiment [7]
5/2	Γ_7	10725	10766
	Γ_6	10472	10410
	Γ_7	10308	10322
7/2	Γ_7	510	588
	Γ_6	498	520
	Γ_6	240	456
	Γ_7	0	0
	Γ_7	$g_{\parallel}$=2.29 $g_{\perp}$=3.81	$g_{\parallel}$=2.414 $g_{\perp}$=3.801

The energy spectrum of a cluster is formed by the crystal field, exchange and resonance interactions [31]. In the case of a rare-earth ion, the spin–orbital interaction and the crystal field of one ion are much more than the exchange, magnetic dipole-dipole, and resonance interactions; because of this, we can consider the Stark levels in one rare earth ion.

Table 7. Energy levels (cm^{-1}) and *g* factors of the hexamer Yb^{3+} clusters in CaF_2

J	Irrep	Calculation		Experiment	
		Yb_6F_{36}	Yb_6F_{37}	[11]	[12]
5/2	Γ_6	10865	10916	10870	10887
	Γ_7	10399	10653	10384	10400
	Γ_7	10261	10183	10204	10209
7/2	Γ_6	677	766	675	686
	Γ_7	495	643	518	531
	Γ_6	108	172	103	110
	Γ_7	0	0	0	0
		$g_{\parallel}$=5.7 $g_{\perp}$=0	$g_{\parallel}$=7.63 $g_{\perp}$=0	$g_{\parallel}$=6.8 $g_{\perp}$=0 [14]	

Thus, after we had calculated the crystal structure, we calculated as well the crystal field at a single Yb^{3+} ion located on axis C_4. In this case, the same parameters G_i of exchange charges were used which were used also in the single centers. The electronic structure of the levels and the *g* factors were calculated for the hexamer Yb_6F_{36} and Yb_6F_{37} clusters.

Table 7 lists the calculated data in comparison with the experimental data. The experimental *g* factors were taken from [14]. The nearest environment of the Yb^{3+} ion in the hexacluster is a distorted antiprism [32].

The experimental energy levels agree well with the calculated levels of the hexacluster Yb_6F_{36} and agree significantly worse with the levels of Yb_6F_{37}. A comparison of the calculated levels with other experiment [12], where the spectrum of CaF_2:Yb^{3+} has been also measured at significant impurity concentrations (to 30 at %), as in the preceding case, gives a good agreement for the cluster Yb_6F_{36}.

Yb^{3+} ion in the hexacluster forms a distorted antiprism [32]. The experimental energy levels agree well with the calculated levels of the hexacluster Yb_6F_{36} and agree significantly worse with the levels of Yb_6F_{37}. A comparison of the calculated levels with other experiment [12], where the spectrum of CaF_2:Yb^{3+} was also measured at significant impurity concentrations (to 30 at %), as in the preceding case, gives a good agreement for the cluster Yb_6F_{36}.

5. Linear cluster Yb–F–Yb–F in BaF_2

The linear cluster Yb–F–Yb–F oriented along a trigonal axis in BaF_2 have been investigated by EPR methods [33, 34]. Ytterbium ions substitute Ba^{2+} ions and interstitial fluorines occupy the empty cubes in fluorine sublattice along the C_3 axis (Fig.2).

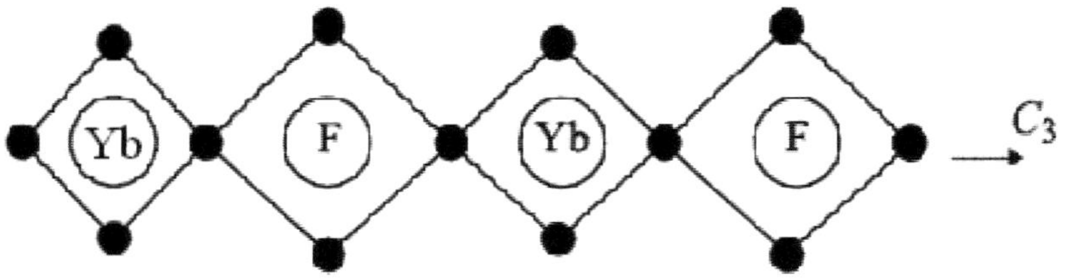

Figure 2. Cluster Yb–F–Yb–F in BaF_2 [33].

Crystal structure of the cluster has been calculated. The calculations predicted that fluorine cubes around Yb are compressed on 1-2% whereas cubes around F extended on 1-2%. The values of g-factors have been calculated for one ytterbium that lie between interstitial fluorines. The calculated values are $g_{\|}=2.12$ and $g_{\perp}=3.87$. The experimental [33,34] values are $g_{\|}=2.25$ and $g_{\perp}=4.02$.

6. Conclusions

The parameters of the exchange charge model G_s,G_σ, and G_π determined from the optical spectra of the cubic Yb^{3+} centers in CaF_2 and CdF_2 allow a successful description of the spectra of other single ytterbium centers. The calculations show that the spectrum of CaF_2:Yb^{3+}, at the impurity concentrations of an order of 1 at % and higher, is due to the Yb_6F_{36} hexamer clusters.

The inclusion of the crystal lattice relaxation during formation of the hexamer cluster shows that the lattice inside the cluster is strongly distorted, and the Yb^{3+}(6) and F(12) ions lie practically on one sphere. However, the distortions quickly damp, and the F(24) fluorine atoms forming the external part of the cluster are displaced by only 0.2Å. Thus, the cluster external sizes (9.4Å) are slightly different from the corresponding distance between the ions in the pure lattice (9Å).

The calculations of crystal structure and g-factors for Yb-F-Yb-F cluster in BaF_2 are in agreement with the experiment. This points to linear cluster formation in BaF_2.

Acknowledgments

This study was supported by the Federal Agency for Education of the Russian Federation (state contract no. P1462 from 04.09.09).

References

[1] Cao C, Qin W, Zhang J, Wang Y, Wang G, Wei G, Zhu P, Wang L and Jin L 2008 *Opt. Commun.* **281** 1716
[2] Ramos C A, Fainstein C and Tovar M 1985 *Phys. Rev.* B **32** 64
[3] Baker J M, Davies E R and Hurrell J P 1968 *Proc. Roy. Soc.* (London) Ser. A **308** 403
[4] Kiro D and Low W 1968 *Phys. Rev. Lett.* **20** 1010
[5] Nazarova O V and Sanadze T I 1977 *Soobshch. Akad. Nauk Gruz. SSR* **87** (2) 329
[6] Berulava B G, Mirianashvili R I, Nazarova O V and Sanadze T I 1977 *Sov. Phys. Solid State* **19** 1033
[7] Petit V, Camy P, Doualan J J, Portier X and Moncorge R 2008 *Phys. Rev.* B **78** 085131
[8] Gerasimov K I and Falin M L 2009 *Phys. Solid State* (St. Petersburg) **51** 721
[9] Kazanskii S A and Ryskin A I 2002 *Fiz. Tverd. Tela* (St. Petersburg) **44** 1356
[10] Falin M L, Gerasimov K I, Latypov V A and Leushin A M 2004 *Appl. Magn. Reson.* **26** 617
[11] Ito M, Gontaudier C, Guyot Y, Lebbou K, Fukuda T and Boulon G 2004 *J. Phys.: Condens. Matter* **16** 1501
[12] Leniec G, Kaczmarek S M and Boulon G 2005 *Proc. SPIE–Int. Soc. Opt. Eng.* **5958** 595825
[13] Ryzhova E A, Molchanov V N, Artyukhov A A, Simonov V I and Sobolev B P 2004 Crystallogr. *Rep.* **49** (4) 591
[14] Kazanskii S A, Ryskin A I, Nikiforov A E, Zaharov A Yu, Ugrumov M Yu and Shakurov G S 2005 *Phys. Rev.* B **72** 014127
[15] Chernyshev V A, Gorlov A D, Mekhonoshin A A, Nikiforov A E, Rokeakh A I, Shashkin S Yu and Zaharov A Yu 1998 *Appl. Magn. Reson.* **14** 37
[16] Malkin B Z 1097 in *Spectroscopy of Solids Containing Rare-Earth Ions*, Ed. by A. A. Kaplyanskii and R. M. Macfarlane (Amsterdam: North_Holland) p. 13.
[17] Larionov A L and Malkin B Z 1975 *Opt. Spectrosc.* **39** (6) 637
[18] Hull R, Parisi J, Osgood R M, Warlimont H, Liu G and Jacquier B 2005 *Spectroscopic Properties of Rare_Earths in Optical Materials* (Berlin: Springer).
[19] Gorlov A D, Chernyshev V A, Ugryumov M Yu and Abrosimov A V 2005 *Phys. Solid State* **47** (8) 1439
[20] Axe J D 1965 *Phys. Rev.* A **139** 1215
[21] Gale J D 1997 *J. Chem. Soc., Faraday Trans.* **93** 629
[22] Al'tshuler S A and Kozyrev B M *Electron Paramagnetic Resonance in Compounds of Transition Elements* (Nauka, Moscow, 1972; Wiley, New York, 1974).
[23] Bumagina L A, Kazakov B N, Malkin B Z and Stolov A L 1977 Sov. Phys. Solid State **19** (4) 683
[24] Reid M F, van Pieterson L and Meijerink A 2002 *J. Alloys Compd.* **344** 240
[25] Sternheimer R M, Blume M and Peierls R F 1968 *Phys. Rev.* **173** (2) 376.
[26] Eremin M V 1989 in *Spectroscopy of Crystals*, Ed. By A. A. Kaplyanskii (Leningrad: Nauka) p. 30
[27] Starostin N V 1975 *Spectroscopy of Crystals*, Ed. By A. A. Kaplyanskii (Moscow: Nauka) p. 12
[28] Newman D J and Ng B 2000 *Crystal Field Handbook* (Cambridge:Cambridge University Press)
[29] van Pieterson L, Reid M F, Wegh R T, Soverna S and Meijerink A 2002 *Phys. Rev.* B **65** 045114
[30] Falin M L, Gerasimov K I, Latypov V A and Leushin A M 2003 *J. Phys.: Condens. Matter* **15** 2833
[31] Mitrofanov V Ya, Nikiforov A E and Cherepanov V I 1985 *Spectroscopy of Exchange-Bonded Complexes in Ionic Crystals* (Moscow: Nauka)
[32] Nikiforov A E, Zakharov A Yu, Ugryumov M Yu, Kazanskii S A, Ryskin A I and Shakurov G S 2005 *Phys. Solid State* **47** (8) 1431
[33] Aminov L K and Kurkin I N 2009 *Phys. Solid State* (St. Petersburg) **51** (4) 741
[34] Aminov L K, Abdulsabirov R Yu, Gafurov M R, Korableva S L, Kurkin I N, Kurzin S P,

Rakhmatullin R M and Ziganshin A G 2005 *Appl. Magn. Reson.* **28** 41

Journal of Physics: Conference Series **324** (2011) 012026 doi:10.1088/1742-6596/324/1/012026

EPR studies of the mechanochemically Er^{3+}-activated fluorite nanoparticles

I A Irisova, A A Rodionov, D A Tayurskii and R V Yusupov

Institute of Physics, Kazan Federal University, Kremlevskaya 18, 420008 Kazan, Russia

E-mail: irisha-kzn@mail.ru

Abstract. In the present work we show with EPR spectroscopy that the simultaneous grinding of the crystalline CaF_2 and ErF_3 salts leads to the mechanochemical doping of the single Er^{3+} ions in the CaF_2 host. Dependence of the EPR spectra intensity on the grain size indicates clearly that the Er^{3+} centers are created at the surface of the particles. Dominant part of the observed paramagnetic centers represent the Er^{3+} ions substituting the regular quasicubic Ca^{2+} lattice site ($g = 6.7$) perturbed by the vicinity of the surface. Fine particles of CaF_2 reveal also the nearly-isotropic EPR signal at $g \sim 1.97$ that originates from the surface defects and can be used for characterization of the obtained samples with respect of the average grain size. Grinding of the of the CaF_2 and ErF_3 mixture in the ratio of 100:1 during 12 hours gives the surface density of the Er^{3+} ions of ~ 0.1 nm^{-2}, or $\sim 10^4$ Er^{3+} ions per the 190-nm size particle.

1. Introduction

Nowadays the nanostructured high-surface area metal fluorides are of increasing importance for different applications like heterogeneous catalysis, thin films, nanoceramics, bioceramics or glasses. Furthermore, investigations of the ionic crystals doped with the rare-earth (RE) ions are actual both for investigations of the mechanisms of the impurity ion and host lattice interaction and for development of the theoretical concepts and practical applications (solid-state lasers, superionic conductors, scintillators etc.) [1].

Electron paramagnetic resonance (EPR) is one of the most informative methods for investigating of the crystals doped with RE ions since the EPR spectra allow determining of the structural properties of the impurity in the lattice, characterization of powder samples with respect to its average particle size, investigating the phenomena occurring on the particle size reduction. Papers [2, 3] notice that till year 2006 only few papers existed on the application of the magnetic resonance spectroscopy methods to studies of the mechanically activated solids and, especially, solid fluorides. This fact motivated our study the fluorides by EPR.

One of the important applications of the RE ions is magnetic resonance imaging (MRI). Contrast agents (CAs) often comprising the gadolinium ions are widely used to improve the sensitivity and diagnostic confidence in MRI [4]. Despite the progress in the design and synthesis of these agents, many are still limited by low relaxivity, low specificity and potential toxicity. In particular, aqueous Gd^{3+} ions are toxic and therefore need to be sequestered using a variety of linear and macrocyclic helates. However, chelation can reduce the contrast enhancement because it decreases the number of coordination sites available for water proton exchange. Paper [5] provides a new approach to solve the above-mentioned problems which is based on confining of the magnevist, gadofullerenes and gadonanotubes inside the nanoporous structure of the silicon particles.

Mechanochemical synthesis can also serve as the perspective method for confining of the RE ions inside the crystal structure. Moreover, mechanochemical synthesis has an important advantage compared with the traditional method of the impurity-doped crystal synthesis (method of the directed crystallization), namely the low temperature, which lifts many limitations associated with different melting points of the components, their vapor pressures, thermal decomposition, and other factors. As a host lattice the CaF_2 compound was chosen since it is one of the main initial components in the mixture for the mechanochemical synthesis of the nanostructured fluoroapatite [6]. The aim of our work was to investigate the mechanoactivated centers of the RE elements in the fine powders of the crystals with fluorite structure. We have studied the fluorite nanoparticles activated by the Er^{3+} ions. Er^{3+} ions were chosen due to the fact that its resonance EPR line occurs at the *g*-factor value different from $g \sim 2$ which gave an opportunity to study explicitly the mechanochemically activated Er^{3+} centers. In our opinion, results the paper may be generalized to the Gd^{3+} ions incorporated to the fluorite nanoparticles in a similar way.

2. Sample preparation

Fluorite fine particles mechanochemically activated with RE ions have been obtained by means of grinding of the mixtures of CaF_2 and RE fluoride high purity crystalline salts. In this paper we studied the samples prepared from the initial mixture of 99 wt.% of CaF_2 and 1 wt.% of ErF_3. The series of the fine powders with various average grain sizes <d> was produced by grinding of the initial mixture for 12 hours in an agate mortar followed by the successive sedimentation in extra-pure isopropyl alcohol. The studied series consisted of the fine powders obtained after the 48, 17, 6 and 2 hours sedimentation.

Obtained samples were characterized with a scanning electron microscope (SEM). Commercial Philips XL 30 microscope was used for this purpose. The images of the 48 h and 2 h sedimentation samples are shown in Figure 1. The corresponding grain size distributions are presented by the histograms shown in Figure 2. In order to find out the average grain size <d>, grain size distributions were fit to the log-normal distribution function

$$f(x) = A\exp\left(-\frac{\left[\ln x - \ln\langle d\rangle\right]^2}{2w^2}\right).$$

Results of the analysis are summarized in Table 1.

Figure 1. Scanning electron microscopy images of the samples obtained by the 48 hours (a) and 2 hours (b) sedimentation.

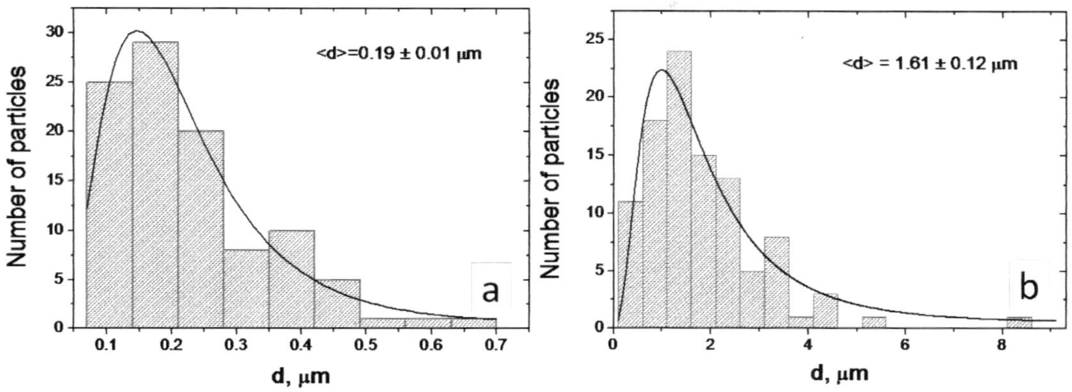

Figure 2. Grain size distributions of the samples obtained by the 48 hours (a) and 2 hours (b) sedimentation.

Table 1. Mean particle sizes for the studied sample series

Sedimentation (h)	Size (μm)
48	0.19 ± 0.01
17	0.29 ± 0.02
6	1.58 ± 0.09
2	1.61 ± 0.12
residue	2.04 ± 0.13

3. Results and discussion

EPR spectra of the obtained samples were measured with a commercial Bruker ESP 300 X-band (~9.5 GHz) spectrometer equipped with the standard TEM_{012} mode rectangular cavity. Temperature of the samples was controlled with the Oxford Instruments ESR9 helium flow cryogenic system.

Effect of grinding on the EPR spectra of CaF_2, ErF_3 and of ($0.99{\cdot}CaF_2 + 0.01{\cdot}ErF_3$) mixture ground in the agate mortar for 45 minutes is shown in Figure 3. The spectra were measured at T = 15 K. Compared with the spectra of the ground for the same time CaF_2 and ErF_3 salts, the spectrum of the ground mixture clearly contains a new component observed at $g \sim 6.71$ (1010 Oe). In the other words, we see that the ground ($0.99{\cdot}CaF_2 + 0.01{\cdot}ErF_3$) mixture is not the same as the mixture of the ground CaF_2 and ErF_3. The origin of the $g \sim 6.71$ component becomes evident if one studies the papers on the EPR spectroscopy of the CaF_2:Er^{3+} crystals [7-11]. This component fits well the properties of the cubic symmetry Er^{3+} ion center.

Another component that is found both in the ground CaF_2 and ($0.99{\cdot}CaF_2 + 0.01{\cdot}ErF_3$) mixture occurs at g ~ 1.97. And, finally, no any pronounced spectral features are found in the EPR spectrum of the ground concentrated ErF_3.

Thus, the mechanochemical activation of the CaF_2 particles with Er^{3+} ions clearly takes place on grinding of the CaF_2 and ErF_3 mixture.

In order to achieve better understanding of the mechanoactivation process we have studied the series of the fine powder samples described in a previous section, that differ essentially in the grain size only. The spectra of the samples measured in the identical conditions with the intensity normalized to the sample mass are shown in Figure 4.

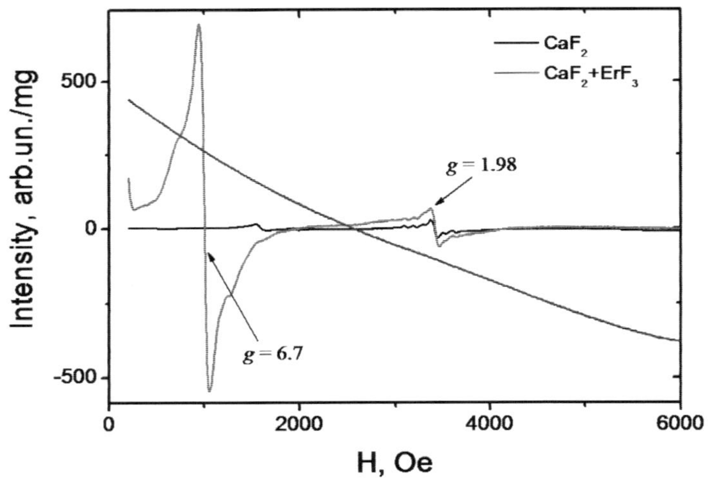

Figure 3. EPR spectra of CaF_2 (black), ErF_3 (blue) and ($0.99{\cdot}CaF_2 + 0.01{\cdot}ErF_3$) mixture ground for 45 minutes; $T = 15$ K.

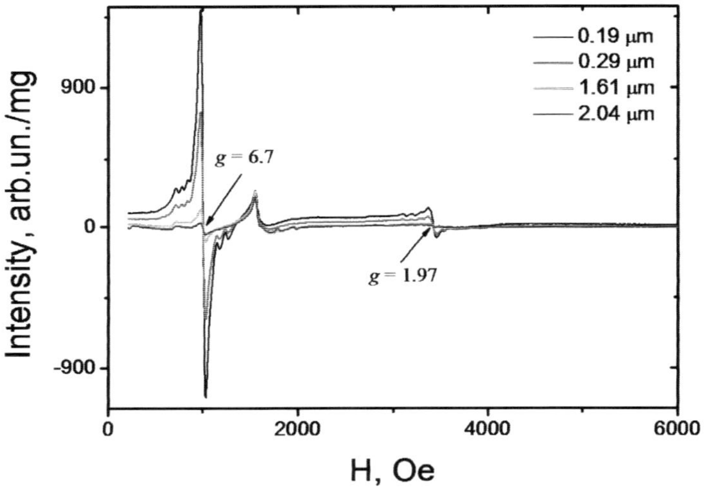

Figure 4. EPR spectra of the various size samples of the ground for 12 hours ($0.99{\cdot}CaF_2 + 0.01{\cdot}ErF_3$) mixture; $T = 15$ K.

Figure 5 depicts the dependencies of the peak-to-peak line intensities for the components found at $g = 6.71$ and $g = 1.97$ on the mean particle size $<d>$. The fits shown in Figure 5 indicate clearly that the intensities of both components dominating in the spectra are proportional to $<d>^{-1}$. As for any shape of the particle its volume and surface are related by $S/D \propto <d>^{-1}$, the described observation shows that both the $g = 6.71$ and $g = 1.97$ features originate from the particle surface. Thus, mechanoactivation of the CaF_2 fine particles with Er^{3+} ions occurs at the particle surface which looks natural. Relation of the $g = 6.71$ spectral feature to Er^{3+} ions is of no doubt: in the spectra of the two smallest grain samples the hyperfine structure is observed for the ^{167}Er isotope ($I = 7/2$, natural abundance 22.9%).

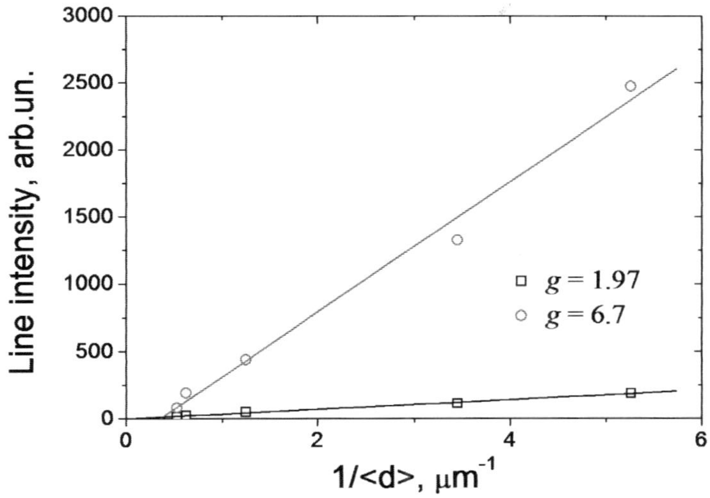

Figure 5. Grain size dependencies of the EPR lines observed at $g = 6.71$ (circles), $g = 1.97$ (squares) and the linear fits of the dependencies.

The component at $g = 1.97$ is also interesting and important. It originates from the particle surface but is not related to erbium ions. This component was observed in the spectrum of the ground CaF_2 in Figure 3. Therefore it is most probably the intrinsic signal from the defects that form at the CaF_2 surface. Then, the intensity of the $g = 1.97$ spectral component can serve as a measure of the mean particle size in the CaF_2 fine powders prepared by means of grinding.

There are two signals caused by contamination of the CaF_2 salt. The first, which is revealed as the six-line hyperfine pattern at $g \sim 2$, obviously originate from the Mn^{2+} ions. The second, which is an asymmetric line at $g \sim 4.3$, clearly scales with the mass of CaF_2; its origin is not clear.

In the range of the g-factor values typical for Er^{3+} centers in CaF_2 additional spectral features are resolved that are especially pronounced for the samples of 1.61 mm and 2.04 mm. Thus, one can clearly see the signals that can be characterized with the g-factor values of 9.78, 4.22, 3.42 and 3.76. None of these values characterize the known Er^{3+} centers found in the CaF_2:Er^{3+} crystals. However, if these, quite well defined centers are localized at the particle surface, the g-factor values may differ from those in the bulk due to the lattice constant relaxation nearby the surface. Note the shift of the g-factor for the surface quasicubic Er^{3+} centers ($g = 6.71$) with respect to its value in the crystal ($g = 6.78$).

Finally, we compare the spectra of the mechanoactivated Er^{3+} centers at the surface of the CaF_2 particles and the doped ones in the CaF_2 single crystal (Figure 6). Obviously, the two spectra essentially reproduce each other, except of the linewidth. This indicates that in the process of mechanoactivation the "cubic" structure doped Er^{3+}-centers are predominantly formed. Linewidth increase can naturally be assigned to the disorder that takes place in the particle volume close to its surface. It is worth mentioning that the substantial value of the Er^{3+}-ion spectrum linewidth in the crystal is related to the comparable terms that occur due to the dipole interaction and non-resolved superhyperfine structure from the neighboring fluorine ions.

The absolute microwave absorption value registered for the Er^{3+}-mechanoactivated CaF_2 particles and CaF_2:Er^{3+} single crystal that was grown from the initial charge that contained 0.8 wt.% of ErF_3 allowed us to estimate the surface density of the mechano-doped Er^{3+} centers. We have found that if one prepares the sample by 12 hours grinding from the mixture of ($0.99 \cdot CaF_2 + 0.01 \cdot ErF_3$), there will be about 0.1 Er^{3+}-ion per nm2, or $\sim 10^4$ Er^{3+}-ions per the particle 190 nm in size.

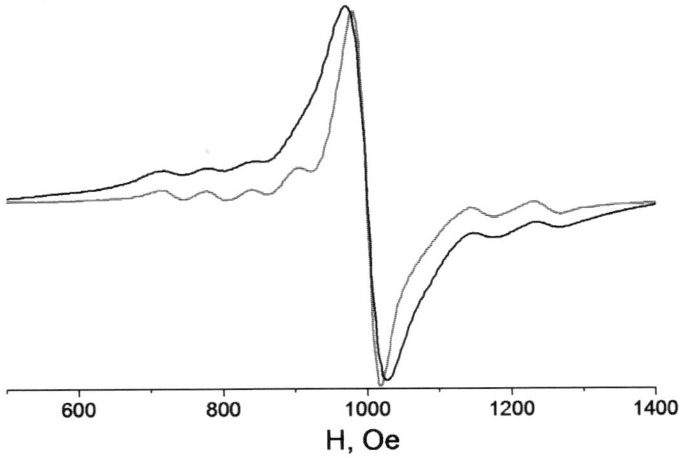

Figure 6. EPR spectra of the CaF_2 nanoparticles (<*d*> = 190 nm) with mechanoactivated Er^{3+} centers and CaF_2:Er^{3+} single crystal with 0.8 wt.% of ErF3 in initial charge; T = 10 K. Resonance frequencies of the two spectra and the peak intensities were adjusted one to the other.

4. Conclusions

Results presented in the paper can be summarized as follows:

- On grinding of the CaF_2 and ErF_3 mixtures mechanochemical activation of the CaF_2 fine particles with Er^{3+} single ion centers occurs.
- In all the so far studied samples of the CaF_2 nanoparticles mechanoactivated with Er^{3+} ions the two most intense EPR lines are observed at g = 6.71 and g = 1.97, with the intensities that scale with the amount of surface ($\propto <d>^{-1}$). This undoubtedly indicates that both lines originate from the surface defects. The line with g = 1.97 is caused by the intrinsic CaF_2 surface defects and thus can be used to estimate the mean grain size of the sample.
- Majority of the mechanochemically doped Er^{3+} centers is represented by the nominally high-symmetry cubic centers of Er^{3+} ions perturbed by the disorder nearby the surface.

Acknowledgement

The authors would like to thank Associate Prof. Sitdikova L.M (Kazan Federal University) for her assistance in SEM observations.

References

[1] Nikcevic I, Jokanovic V, Mitric M, Nedic Z, Makovec D and Uskokovic D 2006 *J. Solid State Chemistry* **179** 1119–1128
[2] Bureau B and Silly G 1999 *J. Phys.: Condens. Matter* **11** L423–L431
[3] Sobolev B. P, Sviridov I. A, Fadeeva V. I, Sulyanov S. N, Sorokin N. I, Zhmurova Z. I, Herrero P, Landa-Canovas A and Rojas R. M. 2005 *J. Crystallography Reports* **50** 478–485
[4] Lauffer R. B 1987 *J. Chem. Rev.* **87** 901–927
[5] Ananta J S et al. 2010 *J. Nature Nanotechnology* **5** 815
[6] Nikcevic I, Jokanovic V, Mitric M, Nedic Z, Makovec D and Uskokovic D 1994 *J. Solid State Chem.* **177** 2565–2574
[7] Rector C V, Pandey B C, Moos H W 1966 *J. Chem. Phys.* **45** 171–179

[8] Ranon U, Low W 1963 *Phys. Rev.* **132** 1609–1611
[9] Baker J M, Hayes W, Jones D A 1959 *Proc. Phys. Soc.* **73** 942–944
[10] Dvir M, Low W 1960 *Proc. Phys. Soc.* **75** 136–137
[11] Weber M J, Bierig K W 1964 *Phys. Rev.* **134** A1492–A1503

Journal of Physics: Conference Series **324** (2011) 012027 doi:10.1088/1742-6596/324/1/012027

Lande factor of the conduction electrons in silicon: temperature dependence

A A Konakov, A A Ezhevskii, A V Soukhorukov, D V Guseinov, S A Popkov and V A Burdov[1]

University of Nizhniy Novgorod, Gagarin ave. 23, Nizhniy Novgorod, 603950 Russia

E-mail: vab3691@yahoo.com

Abstract. Temperature dependence of the electron g-factor in silicon has been investigated both theoretically and experimentally. Theoretical consideration is based on the renormalization of the electron energy in a weak magnetic field by the electron-phonon interaction in the second-order perturbation theory. Interaction of the electron subsystem with the lattice vibrations results in decreasing the conduction electron g-factor. This decreasing was observed experimentally in the electron spin resonance studies for n-Si samples.

1. Introduction

Since pioneer demonstration of the spin injection and detection in Si [1], silicon becomes a perspective material for spintronics due to weak spin-orbit interaction, long spin relaxation time and large spin diffusion length in comparison with typical III-V semiconductors such as gallium arsenide. One of the most important characteristics of conduction electrons in various systems with spin-orbit coupling is the spin relaxation rate. Spin relaxation in silicon mainly occurs through the Elliott [2] and Yafet [3] mechanism, where the spin flip processes are caused by electron-phonon interaction. Quantitative theoretical studies of the Elliott-Yafet spin relaxation in silicon [4] yields the spin relaxation rate proportional to a third power of temperature T. This is in good agreement with the experimental data. The spin relaxation rate can be extracted from the line width of the conduction electron spin resonance (CESR).

Another important characteristic of conduction electrons in magnetic field is their Lande g-factor, which is characterized by the position of the CESR line. First experimental investigations of the conduction electron g-factor were carried out by Wilson and Feher [5]. The measured value for "free" carriers was found to be $g = 1.99875 \pm 0.0001$. In silicon each conduction valley has an axial symmetry, so the g-factor becomes a tensor with two, longitudinal $g_{\parallel}$ and transverse $g_{\perp}$, principal values. The measured value g is isotropic due to cubic symmetry of the Brillouin zone and has the average form over all six conduction band valleys:

$$g = \frac{1}{3} g_{\parallel} + \frac{2}{3} g_{\perp}. \tag{1}$$

[1] To whom any correspondence should be addressed.

In contrast to the spin relaxation rate, there is no information in literature about the temperature dependence of the electron g-factor in silicon. So, in this work we first theoretically discuss and experimentally demonstrate this temperature dependence.

2. Experimental results

We studied the electron g-factor in CESR spectra measurements. We have observed dependence of g-factor of conduction electrons in silicon as a function of temperature and found that in temperature range 80 - 250K they are alike for different donors and their concentrations. The measurements are performed on the spectrometer "Bruker EMX 10/12" using helium cryostat with a system of a temperature control (3.8-300 K) "ER 4112 HV". We use natural silicon samples doped with lithium and phosphorus to make some concentration of electrons in the conduction band. Donor concentrations in both samples were very close to minimize difference in the spin flip process intensity.

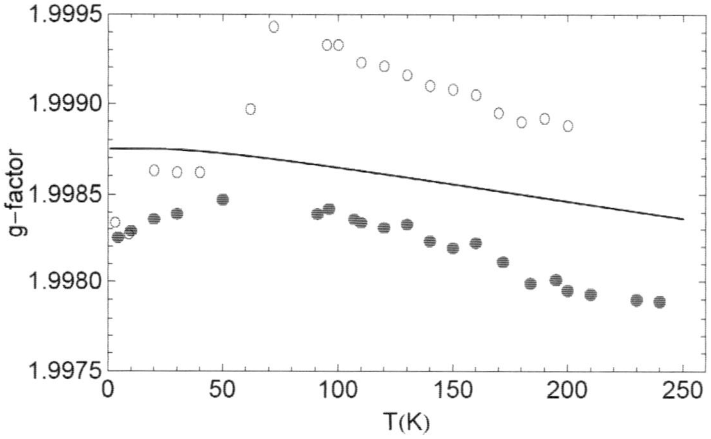

Figure 1. Calculated (solid line), and measured (dots and circles) temperature dependences of the average electron g-factor in Si sample doped with phosphorus (red dots) and lithium (blue circles). Phosphorus and lithium concentrations are $3.7\cdot10^{18}$ cm^{-3} and $3.3\cdot10^{18}$ cm^{-3}, respectively. Calculations were carried out in accordance with Eq. (9). The phonon energy $\hbar\omega$ in Eq. (9) was taken to be 10 meV.

The experimental results are presented in figure 1. As temperature increases from zero to ~80 K, a monotonous rise of the g-factor value takes place. In this temperature range electrons still localized on the donor centres and have the discrete energy spectra, so it can occupy the singlet A_1 or the triplet T_2 and doublet E states. Each of these states can be described by different g-factors, and changing the electron distribution among these states with temperature rising can leads to rising g-factor. Consideration of this effect in more details goes beyond the scope of the present paper. However at higher temperatures the donors become ionized and the electrons can propagate over the sample. In this temperature range one can see approximately linear decrease of the g-factor, which can be explained by the modulation of the lattice spin-orbit interaction by phonons. Below, we discuss some theoretical approach to this problem.

3. Theoretical model

Theoretical study of the g-factor temperature dependence is based on the renormalization of the electron energy in the external magnetic field by the electron-phonon interaction. Within the framework of a many-body picture the renormalization of the excitation energies by lattice vibrations can be defined by the equation [6]

$$\varepsilon(\mathbf{k})=\varepsilon^{(0)}(\mathbf{k})+\mathrm{Re}\,\Sigma^{*}(\varepsilon,\mathbf{k}), \qquad (2)$$

where $\varepsilon^{(0)}(\mathbf{k})$ and $\varepsilon(\mathbf{k})$ are unperturbed and renormalized electron energies, respectively, and $\mathrm{Re}\Sigma^{*}(\varepsilon,\mathbf{k})$ is the self-energy real part that can be presented in covalent semiconductors, such as silicon, using the Rayleigh-Schrödinger perturbation theory [7]:

$$\mathrm{Re}\Sigma^{*}(\varepsilon,\mathbf{k})=\sum_{\mathbf{q}\lambda}\left(n_{\mathbf{q}\lambda}+\frac{1}{2}\mp\frac{1}{2}\right)\frac{\left|V(\mathbf{q})\right|^{2}}{\varepsilon^{(0)}(\mathbf{k})-\varepsilon^{(0)}(\mathbf{k}\pm\mathbf{q})\pm\hbar\omega_{\lambda}(\mathbf{q})},\tag{3}$$

where $V(\mathbf{q})$ is the matrix element describing the electron scattering (from the state $|\mathbf{k}\rangle$ to the $|\mathbf{k}\pm\mathbf{q}\rangle$ state) accompanied by absorption (upper sign) or emission (lower sign) of a phonon with the wave vector $\mathbf{q}$ and polarization λ, $\omega_{\lambda}(\mathbf{q})$ and $n_{\mathbf{q}\lambda}$ are the frequency and the occupation number of phonons. In what follows, we replace the phonon occupation numbers by their equilibrium values describing the Bose-Einstein statistics.

If the spin-orbit interaction is taken into account, the electron states become the states with an effective generalized spin that differs from the standard spin moment. We will denote these states as effective spin-up $|\Uparrow\rangle$ or spin-down $|\Downarrow\rangle$ vectors. In the absence of magnetic field the spin-up and spin-down states have the same energies (the Kramers degeneracy). In the external magnetic field **H** oriented along one of the valleys the electron energy has the additional Zeeman term $\sigma\mu_{\mathrm{B}}g_{i}H/2$, where μ_{B} is the Bohr magneton, g_i is the g-tensor principal value, $i=\perp,\mathrm{P}$ for transverse and longitudinal components, respectively, and $\sigma=\pm1/2$ are the effective-spin projections on the magnetic field. In that case the Zeeman term is also renormalized, and the total electron energy has the form

$$\varepsilon(\mathbf{k})+\sigma\mu_{\mathrm{B}}g_{i}H=\varepsilon^{(0)}(\mathbf{k})+\sigma\mu_{\mathrm{B}}g_{i}^{(0)}H+\sum_{\mathbf{q}\lambda\sigma'}\frac{\left(n_{\mathbf{q}\lambda}+1/2\mp1/2\right)\left|V_{\sigma\sigma'}^{(i)}(\mathbf{q})\right|^{2}}{\varepsilon^{(0)}(\mathbf{k})-\varepsilon^{(0)}(\mathbf{k}\pm\mathbf{q})+(\sigma-\sigma')\mu_{\mathrm{B}}g_{i}^{(0)}H\pm\hbar\omega_{\lambda}(\mathbf{q})},\tag{4}$$

where $g_i^{(0)}$ and g_i are unperturbed and renormalized principal values of the g-tensor, respectively, $V_{\sigma\sigma'}^{(i)}(\mathbf{q})$ is the spin-dependent momentum scattering matrix element for the electron in the i-th valley. Thus, if scattering conserves the spin, only the translatory-motion energy is modified. On the contrary, spin flip processes modify the Zeeman energy. Using standard approach $\mu_{\mathrm{B}}g_{i}H<<\varepsilon(\mathbf{k})$, and neglecting the phonon-absorption process, whose contribution seems to be less compared to the phonon-emission process, one can obtain

$$g_{i}=g_{i}^{(0)}-2g_{i}^{(0)}\sum_{\mathbf{q}\lambda}\frac{\left(n_{\mathbf{q}\lambda}+1\right)\left|V_{\Uparrow\Downarrow}^{(i)}(\mathbf{q})\right|^{2}}{\left(\varepsilon^{(0)}(\mathbf{k})-\varepsilon^{(0)}(\mathbf{k}-\mathbf{q})-\hbar\omega_{\lambda}(\mathbf{q})\right)^{2}}.\tag{5}$$

Here, $V_{\Uparrow\Downarrow}^{(i)}(\mathbf{q})$ is the spin flip matrix element which can be written as [3,8]

$$\left|V_{\Uparrow\Downarrow}^{(i)}(\mathbf{q})\right|=A_{i}q^{2}\sqrt{\hbar/2\rho V\omega_{\lambda}(\mathbf{q})},\tag{6}$$

where ρ and V are the silicon density and the sample volume, respectively, A_i are some constants symbolically referred to as spin deformation potentials.

Considering the zero-temperature limit we should set $n_{\mathbf{q}\lambda}=0$ in Eq. (5). In this case Eq. (5) yields the zero-temperature renormalization of the g-factor by emitting and subsequent absorbing virtual phonon. Evidently, measured low-temperature conduction electron g-factor, such as Wilson and Feher's $g\approx1.9987$, includes this correction. Consequently, we will use the low-temperature experimental values as zeroth approximation:

$$g_{i}(0)=g_{i}^{(0)}-\frac{\hbar A_{i}^{2}g_{i}^{(0)}}{\rho V}\sum_{\mathbf{q}\lambda}\frac{q^{4}}{\omega_{\lambda}(\mathbf{q})\left(\varepsilon^{(0)}(\mathbf{k})-\varepsilon^{(0)}(\mathbf{k}-\mathbf{q})-\hbar\omega_{\lambda}(\mathbf{q})\right)^{2}},\tag{7}$$

while T-dependent correction $\delta g_{i}(T)$ will be defined by

$$\frac{\delta g_i(T)}{g_i^{(0)}} = -\frac{\mathrm{h}A_i^2}{\rho V}\sum_{\mathbf{q}\lambda}\frac{n_{\mathbf{q}\lambda}q^4}{\omega_\lambda(\mathbf{q})\left(\varepsilon^{(0)}(\mathbf{k})-\varepsilon^{(0)}(\mathbf{k}-\mathbf{q})-\mathrm{h}\omega_\lambda(\mathbf{q})\right)^2}. \tag{8}$$

The main contribution to the sum over **q** in Eq. (8) comes from the intervalley phonons connecting different valleys in the electron Brillouin zone. Taking these contributions into account, and averaging over acoustic phonon modes and electron valleys, it is possible to obtain the following estimation for the *T*-dependent correction:

$$\frac{\delta g(T)}{g} : -\frac{15A^2q_0^4 n(\omega)}{\pi \mathrm{h}\rho}\left(\frac{m}{2\mathrm{h}\omega}\right)^{3/2}, \tag{9}$$

where q_0 is a typical wavevector of the intervalley phonon, and $\omega = \omega(q_0)$ is average phonon frequency corresponding to the phonon energy $\mathrm{h}\omega$ whose value may be estimated as 10 – 20 meV. The calculated dependence $\delta g(T)$ is presented in figure 1 along with the experimental data. It is seen that the theoretical estimations agree (at least qualitatively) with the measured values. However, to achieve the quantitative agreement, more rigorous calculations are necessary. In particular, theory should answer the following question: how the donor type influences the Zeeman splitting if the donors are ionized, and the electrons become, in fact, free. Presumably, the spin-orbit interaction and, consequently, the spin-dependent momentum scattering matrix element $V_{\sigma\sigma'}^{(i)}(\mathbf{q})$ depend on the donor type. This dependence has to be taken into account to explain different experimentally observed values of the electron *g*-factor in Si doped with different donors. So far, within the framework of the suggested theoretical treatment, no any specific features of various donors have been considered.

4. Conclusion

Generally speaking, modification of the g-factor is caused by virtual phonon-induced spin flip processes occurring between the initial, intermediate, and final electron states, where the final and initial states completely coincide. This, presumably, implies that the Elliott-Yafet mechanism of the spin relaxation is the main one for the considered system.

Acknowledgments

This work was in part financially supported by the Russian Ministry of Education and Science through the programs RNP and "Kadry", RFBR, and the "Dynasty" Foundation.

References

[1] Appelbaum I, Huang B and Monsma D J 2007 *Nature (London)* **447** 295
[2] Elliott R J 1954 *Phys. Rev.* **96** 266
[3] Yafet Y 1963 *Solid State Physics* ed F Seitz and D Turnbull (New York: Academic) vol 14 p 1
[4] Cheng J L, Wu M W and Fabian J 2010 *Phys. Rev. Lett.* **104** 016601
[5] Wilson D K and Feher G 1961 *Phys. Rev.* **124** 1068
[6] Cardona M 2001 *Phys. Stat. Sol.* (a) **188** 1209
[7] Mahan G D 1990 *Many-particle physics* (New York: Plenum Press)
[8] Pop E, Dutton R W and Goodson K E 2004 *J. Appl. Phys.* **96** 4998

Low temperature adsorption of ^{3}He on silica aerogel surface and its influence on ^{3}He spin kinetics

E M Alakshin[1], R R Gazizulin[1], A V Klochkov[1], V V Kuzmin[1], M S Tagirov[1], D A Tayurskii[1] and N Mulders[2]

[1]Institute of physics, Kazan (Volga Region) Federal University, Kazan, Russia
[2]University of Delaware, Newark, Delaware, USA

E-mail: Slava625@yandex.ru

Abstract. Significant influence of the aerogel surface heterogeneity on the processes of ^{3}He nuclear magnetic relaxation at temperatures 1.5 – 4.2 K is discovered. This influence appears, for instance, in differences of the ^{3}He T_1 relaxation times for small portion of ^{3}He, adsorbed at different temperatures. Binding energy data of ^{3}He and distributions of this energy in two types of aerogel were obtained experimentally. Adsorbed ^{3}He molecules with binding energies 60 – 250 K play supreme role in processes of nuclear magnetic relaxation of gaseous and liquid ^{3}He in aerogel.

1. Introduction

Supreme role of ^{3}He layer adsorbed on an aerogel surface in processes of nuclear magnetic relaxation was studied earlier [1,2]. The spin kinetics of ^{3}He in the silica aerogel was studied above the Fermi temperature of liquid ^{3}He. The magnetic relaxation times T_1 and T_2 of adsorbed, gaseous, and liquid ^{3}He in the 95 % porosity aerogel at a temperature of 1.5 K were obtained by means of pulse nuclear magnetic resonance techniques. It was found that T_1 in all three cases is proportional to the frequency, whereas T_2 is frequency independent. It was proposed that the longitudinal relaxation proceeds due to the exchange motion in the solid adsorbed ^{3}He film. The intrinsic relaxation mechanisms in the liquid and gas phases are much weaker than the relaxation through the adsorbed surface layer. A theoretical model of relaxation in the adsorbed ^{3}He layer has been proposed. This model takes into account the filamentary structure of the aerogel.

In the present work we report experimental data on heterogeneous adsorption potential of ^{3}He in powder silica aerogel and in bulk aerogel (95%) with filamentary structure. The influence of this heterogeneous potential on magnetic relaxation of ^{3}He was also observed.

2. Experimental setup and samples

The aerogel samples were placed in glass tube (pyrex) which was tightly connected with ^{3}He gas handling system. For removing the impurities from the aerogel surface the samples were annealed at temperature 800 K during 30 min under continuous pumping.

The temperature of the NMR cell with aerogel samples has been controlled by ^{4}He vapor pumping and was measured by Allen-Bradley thermometer (10 Ohm resistance at room temperature). The pressure was measured at room temperature part of the gas handling system, using pressure gauge ILMVAC PIZA 111.

The longitudinal magnetization relaxation times T_1 of ^{3}He was measured by the saturation-recovery method using FID signal. The spin-spin relaxation times T_2 was measured by Hahn method. The handmade pulse NMR spectrometer has been used (frequency range 3 – 50 MHz). The pulse NMR spectrometer is equipped by resistive electrical magnet with a magnetic field strength up to 1 T. Two types of aerogel were studied: silica aerogel (95%) with filamentary structure and powder aerogel (EMP-SAP (silica aerogel fine powders) EM-POWER CO. LTD (Korea)). Transmission Electron Microscopy images (TEM) of the samples are presented in figure 1 and figure 2. The particles size of powder aerogel is about 1 - 10 mkm.

Figure 1. TEM of silica aerogel with filamentary structure

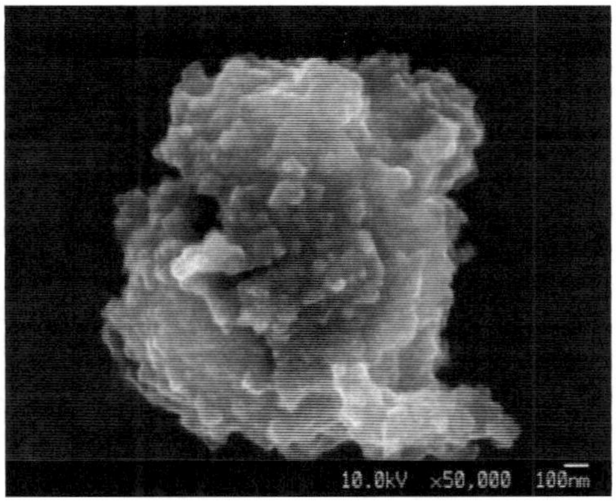

Figure 2. TEM of powder silica aerogel

3. Experiments

Adsorption isotherms of ^{3}He and ^{4}He in an aerogel are presented in figure 3. These isotherms were obtained by following method: at the temperature T = 4.2 K small portions of gaseous ^{3}He were introduced into the cell, which were further thermalized during about 30 min [3] and after that the equilibrium pressure was measured. After each step a new portion of gaseous ^{3}He was introduced. The adsorption capacity of complete layer was measured and was equal to 4 cm^3 for bulk aerogel and 12 cm^3 STP for powder aerogel.

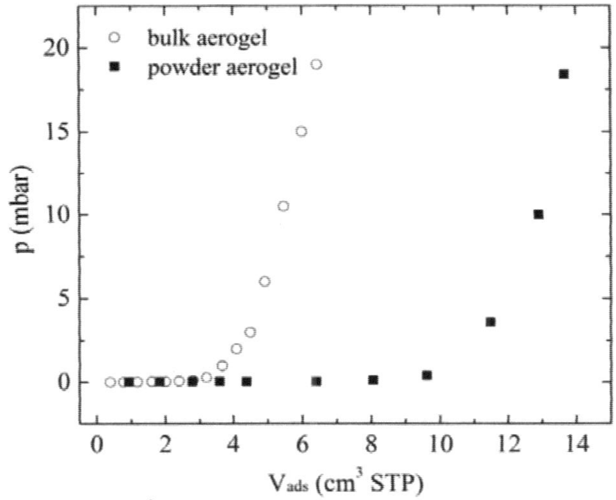

Figure 3. Adsorption isotherms of ^{3}He in bulk and powder aerogel samples at temperature 4.2 K.

In previous works [1,2] the spin kinetics of ^{3}He in the silica aerogel was studied above the Fermi temperature of liquid ^{3}He for complete adsorbed layer. In the figure 4 experimental data of ^{3}He nuclear longitudinal relaxation time T_1 for incomplete adsorbed layer (V_{ads}=0.8 cm^3 STP for bulk aerogel and 2 cm^3 STP for powder aerogel) are presented as a function of the temperature. The adsorbed layer was prepared at two different temperatures (1.5 K and 4.2 K) and as can be seen from figure 4 the behavior in these cases is completely different for both aerogel samples. It can be explained by taking into account heterogeneity of adsorption potential of ^{3}He on aerogel surface since adsorption of ^{3}He at different temperatures in these cases will fill the surface different manner.

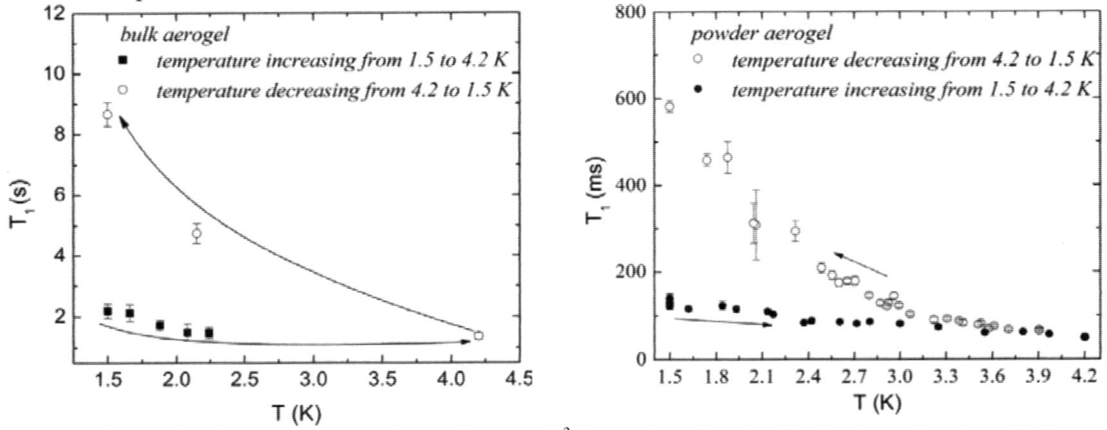

Figure 4. The temperature dependences of ^{3}He nuclear longitudinal relaxation time T_1 for incomplete adsorbed layer. The adsorbed layer was prepared at two different temperatures (1.5 K and 4.2 K).

For obtaining ^{3}He adsorption data and calculation binding energies of ^{3}He on the aerogel surface following method was used. At room temperature whole gas handling system together with NMR cell was filled by gaseous ^{3}He in amount of V(^{3}He) = 12 cm^3 STP (first experiment) and V(^{3}He) = 24 cm^3 STP (second experiment) in case of powder aerogel and 4.7 cm^3 STP in case of bulk aerogel. The whole gas handling system consists of "cold part" (capillary in cryostat and NMR cell) and "hot part" (external capillary, calibrating volume and a pressure gauge). The amount of ^{3}He was chosen, taking into account figure 2. After equilibrium pressure in whole system was achieved the cryostat begun cool down slowly (1 K/min at temperatures above 10 K and 0.05 K/min at low temperatures), using helium flow system. The cooling process leads to pressure decreasing in the system because of two processes: gas cooling in a "cold part" and adsorption on the aerogel surface. The experimental data are presented in figure 5. If we will take into account the redistribution of helium gas between "cold" and "hot" parts of the system during the experiment, the dependence of ^{3}He adsorbed amount versus the temperature could be obtained. In the figure 5 results of ^{3}He adsorption in powder aerogel in case of V(^{3}He) = 24 cm^3 STP are presented.

It can be seen that adsorption continues at lower temperatures, less than 4 K. The case of adsorption experiment with V(^{3}He) = 24 cm^3 and measurements of ^{3}He longitudinal relaxation times T_1 during ^{3}He adsorption experiment in powder aerogel in the temperature range 1.5 – 4.2 K are presented in figures 6 and 7.

Journal of Physics: Conference Series **324** (2011) 012028 doi:10.1088/1742-6596/324/1/012028

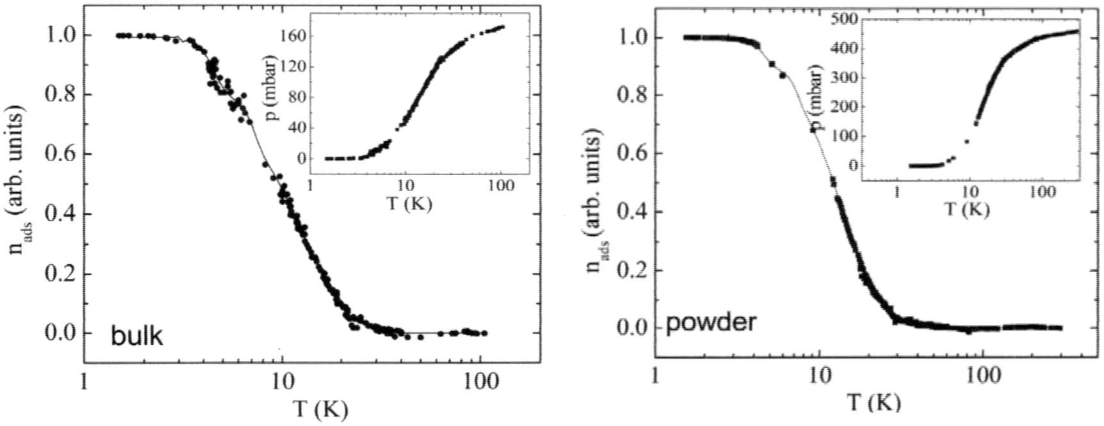

Figure 5. Temperature dependences of ^{3}He amount adsorbed in aerogel: dots – experimental data (V(^{3}He) = 4.7 cm^3 STP -experiment with bulk aerogel, V(^{3}He) = 12 cm^3 STP- experiment with powder aerogel), lines – Langmuir model (see text). On inserts shown the p(T) dependences.

From figure 7 it is clear, that additional amount of ^{3}He in the cell (case of V(^{3}He) = 24cm^3) plays the role of the "additional load" for relaxation process and does not have significant intrinsic relaxation mechanisms. The experimental data can be described, using HR model [4]:

$$T_1 = T_{1s} \cdot N_2(T) / N_1 \qquad (1)$$

where T_{1s} – ^{3}He longitudinal relaxation time in adsorbed (surface) layer. N_2 – amount of ^{3}He molecules in the NMR cell, which depends on temperature. N_1 – amount of ^{3}He molecules in adsorbed layer (V(^{3}He) = 12 cm^3 STP). The solid line in figure 7 represents calculations, taking into account equation (1) and redistribution of gaseous ^{3}He between "cold part" and "hot part" at low temperatures.

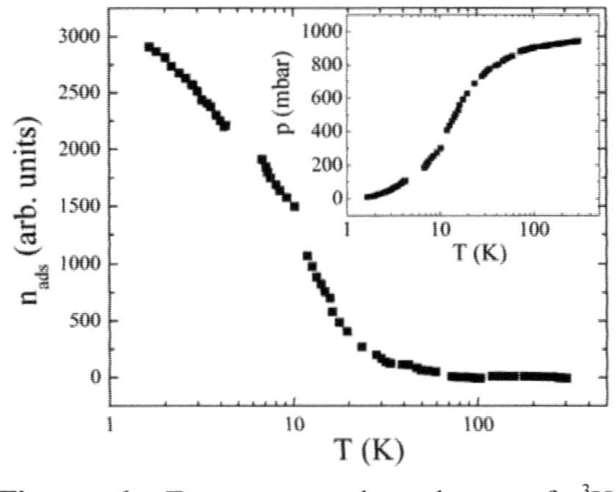

Figure 6. Temperature dependence of ^{3}He amount adsorbed in powder aerogel in case of V(^{3}He) = 24cm^3 STP.). On insert shown the p(T) dependence.

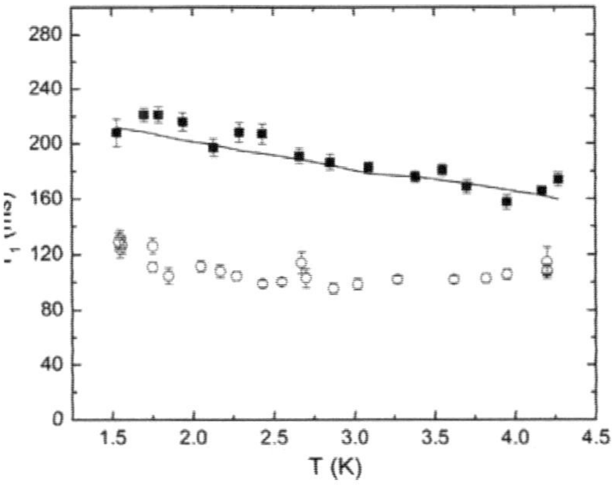

Figure 7. The temperature dependences of ^{3}He longitudinal relaxation time T_1 in powder aerogel: ○ – V(^{3}He) = 12 cm^3 STP, ■ – V(^{3}He) = 24cm^3

The binding energies for adsorption experiments (V(^{3}He) = 4.7 cm^3 STP (bulk aerogel) and V(^{3}He) = 12 cm^3 STP (powder aerogel)) below monolayer capacity can be estimated from Langmuir model [5] by eq.:

$$\theta = p/(p + p^*), \qquad (2)$$

where Θ – degree of layer filling

$$p^* = \alpha T^{5/2} / \langle \exp[-H^S / kT] \rangle , \tag{3}$$

where α – constant, T – temperature, H^S – Hamiltonian of ^{3}He interaction with an aerogel surface. In the case of low temperatures it can be rewritten as:

$$\langle \exp[-H^S / kT] \rangle \approx \exp[-\varepsilon_m / kT] \tag{4}$$

where ε_m – ^{3}He binding energy on aerogel surface.

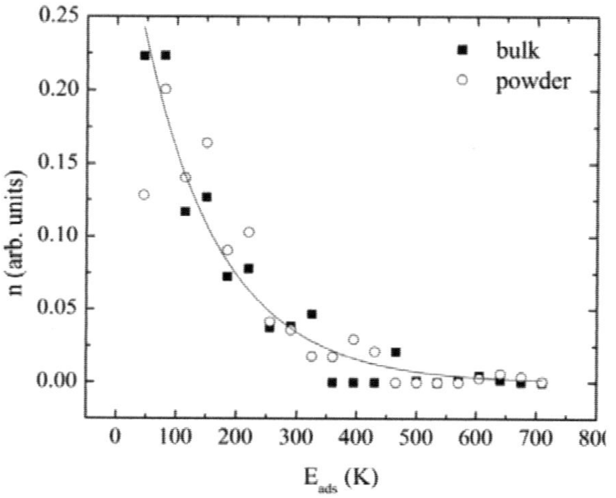

Figure 8. Binding energy distributions of ^{3}He in two types of aerogel

The experimental data on figure 5 was fitted by twenty processes like (2) with binding energy lies in range 45-700K (figure 5). As can be seen from figure 8 observed distributions are similar and can be characterized by exponential function with characteristic parameter 130±11 K.

4. Conclusions

Considering all presented experimental data, significant influence of aerogel surface heterogeneity on the processes of ^{3}He nuclear magnetic relaxation at temperatures 1.5 – 4.2 K was described. For instance, this influence appears in differences of ^{3}He T_1 nuclear relaxation times for small portion of ^{3}He, adsorbed at different temperatures. Binding energy data of ^{3}He on the surface of powder silica aerogel obtained experimentally. It was shown that the binding energy is in the range of 60 – 250 K. Adsorbed ^{3}He molecules with binding energies in range 60-250 K plays dominant role in processes of nuclear magnetic relaxation of gaseous and liquid ^{3}He in aerogel.

Acknowledgements

This work was partially supported by Russian Fund for Basic Research (grant N 09-02-01253) and by the Ministry of Education and Science of the Russian Federation (project no. 13.G25.31.0025).

References

[1] Klochkov A V, Kuzmin V V, Safiullin K R, Tagirov M S, Tayurskii D A and Mulders N 2008 *JETP Lett.* **88** 823
[2] Klochkov A, Kuzmin V, Safiullin K, Tagirov M, Yudin A and Mulders N 2009 *J.Phys.: Conf. Ser.* **150** 032043
[3] Alakshin E M, Gazizulin R R, Klochkov A V, Kuzmin V V, Mulders N, Tagirov M S and Tayurskii D A 2011 *JETP Lett.* **93** 223

[4] Hammel P C and Richardson R C 1984 *Phys. Rev. Lett.* **51** 1441
[5] Steele W A 1970 *J. Low Temp. Phys.* **3** 257

Ab initio simulation of effects of structural singularities in aerogel absorption potential

C. Debras[1), D. Tayurskii[2), B. Minisini[1), Y. Lysogorskiy[2)

[1) Institut Superieur des Materiaux et Mecaniques Avances du Mans, 72000, 44 Av. Bartholdi, Le Mans, France

[2) Institute of Physics, Kazan Federal University, 420008, 18 Kremlevskaya st., Kazan, Russia

E-mail: `yurii.lysogorskii@gmail.com`

Abstract. In the present work simulation of Van der Waals potential between helium atom and part of silica aerogel strand by means of *ab initio* methods was performed. Cell with alpha quartz structure was used as building block of aerogel strand, because it is the most stable structure at low temperature, and only the surface layer of aerogel has been considered. For modeling absorption potential field in plane, summation of potential from individual building blocks has been provided. Two dimensional Van der Waals energy field was calculated for different geometries of aerogel strands. A rather deep potential well has been found in the corner formed due to aerogel strand crossing.

1. Introduction

A material is, by definition, the combination of a matter and a geometry. With the emerging field of nanotechnologies the geometry and the structure of the matter seem to be more important for innovation than the nature of the matter itself. Consequently, the characterization and the comprehension of organization at the molecular scale is the key of future technologies. For more than twenty years [1] the question of aerogels structure appeared as a crucial point to allow us to use practically all features of this extremely porous structure. It is a really very promising material for the functionalized surface technologies. Aerogels are nowadays well known for their unique mechanical, chemical, thermal and optical properties [2], their density is close to only three times that of the air. This amazing combination of properties is mainly due to the unique structure of aerogels, characterized by their continuous solid network.

For studying material with such complex internal geometry very tiny and sensible "probe" should be used. At the role of such probe one can propose quantum liquids like ^{3}He or ^{4}He because atom size of helium is very small and it can penetrate in very narrow nanochannels and nanopores inside aerogel. In superfluid phase these liquids show unusual behavior due to appearance of macroscopic collective wave function of all helium atoms which results in coherent motion. A new "degree of freedom" is emerged. For example in superfluid in addition to pressure oscillation (first sound) one has temperature oscillation [3]. In the case of ^{3}He atoms we have additional magnetic degrees of freedom, relaxation of which inside aerogel porous structure can give us essential information about internal structure of latter.

Aerogels represent a system of inhomogenously distributed impurities for quantum liquids like ^{3}He and ^{4}He. The influence of these impurities on the magnetic relaxation of liquid ^{3}He in normal

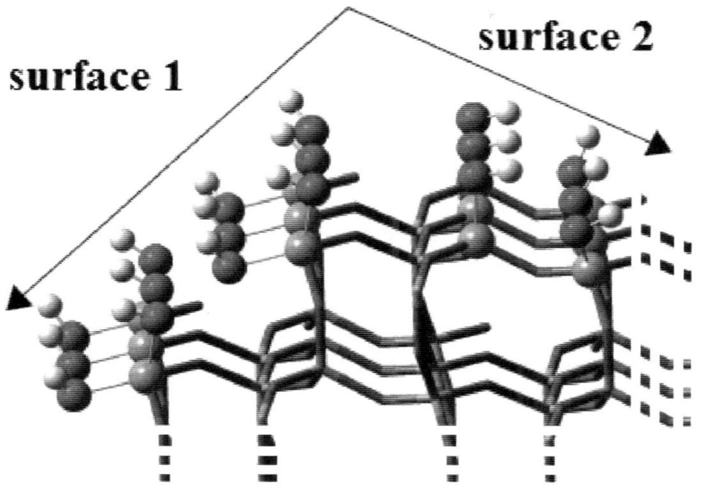

Figure 1. (color online) Two possible functional surfaces with one or two hydroxide groups for one silicon atom. Gray (light gray) - silicon atom, red (dark gray) - oxygen, white - hydrogen.

and superfluid phases was the subject of numerous studies in the recent years, but the complete understanding has not yet been reached. Recently the model of inhomogeneous potential absorption has been proposed to explain the observed temperature and field dependencies of spin-lattice relaxation times of liquid ^{3}He inside aerogel at temperatures about 1.5 K [4].

In the present work simulation of Van der Waals potential between helium atom and part of silica aerogel strand by means of *ab initio* and semi-empirical methods was performed.

2. Aerogel model

The fabrication of aerogel is always divided into three steps : hydrolysis, condensation and supercritical extraction. It is crucial to mention that every step of the fabrication process has an impact on the final aerogel structure. But the global chemical nature of aerogel is still the same. It should be mentioned that aerogel has the fractal structure at a scale length from 10 to 100 nm [5]. The fractal construction takes place during the formation of aerogel because of the percolation phenomena. In present work we will consider silica aerogel, which consist from agglomerated particles of SiO_2.

For running the energy calculation at the nanoscopic scale the quartz α-crystal structure appeared to be the most appropriate. This model does not match perfectly with all possible structure of silica aerogel but it is the most stable for low temperature. But, the entire surface of aerogel is covered by hydroxide groups which determine its hydrophilic nature. One can see on Fig. 1 that different hydroxide group configuration may appear, depending on the cutting direction. If we assume that all reactions are finished, then for computer modelling it is reasonable "to cut" the quartz α cell along the surface 2 because only one hydroxide group will remain.

To minimize the calculation time, we found the two dimension elementary cell on this plane. From this elementary cell, we are able to reconstruct the entire three dimensional structure. From the optimized crystal structure the model of aerogel building block at low temperature has been extracted. The two hydroxide group have been added to represent the remaining hydrophilic layer.

3. *Ab initio* calculations of aerogel-helium potential

The calculation model has been chosen to afford us to calculate really precise interaction for all possible three dimensional geometries. As the first approximation, only the aerogel surface layer has been used because the influence of the lower layers is assumed to be negligible.

At the first stage, Gaussian code [6] has been used to calculate the adsorption potential in one direction. To obtain a more precise adsorption potential field, the total energy has been calculated for three different points, see Fig.2. Some silicon atoms from aerogel building block have been removed to avoid double counting during the calculation of the potential field. From each point, the total energy of both the chosen model and the helium atom has been calculated for range of distances in one direction, see Fig. 2. At least 350 calculation have been performed with Gaussian code using the B3LYP functional and 6-31++G basis set on the interval from 1 to 90 Å. To obtain interaction energy (Van der Waals energy component) the following expression has been used

$$E_{\mathrm{VdW}} = E_{\mathrm{Total}} - E_{\mathrm{Aerogel}} - E_{\mathrm{Helium}}, \tag{1}$$

where E_{Total} denotes energy of system on Fig. 2, E_{Aerogel} and E_{Helium} are energies of the used aerogel building block and helium atom respectively. The three curves obtained are depicted on Fig.3.

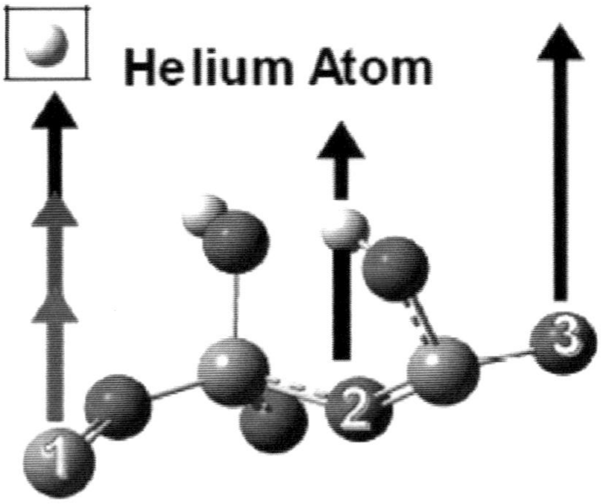

Figure 2. (color online) Calculations of energy have been performed along the direction perpendicular to surface plane for every choosen point

One can note the oscillations of interaction energy for all three reference points in length range from 2 to 25 Å. This seems to be in contrast with results of Ref. [7], where authors obtained non oscillation curve for helium ad-atom adsorption potential in nonplanar geometry. But this publication is completely based on potential curves calculated only for two atoms. In another Ref. [8] one can find results similar to our ones - the calculated with use the Lennard-Jones potential energy of a particle adsorbed by an infinite wall shows oscillations. So the result given by the *ab initio* method can be considered as a reliable result.

4. Two dimensional adsorption potential

It was mentioned above that the used in calculations the block of aerogel is just a piece of a complete aerogel strand. To obtain full adsorption potential in a given point near aerogel strand there exist two possible solutions: using a direct *ab initio* calculations for the whole strand or

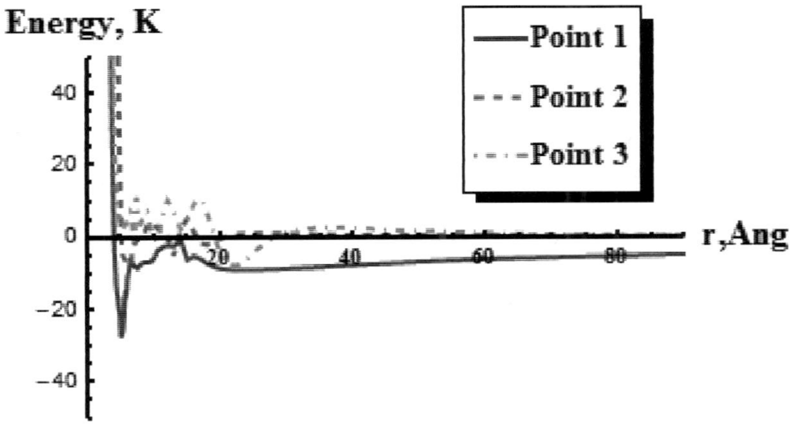

Figure 3. (color online) Van der Waals energy dependencies on distance for three reference point

providing the summation of already calculated potentials for blocks. The second method has been chosen because the calculation time for a complete molecular structure by means of the first method is enormously high.

We have calculated two dimensional Van der Waals energy field for two different types of aerogel strands geometry: two crossing strands and pore inside a silica ring (Fig. 4 and 5). One can observe the appearance of a deep potential wall generated by the geometric singularity - crossing point. At the same time the depth of potential wall in the case of the ring is homogeneous one.

The minimum energy is always located at the distance 20Å from each aerogel strands. To analyze the influence of the strands length and the intersection angle between strands onto near-surface adsorption potential well we have calculated the interaction energy for several models. From these calculations one can conclude that depth of well decreases with length of strand (or number of silica molecules in strands). Close geometry like a pore reduces the depth in comparison to the straight strand. So any irregular geometry with some singularities generates a rather deep potential well just near this singularity, and this observation support the proposed model of inhomogeneous adsorption potential [4].

5. Conclusion

We have provided *ab initio* calculations of the adsorption potential well for helium atom near silica aerogel strands. One can improve calculation accuracy by increasing the number of points and directions for interaction potential. Also it is possible to use the post-Hartree Fock perturbation method MP2 with improved basis set 6-31++G(2d) to describe Van der Waals interactions better. But all these improvements will have an effect on the depth of potential well but not cancel the main result - the appearance of a rather deep potential well for helium atom near geometry singularity (strands intersection, for example). It supports the model of inhomogeneous adsorption potential proposed in [4] to explain very short spin relaxation times for ^{3}He atoms adsorbed on the silica aerogel surface.

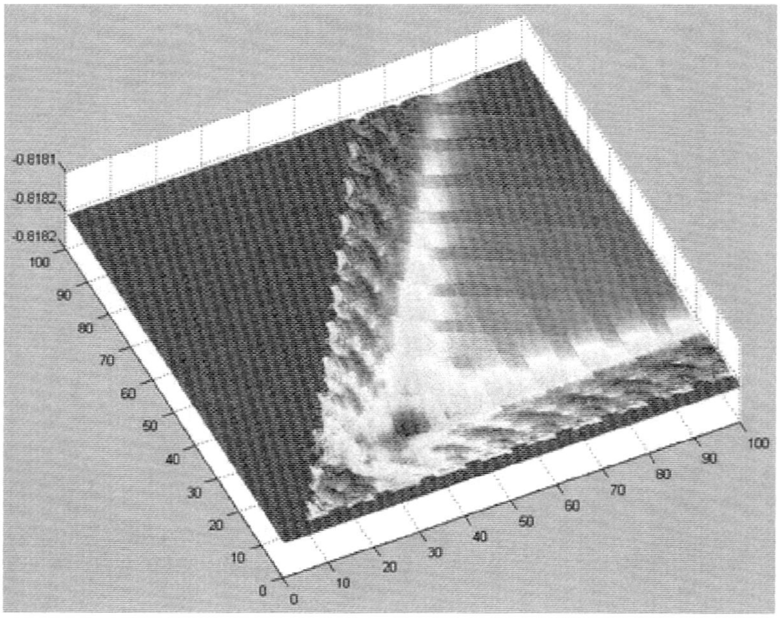

Figure 4. Two dimensional Van der Waals energy field for two intersected strands with the crossing angle $\pi/3$ (x, y in Å; z in a.u.)

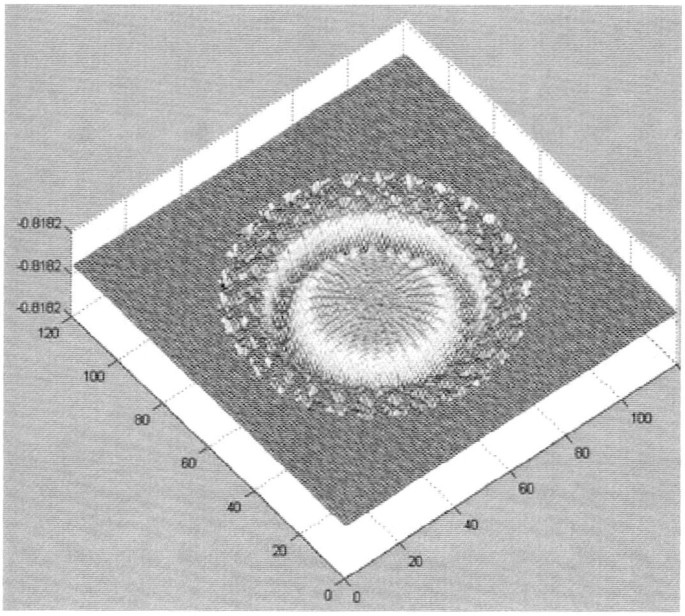

Figure 5. Two dimensional Van der Waals energy field for silica ring with radius 48 Å(x, y in Å; z in a.u.)

Acknowledgments
The work was supported by the Ministry of Science and Education of Russia. Y.L. thanks Region Pays de la Loire for partial individual financial support (MOA119).

References

[1] Vacher R, Woignier T, Phalippou J, Pelous J and Courtens E 1989 *J. Phys. Colloques* **50** 127–131
[2] Aegerter M 2011 *Aerogels Handbook* (Springer Verlag)
[3] Tilley D and Tilley J 1994 *Superfluidity and superconductivity* (Institute of Physics publishing)
[4] Klochkov A V, Kuzmin V V, Safiullin K R, Tagirov M S, Tayurskii D A and Mulders N 2008 *JETP Lett.* **88** 823–827
[5] Porto J and Parpia J 1999 *Phys. Rev. B* **59** 14583
[6] Frisch M J, Trucks G W and Schlegel H B e a Gaussian 03, Revision C.02 Gaussian, Inc., Wallingford, CT, 2004
[7] Hernando A, Hernández E, Mayol R and Pi M 2007 *Phys. Rev. B* **76** 115429
[8] Billes W, Tscheliessnig R and Fischer J 2005 *Act. Biochim. Polon.-English Edition* **52** 685–689

Journal of Physics: Conference Series **324** (2011) 012030 doi:10.1088/1742-6596/324/1/012030

Stress-induced orbital alignment of the Cr^{2+} centers in $KZnF_3$ crystal

I N Gracheva, S I Nikitin and R V Yusupov

Institute of Physics, Kazan (Volga region) Federal University, Kremlevskaya 18, Kazan 420008, Russian Federation

E-mail: ryusupov.kazan@gmail.com

Abstract. Observation of the intense linear dichroism spectrum on the ${}^5E_g \rightarrow {}^5T_{2g}$ transition of Cr^{2+} ions in $KZnF_3$ crystal under uniaxial stress is reported. The model is proposed which assigns the observed spectrum to alignment of the Jahn-Teller Cr^{2+} centers, occurring due to redistribution between the minima of the adiabatic potential that become inequivalent under uniaxial stress applied along the C_4 axis of the crystal. Analysis of the observed dependencies of the dichroism on the applied stress has allowed to estimate the values of the electron-strain interaction constant $V_{ES} = 32900 \pm 1200$ cm^{-1} and the inversion splitting $\delta = 9.2 \pm 1.3$ cm^{-1}, and also to characterize the random strain field.

1. Introduction

Investigations of the impurity centers with an orbitally-degenerate ground state in the high-symmetry crystals were performed intensely in the second half of the XXth century because of the prediction of the Jahn-Teller effect (JTE) and long-lasting unsuccessful attempts of its experimental observation. Nowadays the interest to such objects is defined by the significant role of the orbital ordering, which occurs due to the cooperative JTE, in formation of the unusual electronic properties of the concentrated transition metal compounds, first of all, oxides. Among those unusual phenomena one finds high-temperature superconductivity in cuprates and colossal magnetoresistance in manganites. Properties of the complex transition metal compounds are defined by numerous degrees of freedom such as charge, orbital, exchange etc. Theoretical understanding of these systems behavior demands the explicit knowledge of the corresponding energy scales. One of the approaches to obtain such information is to study the diluted dielectric crystals. Thus, the double exchange in the Cr^{2+}-Cr^{3+} mixed valence pair center has been successfully studied by optical spectroscopy in the $KZnF_3$ crystal doped with chromium ions [1].

It is well known that studies of the impurity centers in the cubic crystals under uniaxial stress can give a lot of information, and, in particular, allow to estimate the vibronic constants [2]. In this paper we present the results of the experimental investigations of the linear dichroism of absorption spectra observed on the ${}^5E_g \rightarrow {}^5T_{2g}$ transition of the Cr^{2+} ions in $KZnF_3$ crystal. This crystal has a cubic perovskite structure; the symmetry is Pm3m. Chromium ions substitute for the Zn^{2+} ion sites and are found thus in a sixfold octahedral surrounding of the fluorine ions. Ground state of the Cr^{2+} ions in octahedral surrounding is the ${}^5E_g$ orbital doublet. In such situation the JTE in the strong-coupling limit is expected [3]. It should be mentioned also that the number of publications on the optical spectroscopy of Cr^{2+} ions both in oxide and fluoride compounds is quite small.

In our studies the JT character of the Cr^{2+} ions ground state is strikingly revealed. The proposed model based on the orbital alignment of the Cr^{2+} centers under uniaxial stress explains all the experimental observations. It is shown that in order to achieve the quantitative agreement of the model predictions with the experimental data it is necessary to take into account the random strain field.

2. Experimental procedure

The samples of $KZnF_3$:Cr crystals were grown by Bridgman-Stockbarger method. For doping the CrF_2 compound was used. Concentration of the chromium ions in the samples was in the range of 0.1 – 1.0 wt.% in the charge. It should be mentioned that all the grown crystals except of the doped Cr^{2+} contained also the Cr^{3+} ions. Sample orientation along the C_4 crystallographic axis was performed via the cleaved (001) crystal planes. Direction of the C_3 axis was found with the DRON-2M X-ray diffractometer. Accuracy of the orientation was ±2°. Typical size of the sample used in our studies was about $2 \times 3 \times 6$ mm^3. Optical absorption spectra were measured with a single-beam spectrometer based on the DFS-12 monochromator with the Ge-photodiode used as a detector.

Linear dichroism of absorption spectra were registered employing a sensitive differential method proposed in [4]. Light from the output of the monochromator passed through a polarizer and then through a superchromatic plate (SCP) that was adjusted at the 45° angle of its optical axis with respect to the axis of the polarizer. After passing through SCP the light becomes linearly pseudo-depolarized, which means that the intensities of the light with any polarization are equal. Rotation of the analyzer placed after SCP leads in the case of the absorption anisotropy in the sample to the intensity modulation of the light transmitted through the sample. Detected signal then would contain the component at the frequency twice that of the analyzer rotation with an amplitude proportional to the linear dichroism. Dichroism value was defined as

$$D = \frac{I_{\parallel}(\lambda) - I_{\perp}(\lambda)}{I_{\parallel}(\lambda) + I_{\perp}(\lambda)} \tag{1}$$

and was calculated using the measured dichroism signal ($I_{\parallel}(\lambda)$ - $I_{\perp}(\lambda)$) and the intensity of the light passed through the sample at zero pressure $I_0(\lambda)$ that were measured keeping all the other experimental conditions identical:

$$D = \frac{I_{\parallel}(\lambda) - I_{\perp}(\lambda)}{2I_0(\lambda)}. \tag{2}$$

Expression (2) is valid if the dichroism is small. Also, if the absorption is weak then the last expression can be rewritten as

$$D = [k_{\perp}(\lambda) - k_{\parallel}(\lambda)]/2, \tag{3}$$

where $k_{\alpha}(\lambda)$ are the absorption coefficients. Investigations were performed in the bath cryostat. Uniaxial strain was created with a pneumatic press, uniaxial stress on the sample had the values up to 100 MPa. Strain magnitudes were calculated using the components of the compliance tensor $S_{11} = 0.821 \cdot 10^{-12}$ cm^2/dyn and $S_{12} = -0.204 \cdot 10^{-12}$ cm^2/dyn [5]. Relative strain value of the E_θ symmetry is $\varepsilon_\theta = (S_{11} - S_{12})/P$, where P is the stress applied along the C_4 axis.

3. Experimental results

Optical absorption and linear dichroism spectra measured at $T = 4.2$ K with stress applied along the C_4 and C_3 axes of the crystal are shown in figure 1. Absorption band with the maximum at 820 nm has a weakly pronounced structure and corresponds to the ${}^5E_g \rightarrow {}^5T_{2g}$ transition of the Cr^{2+} ion. Two narrow zero-phonon lines are observed at the long-wavelength wing of the band. Another wide band with a pronounced structure that has a maximum at 670 nm originates from Cr^{3+} ions and is assigned to the ${}^4A_{2g} \rightarrow {}^4T_{2g}$ transition. When the uniaxial stress is applied along the C_4 principal axis of the crystal the intense linear dichroism signal is observed within the range of the wide vibronic absorption band of the Cr^{2+} ions (figure 1). When the stress of the same value is applied along the C_3 axis, no dichroism signal was found.

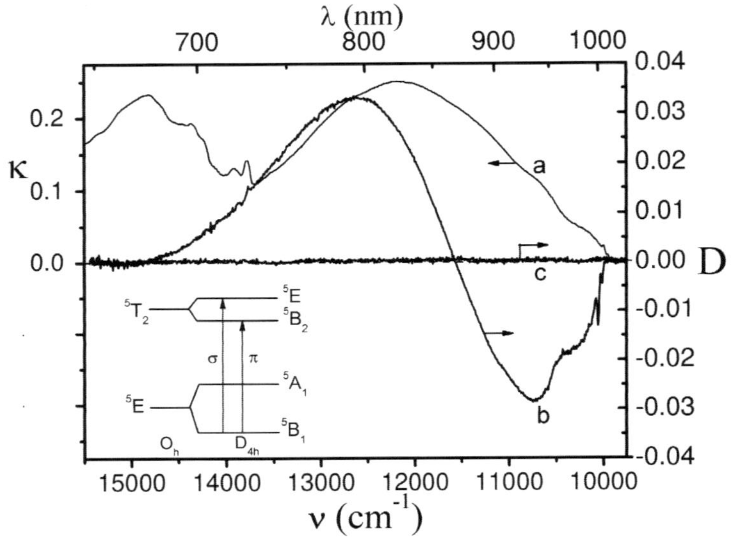

Figure 1. Absorption (a) and the linear dichroism spectra under the stress applied along the C_4 (b) and C_3 (c) axes of the $KZnF_3$:Cr crystal; T = 4.2 K, P = 64 MPa. In the inset the energy level diagram of the Cr^{2+} ion in the crystal field of D_{4h} symmetry and the transition polarizations from the ground $^5B_{1g}$ state are shown.

The shape of the dichroism signal within the wide vibronic band is not modified with the stress value variation. Dependencies of the dichroism signal amplitude on stress at $\lambda = 800$ nm are shown in figure 2. Dependencies on stress measured at different wavelengths within the absorption band of the Cr^{2+} ions at a given temperature have the identical character and differ only in the magnitude.

4. Physical model

Let us consider now the origin of the observed phenomenon. Magnitude of the dichroism signal which is of the same order as the absorption coefficient of the Cr^{2+} ions indicates that an essential fraction of the Cr^{2+} ions is involved. In our opinion, nature of the dichroism signal under uniaxial stress is closely related to the JTE. The ground $^5E_g(t_{2g}^3e_g)$ state of the Cr^{2+} ions interacts strongly with the nearest surrounding distortions of the E_g symmetry. Such $[CrF_6]^{4-}$ cluster distortions are described usually by the two normal modes Q_θ and Q_ε that transform as $(3z^2-r^2)$ and (x^2-y^2), respectively [3]. These modes can be expressed also as $Q_\theta = \rho\cos\varphi$ and $Q_\varepsilon = \rho\sin\varphi$, where φ is a polar angle in the $(Q_\theta, Q_\varepsilon)$ plane counted from the Q_θ axis. The $E\otimes e$ JT problem solution is well known. Account for the linear and the third order ($\propto \rho^3$) terms in the Hamiltonian leads to the adiabatic potential (AP) with the three equivalent minima that correspond to either elongated or compressed along the three C_4 cluster axes. At low temperatures, the particles are localized in the vicinity of the minima, where the system states are described by the vibronic wavefunctions of $\Phi_i = \psi_- \Psi_i(\rho,\varphi)$, $i = X,Y,Z$, where ψ_- is an electronic wavefunction of the lower sheet of AP and $\Psi_i(\rho,\varphi)$ is a vibrational wavefunction localized nearby the points of $\rho = \rho_0$, $\varphi = \varphi_{\min}$ [3]. If the height of the barriers between the minima is finite, a tunneling occurs that significantly modifies the energy spectrum of the system: the three lowest levels are represented by the vibronic doublet E_g and the singlet A_1 (or A_2). Singlet is higher in energy than the doublet by an interval of δ which is called the inversion (or tunneling) splitting [3]. Wavefunctions of the singlet and the doublet in the case of the elongated cluster configurations in the minima of AP are described by the linear combinations

$$\begin{aligned} |A_1\rangle &= \frac{1}{\sqrt{3}}(\Phi_X + \Phi_Y + \Phi_Z), \\ |E\theta\rangle &= \frac{1}{\sqrt{6}}(2\Phi_Z - \Phi_X - \Phi_Y), \\ |E\varepsilon\rangle &= \frac{1}{\sqrt{2}}(\Phi_X - \Phi_Y), \end{aligned} \tag{4}$$

where Φ_i are the vibronic states that correspond to the AP minima [3].

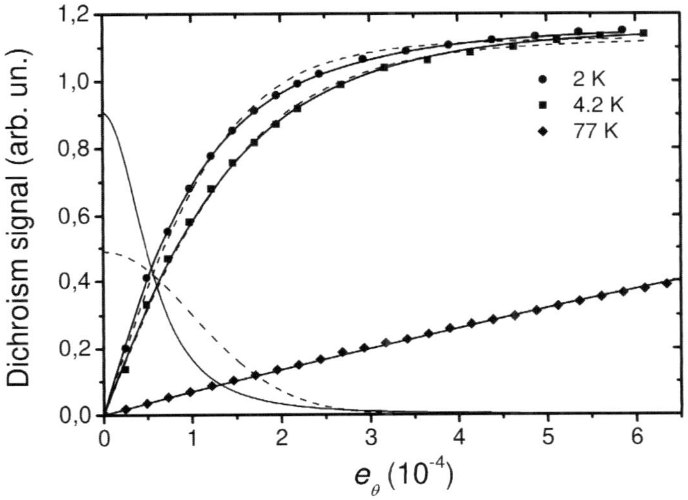

Figure 2. Dependencies of the dichroism signal on stress at the temperatures of 2.0 K, 4.2 K and 77 K, $\lambda = 800$ nm, P||C$_4$. Symbols indicate the experimental data, the lines are the results of the fits: dashed lines – with the use of distribution (9), solid lines – (10). Corresponding random strain distributions are also shown.

Hamiltonian of the system that accounts for the interaction with the E_g-symmetry distortions (described by the ϵ_θ and ϵ_ε components of the strain) in the basis (4) has the form:

$$H = H_0 + H_{ES} = \begin{pmatrix} \delta & 0 & 0 \\ 0 & 0 & 0 \\ 0 & 0 & 0 \end{pmatrix} + \begin{pmatrix} 0 & rV_{ES}\epsilon_\theta & rV_{ES}\epsilon_\varepsilon \\ rV_{ES}\epsilon_\theta & -qV_{ES}\epsilon_\theta & qV_{ES}\epsilon_\varepsilon \\ rV_{ES}\epsilon_\varepsilon & qV_{ES}\epsilon_\varepsilon & qV_{ES}\epsilon_\theta \end{pmatrix}, \tag{5}$$

where q and r are the matrix elements of the type $q = \langle E\varepsilon|U_\theta|E\varepsilon\rangle = -\langle E\theta|U_\theta|E\theta\rangle = \ldots$ and $r = \langle A_1|U_\theta|E\theta\rangle = \langle A_1|U_\varepsilon|E\varepsilon\rangle$, V_{ES} is an electron-strain interaction constant.

Under the uniaxial stress applied along the C$_4$ axis of the crystal, the ensemble of the JT centers becomes at low temperatures essentially anisotropic due to the redistribution of the centers between the inequivalent energy minima. As a consequence, the optical spectra of such system become anisotropic with respect to the direction of the applied stress ($\bar{E} \parallel P$ and $\bar{E} \perp P$). In our opinion, exactly the described situation causes the intense linear dichroism observation within the absorption band of the Cr^{2+} ions in $KZnF_3$ crystal.

In the AP minima the $[CrF_6]^{4-}$ cluster has the tetragonal symmetry. The corresponding energy level pattern is shown in the inset of Fig. 1. Orthogonally polarized transitions from the ground $^5B_{1g}$ state to the ^5E_g and $^5B_{2g}$ sublevels of the excited $^5T_{2g}$ state explain naturally the observed dichroism spectrum structure. Our hypothesis explains also a zero dichroism signal under the stress applied along the C_3 crystalline axis: interaction with such deformation leaves the AP minima equivalent.

The problem of the center redistribution between the AP minima as a function of the applied stress value and the temperature in the case of $P \parallel C_4$ can be solved analytically. Let the stress be applied along the *z*-axis and the light propagates along *x*. We assume also that the light absorption is caused by the electric-dipole transitions and an applied stress does not modify the electronic wavefunctions of the initial and the final states. Then the dichroism signal within the absorption band will be proportional to

$$D \propto \frac{N_Z - N_Y}{N_X + N_Y + N_Z}, \tag{6}$$

where N_i are the populations of the corresponding vibronic states Φ_i. Taking into account the Boltzmann distribution between the energy levels and knowledge of the eigenvectors of the Hamiltonian (5) together with the expressions for the wavefunctions (4) allows to derive the populations of the minima corresponding to Φ_X, Φ_Y and Φ_Z states. As a result, the following expression has been obtained:

$$D \propto \frac{\exp\left(\frac{3\xi - 1}{\tau}\right)\left(\cosh\frac{\Omega}{\tau} + \frac{x\left(1 - 4\sqrt{2}a\right)}{\Omega}\sinh\frac{\Omega}{\tau}\right)}{2\exp\left(\frac{3\xi - 1}{\tau}\right)\cosh\frac{\Omega}{\tau} + 1}, \tag{7}$$

where we introduce the dimensionless parameters $\xi = qV_{ES}\epsilon_\theta / \delta$, $\tau = 2k_B T / \delta$, $\Omega = \sqrt{(4a^2 + 1)\xi^2 + 2\xi + 1}$, $a = r / q$.

The curves $D(\xi, \tau)$, calculated using expression (6), reproduce qualitatively the experimental data. Moreover, each of the dependencies, measured at 2 K and 4.2 K, can be almost ideally fit with (6). However, it has turned out that these two dependencies cannot be fit within the proposed model simultaneously, with the same set of the parameters. Obviously, some additional factors should be taken into account.

One of such factors that had to be accounted for in the majority of the JTE experimental studies is the random strain field present in any real crystal. As we study the cubic symmetry crystal, due both to the orientation degeneracy and the averaging in time between the three axial configurations of the $[CrF_6]^{4-}$ cluster related to the tunneling of each center through the barriers, the dichroism signal is absent at zero applied stress. Random strains also should not violate the high cubic symmetry of the problem. This is why the E_g-symmetry strains were taken as $\epsilon_{\theta s} = \epsilon_s \cos\varphi$ and $\epsilon_{\varepsilon s} = \epsilon_s \sin\varphi$ with the homogeneous distribution over φ in the range of $0 \ldots 2\pi$ and some distribution $f(\epsilon_s)$ that has a maximum at $\epsilon_s = 0$ and drops gradually with ϵ_s increase. When a nonzero stress was applied along the C_4 axis of the crystal, the distortions originating from the random strains and from an applied stress were considered additive. Averaging over the random strain distribution was performed numerically, via evaluation of the integral

$$\langle D(T,\epsilon_\theta)\rangle = \frac{1}{2\pi}\int_0^{2\pi} d\varphi \int_0^{\infty} \epsilon_s D(T,\epsilon_\theta + \epsilon_{\theta s}, \epsilon_{\varepsilon s}) f(\epsilon_s) d\epsilon_s . \tag{8}$$

Here, the function of $D(T,\epsilon_\theta,\epsilon_\varepsilon)$ was defined by (5). The fit of the measured dependencies of the dichroism signal on stress value was performed within an assumption that the JTE in the ground ^5E_g state of the Cr^{2+} ions in $KZnF_3$ crystal is realized in the strong coupling limit ($q = 0.5$, $r = -\sqrt{2}$). Parameters varied during the fitting procedure were V_{ES}, δ, the width of the random strain distribution w and the amplitude.

As far as the shape of the distribution $f(\epsilon_s)$ is concerned, the regularly used one is the Gaussian

$$f(\epsilon_s) = \frac{1}{w\sqrt{\pi/2}} \exp\left(-2\frac{\epsilon_s^2}{w^2}\right). \tag{9}$$

Indeed, from the point of view of the random strain impact modeling, distribution (9) is very nice: it has the steeply vanishing wings ($\propto e^{-x^2}$) and the averaging procedure is performed with no problem. However, physically, following [6, 7], random strains are described by (9) if it originate from the dislocations in the crystal. Otherwise, if random strains occur due to the point defects, the distribution is different [7]:

$$f(\epsilon_s) = \frac{A}{\left(\epsilon_s^2 + w^2\right)^{3/2}}, . \tag{10}$$

where A is a normalizing factor.

We have studied both distributions. Results of the simultaneous fits of the dichroism signal dependencies on the applied stress at 2 K, 4.2 K and 77 K are shown in figure 2 and in Table 1. Both in the figure (visually) and from the table data (χ^2 differs by almost an order of magnitude) it is clearly seen that the use of distribution (10) allows for a much better reproduction of the experimental data. This, in our opinion, indicates unambiguously that the random strain field in the studied sample originates predominantly from the point defects in the crystal.

Table 1. Parameters of the dichroism signal stress dependencies fit obtained with an account of the random strains in the crystal described by the distributions (9) and (10).

Random strain distribution	V_{ES} (cm^{-1})	δ(cm^{-1})	w	χ^2
Gaussian, equation (9)	33600±1600	13±2	(1.44±0.15)·10^{-4}	0.0106
Equation (10)	32900±1200	9.2±1.3	(6.9±0.9)·10^{-5}	0.0027

Thus, we see that an account for the random strains that are inevitably present in the crystal has allowed to fit nicely all the experimental data. Importantly, the kind of experiments we perform is sensitive to the shape of the random strain distribution and such studies are capable to provide one with an additional, both *qualitative and quantitative*, information about the defects in the crystal. It should be mentioned also that the parameter values presented in Table 1 differ for two types of the random strain distributions just slightly, by ~30% maximum, and thus do not depend critically on the shape of the distribution employed.

The width of the random strain distribution is $w = 6.9 \cdot 10^{-5}$ in the relative strain units, which corresponds to the average value of the Cr^{2+} ion interaction with random strains of $qV_{ES}w = 1.1$ cm^{-1}. The linear JT constant V is related in the case of a 6-fold coordination to V_{ES} by $V = \sqrt{3}V_{ES} / 2R$, where R is a distance to the nearest neighbor, $R = a/2$, and $a = 4.055$ Å is the $KZnF_3$ crystal lattice constant. This gives an estimate of $V \approx 1.7$ eV/Å.

6. Conclusions

In summary, we relate an observed at low temperatures intense linear dichroism signal within wide absorption band of the Cr^{2+} ions in $KZnF_3$ crystal under uniaxial stress applied along the C_4 axis of the crystal to the orbital alignment of the Cr^{2+} Jahn-Teller centers. Or, equivalently, the centers are redistributed between the minima of the adiabatic potential that become inequivalent under a stress applied along the C_4 direction. Analysis of the measured dependencies of the dichroism signal on stress allowed to define the values of the electron-strain interaction constant $V_{ES} = 32900 \pm 1200$ cm^{-1}, the inversion splitting $\delta = 9.2 \pm 1.3$ cm^{-1} and the width of the random strain distribution $w = (6.9 \pm 0.9) \cdot 10^{-5}$. Our experiments were found to be sensitive to the shape of the random strain distribution and thus allow to reveal the essential origin of the random strains in the sample.

References

[1] Eremin M V, Nikitin S I, Prosvirnin S Yu, Silkin N I and Yusupov R V 2001 *Solid State Commun.* **117** 297-301
[2] Vinokurov A V, Malkin B Z, Pominov A I and Stolov A L 1986 *Fiz. Tverd. Tela (Leningrad)* **28** 381 [1986 *Sov. Phys. Solid State* **28** 211]
[3] Ham F S 1971 *Phys. Rev.* B **4** 3854-69; Fletcher J R and Stevens K W H 1969 *J. Phys. C* **2** 444-56
[4] Zapasskiĭ V S and Natadze A L 1972 *Opt. Spektrosk.* **32** 1015-19
[5] Burriel R, Bartolome J, Gonzalez D, Navarro R, Ridou C, Rousseau M and Bulou A 1987 *J. Phys. C* **20** 2819-27
[6] Stoneham A M 1969 *Rev. Mod. Phys.* **41** 82-108
[7] Ivanov M A, Mitrofanov V Ya, Falkovskaya L D and Fishman A Ya 1983 *J. Magn. Magn. Mater.* **36** 26

Journal of Physics: Conference Series **324** (2011) 012031 doi:10.1088/1742-6596/324/1/012031

Comparison of Various Correlation Times in Polymer Melts by Molecular Dynamics Simulation

E M Pestryaev

Department of Physics, Ufa State Petroleum Technological University, Kosmonavtov st., 1, 450062, Ufa, Russia

E-mail: physics_usptu@mail.ru

Abstract. This investigation presents data from prolonged molecular dynamics simulations of dense monodisperse polymer melts of linear polymer chains of length between $N = 3$ and $N = 255$. Hear N is number of chain coarse-grained segments. The study aim is to observe the crossover from Rouse-like dynamics for short chains to reptation-like dynamics for long chains and its influence on various translational and orientational correlation times. To address the problem, we calculate a variety of different quantities: standard mean-square displacement of inner monomer and of the chain mass centre, autocorrelation function of chain end-to-end vector, orientational autocorrelation functions of segment and chain end-to-end vector. The last two functions define NMR relaxation in polymer melts. The chain length dependences of all large-scale correlation times reveals clear crossover from non-entangled to entangled dynamics near $N_C \approx 35$. Additional primitive path analysis of the same data gives statistically consistent entanglement value $N_E \approx 14$, i.e., the longest chain investigated is well entangled and displays explicit motional anisotropy. The full data assembly explains qualitatively strong dependence of NMR transverse relaxation time for Rouse-like regime and weak one in reptation regime.

1. Introduction

Phenomenological theory of polymer chains dynamics [1, 2] is used very long time to interpret data of natural [3] and simulated experiments [4-7] and it is undoubted in all basic statements. But quantitative agreement of its statements with experimental data was observed rarely because it had been formulated for ideal chains of infinite length with Gaussian structure. In other words the surrounding chains are treated as a fixed network of impenetrable obstacles. The infinite chain length approximation is very difficult to realize in all kind of experiments. The most part of such experiments is based on the analyses of the mathematical consequences of the fact that chain motion is reptation-like one [1], i.e. it is characterized by strong space anisotropy of chain diffusion along and perpendicular to the contour of a "tube" or "primitive path" which is formed by surrounding chains.

The term "primitive path" had been abstract definition a long time till recent paper [8] in which an algorithm for primitive path finding had been suggested. This algorithm permits to check primary concept of the polymer dynamics, in other words to find quantitative degree of the chain diffusion anisotropy. This concept had been checked some times [9-13] by Monte Carlo simulation and the authors of the works [9-11] had concluded that tube was not needed to observe all time regimes of long chain diffusion. The more, an alternative phenomenological theory of polymer dynamics had been suggested

[11] to explain such the contradiction. The chain diffusional anisotropy revealed in the works [12, 13] was very weak and not enough for reptation regime.

The recent experiments had used complicated methods of samples preparation to create additional artificial tube for real macromolecules. For example, an embedding of real macromolecules into mesoscale cylindrical pores enhanced the space anisotropy of macromolecule diffusion according to the analysis of spin lattice relaxation dispersion [14, 15]. Thermodynamic restriction of the basic notions of the reptation theory had been discovered as the result of these experiments [15].

The aim of this study is scaling analysis of chain diffusion as function of its length and demonstration of simultaneous existence of the tube and indistinct visualization of time diffusive regimes concluded from this constraint. For this purpose the investigation of the crossover from Rouse-like dynamics for short chains to reptation-like dynamics for long chains had been made by protracted molecular dynamics simulation. The crossovers were observed for variety of translational and rotational chain motion each of which is characterized by its proper correlation time. The total collection of the data is applied to explain qualitatively chain length dependence of the NMR transverse relaxation time in polymer melts.

2. Model and simulation technique

A system of spherical particles, i.e., beads or monomers of mass m is modeled in three-dimensional *NVT*, canonical, ensemble [16] to observe chains motion. Non-bonded interaction ε_P between beads on distance r is described by simple truncated and repulsive Lennard-Jones pair potential with length parameter σ [16, 17]:

$$\varepsilon_P(r) = \begin{cases} 4\varepsilon\left[\left(\dfrac{\sigma}{r}\right)^{12} - \left(\dfrac{\sigma}{r}\right)^{6} + \dfrac{1}{4}\right], & r < r_C \\ 0, \quad r \geq r_C, \quad r_C = r_{EQ} = 2^{1\,6}\sigma \end{cases} . \tag{1}$$

Here, $\varepsilon = k_BT$ is depth of potential deep at the equilibrium distance, r_{EQ}, reduced to mean thermal energy, k_BT; r_C is cut-off distance for conventional reducing of computation. Conventional dimensionless set of parameters $m = \varepsilon = \sigma = 1$ defines all other system units including time unit

$$\tau_0 = \sigma \sqrt{\frac{m}{\varepsilon}} = 1 . \tag{2}$$

Diffusion unit is σ^2/τ_0. The cubic volume of the system with side length from 20 to 32 is limited by periodic boundary conditions [16, 17]. Main system parameter is pack coefficient, 0.50, defined as system volume part occupied by beads, the corresponding reduced bead density is 0.96. Such the parameters describe condensed substance in liquid state [17]. Bonded neighbor beads in a chain interact through a finitely extensible non-linear potential [17], which does not prevent their overlap:

$$\varepsilon_B(r) = -15\cdot\left(\frac{r_0}{\sigma}\right)^2\cdot\varepsilon\cdot\ln\left[1-\left(\frac{r}{r_0}\right)^2\right], \qquad r_0 = 1.5\cdot\sigma . \tag{3}$$

Equations (1) and (3) define so-called Kremer-Grest coarse-grained chain with well-known properties for a few lower melt density, 0.85, and altogether with chosen cut-off distance realize for isolated chains good solvent conditions [17]. This coarse-grained chain has not small-scale rotational motion around bonds each of which corresponds to the chain segment. All simulated polymer melts consist of n_{CH} chains with $(N+1)$ beads or N segments, where N is ranged from 3 to 255. General number of beads in the system is $n = (N+1)\, n_{CH}$ and varied from 8 000 to 32 768 for 22 chains of various length. Ratio of the system linear size to the chain mean-square gyration radius was varied from 4 to 20.

Numerical integration of the bead motion equations is made by velocity Verlet algorithm with constant time step $0.01\cdot\tau_0$, the temperature was kept constant by coupling the motion to Nose-Hoover thermostat [12, 18, 19] where the coupling constant is essentially less than unity. The coupling constant smallness implies minor stochastic perturbation of the system dynamics which is originally deterministic one. Additionally the system dynamics is periodically corrected by subtraction of its mass centre velocity from the velocities of all beads [17]. As a result the simulation algorithm permits to observe the system evolution during unlimited long time providing temperature fluctuations no more than 0.001 and stationary Maxwell's distribution of beads velocities in any point of phase trajectory. The last characteristic was checked occasionally, for example, for trajectory of length $2 \cdot 10^8$ integration steps per $5 \cdot 10^6$ steps.

Pre-equilibration of all investigated melts was made before the phase trajectory storage because polymer dynamics theory conclusions are applicable only to the chains in thermodynamically equilibrium conformation. This problem is impotent in a simulation experiment [17, 20] and realized in this study following way. The first, a system of chains of necessary length is designed in crystal state. The second, the system is evolved in Rouse phantom regime till reaching conformational equilibration. The algorithm performance in phantom regime approximately 1000 time more than in excluded volume regime and equilibration is very quick. The third, slow introduction of the increased excluded volume potential is made, taking care that in the beginning the chains can easily cross each other. The increasing excluded volume potential makes local reconstruction of the melt. This process needs beard displacement on its size only, i.e., it is local one and also sufficiently quick. Such the procedure is made for system of one size only one time because equilibrated short chains can be easily got from equilibrated long chains by there breaking with sequential short-time equilibration.

3. Properties of the simulated melt

3.1. Theoretical characteristics of the chain translational and rotational motion

Important quantities used to study the crossover from Rouse to reptation dynamics are the following mean-square displacements [2, 3, 7, 17]:

$$
\begin{aligned}
g_1(t) &= \left\langle [\mathbf{r}_{N/2}(t) - \mathbf{r}_{N/2}(0)]^2 \right\rangle \\
g_{1C}(t) &= \left\langle [\mathbf{r}_{N/2}(t) - \mathbf{r}_{N/2}(0) - \mathbf{r}_{CM}(t) + \mathbf{r}_{CM}(0)]^2 \right\rangle \\
g_{CM}(t) &= \left\langle [\mathbf{r}_{CM}(t) - \mathbf{r}_{CM}(0)]^2 \right\rangle \\
g_2(t) &= \frac{1}{2} \left\langle [\mathbf{r}_1(t) - \mathbf{r}_1(0) + \mathbf{r}_N(t) - \mathbf{r}_N(0)]^2 \right\rangle
\end{aligned}
\tag{4}
$$

Here, g_1 is the mean-square displacement of the middle monomer of a chain (situated at position $\boldsymbol{r}_{N/2}(t)$ at time t), g_{1C} is the same displacement measured relative to the motion of the chain's center of mass ($\boldsymbol{r}_{CM}(t)$ is position of the center of mass at time t), g_{CM} is the mean-square displacement of the chain center of mass, and g_2 is that of the end monomers. To increase statistics g_1 was calculated over 10 % of middle chain beads. Within the framework of the reptation theory for large N the next asymptotic behavior of mean-square displacement may be expected [2, 3, 7, 17]:

$$
g_1(t) \quad \propto \begin{cases} t^{1/2} \cdot N^0, & t << \tau_E \\ t^{1/4} \cdot N^0, & \tau_E << t << \tau_R \\ t^{1/2} \cdot N^{-1/2}, & \tau_R << t << \tau_D \\ t^1 \cdot N^{-2}, & \tau_D << t \end{cases}
\tag{5}
$$

$$
g_{CM}(t) \propto \begin{cases} t^1 \cdot N^{-1}, & t << \tau_E \\ t^{1/2} \cdot N^{-1}, & \tau_E << t << \tau_R \\ t^1 \cdot N^{-2}, & t >> \tau_R \end{cases}
\tag{6}
$$

At early times, a chain does not feel the entanglement constraints imposed by its neighbors. Assuming the Rouse model to provide a realistic description for non-entangled chains, it's expected $g_1(t) \propto t^{1/2}$ and $g_{CM}(t) \propto t$ in this case. If t equals the entanglement time, τ_E, the constraints begin to dominate the polymer dynamics. For larger times, a chain moves as if it were confined in a tube created by its neighbors. It can only slide along the tube axis or primitive path, whereas motion perpendicular to the tube is suppressed. Since the tube can be thought of as an envelope around the chain's random-walk-like configuration, the chain performs Rouse dynamics along a random walk. This leads to $g_1(t) \propto t^{1/4}$ and $g_{CM}(t) \propto t^{1/2}$. If t reaches the Rouse time, τ_R, the chain configuration is relaxed inside the tube. For larger times, the mass center diffuses freely and the middle monomers begin to diffuse out of the tube. The some middle monomers leave the tube completely if t equals the disengagement time, τ_D.

For small N significant corrections of considered regimes must be expected, and all these regimes can be observed clearly only for very large N, where the time scales τ_E, τ_R, and τ_D are well separated, but it is complicated to reach due to the next relations between the characteristic times [2, 3, 7, 17]:

$$\begin{aligned} &\tau_R = Z^2 \tau_E; \\ &\tau_D = 3 \cdot Z \tau_R; \quad Z = (N / N_E) - 1 \end{aligned} \tag{7}$$

Here, Z is the number of entanglements per chain defined by the most impotent and enigmatical parameter of the reptation theory entanglement length, N_E, i.e. mean number of segments between adjacent entanglements along chain. Running ahead, it is useful to notice that the longest simulated chain is characterized by $Z = 17$ entanglements.

The value of g_{1C} plateau is defined by chain mean-square gyration radius R_G^2 and permits to find τ_R as time of intercept of mass centre mean-square displacement and this plateau according to the definition of τ_R [2]:

$$g_{CM}(\tau_R) = g_{1C}(t \to \infty) = R_G^2 \, . \tag{8}$$

Autocorrelation function (**ACF**) of the chain end-to-end vector $\boldsymbol{R}_{EE}$ is used to describe its rotation quantitatively by rotational time τ_{EE}:

$$C_{EE}(t) = \langle \boldsymbol{R}_{EE}(t) \cdot \boldsymbol{R}_{EE}(0) \rangle / \langle [R_{EE}(0)]^2 \rangle \approx \exp(-t / \tau_{EE}). \tag{9}$$

According to [2] τ_{EE} depends on chain length in the same way as Rouse time:

$$\tau_{EE}(N) \propto \tau_R(N) \propto N^x, \quad \begin{cases} x = 2, & N < N_C \\ x = 3, & N > N_C \end{cases} \tag{10}$$

Here, N_C is critical chain length which had come into polymer dynamics from multitude of experiments as quantity bordering non-entangled and entangled behavior of polymer melts. Due to this circumstance there are many statistical data for its ratio to N_E, generalized, for example, in work [21]:

$$N_C \approx (1.6 \div 2.6) \cdot N_E \, . \tag{11}$$

The necessity of the wide region may be additionally proved by the well known fact the physical quantity values obtained by various methods are usually differed and should be averaged [22].

3.2. *The melt structure*

To control the degree of polymer melt equilibrium the squared ratio of the chain end-to-end distance to its gyration radius was used. This ratio is the same for phantom chains and for chains with excluded volume from the theory viewpoint [23]:

$$\langle R_{EE}^2 \rangle \approx N^{2\cdot\nu} b_K^2 \tag{12}$$

$$\langle R_G^2 \rangle \approx \langle R_{EE}^2 \rangle / 6 . \tag{13}$$

Here b_K is the Kuhn segment length, a product of the chain stiffness and mean bond length $\langle b \rangle \approx 0.95 \cdot \sigma$. The last value is a week function of pack coefficient and got as the result of simulation. ν is Flory exponent with theoretical value 0.5 for ideal solution and melt, but may be a slightly more in simulation [7, 23].

3.3. *Orientational mobility in the melt*

A normalized orientational autocorrelation function (**OACF**), *C(t)*, is used as testable experimentally by NMR relaxation. It is defined by the second-degree polynomial, $P_2(t)$, because spins are considered at beads and constant distance from each other [24]:

$$C(t) = \langle P_2(t') \cdot P_2(t'+t) \rangle / \langle [P_2(t')]^2 \rangle , \qquad P_2(t) = [3\cos^2 \theta(t) - 1]/2 \tag{14}$$

Here, θ is angle of inter-spin vector relative to external constant magnetic field. Fluctuations of some inter-spin vector in time define shape and correlation time, τ_i, of the OACF. If OACF is exponential, the τ_i value would be found as

$$\tau_i = \int_0^\infty C_i(t) \cdot dt . \tag{15}$$

Index $i = 1 \div N$ is number of the segments in subunit the fluctuations of which are considered. For example, τ_1 means the segment itself and τ_N describes reorientation of the chain end-to-end vector. Usual shape of the OACF in simulated melt is sum of undefined number of exponents; therefore τ_i was assessed as the OACF decay to $1/e$ of its initial value. Finally, free induction decay is constructed for the case of incomplete motional narrowing of the NMR absorption line [3, 24]:

$$M_i(t) = \exp\left[-\omega_2 \int_0^t (t-\tau) \cdot C_i(\tau) \cdot d\tau \right] . \tag{16}$$

Here, $\omega_2 = 1.5 \cdot 10^{-4}$ is the dimensionless second moment of the NMR line in the absence of the molecular motions. It corresponds to complete motional narrowing condition for τ_1 and does not correspond to one for τ_N for the long chains simulated. Due to non-exponential shape of the free induction decay the effective transverse relaxation time, T_2, was evaluated the same way as τ_i.

3.4. *Characteristics of the chain reptation*

A chain primitive path is the shortest distance between its ends the chain could take without crossing surrounding chains [1] as it is depicted by figure 1. Primitive path is nearly the bent axe of the tube which represents a volume of space that can be explored by a chain without violating the topological constraints. In accordance with this definition Everaers with co-workers [8] had suggested next algorithm to constructing primitive path.

Beads inside a chain are allowed to overlap and decrease in diameter, thus decreasing the contour length of the chain. The intent is to allow chains slowly to shrink their length while maintaining their non-crossability with other chains, until all chains have collapsed onto paths of minimum total length that preserves their original topological interactions. These shrunken chain contours are identified as the primitive paths. In the algorithm, the end beads of all the equilibrated chains are immobilized while the interior beads are allowed to move during concurrent temperature decreasing. Intra-molecular excluded-volume effects are turned off while maintaining intermolecular excluded-volume interactions and chains in the system are shrinking simultaneously. This procedure transforms original

N Kuhn segments of length $\langle b\rangle$ of each chain into primitive segments of length $\langle b_{PP}\rangle$. All original beads are also transformed into primitive beads of the same size $\langle b_{PP}\rangle$ due to the algorithm properties. Hence mean contour length of the primitive path is

$$\langle L_{PP}\rangle = N \cdot \langle b_{PP}\rangle \,. \tag{17}$$

Each chain entanglement makes primitive path bend, the more number of entanglements the more number of the bends, and the longer primitive path, consisting from straight lines of length $N_E\langle b_{PP}\rangle$ between adjacent entanglements. Considering these lines, strands, as Kuhn segments of Gaussian primitive path we get the number of this segments (N/N_E). Finally end-to-end distance of the original chain can be equalized to the primitive path one:

$$\langle R_{EE}^2\rangle = N \cdot N_E^{-1} \cdot \langle (N_E b_{PP})^2\rangle = N_E \cdot N^{-1} \cdot \langle L_{PP}\rangle^2 \,. \tag{18}$$

Hence we are able to find number of beads per entanglement length and primitive path contour length:

$$N_E = N \cdot \langle R_{EE}^2\rangle \cdot \langle L_{PP}\rangle^{-2}, \quad \langle L_{PP}\rangle^2 = N \cdot N_E^{-1} \langle R_{EE}^2\rangle \tag{19}$$

Equation (19) relates all primitive path characteristics to the ones of the original chain. Though its derivation is based on Gaussian statistics for long chain the formula may be applied to the chain of any length. For example, primitive path of short chain without entanglements is straight line and $L_{PP} = R_{EE}$. Substitution of this equality into equation (18) gives result with simple physical significance $N_E = N$. In other words, N_E is monotonically increasing function of the chain length for melt below N_C.

If the chain moves along the tube, there should be substantial memory of the primitive path conformation, at least for some middle chain segments. Thus, the following procedure has been applied in order to extract the longitudinal or axial, g_A, and transverse or normal, g_N, components of the chain motion [2, 9, 13]. Figure 2 schematically depicts two primitive paths, AB and A_1B_1, of the same chain at time 0 and $t < \tau_R$. Let i-th primitive path bead with coordinates $\boldsymbol{r}_i(0)$ has displaced in position $\boldsymbol{r}_i(t)$ at A_1B_1, and the nearest neighbor to its original position is appeared k-th primitive path bead with coordinates $\boldsymbol{r}_k(t)$. The space displacement $\Delta\boldsymbol{r}_i(t)$ may be resolved into axial, $\Delta\boldsymbol{r}_A(t)$, and normal, $\Delta\boldsymbol{r}_N(t)$, components. The axial component is defined as

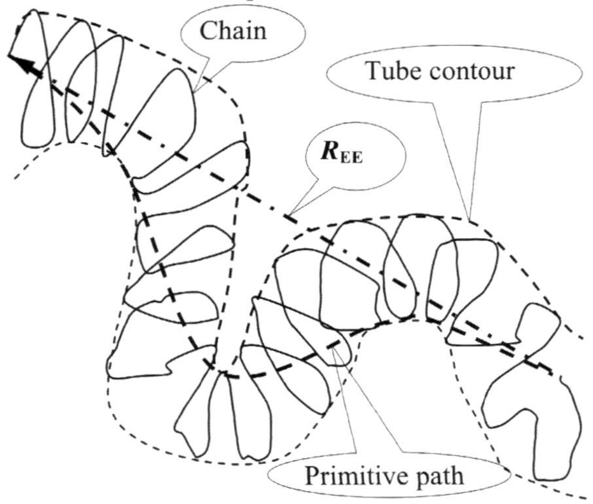

Figure 1. Schematic representation of the chain and its coarse-grained characteristics.

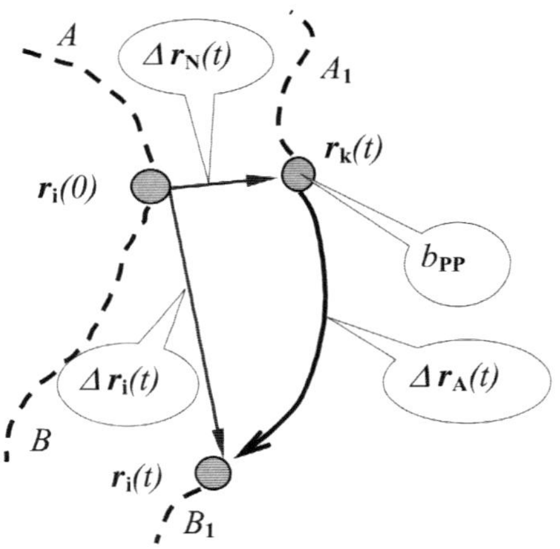

Figure 2. The same primitive path in time moments 0, AB, and t, A_1B_1: resolution of the b_{PP} displacement into axial and normal components.

$$\Delta r_A = |i-k| \cdot \langle b_{PP} \rangle . \tag{20}$$

If difference of the beads number is more than N_E the axial component becomes curvilinear and more than its total space displacement, which is equal to the vector sum

$$\Delta \mathbf{r}_i(t) = \Delta \mathbf{r}_N(t) + \Delta \mathbf{r}_A(t) . \tag{21}$$

Raising to the second power both sides of this equation and suggesting statistical independence of the components [14, 15] we get equation for the middle bead mean-square displacement in the space of the primitive paths:

$$g(t) = g_N(t) + g_A(t) . \tag{22}$$

In according to the definition (19) g_A cannot be more than a quarter of the primitive path squared length, and then it gets plateau and loses its original sense. Combination of the equations (12), (13), and (19) gives the plateau value

$$g_A(t \geq \tau_R) = \left(\frac{L_{PP}}{2} \right)^2 = \frac{3}{2} \frac{N}{N_E} R_G^2 = \frac{3N^{2\nu+1} b_K^2}{2N_E} . \tag{23}$$

Time of the plateau achieving in equation (23) is defined by the chain exit from its tube after τ_R. It is seen the plateau value is higher than g_{1C} plateau, R_G^2, and dependent on chain length approximately quadratically.

Quantitative characteristic of the chain diffusion anisotropy, *a(t)*, is made by the ratio of normal and axial mean-square displacement [9-12]:

$$a(t) = g_N(t) / g_A(t) . \tag{24}$$

Isotropic diffusion is characterized by $a(t) = const = 2$ in region of its definition which is very short in time. When chain diffuses in the tube of diameter d_T [2, 3]

$$d_T^2 = N_E^{2\nu} \langle b_K^2 \rangle = N_E^2 \langle b_{PP} \rangle^2 , \tag{25}$$

its normal displacement is no more than half of the diameter [14] and shouldn't depend on chain length after $N > N_C$. The axial displacement should increase in time till asymptotic value, equation (23), growing with chain length. Substitution of equations (17) and (25) into equation (24) for the moment of exit from the tube, $t = \tau_A$, gives diffusional anisotropy observed:

$$a(\tau_A) = \left(\frac{d_T}{2} \right)^2 / \left(\frac{L_{PP}}{2} \right)^2 = \left(\frac{N_E}{N} \right)^2 = \frac{1}{(Z-1)^2} . \tag{26}$$

It is seen that the anisotropy should be less than unity and decreased as squared number of entanglements per chain. To characterize the *a(t)* minimum the time τ_A was introduced because its physical significance differs from τ_R one. Namely, τ_A means persistent contact time of the Rouse chains with surroundings and the entangled state life time of the reptating chains in the melt.

4. The simulated results and discussion

4.1. Time dependence of the chain diffusional characteristics

Translational and rotational mobility of the chains and their segments in the melt are characterized by typical dependences which are shown in figures 3 and 4 for the chain of 127 segments length. Dynamics of all length chains is observed in the time region $(10^4 \div 10^6)\ \tau_0$ to define the largest characteristic times τ_{EE} and τ_R marked by vertical arrows. Horizontal arrow R_G^2 marks $g_{1C}(t)$ plateau, curve 3 in figure 3, the interception of which with $g_{CM}(t)$, curve 1, gives τ_R. The decay of the end-to-end vector

ACF, curve 1, in figure 4, permits to find τ_{EE} which is close to τ_R for this chain length. Value of τ_E is marked according to the assessment mentioned below such as the horizontal dotted line $(d_T/2)^2$ assessed by equation (25). The intersection of these two lines in figure 3 has to mark the regime change of curve 2, but the changing is unnoticeable. The free diffusive limit for $g_{CM}(t)$, curve 1 in figure 3, is also not achieved therefore chain diffusion coefficient is not defined and discussed.

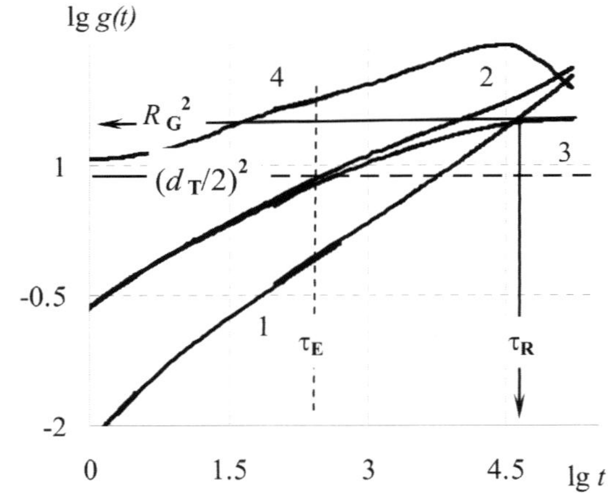

Figure 3. Time dependence of mean-square displacement for chain of length 127 segments: chain mass center (1), middle bead (2), middle bead relative to the mass center (3); ratio of the mean-square displacement of the end bead and the middle bead (4) in linear scale. Arrows R_G^2 and τ_R show their estimation. Dashed lines $(d_T/2)^2$ and τ_E are superposed from other figures.

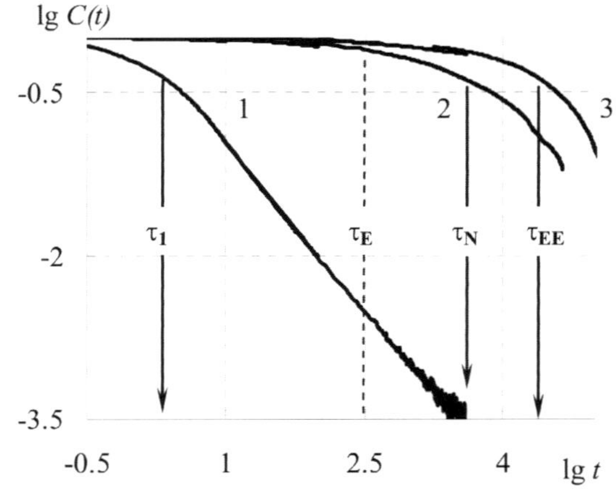

Figure 4. Time dependence of the segmental (1) and chain (2) OACF, the chain end-to-end vector ACF (3) for chain of length 127 segments. Arrows τ_1, τ_N, and τ_{EE} show their estimation. Dashed line τ_E is superposed from other figure.

Curve 4 in figure 3 shows ratio of the mean-square displacement of the end bead, $g_2(t)$, and the middle bead, $g_1(t)$, in linear scale.

The power laws of equation (5) and (6) are observed in figure 3 only approximately because they describe the asymptotic behavior and can be observed clearly only for very large N, where the time scales τ_E, τ_R, and τ_D are well separated. For smaller N, significant deviations have to be expected as it has been demonstrated by a detailed theoretical analysis and simulation [25]. The coarse-grained model is not satisfied this condition, since the subchain units corresponding to various time scales are very close in sizes.

The main information from figure 4 is relation between segmental, curve 1, and chain, curve 2, OACF. It is seen the noticeable decay of the chain OACF is in the time region where the segmental OACF shape is hidden by noise. I.e., contribution of the orientation of the chain as a whole is not impossible to extract from the segmental OACF for the data system discussed.

The collection of the dependences as in figures 3 and 4 for other length chains differs only by values of characteristic times and conservers the same forms. Therefore integrated analysis of all characteristic times will be made latter as chain length dependences.

The similar analysis is made for mean-square displacements in the primitive path space, the example of which is depicted in figure 5 for the same length chain. The axial and normal mean-square displacements of the chain middle monomer are represented by curves 1 and 2, respectively. Both mean-square displacements increases monotonically till τ_R. The axial one approaches plateau after τ_R, the normal one begins to increase faster after this time. These facts are caused by going out of the applicable domain according to the equation (23). The normal mean-square displacement should have plateau due to limitation by the tube radius and its absence is also explained by earlier going out of the applicable domain.

The ratio of the normal mean-square displacement to the axial one, curve 3, is appeared as wide minimum with plateau in its central part. The deepest and the most right point of the plateau is marked by arrows and represents the time, τ_A, and value, $a(\tau_A)$, of the anisotropic diffusion: for this curve $a(\tau_A) \approx 0.1$. The horizontal dashed lines are carried over from other figures to compare the mean-square displacement scales. The axial mean-square displacement plateau, curve 1, is not reached yet, but it is seen that its value should be sufficiently higher R_G^2 in accordance with equation (23).

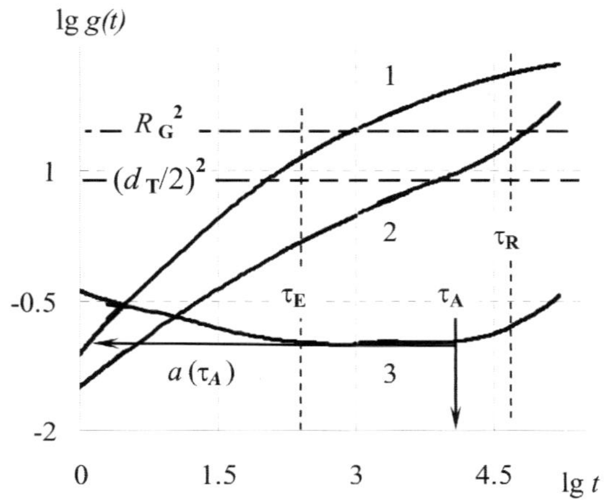

Figure 5. Time dependence of various reptation characteristics for chain of length 127 segments: mean-square displacements of the middle bead relative to the tube in the axial (1) and normal (2) directions, and ratio of the second one to the first one (3). Arrows τ_A and $a(\tau_A)$ show their estimation. Dashed lines R_G^2, $(d_T/2)^2$, τ_E, and τ_R are superposed from other figures.

4.2. Static properties of the chain

Figure 6 depicts dependence of the chain mean-square gyration radius, R_G^2, on the chain length with asymptotic power law $2\nu \approx 1.07$. Data processing according to the equations (12) and (13) gives Kuhn segment length $\langle (b_K)^2 \rangle \approx 1.20$. The large circles along the dependence 1 are mean-square gyration radiuses of the various subunits of the longest chain simulated, 255 segments. The coincidence with the curve demonstrates that the subunit conformation inside the chain repeats the conformation of the same length free chain. It indicates all chains of the simulated melts were in thermodynamic equilibrium what is not always obtainable in simulation of long chains [20].

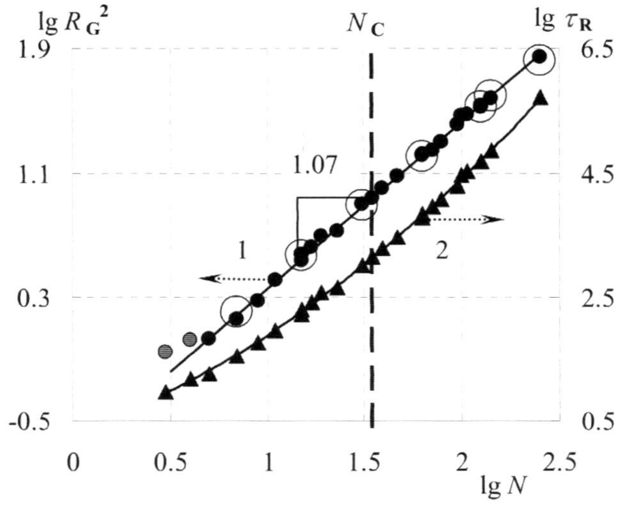

Figure 6. Dependence of mean-square gyration radius (1) and Rouse time (2) on chain length. The large circles are R_G^2 of the various sub-units inside the chain of 256 beads. The trend line power law is ~1.07 without the two lowest grey points. Vertical dashed line superposed from other figure.

The vertical dashed line in figure 6 marks the critical chain length, N_C, which will be got further. The marked N_C value is necessary to demonstrate two facts. The first, the simulated melt static properties are not influenced by chain length. The second, Rouse time dependence does not show sharp flex near of the N_C in spite of the curve ends have other slopes. Since this dependence does not permit to find N_C correctly, this finding will be made further with the help of the scaled dependences [7].

4.3. Scaled dependence of the correlation times on chain length

Ordinate of the scaled dependence is divided by the chain length dependence expected from Rouse model, i.e., N^2 according to equation (10), as it depicted in figure 7 for three large-scale times, τ_R, τ_{EE}, and τ_N. If the original dependence is coincided with theoretical one, the transformed scaling one would be appeared as horizontal line any deviation from which is very noticeable. The scaling introduce some inconvenience – statistical errors become much better visible due to decreasing of the absolute values along scaled ordinate as it is seen from comparison of figures 6 and 7. Though, this defect is easily circumvented with the help of the analytical trend lines which are duplicated for scaled and original dependence.

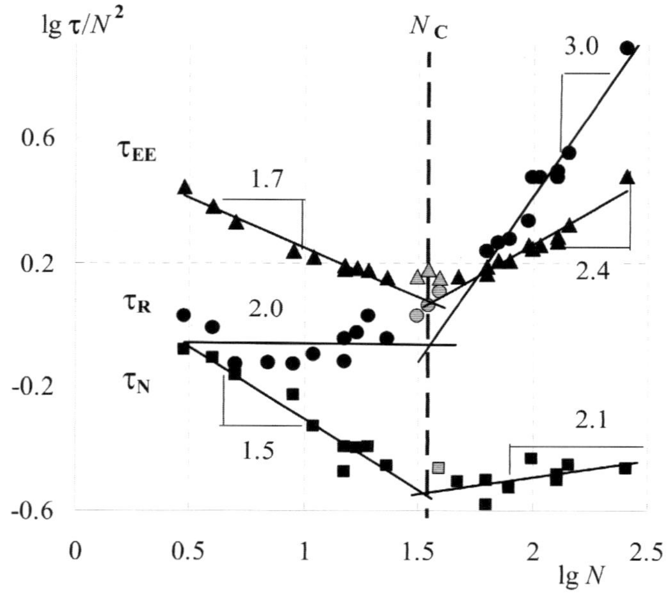

Figure 7. Scaled chain length dependence of the chain rotation correlation time τ_{EE}/N^2, Rouse time τ_R/N^2, and the chain orientational correlation time τ_N/N^2. The vertical dashed line is estimation of N_C as averaged intersection of all trend line pairs. The values near of the right angles are asymptotical power laws defined by the trend lines without including grey points near of N_C.

Asymptotic trend lines of the low chain length and high chain length branch display clearly the change in the chain dynamic behavior. The low chain length part of curve 1 in figure 7, τ_R/N^2, is characterized by large point scattering because short chains do not create coils. In spite of this circumstance this trend line shows power law 2.0, expected for the Rouse regime. The coincidence is probably occasional because the simulated chains have excluded volume, i.e., are not Rouse ones [3, 7]. The long-chain asymptotical branch of this dependence also demonstrates power law 3.0, expected according to equation (9). Intermediate gray points of this and all other dependences are not taken into account for definition of the asymptotical power laws.

Curve 2 in figure 7 for rotational correlation time of the chain end-to-end vector, τ_{EE}/N^2, shows asymptotical power laws which are noticeably deviated from ones of equation (10), 1.7 and 2.4. The points scattering of this dependence is much less than that of curve 1, what is defined by the extracting difference for these two correlation times. In spite of the asymptotical power laws difference the intersection of their trend lines for both dependences are at the same chain length value $N_C = 35\pm1$ within the statistical error. This critical value is marked by vertical dashed line. Thereby, the simulated chains of length more than 34÷36 segments or 35÷37 beads behave themselves as entangled ones. In accordance with equation (11) entanglement length has to be in the interval $N_E = (13 \div 21)$ segments.

Moreover, the lowest dependence in figure 7, the chain orientational correlation time, τ_N/N^2, also displays crossover in the same interval of N_C, but with other values of the asymptotical power law, 1.5 and 2.1.

The N_E value is seemed underestimated one in comparison with Monte Carlo simulation data, ~35 [7-13]. Monte Carlo simulation is traditionally used to investigate dynamical characteristic of various polymer systems due to its much more computational efficiency relative to molecular dynamics simulations. Though, the application of powerful computer during the last years has permitted to use molecular dynamics simulation by many investigators. For example, the similar value of the $N_E \approx 15$ had been obtained in [25]. The dissipative particle dynamics simulation gives even less value, $N_E \approx 13$ [26]. In other words, in spite of the low computational efficiency molecular dynamics simulation permits to study well entangled chains due to less value of the N_E in such simulation.

4.4. Mean-square displacement of the middle and end beads for the different chain lengths

Let's consider the diffusional behavior the middle bead for various length chains, which permits to find entanglement time τ_E. Figure 8 depicts mean-square displacements of the middle chain segments scaled by short-time regime for $t < \tau_E$, when $g_1(t) \sim t^{0.50}$. This scaling transforms original mean-square displacement into increasing one till τ_E and into decreasing one after τ_E according to the equation (5). Furthermore, all curves are collapsed onto a common curve till proper Rouse time of each chain. The collapsed curve maximum defines the entanglement time value, $\tau_E \approx 320$, which was marked in the previous figures.

The sharp lift of each curve is related to the proper Rouse time of each chain. The Rouse chains of 8, 16, and 32 beads in figure 8 have $\tau_R < \tau_E$ and their scaled curves begin to increase before the maximum. The entangled chains of 64, 128, and 256 beads have $\tau_R > \tau_E$ and demonstrate their scaled curves lift much later of τ_E. The dashed line $\sim t^{0.25}$ shows that this regime is clearly visualized only for the longest chain of 256 beads. The curve local minimum of all entangled chains marks transformation of the time regime from $\sim t^{0.25}$ to $\sim t^{0.50}$, and at once to the final regime $\sim t$.

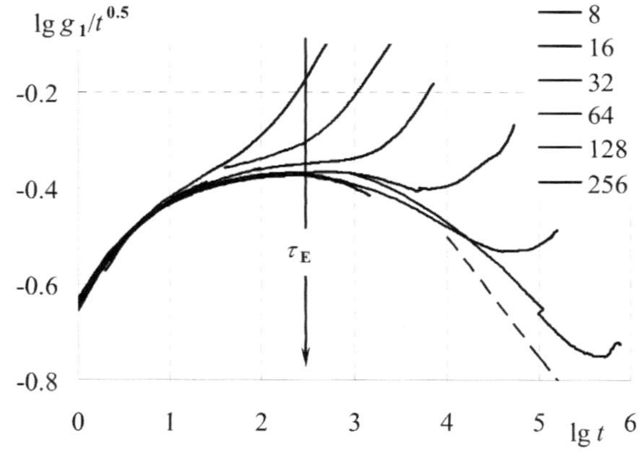

Figure 8. Time dependence of the middle bead mean-square displacement scaled to the regime $\sim t^{0.5}$ for different chain lengths. The quantity of the chain beads is indicated in the column in the same order as the curves. The vertical arrow shows the estimation of the entanglement time, τ_E. The dotted line marks power law –0.25.

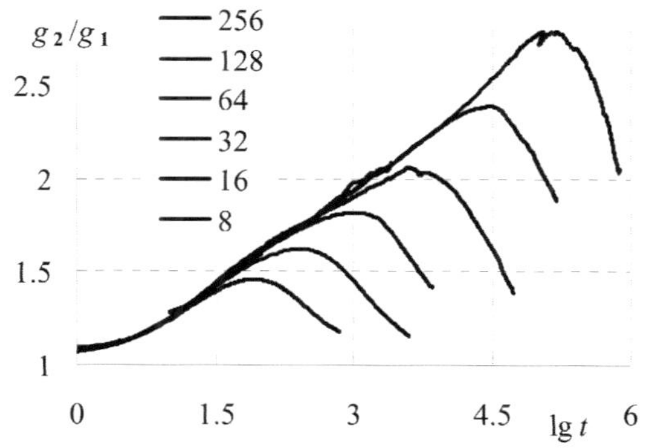

Figure 9. Ratio of the mean-square displacement of the end bead and the middle bead as function of time for different chain lengths. The quantity of the chain beads is indicated in the column in the same order as the curves.

Journal of Physics: Conference Series **324** (2011) 012031 doi:10.1088/1742-6596/324/1/012031

Though the scaled curves in figure 8 are described by equation (5) only in short time intervals which does not permit to identify power laws in these intervals, but the number of the curves twists clearly manifest the presence of these four diffusional regimes in accordance with equation (5).

Figure 9 shows the simulation results for the ratio of the mean-square displacement of the end bead and the middle bead, $g_2(t)/g_1(t)$. The ratio starts around 1.0 at early times, exhibits a maximum at intermediate times and tends to 1.0 for large t. The ratio maximum should be 2.0 at $t \leq \tau_R$, expected for the Rouse chain. For a reptating chain the maximum is still expected at $t \approx \tau_R$, but to be larger than 2.0. For the chain of 64 beads the maximum occurs close to 2.0, i.e., to the value expected from the Rouse model, whereas it is larger for chains of 128 beads, ~2.4, and 256 beads, ~2.9. An increase of the maximum beyond the Rouse limit with growing chain length is predicted theoretically for the reptation and proposed a sensible indicator of reptation dynamics [27]. For a reptating chain the disparity between g_2 and g_1 has to increase because the motion of the inner bead is strongly confined by the tube at intermediate times, whereas end monomers always take part in tube renewal. The data for $N = 128$ and 256 are indicative of such an enhancement beyond the Rouse limit.

Coincidence of the left-hand side of all maxima indicates the identity of the small-scale translational segmental mobility of any chain length till almost its proper Rouse time when the maximum is achieved. This circumstance is preliminary designation of the segmental orientational dynamics identity independently of the chain length as it was also observed in figure 8.

4.5. Dependence of the reptation characteristics on time and chain length

Figure 10 represents the time dependence of the chain diffusional anisotropy, $a(t)$, for the same chain lengths as in previous two figures. In accordance with equation (24) the anisotropy is the more the less $a(t)$ relative to 2.0 corresponding to isotropic diffusion. It tends to zero for infinitely long chain in the time moment of the tube exit. Hence, the simulated chain could be considered as reptating one if $a(\tau_A) << 2$.

$a(t)$ for all chains looks like a minimum which is made deeper and wider with chain length increasing. The minimum depression for Rouse chains, of length 8, 16, and 32, is limited by proper Rouse time, the one for all reptating chains, of length 64, 128, and 256, is continued till τ_E when a middle segment "collides" with tube walls [2]. Further the minimum is continued with constant depth till proper τ_M which means the time of the tube leaving and the border of the $a(t)$ definition region. The minimum of two shortest chains, of length 8 and 16, corresponds to isotropic diffusion, it is close to 2.0. The minimum horizontal part of the reptating chains, of length 64, 128, and 256, falls below 0.3 and for the longest chain is equal to 0.03, and its extension is three decimal orders in time.

The long minimum plateau implies that the curvilinear diffusion regime along the tube axis and the lateral diffusion regime are described by the same power law, $g_A(t) \sim g_N(t) \sim t^{0.31}$, in the plateau time region as it may be found from figure 5. This process of the tube reorganization due to diffusion of surrounding chains in their own tubes is called constraint release [2, 28] or the tube dilation from its middle to its end [29]. Other factor of the plateau extension may be wide distribution of the tube diameters which had been analytically described [30]. The constraint release may be made disable, for example, by intrusion of real macromolecules into mesoscopic pores [14, 15]. However, even the fixed pore constraint does not permit to observe the predicted saturation of $g_N(t)$ due to thermodynamical restriction on reptation [15].

Figure 11 shows the chain length dependence of the entanglement length, $N_E(N)$, and the minimum depth, $a(\tau_A)$, from figure 10. Two vertical dashed lines, N_E and N_C, restrict the crossover region from Rouse dynamics to reptation one. Curve 1, $N_E(N)$, below N_C increases approximately as end-to-end distance in accordance with equation (19). In other words, primitive path of the Rouse chain is straight line due to absence of entanglements. Near of the N_E vertical dashed line the primitive path length is sharply increased relative to the chain length and then curve 1 is smoothly saturated till asymptotical value $N_E \equiv N_E(\infty) \approx 14$ segments, which is reached after N_C.

It is obviously the value N_E could be uniquely defined with the help of equation (19) for a single system of the well entangled chains by time and ensemble averaging of the primitive path length. I.e.,

Journal of Physics: Conference Series **324** (2011) 012031 doi:10.1088/1742-6596/324/1/012031

we needn't investigate the chain length dependences to find N_E and its value is agreed very well with the data of figure 7 by statistical equation (11) according to which $N_E = 13 \div 22$. This coincidence demonstrates significance of such abstract notions as "primitive path" and "tube" for description of the entangled chain diffusion. Moreover, this analysis permits to expand the physical meaning of the chain critical length. It is usually interpreted as one which sharply changes chain dynamics due to the life time of inter-chain contacts becomes comparable with the characteristic time of some experimental method. Though, it is needed to keep in mind such the contacts are present for short chains with excluded volume also and influence on their motion.

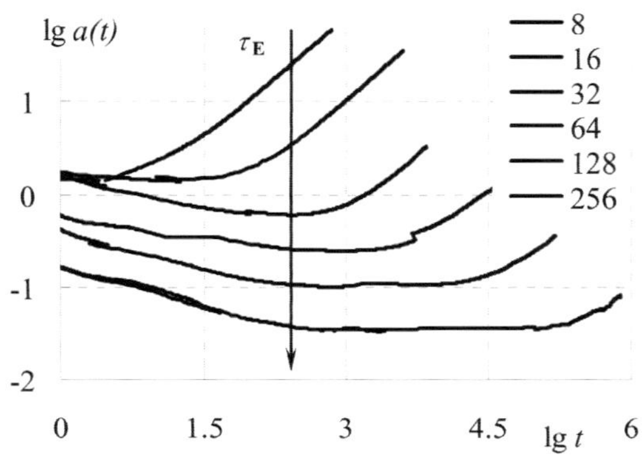

Figure 10. Time dependence of the chain diffusional anisotropy for different chain lengths. The quantity of the chain beads is indicated in the column in the same order as the curves. The vertical arrow marks entanglement time.

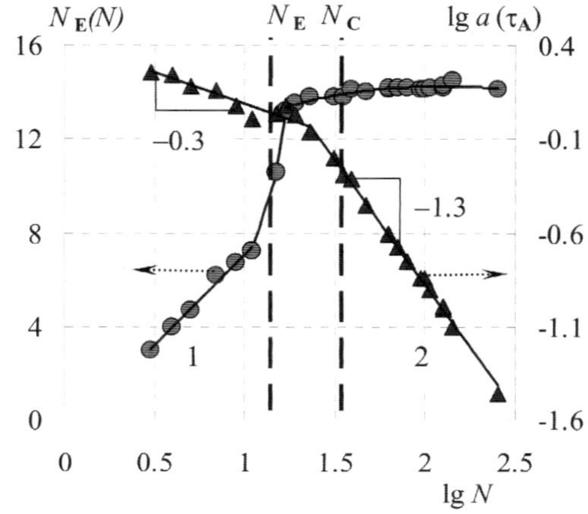

Figure 11. Chain length dependence of the entanglement length (1) and maximal anisotropy of the chain diffusion (2). The values near of the right angles are power laws, defined by the trend lines. The vertical dashed lines mark N_E and N_C.

Such the influence is described by dependence 2 for $a(\tau_A)$ in figure 11. It may be approximated by two power laws with crossover in the region from $N_E \approx 14$ to $N_C \approx 35$ segments. The short chain branch demonstrates that the motional anisotropy is also present and slowly expands for Rouse chains as $\sim N^{-0.30}$. The reptating chain branch is sufficiently faster changed $\sim N^{-1.30}$, but slower than $\sim N^{-2.0}$, expected from equation (26). Nevertheless, the lowest point of the dependence, chain of 255 segments, corresponds to the anisotropy $a(\tau_A) \approx 0.03 \ll 2$ and completely satisfies the basic notion of the reptation theory [1, 2] about neglible lateral chain displacement inside the tube.

Second important conclusion may be made from this dependence. To explain fast increase of the chain life time in permanent surroundings it is unnecessarily to use such the factors as local liquid crystal ordering or specific energetic interaction between chains. The excluded volume and one-dimensionality of the long chain provide the main and alone reason of physical nodes origin in polymer melt. This reason were understandable, if one would have remembered the cage effect in liquid due to which even the vicinity of the small molecules is characterized by finite life time [31].

4.6. NMR transverse relaxation in the simulated polymer melt

It has been demonstrated in the previous sections that the chain dynamics in the simulated melt ranges from Rouse dynamics to reptation one. This demonstration is necessary to reveal the crossover influence on the segmental orientational mobility.

Figure 12 represents the integrated comparison of the different characteristic times as chain length dependences. Only mutual behaviour of all curves may be conclusively converted into NMR transverse relaxation time, the curve T_2, in accordance with equations (14)-(16).

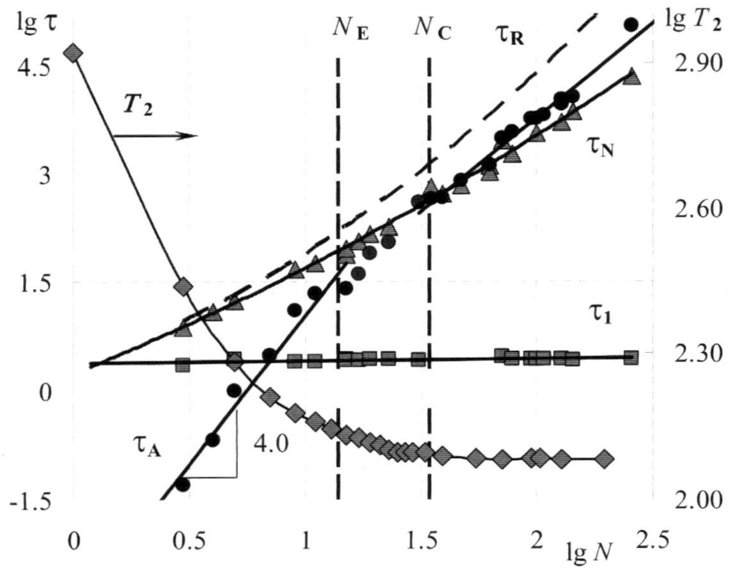

Figure 12. Chain length dependence of the different characteristic times: Rouse time, τ_R (dashed line), time of the anisotropic diffusion, τ_A (circles); chain, τ_N (triangles), and segmental, τ_1 (squares), orientational correlation times; NMR transverse relaxation time, T_2 (rhombs) for subunits inside the chain of length 127 segments. The value 4.0 near of the right angle is power law defined by the trend line. The vertical dashed lines mark N_E and N_C.

It is reasonable to begin from the simplest practically horizontal dependence for the segmental orientational correlation time, τ_1, which defines the most part of the OACF decay in figure 4 and hides information about residual dipole interaction controlling by orientation of the chain as whole. Its behavior is well-known from the multitude of the NMR experiments [3, 30] and indicates independence of the small-scale orientational mobility on chain length.

The highest dashed dependence, τ_R, determines the above mentioned thermodynamical limit for all characteristic large-scale times in the polymer melt [15] and verifies the consistency of simulated data to experiments [3, 30]. The chain end-to-end vector orientational correlation time, τ_N, is partly coincided with Rouse time for short chains, but becomes lower it for reptating chains. The time of anisotropic diffusion, τ_A, is the smallest in the Rouse region, but shows fastest increasing with extraordinary large power law, 4.0. The smallness of the τ_A relative to even τ_1 and moreover τ_N means the isotropy of as small-scale as large-scale orientational mobility of the Rouse chains.

The small-scale segmental mobility anisotropy is appeared when τ_A transcends τ_1 in the region before N_E. It means the segmental orientation becomes a few anisotropic, because it partly relaxes during time of anisotropic diffusion. The appeared residual dipole interaction is influenced only by chain orientation as whole with τ_N. The small-scale mobility anisotropy and residual dipole interaction continue to increase till N_C when τ_A transcends τ_N, later they reach saturation and become constant for reptating chain according to curve 1 in figure 11. Further excess of τ_A over τ_N is already unimportant and unobservable in NMR relaxation because τ_N comes into region of solid-like behavior relative to the constant value of the residual dipole interaction [24].

Exhaustive treatment of the transverse relaxation behavior is outside of this study because it needs great amount of additional simulated data. But qualitative T_2 dependence consisted with above discussion can be found for various subunits inside one of the simulated chains of length 255 segments. Such the dependence on subunit length is inaccessible in NMR experiment, but it gives direct information about influence of the increasing small-scale motional anisotropy on T_2 according to equations (14) and (16). The curve of effective T_2 in figure 12 is got by building of OACF for subunits of length from

1 to 200 segments in suggestion of the same rigid lattice second moment for all of them. Then free induction decay was got by transformation (16).

The curve T_2 for shortest chains shows fast linear decreasing unusual for polymers, because the shortest chains really mimics small molecules with correlation time τ_N in interval of the complete motional narrowing [24]. Then long before N_E the rate of T_2 decreasing is unexplainably changed, at first sight. But in figure 12 the changing is argued by intersection of τ_A and τ_1 dependences. This intersection generates motional anisotropy transforming the molecule into short Rouse chain. From this chain length transverse relaxation is controlled by two motions, but the most part of the second moment as previously is controlled by segmental orientation, and increasing residual part of the second moment is controlled by τ_N. Due to smallness of the last contribution the T_2 begins to decrease very slowly relative to the small molecules. The simplest equation to describe this dependence is different time scale approximation [3] or, so-called, model free theory [34], both of which divide overall OACF into two parts

$$C(t) = (1 - S^2)\, C_1(t/\tau_1) + S^2 C_N(t/\tau_N)\,. \tag{27}$$

Indices 1 and N corresponds, respectively, to the segment and chain orientation. S is local order parameter which is defined by time and ensemble averaging of the second-degree polynomial, $P_2(t)$, in equation (14). S defines the anisotropy of segmental mobility due to onset of entanglements. In accordance with equation (16) the free induction decay is represented by two multipliers, the first of which is constant due to τ_1 constancy, and the second one is slowly decreased due to τ_N increasing. In other words, the decreasing rate of the T_2 curve is caused by appearance of non-zero S value. Transformation of the T_2 curve into horizontal line after N_C is explained by onset of the solid-like for residual dipole interaction defined by S, and increasing τ_N value, which does not satisfy now the condition of narrowing [24].

Another explanation of the total T_2 curve behavior is guess of lower system temperature, then the curve shape is fully coincided with experimental one [3] without any additional suggestion. But it needs additional checking simulation at various model temperatures.

5. Conclusion

The purpose of present paper is to compare behaviour of translational, rotational and orientational correlation times of the chain in dense polymer melt over wide region of the chain lengths, from non-entangled to entangled dynamics. It turned out that molecular dynamics simulation permits to observe crossover from non-entangled to entangled dynamics because the degree of the coarse-graining is more in the denser polymer melt, than it is usually observed in Monte Carlo simulations. This circumstance results in less value of the entanglement length, $N_E \approx 14$, in comparison to usual Monte Carlo values, ~35. In turn, it permitted to simulate well entangled chains with 17 entanglements per longest chain. The well entangled chain demonstrates sufficient diffusional anisotropy as it was predicted by the theory of polymer dynamics [1], but time of this anisotropic diffusion is restricted by thermodynamics reasons [15].

The simulated chain length dependences represent the sufficiently demonstrative system of the data to explain qualitatively the behavior of the nuclear magnetic transverse relaxation time in polymer melts from Rouse till reptation regime.

References

[1] De Gennes P G 1979 *Scaling Concepts in Polymer Physics* (New York - Cornell University Press)
[2] Doi M and Edwards S F 1986 *Theory of polymer dynamics* (New York - Acad Press)
[3] Kimmich R and Fatkullin N 2004 Polymer Chain Dynamics and NMR *Adv. Polym. Sci.* **170** 1-113
[4] Paul W, Binder K, Heermann D W and Kremer K 1991 Dynamics of polymer solutions and

melts. Reptation predictions and scaling of relaxation times *J. Chem. Phys.* **95** 7726-40

[5] Wittmer J, Paul W and Binder K 1992 Rouse and Reptation Dynamics at Finite Temperatures: A Monte Carlo Simulation *Macromolecules* **25** 7211-19

[6] Baschnagel J, Paul W, Tries V and Binder K 1998 Statics and dynamics of bidisperse polymer melts: a monte carlo study of the bond-fluctuation model *Macromolecules* **31** 3856-67

[7] Kreer T, Baschnagel J, Muller M and Binder K 2001 Monte Carlo Simulation of Long Chain Polymer Melts: Crossover from Rouse to reptation dynamics *Macromolecules* **34** 1105-17

[8] Everaers R, Sukumaran S K, Grest G S, Svaneborg C, Sivasubramanian A and Kremer K 2004 Rheology and Microscopic Topology of Entangled Polymeric Liquids *Science* **303** 823-6

[9] Kolinski A, Skolnick J and Yaris R 1987 Does reptation describe the dynamics of entangled, finite length polymer system? A model simulation *J. Chem. Phys.* **86** 1567-85

Skolnick J, Yaris R and Kolinski A 1988 Phenomenological theory of the dynamics of polymer melts. I. Analytic treatment of self-diffusion *J. Chem. Phys.* **88** 1407-17

Skolnick J, Yaris R and Kolinski A 1988 Phenomenological theory of the dynamics of polymer melts. II. Viscoelastic properties *J. Chem. Phys.* **88** 1418-42

[10] Skolnick J, Kolinski A and Yaris R 1987 Monte Carlo studies of the long-time dynamic of dense polymer systems. The failure of the reptation model *Acc. Chem. Res.* **20** 350-6

[11] Skolnick J and Kolinski A 1990 Dynamics of dense polymer systems: computer simulations and analytic theories *Adv. Polym. Sci.* **78** 223-78

[12] Kremer K and Grest G S 1990 Dynamics of entangled linear polymer melts: A molecular-dynaics simulation *J. Chem. Phys.* **92** 5057-86

[13] Shaffer J S 1994 Effect of chain topology on polymer dynamics: bulk melts *J. Chem. Phys.* **101** 4205-13

Shaffer J S 1995 Effect of chain topology on polymer dynamics: configurational relaxation in polymer melts *J. Chem. Phys.* **103** 761-72

[14] Fischer E, Beginn U, Fatkullin N and Kimmich R 2004 Nanoscopic poly(ethylene oxide) strands embedded in semi-interpenetrating methacrylate networks. preparation method and quantitative characterization by field-gradient NMR diffusometry *Macromolecules* **37** 3277-86

Fatkullin N, Fischer E, Mattea C, Beginn U and Kimmich R 2004 Polymer dynamics under nanoscopic constraints: the "corset effect" as revealed by NMR relaxometry and diffusometry *Chem. Phys. Chem.* **5** 884-94

Kausik R, Mattea C, Kimmich R and Fatkullin N 2007 Chain dynamics in mesoscopically confined polymer melts. A field cycling NMR relaxometry study *Eur. Phys. J. Special Topics* 2007 **141** 235-41

[15] Fatkullin N, Kimmich R, Fischer E, Mattea C, Beginn U and Kroutieva M 2004 The confined-to-bulk dynamics transition of polymer melts in naoscopic pores of solid matrices with varying pore diameter *New J. Physics* **6** 46(13), www.njp.org

Fatkullin N F and Kimmich R 2006 Thermodynamic limitations of the reptation model *Macromol. Symp.* **237** 69-72

Fatkullin N, Kausik R and Kimmich R 2007 Molecular diffusion on a time scale between nano- and milliseconds probed by field cycling NMR relaxometry of intermolecular dipolar interactions: application to polymer melts *J. Chem. Phys.* **126** 094904(8)

Fatkullin N and Kimmich R 2010 Anisotropy of the segment mobility versus self- and pair-correlation functions in polymer melts under mesoscopic confinement *Macromolecules* **43** 9821-27

[16] Heerman D W 1986 *Computer simulations methods in theoretical physics* (Berlin – Springer Verlag)

[17] Kremer K 2004 Entangled polymers: from universal aspects to structure property relations *NIC Series* **23** 141-68, www.fz-juelich.de/nic-series

[18] Grest G S and Kremer K 1986 Molecular dynamics simulation for polymers in the presence of a heat bath *Phys. Rev.* A **33** 3628-31
[19] Frenkel D, Smit B 2002 *Undestanding molecular simulation* (New York - Acad Press)
[20] Auhl R and Everaers R, Grest G, Kremer K and Plimpton S J 2003 Equilibration of long chain polymer melt in computer simulations *J. Chem. Phys.* **119** 12718-28
[21] Fetters L J, Lohse D J, Richter D, Witten T A and Zirkel A 1994 Connection between polymer molecular weight, density, chain dimensions and melt viscoelastic properties *Macromolecules* **27** 4639-47
Fetters L J, Lohse D J, Milner S T and Graessley W W 1999 Packing Lengths influence in linear polymer melts on the entanglement, critical, and reptation molecular weights Macromolecules **32** 6847-51
[22] Leon S, van der Vegt N, DelleSite L and Kremer K 2005 Bisphenol A polycarbonate: entanglement analysis from coarse-grained MD simulations *Macromolecules* **38** 8078-92
[23] Grosberg A Yu and Khokhlov A R 1994 *Statistical physics of macromolecules* (New York - AIP)
Teraoka I 2002 *Polymer Solutions: An Introduction to Physical Properties* (New York – Wiley-Interscience)
[24] Abragam A 1961 *The Principles of Nuclear Magnetic Resonance* (London - Oxford University Press)
[25] Ebert U, Schafer L and Baumgartner A 1998 Segment motion in the reptation model of polymer dynamics. I. Analytical Investigation *J. Stat. Phys.* **90** 1325-73
Baumgartner A, Ebert U and Schafer L 1998 Segment motion in the reptation model of polymer dynamics. II. Simulation *J. Stat. Phys.* **90** 1375-400
[26] Tzoumanekas C, Lahmar F, Rousseau B and Theodorou D N 2009 Onset of entanglements revisited. Topological analysis *Macromolecules* **42** 7474-84
Lahmar F, Tzoumanekas C, Theodorou D N and Rousseau B 2009 Onset of entanglements revisited. Dynamical analysis *Macromolecules* **42** 7485-94
[27] Zhou Q and Larson R G 2006 Direct calculation of the tube potential confining entangled polymers *Macromolecules* **39** 6737-43
[28] Klein J 1978 The onset of entangled behavior in semidilute and concentrated polymer solutions *Macromolecules* **11** 852-7
Klein J 1981 Effect of matrix molecular weight on diffusion of labeled molecule in a polymer melt *Macromolecules* **14** 460-1
Klein J 1986 Dynamics of entangled linear, branched, and cyclic polymers *Macromolecules* **19** 105-18
Klein J 1986 Matrix effects on the diffusion of long polymer chains *Macromolecules* **19** 1108-14
[29] Tzoumanekas C and Theodorou D N 2006 Topological analysis of linear polymer melts: a statistical approach *Macromolecules* **39** 4592-604
[30] Kamio K, Moorthi K and Theodorou D N 2007 Coarse-grained end bridging Monte Carlo simulations of poly(ethelene terephtalate) melt *Macromolecules* **40** 710-22
Spyriouni T, Tzoumanekas C, Theodorou D N, Muller-Plate F and Milano G 2007 Coarse-grained and reverse-mapped united-atom Simulations of long-chain atactic polystyrene melts: structure, thermodynamics properties, chain conformation, and entanglements *Macromolecules* **40** 3876-85
[31] Benneman C, Baschnagel J, Paul W and Binder K 1999 Molecular-dynamics simulation of a glassy polymer mMelt: Rouse model and cage effect *Comput. Theoret. Polym. Sci.* **9** 217-26
Baschnagel J, Benneman C and Binder K 2000 Dynamics of a supercooled polymer melt above the mode-coupling critical temperature: cage versus polymer-specific effects *J. Phys. Condens. Matt.* **12** 6365-74
Varnik F and Binder K 2002 Shear viscosity of supercooled polymer melt via non-equilibrium

molecular dynamics simulations *J. Chem. Phys.* **117** 6336-49

Varnik F, Baschnagel J, Binder K and Mareschal M 2003 Confinement effects on the slow dynamics of a supercooled polymer melts: Rouse modes and incoherent scattering function *Eur. Phys. J. E* **12** 167-71

Aichele M, Gebremichael Y, Star FW, Baschnagel J and Glotzer S C 2003 Polymer-specific effects of bulk relaxation and stringlike correlated motion in the dynamics of a supercooled polymer melt *J. Chem. Phys.* **119** 5290-304

Ottochian A, Michele C D and Leporini D 2008 Non-gaussian effects in the cage dynamics of polymers *Philosophical Magazine* **88** 4057-62

Ottochian A, Molin D, Barbieri A and Leporini D 2009 Connectivity effects in the segmental self- and cross-reorientation of unentangled polymer melts *J. Chem. Phys.* **131** 174902(10)

[32] Kariyo S, Gainaru C, Brodin A, Novikov V N, and Rossler E A 2006 From simple liquid to polymer melt: NMR relaxometry study of polybutadiene *Phys. Rev. Lett.* **97** 207803(4)

Kariyo S, Brodin A, Gainaru C, Hermann A, Schik H, Novikov V N and Rossler E A 2008 From simple liquid to polymer melt. Glassy and polymer dynamics studied by fast field cycling NMR relaxometry: low and high molecular weight limit *Macromolecules* **41** 5313-21

Kariyo S, Brodin A, Gainaru C, Hermann A, Hintermeyer J, Schik H, Novikov V N and Rossler E A 2008 From simple liquid to polymer melt. Glassy and polymer dynamics studied by fast field cycling NMR relaxometry: Rouse Rgime *Macromolecules* **41** 5322-32

Brodin A 2008 Segmental versus chain dynamics of linear polymers *J. Chem. Phys.* **128** 104901(8)

Herrmann A, Kariyo S, Elfadl A A, Meier R, Gmeiner J, Novikov V N and Rossler E A 2009 Universal polymer dynamics revealed by field cycling 1H NMR *Macromolecules* **42** 5236-43

Elfadl A A, Kahlau R, Herrmann A, Novikov V N and Rossler E A 2010 From Rouse to fully established entanglement dynamics: A study of polyisoprene by dielectric spectroscopy *Macromolecules* **43** 3340-51

[33] Brereton M G 1989 NMR transverse relaxation function calculated for the Rouse model *Macromolecules* **22** 3667-74

Brereton M G 1990 NMR transverse relaxation function calculated for constrained polymer chains: applications to entanglements and networks *Macromolecules* **23** 1119-31

Brereton M G 1991 An exact expression for the transverse nuclear magnetic resonance relaxation of a dynamic scale invariant polymer chain governed by a single relaxation time *J. Chem. Phys.* **94** 2136-42

Klein P G, Adams C H, Brereton M G, Ries M E, Nicholson T M, Hutchings L R, Richards R W 1998 Rouse and reptation dynamics of linear polybutadiene chains studied by H2 NMR transverse relaxation *Macromolecules* **31** 8871-7

Klein P G and Ries M E 2003 The dynamics and physical structure of polymers obove the glass transition – transverse relaxation studies of linear chains, star polymers and networks *Progr. Nucl. Magn. Res.* **42** 31-52

[34] Lipari G and Szabo A 1982 Model-Free Approach to the Interpretation of Nuclear Magnetic resonance Relaxation in Macromolecules. I. Theory and Range of Validity *J. Am. Chem. Soc.* **104** 4546-59

Lipari G and Szabo A 1982 Model-Free Approach to the Interpretation of Nuclear Magnetic resonance Relaxation in Macromolecules. II. Analysis of Experimental Results *J. Am. Chem. Soc.* **104** 4559-70

Halle B 2009 The physical basis of model-free analysis of NMR relaxation data from proteins and complex fluids *J. Chem. Phys.* **131** 224507(22)

International Conference on Resonances in Condensed Matter: Altshuler100 IOP Publishing
Journal of Physics: Conference Series **324** (2011) 012032 doi:10.1088/1742-6596/324/1/012032

Gaussian fluctuations in an ideal bose-gas – a simple model

A Petrova[1,2], O Nedopekin[2], D Tayurskii[2] and Q A Wang[1]

[1]LUNAM Universite, ISMANS, Laboratoire de Physique Statistique et Systeme Complexes, 44, Avenue F.A. Bartholdi, Le Mans 72000, France

[2]Institute of Physics, Kazan Federal University, Kazan 420008, Russia

E-mail: npetrova131@gmail.com

Abstract. Based on the canonical ensemble, we suggested the simple scheme for taking into account Gaussian fluctuations in a finite system of ideal boson gas. Within framework of scheme we investigated the influence of fluctuations on the particle distribution in Bose -gas for two cases - with taking into account the number of particles in the ground state and without this assumption. The temperature and fluctuation parameter dependences of the modified Bose-Einstein distribution have been determined. Also the dependence of the condensation temperature on the fluctuation distribution parameter has been obtained.

1. Introduction

The experimental possibilities of detecting few and even single atoms confined in magnetic or optical traps with high precision open new perspectives to study the correlations and fluctuations in atomic and molecular gases at the mesoscopic level. Studying fluctuations in statistical systems is important from several points of view. For instance, the stability of a system and its way of reaching the equilibrium state depend drastically on the value of fluctuations [1].

The fluctuations in a Bose-Einstein condensate have been studied in a systematic way in the case of the ideal Bose gas in a box [2-4], and, more recently, in the presence of a harmonic trap [5-7]. The uniform ideal gas is an unstable system, with anomalously strong particle fluctuations. Switching on atomic interactions stabilizes the system, making the fluctuations normal. But further strengthening interactions leads to the condensate depletion. Contrary to the uniform case, the trapped ideal Bose gas is stable, having normal fluctuations. Atomic interactions play only the destructive role, depleting the condensate [8].

The aim of present research is the studying of the overlapping effects coming from Gaussian noise and quantum statistics of the ideal Bose-gas. We investigate the modification of Bose-Einstein distribution due to the thermal fluctuations obeying to the normal distribution and the changes of the dependence of Bose-Einstein condensation temperature on the fluctuation distribution parameter.

In the vicinity of critical points, there always exists a region, where the system experiences strong fluctuations [1]. In finite systems, the phase transition is of crossover type. Around the point of crossover, there can arise strong fluctuations, similarly to those appearing in the vicinity of the critical points. Then the region around a crossover point, where fluctuations are strong, can be called crossover region. Bose-Einstein condensation of trapped atoms is always a crossover. Considering phase transitions under varying temperature, one usually defines the critical region by means of the Ginzburg temperature T_G, such that in the critical region $T_G < T < Tc$ fluctuations are strong, while essentially below T_G, they are suppressed [9]. In general, one may distinguish the longitudinal and transverse fluctuations of an order parameter. The transverse, or directional, fluctuations often survive much longer, when lowering temperature below Tc, than the longitudinal, or size, fluctuations. Therefore the actual critical region is characterized rather by the transverse then longitudinal fluctuations [10].

In Bose-Einstein condensates, one may also distinguish two types of fluctuations. The size fluctuations of the order parameter correspond to density fluctuations. The transverse fluctuations are usually associated with phase fluctuations. In the vicinity of the condensation temperature Tc, both these fluctuations are strong. Lowering temperature below Tc first diminishes the density fluctuations, while the phase fluctuations remain yet strong. Lowering temperature further, one comes to a point T_f, where phase fluctuations are getting suppressed as well. Thus, the crossover region is defined by the temperature interval $T_f < T < Tc$. Density fluctuations always excite current fluctuations. However, there can exist current fluctuations those do not initiate density fluctuations. This explains why the crossover region is mainly defined by the phase (current) fluctuations, but not by the density fluctuations.

The development of physics at the mesoscopic level and its applications (primarily in the field of nanotechnology) is associated with an increase of interest in learning more and more chaotic (including low-dimensional) systems. For such systems, even in thermal equilibrium or close to it the fluctuations of physical quantities are significant. But till now the theoretical description of the fluctuations influence cannot be considered as satisfactory. We should note here that fluctuations cannot be studied thermodynamically at extremely low temperatures or in the case of very rapid changes of their magnitude. Also at sufficient low temperatures pure quantum fluctuations begin to play an important role.

2. The study of the noise effect on the Bose-Einstein distribution

Statistics of weakly interacting Bose gas attracted much attention within recent years due to the experimental realization of Bose-Einstein condensation (BEC) of trapped gases. Statistical physics of Bose-Einstein condensation (BEC) is usually considered by using the grand canonical ensemble (thermodynamic coordinates μ, V, T) in the thermodynamic limit, i.e., for a macroscopic system with the infinite number of particles and volume, but the finite concentration. However, the trapped BEC experiments deal with the finite number of particles ($N \sim 10^2$-10^6), which makes the system mesoscopic. This is the first reason to develop a microscopic theory of BEC in mesoscopic systems by means of the canonical ensemble (thermodynamic coordinates N, V, T) [11].

The grand-canonical description is not appropriate for the analysis of BEC fluctuations, even in the case of mesoscopic ideal Bose gas [12]. For example, the widely used grand-canonical approach yields the grand-canonical catastrophe. The restricted ensemble approach, which fixes the amplitude and the phase of the condensate wave function, is unable to analyze the fluctuation problem at all. To study fluctuations, one should fix only external macroscopic and global topological parameters of the system, such as temperature, pressure, number of particles, superfluid flow pattern, boundary conditions, etc. Such theory is more complex than it was believed commonly, because the result must clearly lead to the dependence on a finite number of particles in the system and its size.

In present work we are interested in the effects of noise on the Bose-Einstein distribution. To study these effects we propose to use the convolution of Bose-Einstein and Gaussian distributions. The main problem consists in the normalization of the convolution of quantum Bose-Einstein statistics and Gaussian function of noise. The standard expression for probability of state with energy ε is

$$n_\alpha(\varepsilon) = \frac{1}{Z}\int_0^\infty n(x) f_\alpha(x-\varepsilon)\,dx\,, \tag{1}$$

where Z is normalization coefficient that could be found from this condition:

$$\int_0^\infty d\varepsilon n(\varepsilon) g(\varepsilon) = 1 \,, \tag{2}$$

$n(\varepsilon)$ stands for the quantum distribution, $g(\varepsilon)$ denotes the density of states, $f_\alpha(x-\varepsilon) = (2\pi\sigma)^{-1/2} \exp\left(-(x-\varepsilon)^2 / 2\sigma^2\right)$ is the usual Gaussian distribution.

This normalization (by means of Z) is straightforward for case without fluctuations. But when fluctuations are introduced we have to take into account the change of chemical potential due to the noise. In self-consistent calculations the normalization will be done by means of the new value of the chemical potential. It is important to note here that the average number of particles in the system is conserved such kind of calculations.

Now we illustrate the proposed procedure for the case of ideal boson-gas, having N particles and confined to the box of volume V. As usually the density of states in 3D space can be written as:

$$\sum_{\vec{k}} \frac{1}{e^{\frac{\varepsilon(\vec{k})-\mu}{\theta}} - 1} \xrightarrow{3D} \frac{V}{(2\pi)^3} \int_0^\infty \frac{1}{e^{\frac{\varepsilon(\vec{k})-\mu}{\theta}} - 1} d^3\vec{k} = \frac{V}{(2\pi)^3} \int_0^\infty d\vec{k}\vec{k}^2 \int_\Omega d\Omega \frac{1}{e^{\frac{\varepsilon(\vec{k})-\mu}{\theta}} - 1} =$$

$$= \frac{V}{(2\pi)^2} \left(\frac{2m}{\hbar^2}\right)^{3/2} \int_0^\infty \frac{\sqrt{\varepsilon}}{e^{\frac{\varepsilon(\vec{k})-\mu}{\theta}} - 1} d\varepsilon, \tag{3}$$

where $\varepsilon(\vec{k}) = \dfrac{\hbar^2 \vec{k}^2}{2m}$. Thus, density of states has the form:

$$g(\varepsilon) = \frac{V}{(2\pi)^2} \left(\frac{2m}{\hbar^2}\right)^{3/2} \sqrt{\varepsilon} = \frac{\sqrt{\varepsilon}}{\varepsilon_0^{3/2}} \,, \tag{4}$$

where

$$\varepsilon_0 = \left(\frac{4\pi^2}{V}\right)^{2/3} \times \frac{\hbar^2}{2m} \tag{5}$$

and m is the mass of free particle.

Using the general expression for Bose-Einstein distribution

$$n(\varepsilon) = \int_0^\infty \frac{g(\varepsilon)}{e^{\frac{\varepsilon-\mu}{kT}} - 1} d\varepsilon \tag{6}$$

one can write the convolution of the Gaussian noise and (6) for the total number of particles in the system in the following form:

$$N = \int_0^\infty \int_0^\infty \frac{f(x-\varepsilon)}{\exp\left(\frac{x-\mu}{kT}\right) - 1} dx g(\varepsilon) d\varepsilon, \tag{7}$$

where

$$f(x-\varepsilon) = \frac{1}{\sigma \sqrt{\frac{\pi}{2}\left(1 + erf\left[\frac{\varepsilon}{\sigma\sqrt{2}}\right]\right)}} \exp\left[-\frac{(x-\varepsilon)^2}{2\sigma^2}\right] \tag{8}$$

represents the probability density of fluctuations with variance σ^2.

The next step is the finding of a new value of the chemical potential under assumption of the conservation of the total particle number (7). It is worth to note that all values of energy x, ε, fluctuation parameter σ, temperature $\theta = kT$ in expression (7) are measured in energetic units. Now we introduce new unit of energy $\varepsilon_0 = \left(4\pi^2 / V\right)^{2/3} \times (\hbar^2 / 2m)$ and will measure all energy units (fluctuation parameter and temperature) by this quantity. The numerical calculations of the fluctuation effects were carried out by use the *Mathematica* software package.

The calculated particle distributions over energy at various temperatures are shown on Figure 1 for the ideal bose-gas with N=100. The analysis of the obtained data allows to make the following conclusions. As the system temperature decreases the probability to find the atoms in the ground state increases. Maximum of the particle's distribution shifts to lower energies with decrease the temperature. The Bose-Einstein distribution diffuses at high temperatures.

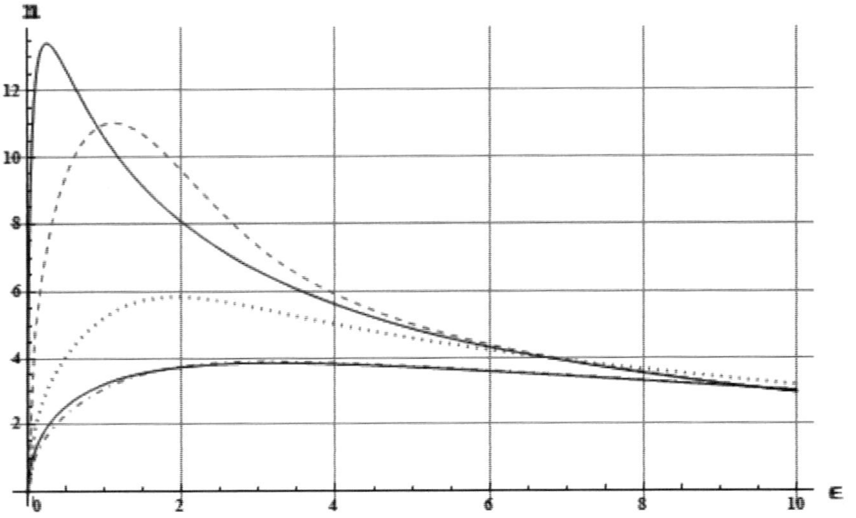

Figure 1. The particle number distributions over energy for different temperatures T. The upper solid line is the usual Bose-Einstein distribution at T=14ε_0, lower solid line is the usual Bose-Einstein distribution at T=20 ε_0 ; dashed, dotted and chain lines show the distributions modified by fluctuations at temperatures $14\varepsilon_0, 17\varepsilon_0, 20\varepsilon_0$ respectively (N=100, σ=1).

In the figure 2 the calculated particle distributions over energy at various values of fluctuation parameter σ are shown.

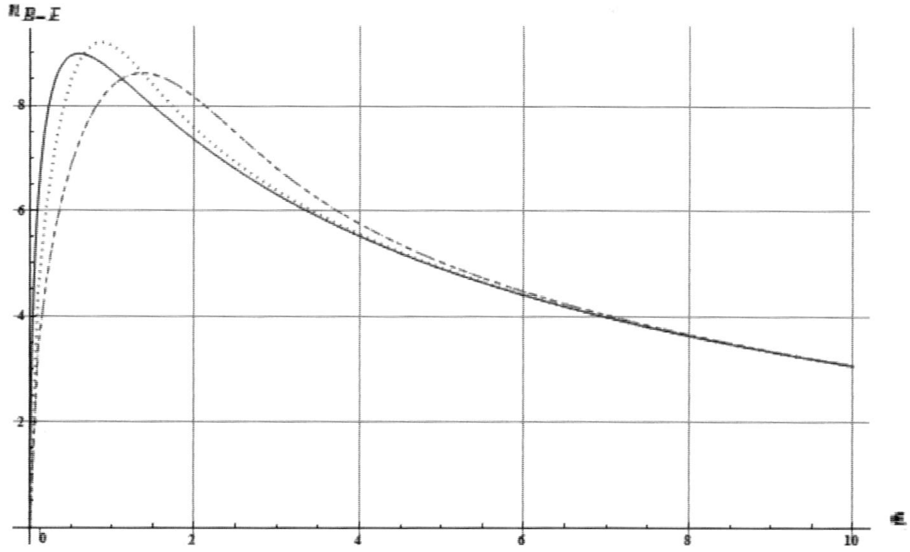

Figure 2. The modified by fluctuations particle distributions over energy for different values σ. The solid, dotted and chain lines of Bose-Einstein distributions at $\sigma = 0.1, 0.5, 1.0$, respectively (N = 100 and T = 15 ε_0). For comparison the black color line shows the distribution of particles without fluctuations.

3. The study of the noise effect on the Bose-Einstein condensation

As noted before the chemical potential of the Bose-gas is always negative, and $\mu\ (\theta = 0) = 0$ (the energy of the particles is measured from zero). This means that as temperature decreases the chemical potential increases (decreases by absolute value) and at a certain temperature θ_0 it reaches zero value and remains equal to zero up to zero temperature. In thermodynamic limit at lowering the temperature in the range closed to $\mu = 0$ the number of particles in all states except the ground one is reduced drastically. This process is called Bose-condensation. To take into account the macroscopic number of particles in the ground state it is usual to separate the large term in the expression for the total number of particles in bose-gas [13]:

$$N = N_0(\theta) + \int_0^\infty \frac{g(\varepsilon)d\varepsilon}{\exp((\varepsilon - \mu)/\theta) - 1} = N_0(\tau) + \frac{\gamma V M^{3/2}}{\sqrt{2}\pi^2\hbar^3}\zeta\left(\frac{3}{2}\right)\Gamma\left(\frac{3}{2}\right)\theta^{3/2}. \quad (9)$$

This expression is used in the standard approach for determining the number of particles in the Bose-Einstein condensate. Let's suppose that there are not fluctuations in the ground state, otherwise they destroy the condensate. In the case with fluctuations the total number of particles in the system is written as follows:

$$N = N_0 + \int_0^\infty \int_0^\infty g(\varepsilon) \frac{f(x-\varepsilon)}{\exp(\frac{x-\mu}{kT}) - 1} dx d\varepsilon, \quad (10)$$

$$N_0 = \frac{1}{\exp(-\frac{\mu}{kT}) - 1}, \quad (11)$$

The chemical potential can be found from (9) and (10) in the case without fluctuation as well as with fluctuations. The corresponding particle distributions over energy for low temperatures with taking into account condensate fraction are shown in figures (3) and (4) (compare with (1) and (2)).

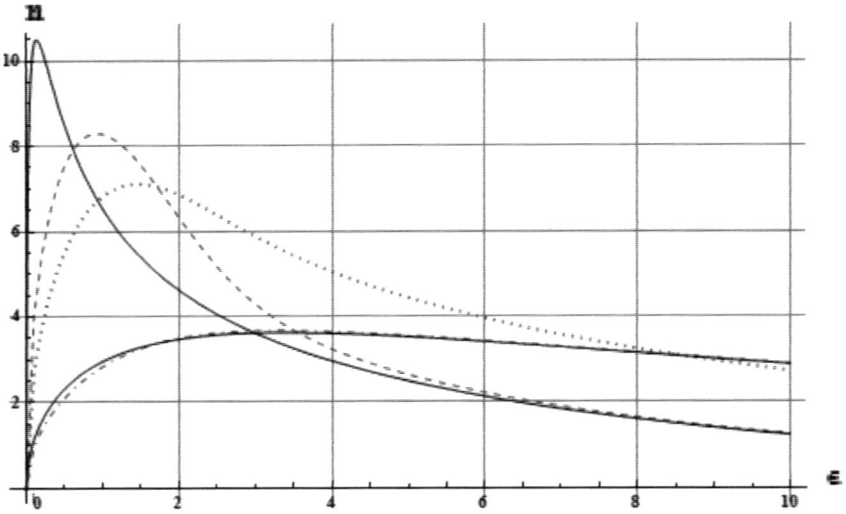

Figure 3. The particle number distributions over energy for different temperatures T. The upper solid line is the usual Bose-Einstein distribution at $T=8\varepsilon_0$, lower solid line is the usual Bose-Einstein distribution at $T=20\,\varepsilon_0$; dashed, dotted and chain lines show the distributions modified by fluctuations at temperatures $8\varepsilon_0, 14\varepsilon_0, 20\varepsilon_0$ respectively (N=100, $\sigma = 1$).

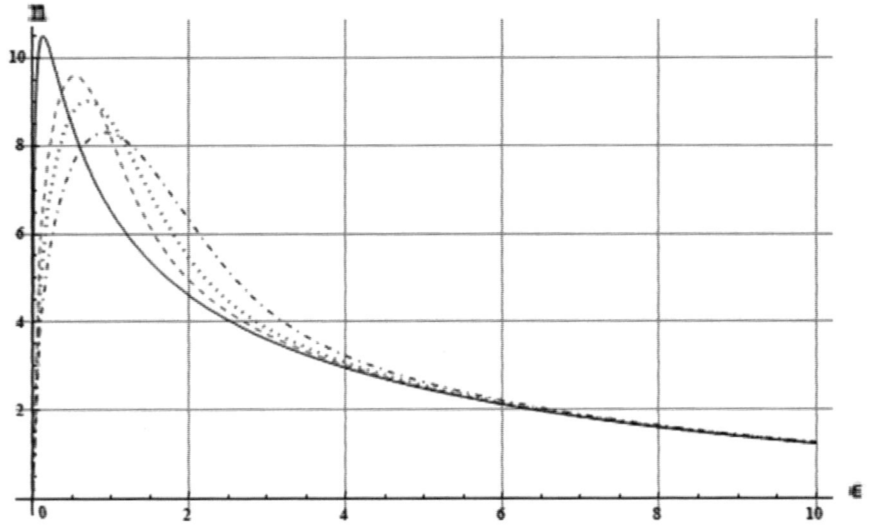

Figure 4. The modified by fluctuations particle distributions over energy for different values σ . The solid, dotted and chain lines of Bose-Einstein distributions at $\sigma = 0.5, 0.7, 1.0$, respectively (N = 100 and T = $8\,\varepsilon_0$). For comparison the black color line shows the distribution of particles without fluctuations.

4. The temperature of Bose- Einstein condensation

The criterion for Bose-Einstein condensation is the equality of chemical potential to zero. When one writes convolution of Bose-Einstein distribution and Gaussian function the logarithmic divergence appears at chemical potential equal to zero. It suggests that for the considering the fluctuation effects we cannot take the chemical potential equal to zero, as it was done for an equilibrium system. For an ideal bose-gas consisting of *N* particles with mass *M* confined in a rigid container of volume *V*, condensation occurs at the temperature [13]:

$$T_C = \frac{h^2}{2\pi kM}\left[\frac{1}{2.612}\frac{N}{V}\right]^{2/3} \tag{12}$$

In the case of fluctuations the total number of particles is written as:

$$N = N_0(\theta) + \int_0^\infty\int_0^\infty \frac{g(\varepsilon)f(\varepsilon - x)}{Exp(\frac{x-\mu}{\theta})-1}d\varepsilon dx$$

$$= N_0(\theta) + \int_0^\infty dxn(x)\int_0^\infty d\varepsilon g(\varepsilon)f(x-\varepsilon) = N_0(\theta) + \int_0^\infty dxn(x)\tilde{g}(x) \quad , \tag{13}$$

$$\int_0^\infty d\varepsilon g(\varepsilon)f(x-\varepsilon) = \tilde{g}(x) \propto \sigma^{1/2}\Gamma(3/4)x^0 + \sigma^{-1/2}\Gamma(5/4)x^1 \quad , \tag{14}$$

$$N \propto N_0(\theta) + \int_0^\infty \frac{\sigma^{1/2}\Gamma(3/4)}{\exp(\frac{x-\mu}{\theta})-1}dx \propto \sigma^{1/2}\Gamma(3/4)\theta\int_0^\infty x^{-1}\,dx \rightarrow \ln x|_0^\infty \tag{15}$$

Function $\tilde{g}(x)$ has the linear dependence like $(ax + b)$, where a and b are the constants. The exponent in Bose-Einstein distribution can be expanded in Taylor series and the first non-vanishing term is proportional to x/θ. Thus, we obtained the logarithmic divergence at $x=0$. If we set $\mu<0$ then the integral will be taken only numerically.

Let's consider the graph method of finding the condensation temperature. We are interested in the dependence of temperature Bose-Einstein condensation on fluctuation parameter [14]. Let's consider the grand canonical ensemble. We separate a very large term, corresponding to number of particles in the ground state:

$$N = N_0 + ZI(\mu), \tag{16}$$

where

$$Z = (\theta)^{3/2} \tag{17}$$

$$I(\mu) = \int_0^\infty\int_0^\infty \sqrt{\varepsilon}\frac{f(x-\varepsilon)}{\exp(x-\frac{\mu}{\theta})-1}dxd\varepsilon \tag{18}$$

All parameters in function $I(\mu)$ are measured in ε_0 units. Further, the equation (16) is divided into two parts. These parts contain the functions dependent on the temperature. This equation (the critical condition of finding the condensation temperature) can be solved graphically:

$$Z(\theta) = \frac{N}{I(\mu,\theta)} \tag{19}$$

The chemical potential in the fluctuation case is not equal to zero because of the divergence of the double integral (7). For each value of the chemical potential from the defined range we can numerically solve equation (19). Thus, we obtained the dependence of the chemical potential on the temperature. Extrapolating this curve to its intersection with the ordinate axis we got the value of the condensation temperature:

$$T_C = \frac{\varepsilon_0 * \theta}{k_b} \tag{20}$$

for the given fluctuation parameter at the point where the chemical potential is equal to zero.

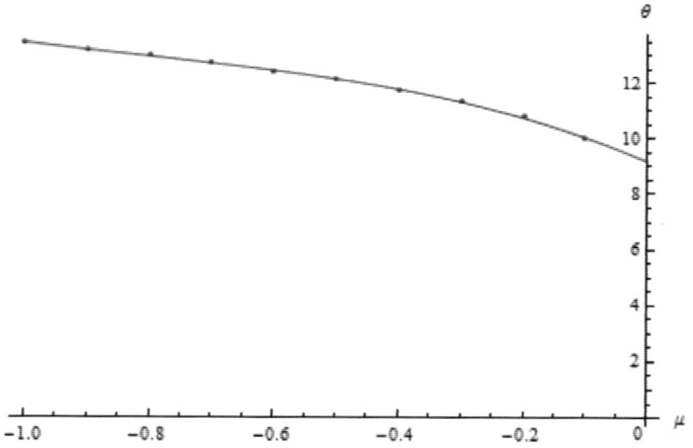

Figure 5. The extrapolated temperature dependence of the chemical potential with taking into account the fluctuations ($\sigma = 1$).

Thus, changing the fluctuation parameter, we can study the condensation temperature changes. The corresponding plot is shown in figure 6. If we extrapolate the curve of condensation temperature dependence on the fluctuation parameter, we shall obtain the value of the condensation temperature in point, where the fluctuation parameter is equal to zero. Following this procedure for the case of ^{4}He atoms we have obtained the condensation temperature equal 3.12 K which is close to the theoretically calculated (3.07 K) [15].

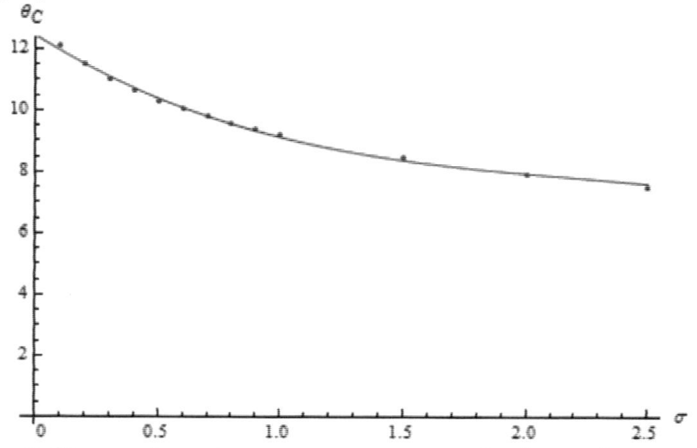

Figure 6. The condensation temperature versus the fluctuation parameter.

A more accurate method for the BEC temperature determination is the finding chemical potential from the condition (in dimensionless parameters normalized to ε_o:

$$N = \frac{1}{e^{-\bar{\mu}} - 1} + \tilde{\theta}^{3/2} \int_0^\infty \int_0^\infty \sqrt{\bar{\varepsilon}} f(\bar{x} - \bar{\varepsilon}) d\bar{\varepsilon} \frac{1}{e^{\bar{x}-\bar{\mu}} - 1} d\bar{x}, \quad (21)$$

In the figure 7 the same data as in figure 6 are shown but in K degrees. Three points obtained by extrapolating the dependence of the chemical potential on the temperature (the value of chemical potential at different temperatures obtained from expression (21)) have been added for σ= 0.1, 1 and 2.5.

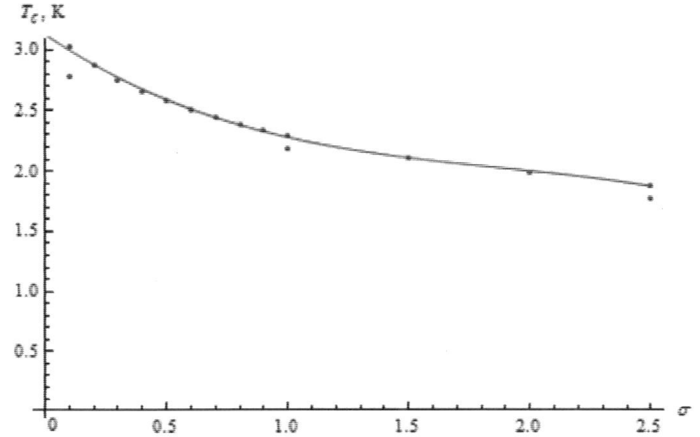

Figure 7. The solid line shows the dependence of the condensation temperature on the fluctuation parameter σ (the dependence of the chemical potential on the temperature obtained by equation (19)). Three points show the dependence of the condensation temperature on the fluctuation parameter σ = 0.1, 1 and 2.5 (the dependence of the chemical potential on the temperature obtained from (21)).

It can be concluded that the temperature of Bose-Einstein condensation decreases as the fluctuation parameter increases. The figure 7 shows that taking into account the fluctuations can give the value of the calculated condensation temperature more close to the experimental observed one (the experimental value 2.17 K for ^{4}He can be obtained at the fluctuation parameter σ approximately equal to 2.4).

Conclusion

We have investigated in detail the effect of fluctuations in the form of Gaussian noise on the different physical quantities of an ideal bose-gas such as the chemical potential and the condensation temperature. The proposed simple model based on the convolution of Gaussian noise and Bose-Einstein distribution allowed to obtain particle distributions over energy for different temperatures and fluctuation parameter values. Using different methods, we have found the condensation temperature as the function of the fluctuations parameters.

Acknowledgement

The work was supported by the Ministry of Science and Education of Russia. A.P. thanks Region Pays de la Loire for partial individual financial support (MOA0020).

References

[1] Yukalov V I 2004 Principal problems in Bose-Einstein condensation of dilute gases *Laser Physics Letters* **1** 435-461
[2] Ziff R M ,Uhlenbeck G E and Kac M 1997, *Phys. Rev.* **32**, 169

[3] Hauge E H 1969 *Physica Norvegica* **4**, 19
[4] Fujiwara I, ter Haar D and Wergeland H 1970 *J. Stat. Phys*.**2**, 329
[5] Politzer H D 1996 *Phys. Rev. A* **54**, 5048
[6] Gajda M and Rzazewski K 1997 *Phys. Rev. Lett.* **78**, 2686; Navez P, Bitouk D, Gajda M, Idziaszek Z and Rzazewski K 1997 *Phys. Rev. Lett.* **79**, 1789
[7] Grossmann S and Holthaus M, Report No. *Cond.mat/* 9709045 and Report No. *Cond.mat/* 9709065
[8] Giorgini S, Pitaevskii L P and Stringari S 1997 Anomalous fuctuations of the condensate in interacting Bose gases *Phys.Rev.Lett.* **80** 5040
[9] Landau L D and Lifshitz E M 1980 *Statistical Physics* (United Kingdom:Pergamon/ Oxford)
[10] Kleinert H 2000 *Phys. Rev. Lett.* **84**, 286
[11] Landau L D and Lifshitz E M 2005 *Statistical physics* vol 5 (Moscow: Fizmatlit, 5th Edition) ed L P Pitaevskii p 616
[12] Dorfman K E, Kim M and Svidzinsky A A 2011 Condensate statistics and thermodynamics of weakly interacting Bose gas: Recursion relation approach *Phys.Rev.A* **83** 033609
[14] Bagnato V, Pritchard D E, Kleppner D 1987 Bose-Einstein condensation in an internal potential *Phys. Rev. A* **35** N 10 4354-58
[15] Wang, Yongli Ma and Jizhou He 2011 Thermodynamics of finite Bose Systems: an exact canonical-ensemble treatment with different traps *J. Low Temp. Phys*.**162** 23–33
[16] Wang Q A , Pezeril M, Le Mehaute A 2000 Nonextensive boson gas and specific heat of ^{4}He superfuid *Physica A* **278** 337-346
[17] Lawani A A, Le Meur J, Tayurskii D, Kaabouchi A El, Nivanen L, Minisini B, Tsobnang F, Pezeril M, Le Mehaute A, Wang Q A 2007 Nonextensive approach to Bose-Einstein condensation of trapped interacting boson gas *J. of Low Temperature Physics* **150**.N3-4 605-611

NMR, high frequency EPR and magnetization studies of $YF_3:Tm^{3+}$ and TmF_3

A V Savinkov[1,3], G S Shakurov[2], S L Korableva[1], A V Dooglav[1], M S Tagirov[1], H Suzuki[3], K Matsumoto[3] and S Abe[3]

[1] Kazan Federal University, 420008 Kazan, Russia
[2] Kazan Physical-Technical Institute, 420029 Kazan, Russia
[3] Kanazawa University, Kakuma-machi, 920-1192 Kanazawa, Japan

E-mail: Andrey.Savinkov@gmail.com

Abstract. Magnetic properties of single crystal and powder samples of thulium fluoride, TmF_3 (orthorhombic *Pnma* space group), and single crystals of YF_3 doped with the Tm^{3+} ions are studied by NMR, high-frequency EPR and dc-magnetometry. It is shown that TmF_3 is a Van Vleck paramagnet. Zero field splitting between two lowest ground state energy levels (ground $^3H_6$ multiplet) of Tm^{3+} ion in TmF_3 crystal lattice is found to be ~6.5 cm^{-1}. The ^{19}F nuclear spin-lattice relaxation in TmF_3 at liquid helium temperatures is driven by the fluctuating magnetic fields created by Tm^{3+} ions occupying the lowest excited singlet.

1. Introduction

The heavy-rare-earth fluorides, from SmF_3 to YbF_3, having an orthorhombic β-YF_3 structure with the space group *Pnma* (62), show a variety of magnetic and optical properties and attract interest for several reasons. The heavy-rare-earth fluorides are model systems for studies of magnetic ordering in rare-earth insulators with competing dipole and exchange interactions. The magnetic ordering was observed in ErF_3 (T_N = 1.05 K) [1], TbF_3 (T_C = 3.95 K) [2, 3], HoF_3 (T_C = 0.53 K) [4], DyF_3 (T_C = 2.55 K) [5, 6] and GdF_3 (T_C = 1.25 K) [7]. Crystals of the heavy-rare-earth fluorides may be also suitable as potential solid-state laser materials and scintillators. Optical spectra of RF_3 and rare earth doped isostructural $YF_3:R^{3+}$ crystals (R = Eu, Tb, Dy, Ho, Er, and Yb) have been measured in Refs. [8-12]. At the present time the PrF_3 and other rare-earth fluorides have attracted more and more attention due to synthesis of the hollow fullerene-like nanoparticles RF_3 (R = Pr, Nd, Sm) [13,14] which can have widespread potential for practical applications.

Fluoride TmF_3 (Tm^{3+} is non-Kramers rare-earth ion, electronic configuration of the partially filled shell $4f^{12}$) has been the subject of quite a few studies in the past. As a result, scanty information exists on magnetic properties of rare-earth ions in both concentrated TmF_3 and diluted $YF_3:Tm^{3+}$ compounds. The Van Vleck paramagnetism often occurs in systems with non-Kramers rare-earth ions (Pr^{3+}, Eu^{3+}, Tb^{3+}, Ho^{3+} and Tm^{3+} as well) in which the crystalline electric field lifts the degeneracy of the *J* multiplets, giving rise to splittings which are typically of the order of 10 – 100 cm^{-1} with the ground state being the electronic singlet. Since conventional EPR microwave frequencies correspond to much smaller energies observation of EPR in Van Vleck paramagnets is rare. The other principal method to study the magnetic properties of Van Vleck paramagnets is a enhanced nuclear magnetic resonance (ENMR). However, often the enhanced nuclear magnetic moment of the rare earth ion is so big that the nuclear dipole-dipole and exchange interactions make ENMR observation difficult, at least by the pulsed NMR techniques. In this report we present magnetic properties of thulium fluoride (TmF_3) studied by high-frequency EPR, ^{19}F NMR methods and dc-magnetometry.

2. Experiment

The TmF_3 and YF_3 : 0.5 at % Tm^{3+} single crystals were grown by the Bridgman method in carbon crucibles in the atmosphere of the high purity argon at a pulling rate of 1 mm/h. High purity the TmF_3 and YF_3 powders (99,99% grade) were used as starting materials. Additionally, the growth atmosphere was fluorinated by burning the tetra-fluorine-ethylene. The TmF_3 single crystals were grown for magnetometry measurements and the YF_3 : 0.5 at % Tm^{3+} single crystal was grown for measurements by the high-frequency EPR. All single crystals have a needle shape. The samples were oriented in the sample holder using the X-ray diffraction pattern with the accuracy of 3^0. The TmF_3 powder for NMR measurements was prepared by grinding of TmF_3 single crystals in agate mortar.

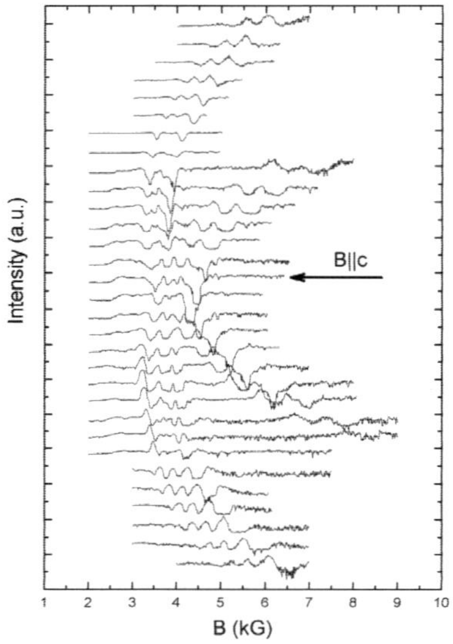

Figure 1. High frequency EPR spectra measured for YF_3:Tm^{3+} (0.5%) single crystal in the field $B_0 \perp b$ (angular step in ac-plane is 5^0) at frequency 207 GHz.

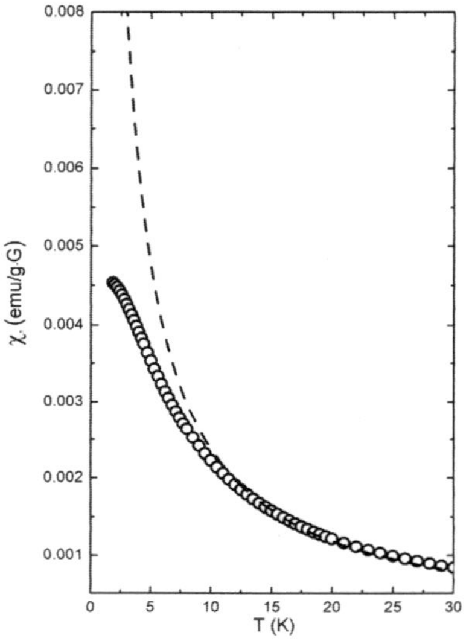

Figure 2. Measured temperature dependence of the static magnetic susceptibility in TmF_3 powder (open symbols). Curie's Law simulation is given by dashed line.

The dc-SQUID magnetometer MPMS (Quantum Design) was used for measurements of the magnetic susceptibility in the temperature range 1.8 – 300 K at applied magnetic field 0.01 T. High frequency EPR spectra of Tm^{3+} ion in diluted YF_3:Tm^{3+} (0.5% Tm) single crystal were taken with high-frequency tunable EPR spectrometer [15] in magnetic fields from 0 to 0.9 T at temperature 4.2 K. In the high frequency EPR measurements radiofrequency magnetic field B_1 was applied parallel to static magnetic field B_0, i.e. $B_1 \| B_0$. For the detection of the ^{19}F NMR signal a standard NMR pulse spectrometer was used employing spin-echo technique and echo recording in quadrature. The NMR powder spectra of ^{19}F were taken in TmF_3 powder at T = 4.2 K and 1.5 K as the magnetic field dependence of the spin echo intensity at a frequency 16.5 MHz. The measurements were performed in magnetic field B_0 from 0 to 1.05 T. The ^{19}F nuclear longitudinal relaxation time T_1 was measured by plotting the ^{19}F nuclear spin-echo intensity as a function of the time between a saturating and $\pi/2$–π probing pulses; the nuclear transverse relaxation time T_2 was measured by plotting the ^{19}F spin-echo intensity as a function of the interval time between $\pi/2$ and π pulses.

3. Experimental results and discussion

Figure 1 shows high frequency EPR spectra for YF_3:Tm^{3+} (0.5%) single crystal in the field $B_0 \perp b$ (angular step in *ac*-plane was 5^0) at frequency 207 GHz and T = 4.2 K. The hyperfine structure of the EPR spectra consisting of two resonant lines was observed. This gives evidence, that the lines belong to Tm^{3+} ions (^{169}Tm, 100 % natural abundance, I = 1/2). The angular dependence of the EPR spectra revealed the occurrence of two magnetically non-equivalent sites of Tm^{3+} ions in TmF_3 crystal structure. High frequency EPR spectrum in the field directed in the *ab*-plane wasn't observed because the EPR signals were out of magnetic field range of our magnet.

Dependence of the resonance transitions on the magnetic field implies that the transitions occur between two lowest singlet states of Tm^{3+}. The EPR spectra are described by the effective spin Hamiltonian $H = g\mu_B S_z B_{0z} + \Delta S_x$, where Δ is the zero-field splitting (actually, the energy gap between two lowest singlets of Tm^{3+}), μ_B is the Bohr magneton, B_{0z} is the component of the external magnetic field along the local *z*-axis. The EPR experiment shows that these two lowest singlets of the $^3H_6$ ground multiplet of Tm^{3+} ion are separated by the energy gap Δ ~ 195 GHz ~ 6.5 cm^{-1}. Such structure of electron energy levels is typical for Van-Vleck paramagnets. The magnetic moments corresponding to this quasi-doublet (two singlets) are determined by the g-factor ~13.5 and lie in the *ac*-plane along the directions declined by ~ ±23^0 from the *c*-axis. Defined value of the g-factor is close to that of the free Tm^{3+} ion. It is also evidenced that observed EPR signals belong to Tm^{3+} ions.

Some of EPR lines were distorted and split, weak satellite signals were observed. The extra lines can appear due to forbidden electron-nuclear transitions ($\Delta M_I \neq 0$). Wave functions of Tm^{3+} lowest singlets depend on the magnetic field strength, therefore probabilities of transitions are also field dependent. Besides, the strong dispersion of microwave radiation in the needle-shaped crystals occurs near the resonance, this can change the refractive index of the sample and thus change the microwave power reflected to a detector.

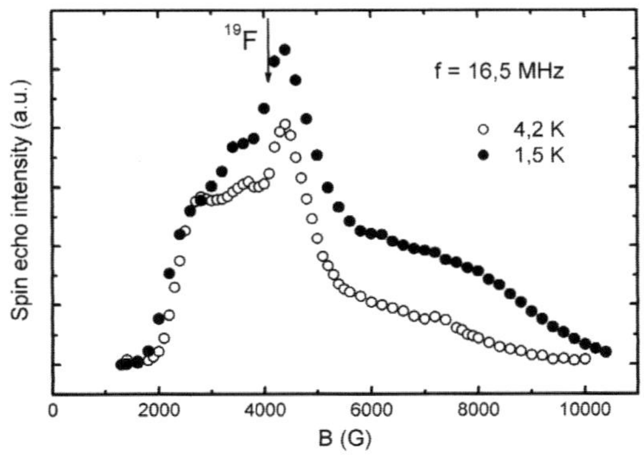

Figure 3. The ^{19}F NMR spectra measured in TmF_3 powder at T = 1.5 K (solid symbols) and 4.2 K (open symbols). Arrow points to the "free" ^{19}F NMR line position.

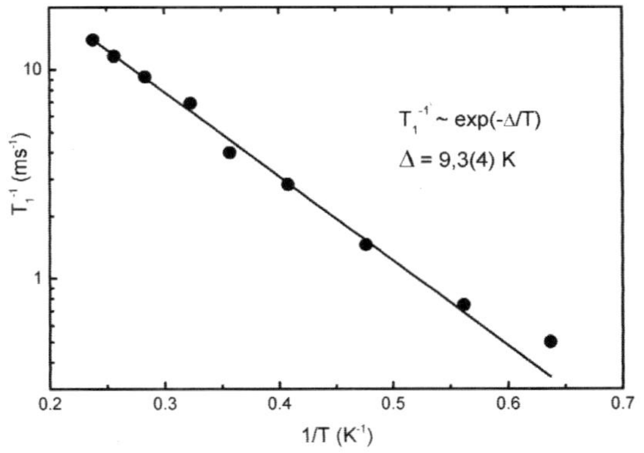

Figure 4. Temperature dependence of ^{19}F nuclear longitudinal relaxation rate. Solid line represents the result of fitting by the function $T_1^{-1} = A\exp(-\Delta/T)$.

Temperature dependence of the static magnetic susceptibility in TmF_3 powder is shown in Fig.2. One can see that the susceptibility deviates from the Curie law at temperatures below ~15 K. At temperatures below ~2 K the susceptibility tends to a constant value implying that only the ground state singlet is populated at T < 2 K. Such temperature behavior of dc-susceptibility is typical for Van-Vleck paramagnets.

Figure 3 shows the NMR powder spectra of ^{19}F (I = 1/2) for TmF_3 powder at T = 4.2 K and 1.5 K at a frequency 16.5 MHz. At low temperatures in the external magnetic field all fluorine nuclei are additionally subject to the internal quasi-static magnetic field (dipole and transferred) created by the adjacent Tm^{3+} electronic magnetic moments induced by the external field. This internal field depends on the direction of the external field in respect to the crystal axes. In general four magnetically nonequivalent sites of ^{19}F are possible in TmF_3 structure, so in single crystal one should observe four orientation-dependent ^{19}F NMR lines. These lines should be broad due to the distribution of the internal magnetic field. In powder sample the ^{19}F NMR spectrum should also be additionally broadened due to averaging of crystallites orientations in respect to the external magnetic field. This can be observed in the experimental spectrum of Fig.3 extending from about 2 to 10 kG. Such broad spectrum evidences on rather strong local magnetic fields on the ^{19}F sites due to dipole and transferred super-hyperfine interactions between the Tm^{3+} ions and fluorine nuclei.

The measured value of the transverse relaxation time T_2 of ^{19}F was about 18 μs at $T = 4.2$ K and ~63 μs at 1.5 K for all over the spectrum. Figure 4 shows the longitudinal relaxation rate T_1^{-1} of ^{19}F nuclei in the magnetic field 4.3 kG in the temperature range 1.5 – 4.2 K. The $T_1^{-1}(T)$ behavior is clearly exponential and can be well described by the function $T_1^{-1} \sim \exp(\Delta/T)$ with $\Delta = 9.3(4)$ K = 6.5(3) cm^{-1}. Such temperature dependence of the nuclear longitudinal relaxation for ligands with $I = 1/2$ is typical for Van-Vleck paramagnets [16] and is governed by the fluctuating magnetic field created by the Van-Vleck ion at ligands sites at temperatures when the excited electronic energy levels become populated. The parameter Δ in this dependence is equal to the splitting between the ground and first excited states of the Van-Vleck ion. Thus, the splitting between two lowest electron states of Tm^{3+} ion in TmF_3 obtained from the high frequency EPR is confirmed by ^{19}F NMR measurements.

Our attempts to observe ^{169}Tm NMR signal at helium temperatures were not successful, most likely due to extremely short transverse relaxation time of ^{169}Tm nuclei.

4. Conclusion

The results of the present work give unambiguous evidence that TmF_3 is a Van Vleck paramagnet. Both ^{19}F NMR and Tm^{3+} high frequency EPR measurements give the same splitting between the ground state and lowest excited level of Tm^{3+} ion: Δ~6.5 cm^{-1}.

References

[1] Kramer K, Romstedt H, Gudel H U, Fisher P, Murasik A and Fernandez-Diaz M T 1996 *Eur. J. Solid State Inorg. Chem.* **33** 273
[2] Holmes L, Guggenheim H J and Hull G W 1970 *Solid State Commun.* **8** 2005
[3] Holmes L and Guggenheim H J 1971 *J. physique, Colloque* **C1** 32
[4] Bleaney B, Gregg J F, Hill R W, Lazzouni M, Leask M J M and Wells M R 1988 *J. Phys. C: Solid State Phys.* **21** 2721
[5] Holmes L M, Hulliger F, Guggenheim H J and Maita J P 1974 *Phys. Lett. A* **50** 163
[6] Savinkov A V, Korableva S L, Rodionov A A, Kurkin I N, Malkin B Z, Tagirov M S, Suzuki H, Matsumoto K and Abe S 2008 *J. Phys.: Condens. Matter.* **20** 485220
[7] Katila T E, Typpi V K, Shenoy G K and Niinistö L 1972 *Solid State Comm.* **11** 1147.
[8] Caspers H H, Rast H E and J.L. Fry J L 1967 *J. Chem. Phys.* **47** 4505
[9] Kuroda H, Shionoja S and Kushida T 1972 *J. Phys. Soc. Japan* **33** 125
[10] Bumagina L A, Kazakov B N, Malkin B Z and Stolov A L 1977 *Fiz. Tverd. Tela* **19** 1073 (*Sov. Phys.Solid State* **19** 624).
[11] Davidova M P, Kazakov B N and. Stolov A L 1978 *Fiz. Tverd. Tela* **20** 2391 (Sov. *Phys. Solid State* **20** 1378).
[12] Sharma K K, Spedding F H and Blinde D R 1981 *Phys. Rev. B* **24** 82
[13] Ma L, Chen W-X, Zheng Yi-F, Zhao J and Xu Zh 2007 *Materials Letters* **61** 2765
[14] Wang X and Li Y D 2003 *Chem. Eur. J.* **9** 5627
[15] Tarasov V F and Shakurov G S 1991 *Appl. Magn. Res.* **2**, 571
[16] Aminov L K and Teplov M A 1985 *Sov. Phys. Usp.* **28** 762

^{19}F NMR study of $LiTbF_4$ single crystals

I V Romanova, A V Egorov, S L Korableva, B Z Malkin and M S Tagirov

Kazan Federal University, Kremlevskaya 18, 420008 Kazan, Russia

E-mail: Irina.Choustova@ksu.ru

Abstract. The angular dependences of ^{19}F NMR spectra have been measured in the external magnetic field of 0.5 T oriented in the basis plane of $LiTbF_4$ at the room temperature. We have obtained the constants of transferred hyperfine interaction and the corrected set of crystal field parameters for the Tb^{3+} ions in $LiTbF_4$. The results of simulations of the magnetization in high pulsed magnetic fields with taking into account magnetoelastic interactions agree satisfactorily with experimental data presented in the literature.

1. Introduction

Double lithium rare-earth fluorides crystallizing in the tetragonal scheelite ($CaWO_4$) structure (C_{4h}^6) attract much attention as model objects in physics of magnetism [1]. The unit cell of $LiRF_4$ contains two magnetically equivalent lanthanide R^{3+} ions at sites with the S_4 point symmetry. $LiTbF_4$ is a dipolar Ising-like ferromagnet (T_c=2.89 K) [2]. Earlier ^{19}F NMR in this material at 295 K was studied by Hansen et al. [3], and they determined parameters of the transferred hyperfine interactions between the Tb^{3+} ions and the ^{19}F and ^{7}Li nuclei. The sets of transferred hyperfine interaction constants were obtained from fitting the calculated NMR spectra to the experimental data. The local magnetic fields induced by Tb^{3+} magnetic moments at ^{19}F nuclei were calculated using bulk magnetic susceptibilities of $LiTbF_4$ measured in [4, 5].

In general case, temperature variations of local magnetic fields in a paramagnetic crystal can be calculated if parameters of the crystal field affecting paramagnetic ions are known. However, different sets of the crystal field parameters have been introduced in the literature to describe magnetic and spectral properties of $LiTbF_4$ [6, 7] and terbium doped isomorphic $LiYF_4$ crystals [8, 9]. Earlier we studied temperature and magnetic field dependences of the magnetization of $LiRF_4$ single crystals (R=Dy, Ho, Tb) [10, 11] taking into account magnetoelastic interactions and found corrected sets of the crystal field parameters for Tb^{3+}, Dy^{3+} and Ho^{3+} ions. It should be noted that in theoretical studies of $LiTbF_4$ spectral properties carried out earlier, as well as in our previous work [11], crystal field energies of the Tb^{3+} ion were calculated by making use of the ion Hamiltonian acting in the truncated space of electronic states belonging to the ground $4f^8$ configuration. This resulted in remarkable differences between some crystal field parameters obtained for stoichiometric $LiTbF_4$ [6, 7, 11] and for the impurity Tb^{3+} ions in $LiYF_4$ [8, 9]. To determine unambiguously the total set of seven crystal field parameters (in crystallographic frame) for rare earth ions at the sites with the S_4 symmetry, it is necessary to analyze responses of the ions on large enough direction dependent perturbations (in particular, on external magnetic fields). In the present work we use additional information about the local magnetic fields in $LiTbF_4$ obtained from ^{19}F NMR spectra and the results of the magnetization measurements in high pulsed magnetic fields [12], as well as more precise technique of calculations of the terbium magnetic moments, in order to validate different parameters of the microscopic model describing static and dynamic magnetic properties of $LiTbF_4$. Theoretical grounds for simulations of the local magnetic fields are considered in section 2 of the present work, and in section 3 we analyze the results of ^{19}F NMR measurements.

2. Magnetic moment of the Tb^{3+} ion in $LiTbF_4$ in external magnetic field

To calculate the angular dependence of the ^{19}F NMR spectra, we need to know local magnetic fields affecting the ^{19}F nuclei. Local fields depend on the average values of the Tb^{3+} magnetic moments. The magnetic moments of the Tb^{3+} ions are considered here with taking into account interactions of the Tb^{3+} ion with the static crystal field and lattice deformations and magnetic (dipole-dipole and exchange) interactions between the Tb^{3+} ions.

In the presence of an applied magnetic field $\boldsymbol{B}$ (below a direction of $\boldsymbol{B}$ relative to the crystallographic axes is specified by spherical coordinates θ (the angle between $\boldsymbol{B}$ and the c-axis) and φ (the angle between the $\boldsymbol{B}$ projection onto the ab-plane and the [100] axis)), we write down the Hamiltonian of a single Tb^{3+} ion in the following form (the nuclear Zeeman energy and the hyperfine interaction are neglected):

$$H = H_0 + H_{cf} - \mu_B \boldsymbol{B}_{loc}(\boldsymbol{L} + 2\boldsymbol{S}) + \sum_{\alpha\beta} V'_{\alpha\beta} e_{\alpha\beta} + \sum_{\alpha,s} V''_{\alpha}(s) w_{\alpha}(s) . \tag{1}$$

The Hamiltonian (1) operates in the total space of 3003 states of the electronic $4f^8$ configuration, the first term is the free ion Hamiltonian, the second term is the crystal field Hamiltonian:

$$H_{cf} = B_2^0 O_2^0 + B_4^0 O_4^0 + B_4^4 O_4^4 + B_4^{-4} \Omega_4^4 + B_6^0 O_6^0 + B_6^4 O_6^4 + B_6^{-4} \Omega_6^4 \tag{2}$$

determined in the crystallographic system of coordinates by the set of seven crystal field parameters B_p^k (O_p^k and Ω_p^k are the linear combinations of spherical tensor operators similar to the corresponding Stevens operators [13]). The third term in (1) is the electronic Zeeman energy (μ_B is the Bohr magneton, $\boldsymbol{L}$ and $\boldsymbol{S}$ are the electronic orbital and spin moments, respectively, and $\boldsymbol{B}_{loc}$ is the local magnetic field defined below). The last two terms define linear interactions of rare earth ions with the homogeneous macro- and microdeformations, respectively, where $\hat{\boldsymbol{e}}$ is the deformation tensor, and $\boldsymbol{w}(s)$ is the vector of the s-sublattice displacement. The electronic operators $V'_{\alpha\beta}$ and $V''_{\alpha}(s)$ can be represented, similar to the crystal field energy, through the linear combinations of spherical tensor operators with the parameters which have been calculated earlier in the framework of the exchange charge model [14-16]. Below we consider the electron-deformation interaction (last two terms in (1)) as a perturbation and take into account the corresponding contributions of the first order in deformation parameters $\hat{\boldsymbol{e}}$ and $\boldsymbol{w}(s)$ into the free energy of the magnetic subsystem.

The energy of a deformed crystal lattice (per unit cell with the volume v_0) is written as [16]

$$E_{lat} = \frac{v_0}{2} \hat{\boldsymbol{e}} \hat{\boldsymbol{C}}' \hat{\boldsymbol{e}} + \sum_s \boldsymbol{w}(s) \hat{\boldsymbol{b}}(s) \hat{\boldsymbol{e}} + \frac{1}{2} \sum_{ss'} \boldsymbol{w}(s) \hat{\boldsymbol{a}}(s,s') \boldsymbol{w}(s') \tag{3}$$

where $\hat{\boldsymbol{a}}$ is the dynamic matrix of the lattice at the Brillouin zone center, and the tensor of elastic constants is $\hat{\boldsymbol{C}} = \hat{\boldsymbol{C}}' - \sum_{ss'} \hat{\boldsymbol{b}}(s) \hat{\boldsymbol{a}}^{-1}(s,s') \hat{\boldsymbol{b}}(s') / v_0$. Taking into account the equilibrium conditions for the coupled paramagnetic ions and the elastic lattice, we obtain the lattice macrodeformation induced by the magnetic field

$$\hat{\boldsymbol{e}}(\boldsymbol{B}) = -\frac{n}{v_0} \hat{\boldsymbol{S}} [\langle \boldsymbol{V}' \rangle_B - \langle \boldsymbol{V}' \rangle_0] \tag{4}$$

and the sublattice displacements

$$\Delta \boldsymbol{w}(\boldsymbol{B}) = \boldsymbol{w}(\boldsymbol{B}) + \hat{\boldsymbol{a}}^{-1} \hat{\boldsymbol{b}} \hat{\boldsymbol{e}}(\boldsymbol{B}) = -n \hat{\boldsymbol{a}}^{-1} [\langle \boldsymbol{V}'' \rangle_B - \langle \boldsymbol{V}'' \rangle_0] . \tag{5}$$

Here $n = 2$ is the number of rare earth ions in the unit cell, $\hat{\boldsymbol{S}} = \hat{\boldsymbol{C}}^{-1}$ is the compliance tensor of the lattice, and angular brackets $<\ldots>_B$ and $<\ldots>_0$ indicate thermal averages over the eigenstates of the

Hamiltonian (1) for $\boldsymbol{B}\neq 0$ and $\boldsymbol{B}=0$, respectively. Operators $\boldsymbol{V} = \boldsymbol{V}' - \hat{\boldsymbol{b}}(s)\hat{\boldsymbol{a}}^{-1}(s,s')\boldsymbol{V}''(s')$ in (4) equal to operators $\boldsymbol{V}'$ renormalized due to linear coupling between macro- and microdeformations.

Table 1. Coupling constants (in cm^{-1}) of the electron-deformation interaction [16, 17].

	$a_{\lambda,2}^{0}$	$a_{\lambda,4}^{0}$	$a_{\lambda,4}^{4}$	$a_{\lambda,4}^{-4}$	$a_{\lambda,6}^{0}$	$a_{\lambda,6}^{4}$	$a_{\lambda,6}^{-4}$	
A_g^1	350	140	2300	1550	90	700	300	
A_g^2	-900	500	2400	795	-50	1300	1100	
	$b_{\lambda,2}^{2}$	$b_{\lambda,2}^{-2}$	$b_{\lambda,4}^{2}$	$b_{\lambda,4}^{-2}$	$b_{\lambda,6}^{2}$	$b_{\lambda,6}^{-2}$	$b_{\lambda,6}^{6}$	$b_{\lambda,6}^{-6}$
B_g^1	1644	1846	-454	1885	188	-543	-858	-738
B_g^2	3814	-836	-1532	1424	-243	-658	-1444	-1245

To take into account magnetic dipole-dipole and exchange interactions between the rare earth ions, we use the mean field approximation. The local field in the spherical sample equals

$$\boldsymbol{B}_{loc} = \boldsymbol{B} + (\hat{\boldsymbol{Q}} + \hat{\boldsymbol{\lambda}} - N\hat{\boldsymbol{I}})\boldsymbol{M} \qquad (6)$$

($\hat{\boldsymbol{Q}}$ is the tensor of dipole lattice sums, the tensor $\hat{\boldsymbol{\lambda}}$ determines the exchange field, the demagnetizing factor $N=4\pi/3$, $\hat{\boldsymbol{I}}$ is the unit tensor, $\boldsymbol{M}$ is the magnetization). We substitute expressions (4) and (5) for $\hat{\boldsymbol{e}}$ and $\boldsymbol{w}(s)$, respectively, in (1) and thus obtain the effective self-consistent single-ion Hamiltonian $H_{eff}(\boldsymbol{B},\hat{\boldsymbol{e}}(\boldsymbol{B}),\boldsymbol{w}(\boldsymbol{B}),\boldsymbol{M}(\boldsymbol{B}))$ parametrically dependent on the magnetization and renormalized by the electron-deformation interactions. Additional terms, depending on the external magnetic field and the temperature, change the response of rare earth ions on the field $\boldsymbol{B}$.

Table 2. Crystal field parameters B_p^k (cm^{-1}) for the R^{3+} ions in $LiRF_4$ (R=Ho, Dy, Tb) crystals.

B_p^k	$LiTbF_4$ This work	$LiTbF_4$ [8]	$LiDyF_4$ [10]	$LiHoF_4$ [10]
B_2^0	200	200	170	219.7
B_4^0	-100.5	-100	-85	-87.3
B_6^0	-2	-3.6	-4.2	-3.55
B_4^4	-694	-1103	-721	-710
B_4^{-4}	-829	0	-661	-612
B_6^4	-435	-529	-390	-387
B_6^{-4}	-283	0	-248	-253.7

Calculations are essentially simplified when making use of symmetry properties of a system. We worked with linear combinations of the deformation tensor $e(A_g^1)=e_{zz}$, $e(A_g^2)=(e_{xx}+e_{yy})/2$, $e(B_g^1)=e_{xx}-e_{yy}$, $e(B_g^2)=e_{xy}$, $e_1(E_g)=e_{xz}$, $e_2(E_g)=e_{yz}$) and the sublattice displacements corresponding to irreducible representations $\Gamma=A_g$, B_g, E_g of the lattice factor group C_{4h}. In particular, the magnetic field $\boldsymbol{B}$ directed along the crystal symmetry c-axis brings about only totally symmetric A_g deformations, but the field in the ab-plane induces A_g and rhombic B_g deformations. The corresponding internal A_g and B_g deformations are described by three and five independent linear combinations of the sublattice displacements, respectively. We can write the electron-deformation interaction in the basis of symmetrical strains as follows

$$H_{ed} = \sum_{\Gamma\lambda}[\sum_k V_k(\Gamma^\lambda)e_k(\Gamma^\lambda) + \sum_p V_p''(\Gamma^\lambda)\Delta w_p(\Gamma^\lambda)], \qquad (7)$$

where $e_k(\Gamma^\lambda)$ and $\Delta w_p(\Gamma^\lambda)$ are independent variables. In case of coupling with A_g and B_g strains, the electronic operators equal

$$V(A_g^\lambda) = a_{\lambda,2}^0 O_2^0 + a_{\lambda,4}^0 O_4^0 + a_{\lambda,4}^4 O_4^4 + a_{\lambda,4}^{-4} O_4^{-4} + a_{\lambda,6}^0 O_6^0 + a_{\lambda,6}^4 O_6^4 + a_{\lambda,6}^{-4} O_6^{-4}, \quad (8)$$

$$V(B_g^\lambda) = b_{\lambda,2}^2 O_2^2 + b_{\lambda,2}^{-2} O_2^{-2} + b_{\lambda,4}^2 O_4^2 + b_{\lambda,2}^{-2} O_4^{-2} + b_{\lambda,6}^2 O_6^2 + b_{\lambda,6}^{-2} O_6^{-2} + b_{\lambda,6}^6 O_6^6 + b_{\lambda,6}^{-6} O_6^{-6}. \quad (9)$$

The coupling constants $a_{\lambda,k}^p$ and $b_{\lambda,k}^p$ determined from piezospectroscopic measurements of the rare earth doped $LiYF_4$ crystals are given in table 1. We used in calculations the compliance constants of $LiYF_4$: $S(A_g11) = 7.78 \cdot 10^{-12}$, $S(A_g12) = -2.33 \cdot 10^{-12}$, $S(A_g22) = 3.36 \cdot 10^{-12}$, $S(B_g11) = 37.1 \cdot 10^{-12}$, $S(B_g12) = 8.0 \cdot 10^{-12}$, $S(B_g22) = 14.4 \cdot 10^{-12}$ (in units of m^3/J) [18], components of tensors $\hat{\boldsymbol{a}}$ and $\hat{\boldsymbol{b}}$ were calculated using parameters of a rigid ion model of the $LiYF_4$ lattice dynamics [19].

Components of the magnetic moment of a rare earth ion satisfy to self-consistent equations

$$m_\alpha = \sum_i \mu_B < i \,|\, L_\alpha + 2S_\alpha \,|\, i > e^{-E_i(\boldsymbol{M})/k_B T} \left(\sum_i e^{-E_i(\boldsymbol{M})/kT}\right)^{-1}, \quad (10)$$

where $E_i(\boldsymbol{M})$ are the crystal energies (eigenvalues of the Hamiltonian H_{eff}), T – temperature, k_B is the Boltzman constant. To obtain crystal field energies and the magnetization $\boldsymbol{M} = n\boldsymbol{m}/v_0$, the following actions are performed: the matrix of the effective Hamiltonian with $\boldsymbol{M}$=0, $\hat{\boldsymbol{e}}$ =0, $\boldsymbol{\Delta w}$=0 is diagonalized, and the macro- ($\hat{\boldsymbol{e}}$ ($\boldsymbol{B}$)) and microdeformations ($\boldsymbol{\Delta w}(\boldsymbol{B})$), and components (10) of the magnetic moment are calculated. At the next step, the obtained values of $\boldsymbol{M}$, $\hat{\boldsymbol{e}}$, $\boldsymbol{\Delta w}$ are substituted into the Hamiltonian H_{eff}, and the procedure is repeated (up to seven times) to get a steady solution.

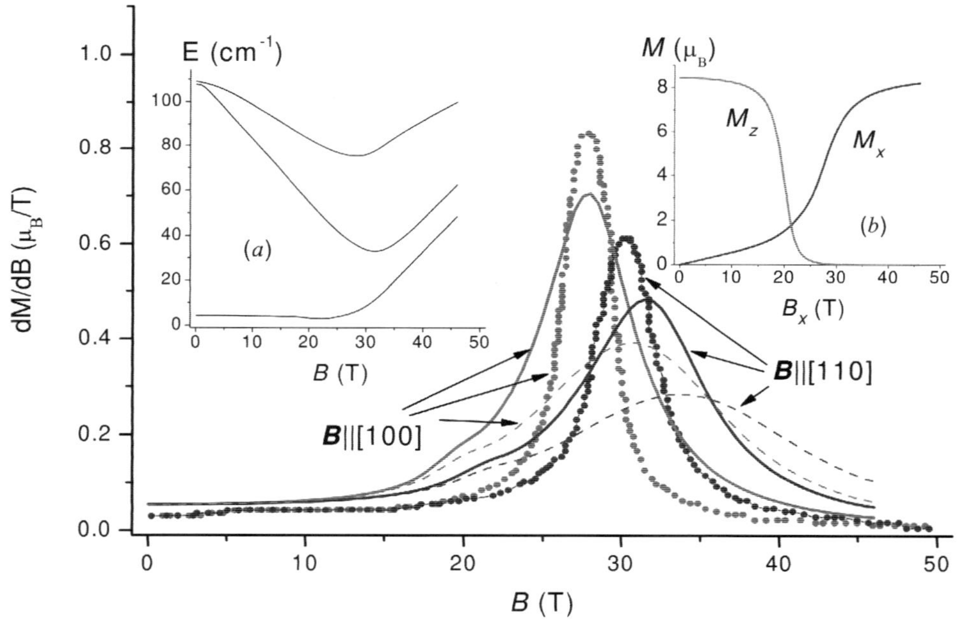

Figure 1. Differential magnetic susceptibilities of $LiTbF_4$ single crystals in magnetic fields $\boldsymbol{B}$||[100] and $\boldsymbol{B}$||[110] (in red and blue, respectively), T=1.4 K. Symbols – experimental data digitized from Ref. [12], solid and dashed curves represent the results of calculations carried out with or without taking into account the magnetoelastic interactions, respectively. Inset (*a*) – calculated energies of the three lower sublevels of the $^7F_6$ multiplet relative to the ground state vs. the magnetic field $\boldsymbol{B}$||[110]. Inset (b) – calculated magnetization components along (induced moment M_x) and normal (spontaneous moment M_z) to the external field $\boldsymbol{B}$||[100] vs. the field intensity.

Journal of Physics: Conference Series **324** (2011) 012034 doi:10.1088/1742-6596/324/1/012034

Table 3. Energy levels (cm^{-1}) of the Tb^{3+} ion in $LiTbF_4$ and $LiYF_4$.

Multiplet, symmetry		Experimental energy levels		Calculated energies
		$LiTbF_4$ [6, 20]	$LiYF_4$:Tb [8,21]	
$^7F_6$	Γ_2	0	0	0
	Γ_2	1.0	0.9	0.97
	$\Gamma_{3,4}$	107	98	107
	Γ_1	124		114
	Γ_2	136		128
	$\Gamma_{3,4}$	166		169
	Γ_1	217		220
	Γ_1			344
	$\Gamma_{3,4}$			355
	Γ_2		360	363
$^7F_5$	Γ_2		2110	2120
	$\Gamma_{3,4}$			2133
	Γ_1			2169
	Γ_1			2174
	$\Gamma_{3,4}$		2196	2199
	Γ_1			2353
	$\Gamma_{3,4}$			2391
	Γ_2		2400	2401
$^7F_4$	Γ_1			3348
	$\Gamma_{3,4}$			3385
	Γ_2			3413
	Γ_2			3524
	Γ_1			3526
	$\Gamma_{3,4}$			3600
	Γ_1			3784
$^7F_3$	Γ_2	4333		4337
	$\Gamma_{3,4}$	4406		4406
	Γ_1	4473		4468
	$\Gamma_{3,4}$	4524		4509
	Γ_2	4532		4517
$^7F_2$	Γ_1	5035		5033
	Γ_2	5067		5071
	$\Gamma_{3,4}$	5285		5274
	Γ_2	5385		5366
$^7F_1$	Γ_1	5592		5592
	$\Gamma_{3,4}$	5690		5691
$^7F_0$	Γ_1	5917		5874
$^5D_4$	Γ_1	20561	20554	20567
	$\Gamma_{3,4}$	20571	20559	20559
	Γ_2			20569
	Γ_1	20580	20568	20590
	Γ_2			20626
	$\Gamma_{3,4}$	20641	20627	20640
	Γ_1	20654	20645	20661

Crystal field parameters (see table 2) were determined by fitting the calculated crystal field energies to the energy levels determined from the optical and EPR spectra [6, 20] of $LiTbF_4$. As it is seen in table 2, our parameters B_2^0, B_4^0, B_6^0, $[(B_4^4)^2+(B_4^{-4})^2]^{1/2}$=1081 cm^{-1} and $[(B_6^4)^2+(B_6^{-4})^2]^{1/2}$=519 cm^{-1} differ only slightly from corresponding parameters determined in [8] for the impurity Tb^{3+} ions in $LiYF_4$ (note that in [8] instead of the point S_4 symmetry, the D_{2d} symmetry was assumed). The B_2^0 parameter is fixed by the measured splitting of the $^7F_1$ multiplet. The calculated energies of sublevels of the $^7F_6$, $^7F_3$ and $^7F_2$ multiplets (see table 3) agree satisfactorily with the measured energies, however, large differences remain between the results of calculations and the experimental data for the $^5D_4$ multiplet. In particular, reversed positions of the two lower $^5D_4$ sublevels $\Gamma_{3,4}$ and Γ_1 are predicted by our set of parameters, as well as by all other published sets [6-9, 11].

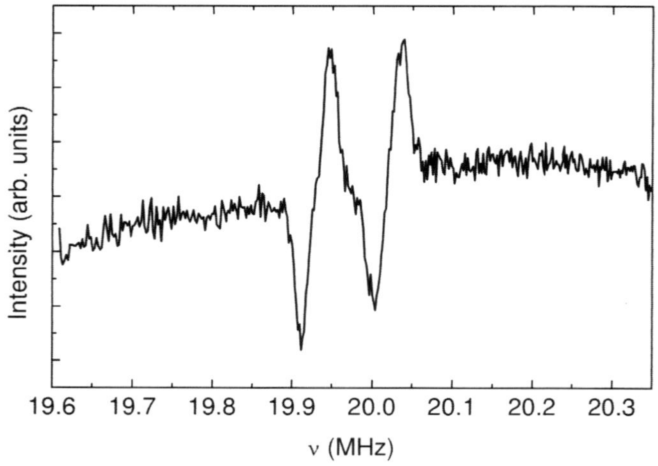

Figure 2. ^{19}F NMR spectrum of $LiTbF_4$ in the magnetic field B=0.5 T at room temperature ($\boldsymbol{B}\perp$c, φ=50°).

To determine ratios of the parameters B_p^4 and B_p^{-4} (p=4 and 6), we analyzed presented in [12] dependences of the longitudinal magnetization in $LiTbF_4$ on the external pulsed magnetic fields applied to the oriented samples (rectangular rods with dimensions of 1.5×1.5×2.5 mm^3) in the *ab*-plane along the [100] and [110] directions (coinciding with the long edge of a rod) at the temperature 1.4 K. The magnetic field varied from zero up to 50 T during 20 μs, and we calculated the magnetization assuming, similarly to the authors of [12], the adiabatic process. Changes of the temperature ΔT of the magnetic subsystem induced by the increase of the field with steps $\Delta\boldsymbol{B}$ were calculated using the expression

$$\Delta T=-T\frac{<H_{eff}(\boldsymbol{B})\boldsymbol{m}>-<H_{eff}(\boldsymbol{B})><\boldsymbol{m}>}{<H_{eff}(\boldsymbol{B})^2>-<H_{eff}(\boldsymbol{B})>^2}\Delta\boldsymbol{B}\,. \tag{11}$$

There are two terbium ions, Tb1 and Tb2, with coordinates $\boldsymbol{r}$(Tb1)=(0, 0, c/2), $\boldsymbol{r}$(Tb2)=(0, a/2, -c/4) in the unit cell. According to neutron scattering data [2] at 100 K, the lattice constants of $LiTbF_4$ equal a=0.5181, c=1.0873 nm. In the magnetically ordered state, the local field at the terbium ion Tbη (η=1, 2) is written as follows

$$B_{loc,\alpha}(Tb\eta)=B_\alpha+\sum_\beta[\sum_{\eta'=1,2}Q_{Tb\eta,Tb\eta'}(\alpha\beta)+2\frac{4\pi}{3v}(\lambda-N_\alpha)\delta_{\alpha\beta}]m_\beta(Tb)\,. \tag{12}$$

Here the demagnetizing factors of the rods are N_x=N_y=0.68 [22], N_z=0 (for the domains in the system with the dominant dipole-dipole interactions), the lattice sums $Q_{Tb\eta,Tb\eta}(xx)=Q_{Tb\eta,Tb\eta}(yy)=1.18053$,

$Q_{Tb\eta,Tb\eta}(zz)=0.63894$, $Q_{Tb1,Tb2}(xx)=Q_{Tb1,Tb2}(yy)=0.41738$, $Q_{Tb1,Tb2}(zz)=2.16525$ (in units of $4\pi/3v$) have been computed by the Ewald method, the value of the isotropic exchange (antiferromagnetic) field constant λ=-0.316 have been determined previously in [11]. To account for the ordering of magnetic moments along the *c*-axis, a small initial external field of 0.01 T has been introduced. With the increasing magnetic field normal to the magnetic easy axis [001], the gap between the lower sublevel of the first excited Γ_{34} state and the ground quasi-doublet quickly diminishes (see Inset (*a*) in figure 1), and the peaks in the differential magnetic susceptibility are observed at the anticrossing point. As an example, evolutions of the spontaneous magnetic moment with the increasing transversal external field and of the induced moment along the field are shown in Inset (b) in figure 1.

As it is seen in figure 1, the differential susceptibilities computed with our set of the crystal field parameters match satisfactorily experimental data. However, the widths of the calculated peaks are much broader than the measured ones. It is evident that more elaborated study of the quantum phase transition in $LiTbF_4$ in the transversal magnetic fields beyond the mean field approximation is necessary. A comparison of the solid and dashed curves in figure 1 shows that magnetoelastic interactions contribute substantially into the magnetization if magnetic field is strong enough.

3. Experimental results and simulations of the NMR frequencies

The unit cell of $LiRF_4$ contains eight fluorine ions. Each of the two magnetically non-equivalent subsystems (labeled below by s=1, 2) of ^{19}F nuclei in the external magnetic field lying in the *ab*-plane involve four F^- sublattices with the following basis vectors $\mathbf{r}_k(s)$: $\mathbf{r}_1(1)=(ax, ay, cz)$, $\mathbf{r}_2(1)=(-ax,- ay, cz)$, $\mathbf{r}_3(1)=(ax, a(y\text{-}0.5), c(0.25\text{-}z))$, $\mathbf{r}_4(1)=(-ax, -a(y\text{-}0.5), c(0.25\text{-}z))$; $\mathbf{r}_1(2)=(-ay, ax,-cz)$, $\mathbf{r}_2(2)=(ay, -ax,-cz)$, $\mathbf{r}_3(2)=(-a(y\text{-}0.5), ax, -c(0.25\text{-}z))$, $\mathbf{r}_4(2)=(a(y\text{-}0.5), -ax, -c(0.25\text{-}z))$. At 295 K, the lattice structure constants of $LiTbF_4$ equal a=0.5192, c=1.0875 nm, x=0.2802, y=0.1619, z=0.0810 [2].

Single crystal of $LiTbF_4$ was grown using Bridgman-Stockbarger method. After axes definition by using X-ray diffractometer, it was shaped as a sphere to obtain the homogeneous demagnetizing field. The angular dependence of ^{19}F NMR spectrum was obtained by using homebuilt CW NMR spectrometer with a frequency sweep. The external magnetic field of 0.5 T was oriented in the basis plane of $LiTbF_4$. An example of the recorded spectrum is shown in figure 2 where two lines correspond to magnetically non-equivalent fluorine sites. The measured angular dependences of the resonance frequencies are represented in figure 3.

Table 4. Dipole lattice sums (in units of $4\pi/3v$)

$\alpha\beta$	*xx*	*xy*	*xz*	*yy*	*yz*	*zz*
$Q_{Tb1,F_1(1)}(\alpha\beta)$	0.1971	2.8226	-1.7445	3.8486	-2.3506	-1.0458
$Q_{Tb1,F_3(1)}(\alpha\beta)$	0.1925	-1.3327	-3.1669	-0.7825	2.4262	3.5900

According to expression (6), the local magnetic fields at the ^{19}F nuclei in $LiTbF_4$ in the paramagnetic phase equal (in the crystallographic frame)

$$B_x(F_k(s)) = B_{loc,x}(ks) = B_x + [Q_{Tb1,F_k(s)}(xx) + Q_{Tb2,F_k(s)}(xx) + \varepsilon_{xx}(ks) - 2\frac{4\pi}{3v}N]m_x + [Q_{Tb1,F_k(s)}(xy) + Q_{Tb2,F_k(s)}(xy) + \varepsilon_{xy}(ks)]m_y, \tag{13}$$

$$B_y(F_k(s)) = B_{loc,y}(ks) = B_y + [Q_{Tb1,F_k(s)}(xy) + Q_{Tb2,F_k(s)}(xy) + \varepsilon_{yx}(ks)]m_x + [Q_{Tb1,F_k(s)}(yy) + Q_{Tb2,F_k(s)}(yy) + \varepsilon_{yy}(ks) - 2\frac{4\pi}{3v}N]m_y. \tag{14}$$

The independent dipole lattice sums have been calculated by the Ewald method and are given in table 4. Components of tranferred hyperfine interaction tensors $\varepsilon_{\alpha\beta}(ks)$ have been considered as the fitting parameters. Note that tensors $\varepsilon_{\alpha\beta}(ks)$ contain contributions due to the transferred hyperfine interactions between a ^{19}F nucleus and a couple of the nearest neighbor terbium ions, Tb1 and Tb2. The dipole sums and tensors $\hat{\varepsilon}(ks)$ satisfy the symmetry relations, in particular, $Q_{Tb1,F_1(2)}(xx)=Q_{Tb1,F_1(1)}(yy)$, $Q_{Tb1,F_1(2)}(yy)=Q_{Tb1,F_1(1)}(xx)$, $Q_{Tb1,F_1(2)}(xy)=-Q_{Tb1,F_1(1)}(xy)$, $\varepsilon_{xx}(12)=\varepsilon_{yy}(11)$, $\varepsilon_{yy}(12)=\varepsilon_{xx}(11)$, $\varepsilon_{xy}(12)=-\varepsilon_{xy}(11)$, $Q_{Tb2,F_1(1)}(\alpha\beta)=Q_{Tb1,F_3(1)}(\alpha\beta)$, $Q_{Tb2,F_3(1)}(\alpha\beta)=Q_{Tb1,F_1(1)}(\alpha\beta)$ for $\alpha,\beta=x,y$.

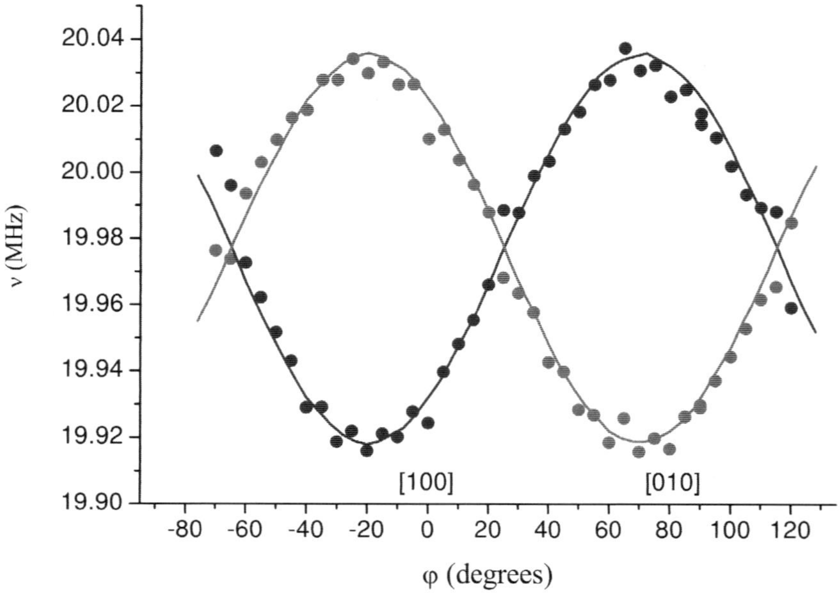

Figure 3. The angular dependence of ^{19}F NMR spectra in $LiTbF_4$ in the magnetic field of 0.5 T (***B***⊥c), *T*=295 K. The circles are the experimental points, solid curves represent the results of calculations where the transferred hyperfine interaction parameters obtained in the present work have been used.

Three independent parameters, $\varepsilon_{xx}(11)$=-1.743, $\varepsilon_{xy}(11)$=-0.279 and $\varepsilon_{yy}(11)$=-1.481 (in units of $4\pi/3v$) were found from fitting the calculated resonance frequencies of ^{19}F nuclei $\nu(s)=\gamma_F[B_{loc,x}(1s)^2+B_{loc,y}(1s)^2]^{1/2}$ (here $\gamma_F/2\pi$=40.07 MHz/T is the fluorine gyromagnetic ratio) to the results of measurements. The obtained hyperfine interaction constants are compared with the ones presented in [2] in table 5.

Table 5. Tranferred hyperfine interaction constants for $LiTbF_4$, g/cm^3 (*M* is the mass per formula unit).

	This work, accuracy ±1 g/cm^3	Ref. [2], accuracy ±3 g/cm^3
$M\varepsilon_{xx}(11)$	-22.38	-18.15
$M\varepsilon_{xy}(11)$	-4.35	0.11
$M\varepsilon_{yy}(11)$	-17.22	-19.49

4. Conclusion

In the present work corrected set of the crystal field parameters for the Tb^{3+} ions in $LiTbF_4$ have been obtained from the analysis of the literature data on the crystal field energies and the differential magnetic susceptibilities in high pulsed magnetic fields. We argue that magnetoelastic interactions contribute remarkably into formation of the magnetization in high magnetic fields. The obtained crystal field parameters correlate with the parameters determined earlier for isomorphic rare earth compounds.

Angular dependence of ^{19}F NMR spectra in the magnetic field perpendicular to the crystal *c*-axis has been measured and the corrected set of tranferred hyperfine interaction constants for $LiTbF_4$ has been determined. Diagonal components of the transferred hyperfine interaction tensor are similar to, but the nondiagonal component is much larger than corresponding parameters obtained earlier in [2]. The constants of the transferred hyperfine interaction and of the magnetic dipole-dipole interaction between the fluorine nuclei and the terbium ions in $LiTbF_4$ have comparable magnitudes (similar conclusions were obtained in the studies of superhyperfine structures of EPR spectra in dilute paramagnets $LiYF_4$:Yb^{3+} [23], $LiYF_4$:Nd^{3+} [24]), but different signs.

Acknowledgements

The authors are grateful to V. A. Shustov for the X-Ray orientation of the samples. This work was partially supported by the Ministry of Education and Science of the Russian Federation (project no. 13.G25.31.0025) and by the RFBR grant №09-02-00930.

References

[1] Aminov L K, Malkin B Z and Teplov M A 1996 *Handbook on the Physics and Chemistry of Rare Earths,* vol.22 (Amsterdam: North-Holland)
[2] Als-Nielsen J, Holmes L M, Larsen F K and Guggenheim H J 1975 *Phys. Rev.* B **12** 191
[3] Hansen P E and Nevald R 1977 *Phys. Rev.* B **16** 146
[4] Hansen P E, Johanson T and Nevald R 1975 *Phys. Rev.* B **12** 5315
[5] Nevald R and Hansen P E 1977 *Physica* B **86-88** 1443
[6] Christensen H P 1978 *Phys. Rev.* B **17** 4060
[7] Krotov V I, Malkin B Z and Mittelman A A 1982 *Physics of the Solid State* (St Petersburg) **24** 542
[8] Liu G K, Carnall W T, Jones R P, Cone R L and Huang J 1994 *J. Alloys Compounds* **207-208** 69
[9] Liu G K 2005 *Spectroscopic Properties of Rare Earths in Optical Materials* Eds Guokui Liu and Bernard Jacquier (Springer Series in Materials Science, vol. 83) pp 1-89
[10] Romanova I V, Abdulsabirov R Yu, Korableva S L, Malkin B Z, Mukhamedshin I R, Suzuki H and Tagirov M S 2006 *Magn. Resonance in Solids EJ* **8** 1
[11] Romanova I V, Malkin B Z, Mukhamedshin I R, Suzuki H and Tagirov M S 2002 *Physics of the Solid State* (St Petersburg) **44** 1475
[12] Kazei Z A, Snegirev V V, Abdulsabirov R Yu, Korableva S L, Broto J M and Rakoto H 2006 *Proceedings of 34 Conference on Low Temperature Physics*, part 1, p. 13
[13] Altshuler S A and Kozyrev B M 1972 *Electron paramagnetic resonance* (Moscow: Nauka)
[14] Abdulsabirov R Yu, Kazantsev A A, Korableva S L, Malkin B Z, Nikitin S I and Stolov A L 2006 *J. Lumin.* **117** 225
[15] Abdulsabirov R Yu, Kazantsev A A, Korableva S L, Malkin B Z, Nikitin S I, Stolov A L, Tagirov M S, Tayurskii D A and van Tol J 2002 *SPIE Proceedings* **4766** 59
[16] Aminov L K and Malkin B Z 2008 *Dynamics and kinetics of electronic and spin excitations in paramagnetic crystals (*Kazan: Kazan State University)
[17] Bertaina S, Barbara B, Giraud R, Malkin B Z, Vanuynin M V, Pominov A I, Stolov A L and Tkachuk A M 2006 *Phys. Rev.* B **74** 184421
[18] Blanchfield P and Saunders G A 1979 *J. Phys.* C **12** 4673

[19] Salaun S, Bulou A, Rousseau M, Hennion B and Gesland G Y 1997 *J. Phys.:Condens. Matter* **9** 6957
[20] Magarino J, Tuchendler J, Beauvillain P and Laursen I 1980 *Phys. Rev.* B **21** 18
[21] Liu G K, Huang J, Cone R L and Jacquier B 1988 *Phys. Rev.* B **38** 11061
[22] Sato M and Ishii Y 1989 *J. Appl. Phys.* **66** 983
[23] Aminov L K, Ershova A A, Zverev D G, Korableva S L, Kurkin I N and Malkin B Z 2008 *Appl. Magn. Res.* **33** 351
[24] Aminov L K, Ershova A A, Korableva S L, Kurkin I N, Malkin B Z and Rodionov A A 2011 *Physics of the Solid State* (St Petersburg) **53** 2129

Quadratic non-Condon effect in optical spectra of impurity paramagnetic centers in dielectric crystals

R Yu Yunusov[1] and O V Solovyev

Kazan (Volga Region) Federal University, 420008, Kremlevskaya Street 18, Kazan, Russia

E-mail: thor@chelny.com

Abstract. Analytical expressions for the absorption and luminescence form functions of impurity paramagnetic centers in dielectric crystals at zero temperature are derived in the adiabatic approximation taking into account the quadratic non-Condon effect. It is proved that, if the optical transition is forbidden due to symmetry selection rules, the non-Condon absorption and luminescence spectra are not mirror symmetric and can contain a zero-phonon line, contrary to the case of linear non-Condon effect. Conditions under which the zero-phonon line is contained in the optical spectra of a symmetry-forbidden transition are determined.

1. Introduction

The purpose of this theoretical study is to derive and analyze formulas for calculating the form of vibrational optical spectra of weakly allowed transitions in impurity centers of dielectric crystals. For a correct description of the spectra of weakly allowed transitions, it is necessary to go beyond the Condon approximation. In [1] we derived analytical expressions for non-Condon absorption and luminescence spectra of impurity centers for the case of zero temperature taking into account linear non-Condon effect. In the present paper we consider linear and quadratic non-Condon effects.

The statement of the problem is determined by the necessity of interpreting the interconfigurational $4f^n - 4f^{n-1}5d$ spectra of absorption and luminescence of heavy rare-earth ions ($n > 7$) for which transitions between the ground state of the $4f^n$ electronic configuration and the lowest-energy state of the $4f^{n-1}5d$ electronic configuration are spin-forbidden.

In this study, we use the following simplifying assumptions: (1) the host crystal is transparent in the spectral region under investigation; (2) the concentration of impurity optical centers is low, and hence the cooperative effects due to the interaction between these centers can be ignored. We consider the following Hamiltonian of the electron–vibrational system [2]:

$$H = H_e + H_{int} + H_{vib}. \qquad (1)$$

Here H_e is the Hamiltonian of optical electrons of the impurity paramagnetic center, which includes the energy of interaction of the electrons with a static crystal field (the set of coordinates of the optical electrons of the impurity ion with respect to its nucleus is designated as **r**). The Hamiltonian H_{vib} of the vibrational subsystem is considered in the harmonic approximation:

[1] To whom any correspondence should be addressed.

$H_{\mathrm{vib}} = \sum_{\chi} \hbar\omega_{\chi}(q_{\chi}^2 - \partial^2/\partial q_{\chi}^2)/2$, where q_{χ} stands for the real dimensionless normal coordinates of the vibrational subsystem (the set of normal coordinates is designated as q) and ω_{χ} are the corresponding frequencies. The mechanism of relaxation of the vibrational subsystem is postulated, and the relaxation time is supposed to be short in comparison with the lifetime of excited electronic states. We consider the electron-vibrational interaction Hamiltonian H_{int} linear in the normal coordinates of the vibrational subsystem: $H_{\mathrm{int}} = \sum_{\chi} v_{\chi} q_{\chi}$, where v_{χ} are the electronic Hermitian operators, which have dimension of energy.

2. The non-Condon absorption and luminescence generating functions

This work is based on the adiabatic approximation. Let us assume that the eigenvalues and eigenfunctions of the Hamiltonian H_{e} are known: $H_{\mathrm{e}}\psi_n^0(\boldsymbol{r}) = E_n^0\psi_n^0(\boldsymbol{r})$. Let $\psi_n(\boldsymbol{r},q)$ (in what follows, we will also use the notation $|n>$, $\psi_n^0(\boldsymbol{r}) = |n^0>$) be the nth solution of the electronic Schrödinger equation for a fixed configuration of the vibrational subsystem:

$$(H_{\mathrm{e}} + H_{\mathrm{int}} + \sum_{\chi} \frac{\hbar\omega_{\chi}}{2} q_{\chi}^2)\psi_n(\boldsymbol{r},q) = U_n(q)\psi_n(\boldsymbol{r},q). \tag{2}$$

Equation (2) will be solved using the perturbation theory with the Hamiltonian H_{int} taken as perturbation (the Kubo–Toyodzawa approach [3]). We find the electronic wave function $\psi_n(\boldsymbol{r},q)$ to the second order in the perturbation:

$$\psi_n^2(\boldsymbol{r},q) = \psi_n^0(\boldsymbol{r}) + \sum_{c\neq n} \frac{<c^0|H_{\mathrm{int}}|n^0>}{E_n^0 - E_c^0}\psi_c^0(\boldsymbol{r}) - \psi_n^0(\boldsymbol{r}) \cdot \sum_{c\neq n} \frac{|<c^0|H_{\mathrm{int}}|n^0>|^2}{2(E_n^0 - E_c^0)^2}$$
$$+ \sum_{c\neq n}\left(\sum_{k\neq n} \frac{<c^0|H_{\mathrm{int}}|k^0><k^0|H_{\mathrm{int}}|n^0>}{(E_n^0 - E_k^0)(E_n^0 - E_c^0)} - \frac{<n^0|H_{\mathrm{int}}|n^0><c^0|H_{\mathrm{int}}|n^0>}{(E_n^0 - E_c^0)^2}\right)\psi_c^0(\boldsymbol{r}). \tag{3}$$

After substituting the adiabatic potential $U_n(q)$ into the equation for the vibrational subsystem and considering the latter to the first order in the perturbation H_{int}, we obtain again the Hamiltonian of the system of independent harmonic oscillators with the same vibration frequencies but with the displaced equilibrium positions and denote it as H_{vib}^n:

$$U_n(q) - \sum_{\chi} \hbar\omega_{\chi}/2 \cdot \partial^2/\partial q_{\chi}^2 = E_n^0 - \sum_{\chi} \hbar\omega_{\chi} q_{\chi n}^2/2 + H_{\mathrm{vib}}^n, \tag{4}$$

where the new equilibrium positions are given by $q_{\chi n} = -<n^0|v_{\chi}|n^0>/\hbar\omega_{\chi}$.

Then, we consider the transition from the electron state a to the electron state b upon absorption of an electromagnetic radiation quantum by the impurity–vibrational system. Let d be the projection of the electric dipole moment of the impurity center onto the direction of the polarization of the photon (the expressions given below can be easily generalized to the case of multipole radiation). The dependence of the absorption coefficient on the frequency is determined by the form function [2]

$$F_{ab}^{(\mathrm{abs})}(\Omega) = \int_{-\infty}^{+\infty} I_{ab}^{(\mathrm{abs})}(t)\exp(-i\Omega t)dt, \tag{5}$$

where $I_{ab}^{(\mathrm{abs})}(t)$ is the absorption generating function, which in the adiabatic approximation is represented by the expression [2]

$$I_{ab}^{(\mathrm{abs})}(t) = \left\langle <a|d|b> \exp\left(itH_{\mathrm{vib}}^{b}/\hbar\right) <b|d|a> \exp\left(-itH_{\mathrm{vib}}^{a}/\hbar\right)\right\rangle_a \exp(i\Omega_{ba}t), \quad (6)$$

where $\langle ... \rangle_a$ means temperature averaging over the states of the vibrational subsystem in the electronic state a and

$$\Omega_{ba} = (E_b^0 - E_a^0)/\hbar + \sum_{\chi} \omega_{\chi}(q_{\chi a}^2 - q_{\chi b}^2)/2 \quad (7)$$

is the frequency of the zero-phonon transition.

We neglect the electron-vibrational interaction for the lower energy electronic state, involved in optical transition. In practice, this approximation is fulfilled with good accuracy in the case of interconfigurational optical transitions, for example, $4f^n - 4f^{n-1}5d$ transitions. Therefore, we can set $|a>=|a^0>$ and $q_{\chi a} = 0$.

Then, it is necessary to choose the approximation in which the wave function of the excited electronic state b is calculated. In the Condon approximation, the static electronic wave functions are considered at fixed vibrational coordinates ($q_{\chi} = 0$, as the Kubo–Toyodzawa approach implies). In this case, the matrix element of the operator d is factored outside the sign of averaging over the states of the vibrational subsystem in equation (6). Setting $|b>=|b^0>$ and introducing the notation $<a^0|d|b^0>=d_{ab}^0$, the following expression for the Condon absorption generating function can be obtained [4]:

$$I_{ab}^{\mathrm{cond\,(abs)}}(t) = |d_{ab}^0|^2 \exp\left(i\Omega_{ba}t - \sum_{\chi}\frac{1}{2}q_{\chi b}^2 + \sum_{\chi}\frac{1}{2}q_{\chi b}^2 \exp(i\omega_{\chi}t) \right) = |d_{ab}^0|^2 \, I_{ab}^{0\,(\mathrm{abs})}(t). \quad (8)$$

Here we introduced the normalized Condon generating function of absorption $I_{ab}^{0\,(\mathrm{abs})}(t)$, which corresponds to the Condon form function of absorption $F_{ab}^{0\,(\mathrm{abs})}(\Omega)$ normalized by 2π. In the Condon approximation, the absorption and luminescence form functions are mirror symmetric [2]:

$$I_{ba}^{0\,(\mathrm{lum})}(t) = I_{ab}^{0\,(\mathrm{abs})}(-t)\cdot \exp(2i\Omega_{ba}t) \quad \text{and} \quad F_{ba}^{0\,(\mathrm{lum})}(\Omega) = F_{ab}^{0\,(\mathrm{abs})}(2\Omega_{ba} - \Omega). \quad (9)$$

For a spin-forbidden or symmetry-forbidden electronic transition $a \leftrightarrow b$ the matrix element d_{ab}^0 is small or equal to zero, and it is necessary to go beyond the Condon approximation. The forbidden transition can become allowed through borrowing the intensity from allowed transitions via electron-vibrational interaction; this mechanism was proposed by Herzberg and Teller [5].

Let us consider the dynamic wave function (3) for the electronic state b and the corresponding matrix element of the electric dipole moment, containing terms linear and quadratic in the normal coordinates:

$$<a|d|b>= d_{ab}^0 + \sum_{\chi} \gamma_{\chi}^{*} q_{\chi} + \sum_{\chi_1\chi_2} B_{\chi_1\chi_2} q_{\chi_1} q_{\chi_2}, \quad (10)$$

$$\gamma_{\chi} = \sum_{c \neq b} \frac{<b^0|v_{\chi}|c^0>}{(E_b^0 - E_c^0)} d_{ca}^0, \quad (11)$$

$$B_{\chi_1\chi_2} = -d_{ab}^0 \sum_{c\neq b} \frac{<c^0|v_{\chi_1}|b^0><b^0|v_{\chi_2}|c^0>}{2(E_n^0 - E_c^0)^2} + \sum_{c\neq b} d_{ac}^0 \left(\sum_{k\neq b} \frac{<c^0|v_{\chi_1}|k^0><k^0|v_{\chi_2}|b^0>}{(E_b^0 - E_k^0)(E_b^0 - E_c^0)} \right.$$

$$-\frac{<b^0|v_{\chi_1}|b^0><c^0|v_{\chi_2}|b^0>}{(E_b^0-E_c^0)^2}\Bigg). \qquad (12)$$

We will mark the generating functions and form functions calculated within the established approximation by the index 2. We substituted (10) into equation (6) and the analogous equation for luminescence and calculated the temperature averages (for the case of zero temperature) over the states of the vibrational subsystem, using the representation of the second quantization and the Feynman's operator calculus [4, 6]. In all further formulas the symmetrized coefficients $B^{\text{sym}}_{\chi_1\chi_2}=(B_{\chi_1\chi_2}+B_{\chi_2\chi_1})/2$ are used, with the notation «*sym*» omitted for brevity. We obtained the following formulas for the generating functions:

$$I^{(2)}(t)/I^{(0)}(t)=\left(\sum_{\chi_1\chi_2}\frac{1}{2}|B_{\chi_1\chi_2}|^2\exp(\pm i(\omega_{\chi_1}+\omega_{\chi_2})t)\right)\mp\left(\sum_{\chi_1\chi_2}\text{Re}(\gamma_{\chi_1}B_{\chi_1\chi_2})\,q_{\chi_2 b}\exp(\pm i\omega_{\chi_1}t)(\exp(\pm i\omega_{\chi_2}t)\mp 1)\right)$$

$$+\left(\sum_{\chi_1\chi_2\chi_3}\frac{1}{2}B_{\chi_1\chi_2}B^*_{\chi_1\chi_3}q_{\chi_2 b}q_{\chi_3 b}\exp(\pm i\omega_{\chi_1}t)(\exp(\pm i\omega_{\chi_2}t)\mp 1)(\exp(\pm i\omega_{\chi_3}t)\mp 1)\right)$$

$$+\sum_{\chi}\frac{1}{2}\left|\gamma_\chi\right|^2\exp(\pm i\omega_\chi t)+\left|d^0_{ab}\mp\sum_{\chi}\frac{1}{2}\gamma^*_\chi q_{\chi b}(\exp(\pm i\omega_\chi t)\mp 1)\right.$$

$$\left.+\sum_{\chi_1\chi_2}\frac{1}{4}B_{\chi_1\chi_2}q_{\chi_1 b}q_{\chi_2 b}(\exp(\pm i\omega_{\chi_1}t)\mp 1)(\exp(\pm i\omega_{\chi_2}t)\mp 1)+\sum_{\chi}\frac{1}{2}B_{\chi\chi}\right|^2. \qquad (13)$$

In this formal expressions modulus concerns complex constants, but not exponential functions of time; upper and lower signs are taken for absorption and luminescence, respectively. The mirror symmetry of absorption and luminescence breaks due to the terms included in (13) with different signs for absorption and luminescence. Physical cause of the asymmetry lies in the fact that transitions between the same vibrational sublevels have different weights in absorption and luminescence spectra, being temperature averaged. If we put all coefficients $B_{\chi_1\chi_2}$ equal to zero, (13) transforms to the formulas for the linear non-Condon effect, which we obtained in [1]. If we put also coefficients γ_χ equal to zero, we obtain the Condon case (see (8)).

Note, that the obtained formulas (13) will be also valid if we take into consideration quadratic electron-vibrational interaction in the first order of the perturbation theory for the electronic wave function; this will only alter the values of $B_{\chi_1\chi_2}$ constants (see (12)).

3. The non-Condon absorption and luminescence form functions

Let's find the Fourier transforms of the obtained generating functions $I^{2\,(\text{abs})}_{ab}(t)$ and $I^{2\,(\text{lum})}_{ba}(t)$ (see (5), (13)). It is convenient to present the explicit formulas for the absorption and luminescence form functions in terms of convolution operators. For the distribution $A(\omega)$ determined for positive values of the argument, we define the convolution operators ρ^-_A and ρ^+_A, which, by acting on a function $\vartheta(\Omega)$, give its convolutions with $A(\omega)$ [1]:

$$\rho^{\pm}_A[\vartheta(\Omega)]=\int_0^{+\infty}A(\omega)\vartheta(\Omega\pm\omega)d\omega. \qquad (14)$$

By definition, we assume that $\rho^{\pm\,*}_{\xi}=\rho^{\pm}_{\xi*}$. Similarly, we introduce the convolution operators with the distributions of two and three frequencies:

$$\rho^{\pm}_{A(\omega_1,\omega_2)}[\vartheta(\Omega)] = \int_0^{+\infty} d\omega_1 \int_0^{+\infty} d\omega_2 A(\omega_1,\omega_2)\vartheta(\Omega \pm \omega_1 \pm \omega_2), \tag{15}$$

$$\rho^{\pm}_{A(\omega_1,\omega_2,\omega_3)}[\vartheta(\Omega)] = \int_0^{+\infty} d\omega_1 \int_0^{+\infty} d\omega_2 \int_0^{+\infty} d\omega_3 A(\omega_1,\omega_2,\omega_3)\vartheta(\Omega \pm \omega_1 \pm \omega_2 \mp \omega_3). \tag{16}$$

The following equality can be proved for an arbitrary function $f(x)$, which can be expanded in the Maclaurin's series:

$$\int_{-\infty}^{+\infty} f\left(\int_0^{+\infty} A(\omega)\exp(\pm i\omega t)d\omega\right)\cdot\Phi(t)\cdot\exp(-i\Omega t)dt = f(\rho_A^{\mp})\left[\int_{-\infty}^{+\infty}\Phi(t)\cdot\exp(-i\Omega t)dt\right]. \tag{17}$$

Analogous equalities involving the distributions $A(\omega_1,\omega_2)$ and $A(\omega_1,\omega_2,\omega_3)$ can be written.

Utilizing (5), (8), (9), (17) one can write the following expression for the normalized Condon absorption and luminescence form functions:

$$F^{0\,(\mathrm{abs})}_{\ (\mathrm{lum})}(\Omega) = 2\pi\exp(-\int L(\omega)d\omega + \rho_L^{\mp})\,[\delta(\Omega-\Omega_{ba})], \tag{18}$$

where the upper and lower signs are taken for absorption and luminescence, respectively; integrals here and further are taken in the limits from zero to the maximum frequency of vibrations; spectral distribution $L(\omega)$ is established as

$$L(\omega) = \sum_{\chi} q^2_{\chi b}\delta(\omega-\omega_\chi)/2. \tag{19}$$

The integral $\int L(\omega)d\omega$ has the meaning of the Huang-Rhys parameter of the transition.

With the use of (5), (13), (17) we obtained the following expression for the non-Condon form functions of absorption (upper signs) and luminescence (lower signs) at zero temperature:

$$F^{2\,(\mathrm{abs})}_{\ (\mathrm{lum})}(\Omega) = (\rho_A^{\mp} + \rho_P^{\mp} + \rho_{P^*}^{\mp} \mp \rho_Q^{\mp} \mp \rho_{Q^*}^{\mp} + \rho_G^{\mp} \mp \rho_M^{\mp} \mp \rho_{M^*}^{\mp} + \rho_\eta^{\mp} + \rho_f^{\mp}$$

$$+\left|d^0_{ab} + \sum_{\chi}\frac{1}{2}\gamma^*_\chi q_{\chi b} \pm \rho_\varphi^{\mp} + \rho_N^{\mp} \mp 2\rho_\zeta^{\mp} + \sum_{\chi_1\chi_2}\frac{1}{4}B_{\chi_1\chi_2}q_{\chi_1 b}q_{\chi_2 b} + \sum_{\chi}\frac{1}{2}B_{\chi\chi}\right|^2)[F^{0(\mathrm{abs})}_{\ (\mathrm{lum})}(\Omega)]. \tag{20}$$

Here the following distributions were introduced (the distributions $f(\omega)$ and $\varphi(\omega)$, which correspond to the linear non-Condon effect, have been established in [1]):

$$A(\omega_1,\omega_2) = \sum_{\chi_1\chi_2}\frac{1}{2}|B_{\chi_1\chi_2}|^2\,\delta(\omega_1-\omega_{\chi_1})\delta(\omega_2-\omega_{\chi_2}),$$

$$P(\omega_1,\omega_2) = -\sum_{\chi_1\chi_2}\frac{1}{2}\gamma^*_{\chi_1}B^*_{\chi_1\chi_2}q_{\chi_2 b}\delta(\omega_1-\omega_{\chi_1})\delta(\omega_2-\omega_{\chi_2}), \qquad Q(\omega_1) = \int P(\omega_1,\omega_2)d\omega_2,$$

$$G(\omega_1,\omega_2,\omega_3) = \sum_{\chi_1\chi_2\chi_3}\frac{1}{2}B_{\chi_1\chi_2}B^*_{\chi_1\chi_3}q_{\chi_2 b}q_{\chi_3 b}\delta(\omega_1-\omega_{\chi_1})\delta(\omega_2-\omega_{\chi_2})\delta(\omega_3-\omega_{\chi_3}),$$

$$M(\omega_1,\omega_2) = \int G(\omega_1,\omega_2,\omega_3)d\omega_3, \qquad \eta(\omega_1) = \int M(\omega_1,\omega_2)d\omega_2,$$

$$f(\omega)=\sum_{\chi}\frac{1}{2}|\gamma_{\chi}|^2\,\delta(\omega-\omega_{\chi}),\qquad \varphi(\omega)=-\sum_{\chi}\frac{1}{2}\gamma_{\chi}^{*}q_{\chi b}\delta(\omega-\omega_{\chi}),$$

$$N(\omega_1,\omega_2)=\sum_{\chi_1\chi_2}\frac{1}{4}B_{\chi_1\chi_2}q_{\chi_1 b}q_{\chi_2 b}\delta(\omega_1-\omega_{\chi_1})\delta(\omega_2-\omega_{\chi_2}),$$

$$\zeta(\omega_1)=\int N(\omega_1,\omega_2)d\omega_2. \qquad (21)$$

4. Computational formulas for the non-Condon form functions

For computational reasons it is convenient to express the sums over the normal coordinates, appearing in (7), (19), (21), through the advanced Green's functions for the vibrational subsystem. In [1] we introduced a complex function of the positive frequency with the indices corresponding to four electron states b_i, $i=1..4$, which are the eigenstates of the Hamiltonian H_e:

$$D_{b_1b_2b_3b_4}(\omega)=\operatorname{Im}G^{a}_{<b_1^0|H_{\text{int}}|b_2^0>,\,<b_3^0|H_{\text{int}}|b_4^0>^*}(\omega). \qquad (22)$$

Here and further the sign of the imaginary part should be assigned to the advanced Green's function [7] for the Hermitian vibrational operators; the complex coefficients, i.e., the matrix elements of the electronic operators, are factored outside the sign of the imaginary part. The properties of symmetry of the introduced function are obvious: $D_{b_1b_2b_3b_4}(\omega)=D_{b_4b_3b_2b_1}(\omega)=D^{*}_{b_3b_4b_1b_2}(\omega)$. The required sums over the normal coordinates can be expressed through the $D_{b_1b_2b_3b_4}(\omega)$ functions with the use of the following equation, which can be proved for an arbitrary function $f(\omega)$:

$$\sum_{\chi}<b_1^0|v_{\chi}|b_2^0><b_3^0|v_{\chi}|b_4^0>^*\,f(\omega_{\chi})\delta(\omega-\omega_{\chi})=\frac{2\hbar}{\pi}D_{b_1b_2b_3b_4}(\omega)f(\omega). \qquad (23)$$

For example, the distribution $L(\omega)$ (19), which defines the Condon band shape, with the use of (23) can be written as $L(\omega)=\frac{1}{\pi\hbar}\frac{D_{bbbb}(\omega)}{\omega^2}$.

Introducing $B_{\chi_1\chi_2}=\sum_{b_1b_2b_3b_4}C_{b_1b_2b_3b_4}<b_1^0|v_{\chi_1}|b_2^0><b_3^0|v_{\chi_2}|b_4^0>$, where coefficients $C_{b_1b_2b_3b_4}$ can be determined from the expression (12), symmetrized upon indices χ_1, χ_2, we can rewrite the distributions (21), involved in the non-Condon form-functions, in the following form

$$A(\omega_1,\omega_2)=\sum_{b_1\ldots b_8}\frac{2\hbar^2}{\pi^2}C_{b_1b_2b_3b_4}C^{*}_{b_5b_6b_7b_8}D_{b_1b_2b_5b_6}(\omega_1)D_{b_3b_4b_7b_8}(\omega_2),$$

$$P(\omega_1,\omega_2)=\sum_{b_1\ldots b_4}\sum_{c\neq b}\frac{2\hbar}{\pi^2}\frac{d^0_{ac}}{E_b^0-E_c^0}C^{*}_{b_1b_2b_3b_4}D_{cbb_1b_2}(\omega_1)\frac{D_{bbb_3b_4}(\omega_2)}{\omega_2},$$

$$G(\omega_1,\omega_2,\omega_3)=\sum_{b_1\ldots b_8}\frac{4\hbar}{\pi^3}C_{b_1b_2b_3b_4}C^{*}_{b_5b_6b_7b_8}D_{b_1b_2b_5b_6}(\omega_1)\frac{D_{b_3b_4bb}(\omega_2)}{\omega_2}\frac{D_{bbb_7b_8}(\omega_3)}{\omega_3},$$

$$f(\omega)=\sum_{c,c'\neq b}\frac{\hbar}{\pi}\frac{d^0_{ac}\,d^{0\,*}_{ac'}}{(E_b^0-E_c^0)(E_b^0-E_{c'}^0)}D_{cbc'b}(\omega),$$

$$\varphi(\omega)=\sum_{c\neq b}\frac{1}{\pi}\frac{d^0_{ac}}{E^0_b-E^0_c}\frac{D_{cbbb}(\omega)}{\omega},$$

$$N(\omega_1,\omega_2)=\sum_{b_1\ldots b_4}\frac{1}{\pi^2}C_{b_1b_2b_3b_4}\frac{D_{b_1b_2bb}(\omega_1)}{\omega_1}\frac{D_{b_3b_4bb}(\omega_2)}{\omega_2}. \qquad (24)$$

5. Group-theoretical analysis of the properties of the non-Condon band

Let G_0 be the point group of symmetry of the Hamiltonian H_e. The electron-vibrational interaction Hamiltonian H_{int} can be expressed in terms of symmetrized displacements $Q_{\Gamma\gamma}$ of the ions of the crystal lattice, which are transformed according to the irreducible representations of the group G_0 (γ is the row of the irreducible representation Γ). The electron functions of the level, containing the state b, are also transformed according to a certain irreducible representation of the group G_0; we will denote this representation as Γ^b (for brevity, the level itself will also be denoted as Γ^b). In the adiabatic approximation, we take into account only interaction with the vibrations that are adiabatic for this level. These are the totally symmetric vibrations, which are transformed according to the identical representation Γ_1, and the vibrations transformed according to the representations Γ that are not contained in $\left[\Gamma^b\times\Gamma^b\right]^K$. Here K denotes the symmetric part of the direct product of the representation by itself if H_e is the Hamiltonian of the even number of electrons and the antisymmetric part if H_e is the Hamiltonian of the odd number of electrons.

Let's consider the obtained non-Condon form-functions of absorption and luminescence (20) for a symmetry-forbidden transition $a\leftrightarrow b$, for which the electric dipole moment matrix element vanishes in Condon approximation: $d^0_{ab}=0$. In [1] we showed that in this case the distribution $\varphi(\omega)$ and the quantity $\sum_\chi\frac{1}{2}\gamma^*_\chi q_{\chi b}$ (in fact, the integral of $\varphi(\omega)$ with the opposite sign) are equal to zero. Now we will show that for the symmetry-forbidden transition $a\leftrightarrow b$ the distribution $N(\omega_1,\omega_2)$ (see (24)) is also equal to zero. Substituting the explicit expressions of the $C_{b_1b_2b_3b_4}$ coefficients in (24) we obtain

$$N(\omega_1,\omega_2)=\sum_{c\neq b}\sum_{k\neq b}\frac{1}{2\pi^2}\frac{d^0_{ac}}{(E^0_b-E^0_k)(E^0_b-E^0_c)}\left(\frac{D_{ckbb}(\omega_1)}{\omega_1}\frac{D_{kbbb}(\omega_2)}{\omega_2}+\frac{D_{kbbb}(\omega_1)}{\omega_1}\frac{D_{ckbb}(\omega_2)}{\omega_2}\right)$$
$$-\sum_{c\neq b}\frac{1}{2\pi^2}\frac{d^0_{ac}}{(E^0_b-E^0_c)^2}\left(\frac{D_{bbbb}(\omega_1)}{\omega_1}\frac{D_{cbbb}(\omega_2)}{\omega_2}+\frac{D_{cbbb}(\omega_1)}{\omega_1}\frac{D_{bbbb}(\omega_2)}{\omega_2}\right). \qquad (25)$$

Let us consider the second line in (25). Since the imaginary parts of the crystal Green's functions are diagonal in the indices of the irreducible representations and only interaction with the totally symmetric vibrations contribute to the matrix element $<b^0|H_{int}|b^0>$, the sum over the states c for these terms is limited by states of the same symmetry as the state b. Thus, the terms in the second line in (25) give zero contribution to the distribution $N(\omega_1,\omega_2)$. Now, consider the first line in (25). From similar reasoning it turns out that the sum over the states k is limited by states of the same symmetry as the state b, and the sum over the states c for these terms is limited by states of the same symmetry as the state k. Hence the terms in the first line in (25) also give zero contribution to the distribution $N(\omega_1,\omega_2)$.

The distribution $\zeta(\omega)$ and the quantity $\sum_{\chi_1\chi_2}\frac{1}{4}B_{\chi_1\chi_2}q_{\chi_1 b}q_{\chi_2 b}$, being the integrals of the distribution $N(\omega_1,\omega_2)$, are also equal to zero for a symmetry-forbidden transition $a\leftrightarrow b$. It can be seen that the remaining distributions, involved in the non-Condon form functions (20), in general case do not vanish for a symmetry-forbidden transition. Given the above, the formulas (20) for the non-Condon form functions of a symmetry-forbidden transition reduce to simpler expressions:

$$F^{2\,(\text{abs})}_{\,(\text{lum})}(\Omega)=(\rho_A^{\mp}+\rho_P^{\mp}+\rho_{P^*}^{\mp}\mp\rho_Q^{\mp}\mp\rho_{Q^*}^{\mp}+\rho_G^{\mp}\mp\rho_M^{\mp}\mp\rho_{M^*}^{\mp}+\rho_\eta^{\mp}+\rho_f^{\mp}+\left|\sum_\chi\frac{1}{2}B_{\chi\chi}\right|^2)[F^{0(\text{abs})}_{\ (\text{lum})}(\Omega)]. \tag{26}$$

As follows from (26), the form functions $F^{2(\text{abs})}(\Omega)$ and $F^{2}_{(\text{lum})}(\Omega)$ of a symmetry-forbidden transition, contrary to the case of linear non-Condon effect [1], are not mirror-symmetric, and can contain a zero-phonon line. The only term in (26) that causes a zero-phonon line in the spectra is $\left|\sum_\chi\frac{1}{2}B_{\chi\chi}\right|^2$, since all other terms give convolutions of the normalized Condon form function with some distributions. Let's determine the conditions under which the quantity $\sum_\chi\frac{1}{2}B_{\chi\chi}$ is nonzero for a symmetry-forbidden transition. We express the quantity $\sum_\chi\frac{1}{2}B_{\chi\chi}$ through the Green's functions of the vibrational subsystem (see (12), (23)):

$$\sum_\chi\frac{1}{2}B_{\chi\chi}=\sum_{c\neq b}\sum_{k\neq b}\frac{\hbar}{\pi}\frac{d^0_{ac}}{(E^0_b-E^0_k)(E^0_b-E^0_c)}\int \operatorname{Im}G^a_{<c^0|H_{\text{int}}|k^0>,<k^0|H_{\text{int}}|b^0>}(\omega)d\omega$$

$$-\sum_{c\neq b}\frac{\hbar}{\pi}\frac{d^0_{ac}}{(E^0_b-E^0_c)^2}\int \operatorname{Im}G^a_{<c^0|H_{\text{int}}|b^0>,<b^0|H_{\text{int}}|b^0>}(\omega)d\omega\,. \tag{27}$$

The second term in (27) vanishes similarly to the second line in the expression (25) for the $N(\omega_1,\omega_2)$ distribution. Now let's determine the conditions under which the first term in (27) does not vanish. We should find such a state c (Γ^c is not equivalent to Γ^b) that satisfies the following conditions:

(1) $\Gamma^d\in\Gamma^a\times\Gamma^c$, i.e., the mixing of the state c with the state b can allow the transition $a\leftrightarrow b$;

(2) In the expansion of the electron-vibrational interaction Hamiltonian H_{int} over the symmetrized ionic displacements there is such a displacement $Q_{\Gamma\gamma}$ that

(a) $\Gamma\notin\left[\Gamma^b\times\Gamma^b\right]^K$, i.e., the vibration is adiabatic for the level Γ^b (but not totally symmetric);

(b) $\Gamma\in\Gamma^b\times\Gamma^k$ and $\Gamma\in\Gamma^c\times\Gamma^k$, i.e., interaction with this vibration mixes both states c and k and states k and b.

Thus, the zero-phonon line can be allowed in the spectra of a symmetry-forbidden transition via mechanism of a clearly quadratic nature – mixing, induced by interaction with vibrations, of the states b and c through the third state k; if the aforementioned conditions cannot be satisfied simultaneously, the zero-phonon lines will be absent in spectra within the established approximations.

6. Conclusions

In the present paper we develop theory of quadratic non-Condon effect in optical spectra of impurity ions at zero temperature within the adiabatic approximation. This effect has not been considered earlier in literature. The results obtained can be used in further investigation of mechanisms of optical spectra formation in the case of the Condon approximation violation, in modeling and interpretation of interconfigurational $4f^n - 4f^{n-1}5d$ absorption and luminescence spectra of impurity rare earth ions, which is necessary for predicting characteristics of potential phosphors and scintillators in the vacuum ultraviolet region of the spectrum of electromagnetic radiation. The strength of the quadratic non-Condon effect in comparison with the linear non-Condon effect will be estimated in our future investigation of the $LiYF_4:Lu^{3+}$ crystal optical spectra, corresponding to symmetry-forbidden $4f^{14} - 4f^{13}5d$ transitions in the Lu^{3+} ion.

Acknowledgements
This work was supported by the RFBR Grant 09-02-00930.

References

[1] Solovyev O V 2010 *Phys. Solid State* **52** 728
[2] Perlin Yu E and Tsukerblat B S 1974 *Effects of Electronic–Vibrational Interaction in the Optical Spectra of Impurity Paramagnetic Ions* (Kishinev: Shtiintsa)
[3] Kubo R and Toyodzawa Y 1955 *Prog. Theor. Phys.* **13** 160
[4] Lax M 1952 *J. Chem. Phys.* **20** 1752
[5] Herzberg G and Teller E 1933 *Z. Phys. Chem. Abt.* B **21** 410
[6] Rickayzen G 1957 *Proc. R. Soc. London, Ser.* A **241** 480
[7] Böttger H 1983 *Principles of the Theory of Lattice Dynamics* (Berlin: Akademie-Verlag)

Journal of Physics: Conference Series **324** (2011) 012036 doi:10.1088/1742-6596/324/1/012036

Simulations of magnetic and magnetoelastic properties of $Tb_2Ti_2O_7$ in paramagnetic phase

V V Klekovkina[1], A R Zakirov[1], B Z Malkin[1] and L A Kasatkina[2]

[1] Kazan Federal University, Kazan 420008, Russia
[2] Kazan State Technical University, Kazan 420111, Russia

E-mail: Vera.Klekovkina@gmail.com

Abstract. Magnetic and magnetoelastic properties of terbium titanate pyrochlore in paramagnetic phase are simulated. The magnetic field and temperature dependences of magnetization and forced magnetostriction in $Tb_2Ti_2O_7$ single crystals and polycrystalline samples are calculated in the framework of exchange charge model of crystal field theory and a mean field approximation. The set of electron-deformation coupling constants has been determined. Variations of elastic constants with temperature and applied magnetic field are discussed. Additional strong softening of the crystal lattice at liquid helium temperatures in the magnetic field directed along the rhombic symmetry axis is predicted.

1. Introduction

Rare earth titanates $R_2Ti_2O_7$ crystallizing in the pyrochlore lattice (the rare earth ions R^{3+} form a sublattice of corner sharing regular tetrahedrons) are extensively studied during last ten years (for a review, see [1]). These compounds refer to the class of geometrically frustrated magnets and exhibit a wide variety of magnetic behaviors at low temperatures such as spin liquids (R=Tb), spin ices (R=Dy, Ho) or non-collinear long-range orders (R=Gd, Er) [2]. Thus $Tb_2Ti_2O_7$ is especially interesting since the origin of its spin-liquid ground state remains mysterious for many years.

The titled compound does not show any magnetic long-range order down to 50 mK and remains dynamic with short range correlations [3-4]. Theoretical investigation within a random phase approximation showed that it should undergo a transition into a magnetic long-range ordered state at temperature about 1.8 K [5]. However, magnetic properties of rare earth compounds can be strongly affected by magnetoelastic interactions. In particular, these interactions bring crystal lattice deformations (parastriction) in paramagnetic phase in presence of an external magnetic field. A cooperative Jahn-Teller phase transition (spontaneous lowering of crystal lattice symmetry due to interactions between the rare earth ions through fields of static and dynamic deformations [6]) can suppress the magnetic ordering.

Experimental investigations of parastriction and anomalous elastic properties of $Tb_2Ti_2O_7$ at low temperatures were performed over twenty years ago in [7-10] where a theoretical interpretation of the measured magnetic field and temperature dependences of the lattice constants and the Young's modulus was presented as well. At liquid helium temperatures and in the magnetic field of 5 T, the relative change in length along the applied field direction in polycrystalline and single crystal samples of $Tb_2Ti_2O_7$ reaches giant value of $0.5 \cdot 10^{-3}$ [8] that is at least two orders of magnitude larger than in typical paramagnets. The single ion mechanism of the parastriction [11] usually dominates in insulating rare earth compounds, and the field induced lattice strains depend on the crystal field energies of a rare earth ion, electron-deformation coupling constants and elastic constants of a lattice. The authors of [8, 10] determined correctly the ground state of the Tb^{3+} ions and predicted a relatively small gap of 16 K between the ground and the first excited crystal field doublets. The results of measurements were reproduced satisfactorily in terms of a purely single-ion origin with making use of several phenomenological parameters of magnetoelastic coupling, and the temperature $T_C \sim 0.1$ K of the cooperative Jahn-Teller structural transition was estimated from calculations [9, 10]. High-resolution neutron scattering data do not show any signs of symmetry breaking down to temperatures of 0.4-0.1 K [12].

Recent high-resolution x-ray diffraction measurements of $Tb_2Ti_2O_7$ single crystal revealed significant effects of the magnetoelastic coupling on properties of the crystal lattice at low temperatures in pulsed magnetic fields [13]. Essential softening of elastic constants at low temperatures was found from ultrasonic measurements [14, 15].

In the present work, our goal is to construct the model of magnetic and magnetoelastic interactions in $Tb_2Ti_2O_7$ that would allow us to describe satisfactorily the observed temperature and field dependences of the parastriction and elastic constants in this fascinating compound.

The paper is organized as follows. In section 2, we discuss the theoretical background and give values of model parameters. In section 3, the calculated bulk magnetization, individual Tb^{3+} magnetic moments in external magnetic fields, the temperature and field dependences of the parastriction and elastic constants are compared with experimental data available from literature. Finally, conclusions are given in section 4.

2. Theoretical background and model parameters

2.1. Structural properties

The $Tb_2Ti_2O_7$ crystal lattice structure is isometric with space group $Fd3m$ (O_h^7, No. 227). The Tb^{3+} and Ti^{4+} ions occupy $16d$ $(1/2, 1/2, 1/2)$ and $16c$ $(0, 0, 0)$ sites, respectively, and the oxygen ions are in $8b$ $(3/8, 3/8, 3/8)$ (we will designate it as O1) and $48f$ $(x, 1/8, 1/8)$ (O2) Wyckoff positions. The lattice structure is completely described by the lattice constant a =1.01694 nm and the parameter x=0.3287 [16]. Geometrically, the Tb^{3+} ion is in a strongly distorted cubic polyhedron formed by two O1 and six O2 ions (local point symmetry is trigonal $\bar{3}m$ (D_{3d})).

Crystallographic Cartesian coordinates (in $a/8$ units) of the four nonequivalent positions of rare earth ions in the unit cell are: $\boldsymbol{r}_1$ =(1,1,1), $\boldsymbol{r}_2$ =(-1,-1,1), $\boldsymbol{r}_3$ =(-1,1,-1), $\boldsymbol{r}_4$ =(1,-1,-1). Below we use local coordinate systems for rare earth ions in the sublattices specified by vectors $\boldsymbol{r}_n$ connected to the crystallographic frame by the following rotation matrices, respectively:

$$R_1 = \begin{pmatrix} \frac{-1}{\sqrt{6}} & \frac{-1}{\sqrt{6}} & \sqrt{\frac{2}{3}} \\ \frac{1}{\sqrt{2}} & \frac{-1}{\sqrt{2}} & 0 \\ \frac{1}{\sqrt{3}} & \frac{1}{\sqrt{3}} & \frac{1}{\sqrt{3}} \end{pmatrix} \quad R_2 = \begin{pmatrix} \frac{1}{\sqrt{6}} & \frac{1}{\sqrt{6}} & \sqrt{\frac{2}{3}} \\ \frac{-1}{\sqrt{2}} & \frac{1}{\sqrt{2}} & 0 \\ \frac{-1}{\sqrt{3}} & \frac{-1}{\sqrt{3}} & \frac{1}{\sqrt{3}} \end{pmatrix} \quad R_3 = \begin{pmatrix} \frac{-1}{\sqrt{6}} & \frac{1}{\sqrt{6}} & \sqrt{\frac{2}{3}} \\ \frac{-1}{\sqrt{2}} & \frac{-1}{\sqrt{2}} & 0 \\ \frac{1}{\sqrt{3}} & \frac{-1}{\sqrt{3}} & \frac{1}{\sqrt{3}} \end{pmatrix} \quad R_4 = \begin{pmatrix} \frac{1}{\sqrt{6}} & \frac{-1}{\sqrt{6}} & \sqrt{\frac{2}{3}} \\ \frac{1}{\sqrt{2}} & \frac{1}{\sqrt{2}} & 0 \\ \frac{-1}{\sqrt{3}} & \frac{1}{\sqrt{3}} & \frac{1}{\sqrt{3}} \end{pmatrix} \qquad (1)$$

We will mark all quantities in local coordinate systems by prime.

2.2. Crystal field parameters

Hamiltonian of a rare earth ion in the external magnetic field $\boldsymbol{B}$ can be presented in the form:

$$H = H_0 + H_{CF} + H_Z + H_{el-def}\,. \qquad (2)$$

Here H_0 is the Hamiltonian of a free Tb^{3+} ion, operating in the total space of 3003 states of the electronic $4f^8$ configuration, H_{CF} is the crystal field Hamiltonian, $H_{el\text{-}def}$ is the Hamiltonian of the electron-deformation interaction specified below, $H_Z = \mu_B \boldsymbol{B}_{eff}(\boldsymbol{L}+2\boldsymbol{S})$ is the electronic Zeeman energy, μ_B is the Bohr magneton, $\boldsymbol{L}$ and $\boldsymbol{S}$ are the electronic orbital and spin moments, respectively. Anisotropic exchange and magnetic dipolar interactions between the Tb^{3+} ions were taken into account by making use of the mean field approximation. The effective magnetic field experienced by the Tb^{3+} ions belonging to the n-sublattice (n=1-4) is written as [17]:

$$\boldsymbol{B}_{eff}(n) = \boldsymbol{B} - \boldsymbol{B}_d + \sum_{n'} \boldsymbol{Q}(n,n')\boldsymbol{\mathcal{M}}(n')$$
$$+\lambda_{\parallel}\sum_p (\boldsymbol{r}^-_{np}\cdot\boldsymbol{\mathcal{M}}(p))\boldsymbol{r}^-_{np}/(\boldsymbol{r}^-_{np})^2 + \lambda_{\perp 1}\sum_p (\boldsymbol{r}^+_{np}\cdot\boldsymbol{\mathcal{M}}(p))\boldsymbol{r}^+_{np}/(\boldsymbol{r}^+_{np})^2 + \lambda_{\perp 2}\sum_p (\boldsymbol{R}_{np}\cdot\boldsymbol{\mathcal{M}}(p))\boldsymbol{R}_{np}/(\boldsymbol{R}_{np})^2. \quad (3)$$

Here $\boldsymbol{\mathcal{M}}(n)$ is the thermal average of the magnetic moment $\boldsymbol{m} = -\mu_B(\boldsymbol{L}+2\boldsymbol{S})$ of the Tb^{3+} ion with the basis vector $\boldsymbol{r}_n$ of the corresponding sublattice, n and p=1-6 number four magnetic ions in the unit cell and their nearest neighbors, respectively, $\boldsymbol{r}^{\pm}_{np} = \boldsymbol{r}_n \pm \boldsymbol{r}_p$, $\boldsymbol{R}_{np} = [\boldsymbol{r}_n \times \boldsymbol{r}_p]$. The second line in (3) corresponds to the exchange field defined by three independent parameters. Values of these parameters, $\lambda_{\parallel}$=-0.079 T/μ_B, $\lambda_{\perp 1} = \lambda_{\perp 2}$=-0.06 T/$\mu_B$, were determined in [17] by fitting the simulated dc-susceptibility of $Tb_2Ti_2O_7$ to experimental data. Components of the tensor $\boldsymbol{Q}$ are the corresponding dipole lattice sums. In particular, the nonzero components of this tensor are $Q_{\alpha\alpha}(n,n)=1$, $Q_{xx}(1,2)=Q_{yy}(1,2)=2.0346$, $Q_{xy}(1,2)=3.4522$, $Q_{zz}(1,2)=-1.0692$ in units of $4\pi/3v$ where $v=a^3/4$ is the unit cell volume (other components can be obtained using symmetry properties of the lattice). $\boldsymbol{B}_d$ is the demagnetizing field, in the present work we consider homogeneous demagnetizing fields collinear to the external field:

$$\boldsymbol{B}_d = \frac{4\pi N}{3vB^2}\sum_n (\boldsymbol{\mathcal{M}}(n)\cdot\boldsymbol{B})\boldsymbol{B}. \quad (4)$$

Here N is the demagnetizing factor (N=1 for a sample of spherical shape).

The crystal field Hamiltonian H_{CF} for rare earth ions in the pyrochlore lattice is determined by six crystal field parameters B^k_p. In the local Cartesian systems of coordinates with the z'-axes directed along the corresponding crystal ternary axes and the local x'-axes lying in the plane containing the crystal z-axis and the basis vector $\boldsymbol{r}_n$, the crystal field Hamiltonian H_{CF} has the following form:

$$H_{CF} = B^0_2 O^0_2 + B^0_4 O^0_4 + B^0_6 O^0_6 + B^3_4 O^3_4 + B^3_6 O^3_6 + B^6_6 O^6_6 \quad (5)$$

where operators O^k_p ($k \le p$) are similar to Stevens' operators [18] and can be written as linear combinations of spherical tensor operators $C^{(p)}_k$ [19]. In particular, $O^0_p = a_{p0}C^{(p)}_0$, $O^{2k-1}_p = -a_{p\ 2k-1}(C^{(p)}_{2k-1} - C^{(p)}_{-2k+1})$, $O^{-2k+1}_p = ia_{p\ 2k-1}(C^{(p)}_{2k-1} + C^{(p)}_{-2k+1})$, $O^{2k}_p = a_{p\ 2k}(C^{(p)}_{2k} + C^{(p)}_{-2k})$, $O^{-2k}_p = -ia_{p\ 2k}(C^{(p)}_{2k} - C^{(p)}_{-2k})$ for k=1, 2, 3; numerical factors equal $a_{20}=2$; $a_{21}=1/\sqrt{6}$; $a_{22}=2/\sqrt{6}$; $a_{40}=8$; $a_{41}=2/\sqrt{5}$; $a_{42}=4/\sqrt{10}$; $a_{43}=2/\sqrt{35}$; $a_{44}=8/\sqrt{70}$; $a_{60}=16$; $a_{61}=8/\sqrt{42}$; $a_{62}=16/\sqrt{105}$; $a_{63}=8/\sqrt{105}$; $a_{64}=16/\sqrt{126}$; $a_{65}=8/\sqrt{693}$; $a_{66}=16/\sqrt{231}$.

The parameters B^k_p of the crystal field were calculated as sums of Coulomb and exchange contributions in exchange charge approximation [19]:

$$B^k_p = B^k_{pq} + B^k_{pS}. \quad (6)$$

The Coulomb and exchange contributions are determined by the following formulae [19], respectively:

$$B^k_{pq} = -K^k_p \sum_{Ls} e^2 q_s (1-\sigma_p) \langle r^p \rangle O^k_p(\Theta_{Ls}\Phi_{Ls}) / R_{Ls}{}^{p+1} , \tag{7}$$

$$B^k_{pS} = K^k_p \sum_{\nu} \frac{2(2p+1)}{7} \frac{e^2}{R_\nu} S_p(R_\nu) O^k_p(\Theta_\nu \Phi_\nu) , \tag{8}$$

Here $K^0_p = (1/a_{p0})^2$, $K^k_p = \frac{1}{2}(1/a_{pk})^2$ for $k \neq 0$, R_{Ls}, Θ_{Ls}, Φ_{Ls} are spherical coordinates of the ion belonging to the s-sublattice at the unit cell L, R_ν, Θ_ν, Φ_ν are the same for the ligand, eq_s are the ion charges, σ_p are the shielding factors, $\langle r^p \rangle$ are the moments of the 4f electron space distribution, e is the proton charge, $S_p(R_\nu)$ are bilinear forms constructed from the overlap integrals of 4f wave functions and oxygen 2s-, 2 $p\sigma$-, 2 $p\pi$-functions with dimensionless phenomenological model parameters G_s, G_σ, G_π :

$$S_p(R_\nu) = G_s(S_s(R_\nu))^2 + G_\sigma(S_\sigma(R_\nu))^2 + [2 - p(p+1)/12] G_\pi (S_\pi(R_\nu))^2 . \tag{9}$$

Table 1. Crystal field parameters and energy levels E of the Tb^{3+} ground multiplet $^7F_6$.

p	k	B^k_p (cm^{-1}) This work	[17]	[20]	Symmetry	E (cm^{-1}) Calculated	Measured [21]
2	0	218.5	219	220	Γ_3	0	0
4	0	320.6	319.4	317	Γ_3	11.1	12.1
4	3	-2188.7	-2188.9	-2174	Γ_2	77.4	84.3
6	0	51.2	52.6	53	Γ_1	121.1	118.8
6	3	777	807	807	Γ_3	282.9	
6	6	809	779	807	Γ_2	320.6	
					Γ_1	323.9	
					Γ_3	440.2	
					Γ_1	512.7	

Numerical values of the overlap integrals and the shielding factor σ_2 were taken from [20]. Values of charges eq_s and parameters G_α were chosen so that the crystal field parameters would be close to ones from [17, 20] that agree satisfactorily with the available information on crystal field energies of the Tb^{3+} ions in $Tb_2Ti_2O_7$ [21]. The Coulomb contributions to the crystal field parameters were calculated using effective point charges 2.59 (Tb^{3+}), 3.39 (Ti^{4+}), -1.64 ($O1^{2-}$) and -1.72 ($O2^{2-}$) in units of the proton charge. The exchange charge contributions were calculated using values of the parameters G_α independent on the type of bonding $G_\sigma = G_s = G_\pi$ =6.52 for O1 ligands and $G_\sigma = G_s = G_\pi$ /0.67=10.38 for O2 ligands.

The obtained set of crystal field parameters is compared with the literature data in table 1. The single ion Hamiltonian $H_s = H_0 + H_{CF}$ was diagonalized numerically. The lowest 13 crystal field energies of the ground $^7F_6$ multiplet and the corresponding irreducible representations of the D_{3d} point symmetry group are presented in table 1.

2.3. Electron-deformation interaction

The Hamiltonian of the linear electron-deformation interaction can be written as follows [19]:

$$H_{el-def} = \sum_{pk,\,\alpha\beta} B^k_{p,\alpha\beta} e'_{\alpha\beta} O^k_p = \sum_{pk,\,ij\nu} B^k_{p,\nu}(\Gamma^j_i) e'_\nu(\Gamma^j_i) O^k_p . \tag{10}$$

Journal of Physics: Conference Series **324** (2011) 012036 doi:10.1088/1742-6596/324/1/012036

Here $e'_{\alpha\beta}$ are components of the macroscopic strain tensor; i, j and ν are type, number and row of an irreducible representation Γ of the symmetry group D_{3d}, correspondingly. The symmetrized linear combinations of the components of the strain tensor are:

$$e'(\Gamma_1^1) = e'_{xx} + e'_{yy};\quad e'(\Gamma_1^2) = e'_{zz}; \tag{11}$$

$$e'_1(\Gamma_3^1) = e'_{xx} - e'_{yy};\quad e'_2(\Gamma_3^1) = 2e'_{xy}; \tag{12}$$

$$e'_1(\Gamma_3^2) = 2e'_{xz};\qquad e'_2(\Gamma_3^2) = -2e'_{yz}. \tag{13}$$

Table 2. Parameters of the electron–deformation interaction $B^k_{p,\lambda}(\Gamma^j_i)$ (cm^{-1}).

pk	$B^k_p(\Gamma_1^1)$	$B^k_p(\Gamma_1^2)$	pk	$B^k_{p,1}(\Gamma_3^1)$**	$B^k_{p,1}(\Gamma_3^2)$**	pk	$B^k_{p,1}(\Gamma_3^1)$**	$B^k_{p,1}(\Gamma_3^2)$**
2 0	2320 (3770)*	-6160 (-3960)*	2 1	3950 (-1215)*	6440 (5540)*	2 2	-2110 (-2910)*	1976 (2480)*
4 0	-230	-1860	4 1	-2450	2585	4 2	665	-1434
4 3	10050	-2610	6 1	1228	1160	6 2	-133	666
6 0	27	-391	4 4	-4810	1120	6 5	2174	-138
6 3	-3595	79	6 4	813	-659			
6 6	-3500	-428						

*- final values of coupling constants determined from the fitting procedure

**- $B^{-k}_{p,2}(\Gamma_3^j) = -B^k_{p,1}(\Gamma_3^j)$ for k=1, 4; $B^{-k}_{p,2}(\Gamma_3^j) = B^k_{p,1}(\Gamma_3^j)$ for k=2, 5

Parameters $B^k_{p,\alpha\beta}$ are related to the crystal field parameters by the following expression:

$$B^k_{p,\alpha\beta} = \frac{1}{2}\sum_{Ls}\left[X'_\alpha(Ls)\frac{\partial B^k_p}{\partial X'_\beta(Ls)} + X'_\beta(Ls)\frac{\partial B^k_p}{\partial X'_\alpha(Ls)}\right] \tag{14}$$

where $X'_\alpha(Ls)$ are the ion coordinates. It should be noted that we neglect additional terms in (10) caused by the interaction of the rare earth ions with sublattice displacements [19]. Nonzero electron-deformation parameters calculated according to (14) are presented in table 2.

Deformation of a crystal lattice in the magnetic field reduces free energy of a crystal and changes magnetic moments of rare earth ions. Parastriction in $Tb_2Ti_2O_7$ accept giant values [7, 8], and one may expect that the corresponding contributions into the magnetization are essential. The free energy per unit volume of the homogeneously deformed cubic lattice of $Tb_2Ti_2O_7$ equals

$$F_{lat} = \frac{1}{2}C_{\alpha\beta\gamma\delta}e_{\alpha\beta}e_{\gamma\delta} = \frac{1}{2}C(A_{1g})e(A_{1g})^2 + \frac{1}{2}C(E_g)\sum_{\lambda=1,2}e_\lambda(E_g)^2 + \frac{1}{2}C(F_{2g})\sum_{\lambda=1,2,3}e_\lambda(F_{2g})^2 \tag{15}$$

where linear combinations of the components of the strain tensor in the crystallographic frame

$$e(A_{1g}) = e_{xx} + e_{yy} + e_{zz}, \tag{16}$$

$$e_1(E_g) = e_{xx} - e_{yy},\ e_2(E_g) = (2e_{zz} - e_{xx} - e_{yy})/\sqrt{3}, \tag{17}$$

$$e_1(F_{2g}) = e_{xy},\ e_2(F_{2g}) = e_{yz},\ e_3(F_{2g}) = e_{zx} \tag{18}$$

transform according to irreducible representations A_{1g}, E_g, F_{2g} of the cubic symmetry group O_h, and $C_{\alpha\beta\gamma\delta}$ ($C_{\alpha\alpha\alpha\alpha}$=$C_{11}$, $C_{\alpha\alpha\beta\beta}$=$C_{12}$, $C_{\alpha\beta\alpha\beta}$=$C_{44}$) are the elastic constants which have been measured recently in [14, 15] in the temperature range from 1.5 to 300 K. The symmetrized elastic moduli $C(A_{1g}) = (C_{11} + 2C_{12})/3$, $C(E_g) = (C_{11} - C_{12})/2$ and $C(F_{2g}) = 4C_{44}$ refer to full symmetrical, tetragonal and trigonal deformations, respectively. For each rare earth n-sulattice, components of the strain tensor in the local and crystallographic frames are connected by the transformations

$$e'^{(n)}_{\alpha\beta} = \sum_{\gamma\delta} R_{n,\alpha\gamma} R_{n,\beta\delta} e_{\gamma\delta} . \qquad (19)$$

Substitution of (19) into (10) allows us to obtain the free energy of the magnetic subsystem $F_M(\boldsymbol{B}, e_{\alpha\beta}) = -\frac{k_B T}{v}\sum_{n=1}^{4}\ln\{\mathrm{Tr}\exp[-H_n(\boldsymbol{B}, e_{\alpha\beta})/k_B T]\}$ as a function of external magnetic field and macroscopic strains (here k_B is the Boltzman constant, H_n is the Hamiltonian (2) of the Tb^{3+} ion belonging to the n-sublattice). From the minimum conditions of the total free energy of a crystal, $\partial(F_{lat} + F_M)/\partial e_{\alpha\beta} = 0$, and the definition of the magnetization $M_\alpha = \sum_n \mu_\alpha(n)/v = -\partial F_M/\partial B_\alpha$, we obtain the following system of self-consistent equations for strains and magnetic moments of the Tb^{3+} ions induced by the external magnetic field:

$$e_\upsilon(\Gamma_i^j) = -\frac{1}{C(\Gamma_i^j)v}\sum_{n,pk,i'j'\upsilon'} B^k_{p,\upsilon'}(\Gamma^{j'}_{i'})\left(\left\langle O_p^k\right\rangle_n - \left\langle O_p^k\right\rangle_{n,\boldsymbol{B}=0}\right) T^{i'j'\upsilon'}_{ij\upsilon}(n), \qquad (20)$$

$$\mathcal{M}(n) = -\mu_B\left\langle \boldsymbol{L} + 2\boldsymbol{S}\right\rangle_n = -\mu_B \frac{\mathrm{Tr}[(\boldsymbol{L}+2\boldsymbol{S})\exp(-H_n(\boldsymbol{B},\{\mathcal{M}, e(\Gamma)\})/k_B T)]}{\mathrm{Tr}\exp[-H_n(\boldsymbol{B},\{\mathcal{M}, e(\Gamma)\})/k_B T]} . \qquad (21)$$

Here $T^{i'j'\upsilon'}_{ij\upsilon}(n) = \partial e'^{(n)}_{\upsilon'}(\Gamma^{j'}_{i'})/\partial e_\upsilon(\Gamma_i^j)$, $\langle ...\rangle_n$ and $\langle ...\rangle_{n,\boldsymbol{B}=0}$ are the thermal averages with the Hamiltonian of the Tb^{3+} ion at the site n in the magnetic field $\boldsymbol{B}$ and in zero field, respectively. The system of equations (20), (21) was solved using the method of successive approximations for each value of the field $\boldsymbol{B}$ and temperature T.

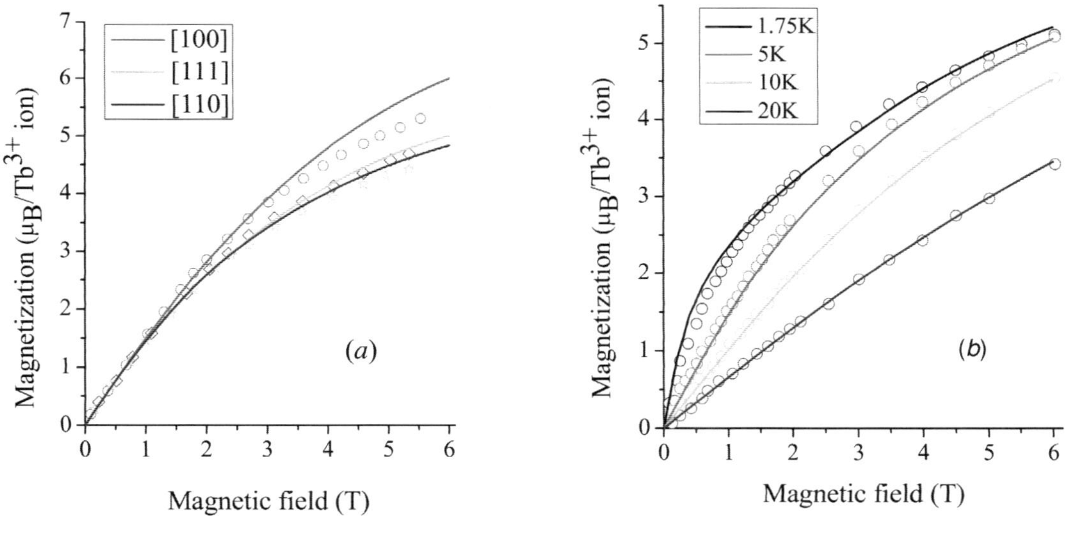

Figure 1. Measured (symbols) and calculated (solid curves) field dependences of the magnetization of $Tb_2Ti_2O_7$ single crystals for magnetic fields directed along the crystallographic axes at T=4.8 K (*a*) and of the magnetization of polycrystalline samples at different temperatures (*b*). Experimental points are digitized from [8] (*a*) and [25] (*b*).

2.4. Elastic constants

The contributions of the magnetic subsystem into the elastic constants were obtained from computations of internal stresses induced by small but finite deformations ($e_{0\upsilon}(\Gamma_i^j)$~$10^{-6}$):

$$\Delta C(\Gamma_i^j)=\sum_n \frac{\sum_{pk,i'j'\upsilon'} B_{p,\upsilon'}^k(\Gamma_{i'}^{j'})T_{ij\upsilon}^{i'j'\upsilon'}(n)\left[\langle O_p^k\rangle_{n,e_0}-\langle O_p^k\rangle_{n,e_0=0}\right]}{e_{0\upsilon}(\Gamma_i^j)v}. \tag{22}$$

Here $\langle ...\rangle_{n,e_0}$ is the thermal average with the Hamiltonian H_n containing the corresponding energy of electron-deformation interaction $\sum_{pk,\,i'j'v'} B_{p,v'}^k(\Gamma_{i'}^{j'})T_{ijv}^{i'j'v'}(n)e_{0v}(\Gamma_i^j)O_p^k$. Note that we neglect second order terms in the electron-deformation interaction and intersite interactions via the phonon field.

2.5. Thermal expansion

Due to changes of populations of the crystal field sublevels of the ground multiplet with temperature, the Tb^{3+} ions contribute into the thermal expansion of the crystal lattice. The corresponding difference between the lattice constant values at temperatures T_1 and T_2 equals $\delta a(T_1,T_2)=\Delta a(T_1)-\Delta a(T_2)$ where

$$\Delta a(T)=-\frac{4a}{9C(A_{1g})v}\sum_{pk}[2B_p^k(\Gamma_1^1)+B_p^k(\Gamma_1^2)]\frac{\mathrm{Tr}[O_p^k\exp(-H_s/k_BT)]}{\mathrm{Tr}[\exp(-H_s/k_BT)]}. \tag{23}$$

2.6. Parastriction

Forced magnetostriction arises in a paramagnet due to redistribution of electronic density of paramagnetic ions in the external magnetic field that results in displacements of the ions in the local environment of each paramagnetic centre from their equilibrium positions, thus linear dimensions of the sample change.

Relative elongation (contraction) of uniformly strained crystal lattice along the unit vector ***d*** equals

$$\Delta l/l=\sum_{\alpha\beta} e_{\alpha\beta}d_\alpha d_\beta . \tag{24}$$

In particular, we obtain from (24) the longitudinal parastriction (***B***||***d***) of a cubic crystal in the following form:

$$(\Delta l/l)_{\parallel}=\frac{1}{3}e(A_{1g})+e_1(E_g)(n_x^2-n_y^2)+\frac{\sqrt{3}}{6}e_2(E_g)(3n_z^2-1)+ \\ 2e_1(F_{2g})n_xn_y+2e_2(F_{2g})n_yn_z+2e_3(F_{2g})n_xn_z \tag{25}$$

where n_α are the directional cosines of the field ***B*** in the crystallographic frame. In general case, the transversal parastriction depends on the vector ***d*** direction in the plane normal to ***B***, however, it is isotropic for magnetic fields directed along the C_3 and C_4 symmetry axes:

$$\boldsymbol{B}\|[001]\quad (\Delta l/l)_{\parallel}=\frac{1}{3}e(A_{1g})+\frac{1}{\sqrt{3}}e_2(E_g),\ (\Delta l/l)_{\perp}=\frac{1}{3}e(A_{1g})-\frac{1}{2\sqrt{3}}e_2(E_g); \tag{26}$$

$$\boldsymbol{B}\|[111]\quad (\Delta l/l)_{\parallel}=\frac{1}{3}e(A_{1g})+2e_1(F_{2g}),\quad (\Delta l/l)_{\perp}=\frac{1}{3}e(A_{1g})-e_1(F_{2g}). \tag{27}$$

The transversal parastriction of a single crystal in the magnetic field directed along the C_2 symmetry axis is anisotropic:

$$\boldsymbol{B}\|[110]\quad (\Delta l/l)_{\parallel}=\frac{1}{3}e(A_{1g})-\frac{1}{2\sqrt{3}}e_2(E_g)+e_1(F_{2g}), \tag{28}$$

$$(\Delta l/l)_{\perp 1}=(\Delta l/l)_{[0\ 0\ 1]}=\frac{1}{3}e(A_{1g})+\frac{1}{\sqrt{3}}e_2(E_g), \tag{29}$$

$$(\Delta l/l)_{\perp 2}=(\Delta l/l)_{[1\ -1\ 0]}=\frac{1}{3}e(A_{1g})-\frac{1}{2\sqrt{3}}e_2(E_g)-e_1(F_{2g}). \tag{30}$$

It is possible to present the strain tensor by power series in components of the magnetic field. In particular, $e(A_{1g}) = f(B,T) + b(T)(B_x^4 + B_y^4 + B_z^4) + \dots$, thus the full symmetrical strain, similarly to low symmetry strains, depends not only on the magnitude of the field, but on its direction as well. For a powder sample, the expression (25) should be averaged over magnetic field directions.

3. Results of simulations and discussion

Due to a phenomenological character of the crystal field model, the obtained set of electron-deformation coupling constants can be used only for the initial estimations of the parastriction. Six parameters from the total set of thirty, $B^k_{2,\lambda}(\Gamma^j_i)$, which define changes of the quadrupolar component of the crystal field in the deformed crystal lattice, have been varied to reproduce available data on the parastriction in single crystals [8, 13]. The necessity of introducing large corrections of the calculated parameters $B^k_{2,\lambda}(\Gamma^j_i)$ most likely steams from the neglect of the sublattice displacements in the deformed crystal lattice. The final values of the varied coupling constants (see table 2) depend essentially on the values of the elastic moduli used in the simulations. The measured field and temperature dependences of the elastic constants of $Tb_2Ti_2O_7$ presented in [14] and [15] are qualitatively the same, but numerical values are quite different. The results of ultrasonic measurements in [14] do not agree with the previously measured bulk modulus B_0 in the x-ray synchrotron diffraction studies [22] of the crystal structure under pressure (values of $(C_{11}+2C_{12})/3$ in [14] are about three times less than B_0 in [22]). Thus we have used in the present work the elastic constants measured in [15].

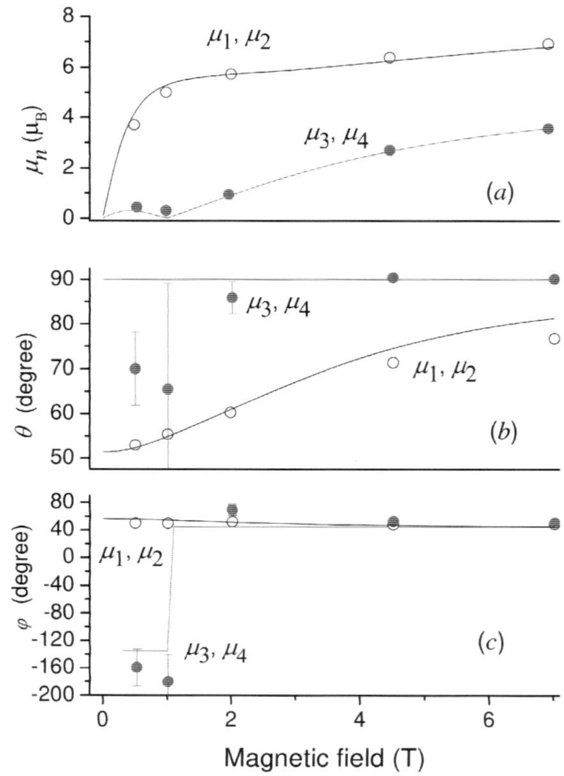

Figure 2. Field dependences of the Tb^{3+} magnetic moments for magnetic field ***B*** $\parallel$ [110] at temperature 1.6 K: calculated (solid curves) magnetic moments (*a*), angles θ (*b*) and φ (*c*) between a moment and the [001] and [100] axes, respectively. Experimental data (symbols) are digitized from [26].

We calculated dependences of the magnetization and individual terbium magnetic moments on the external field and temperature to test our model. In arbitrary directed magnetic field, the Tb^{3+} ions occupy four nonequivalent sites. Four terbium sublattices are equivalent for magnetic field parallel to a tetragonal axis of the lattice, but there are two nonequivalent sites with different magnetic moments if the external field is parallel to trigonal or rhombic axes. Figure 1(*a*) shows calculated field dependences of the magnetization of a single crystal for magnetic fields directed along C_4, C_3 and C_2 symmetry axes at temperature 4.8 K. From a comparison of the results of calculations with experimental data, it follows that neglect of magnetic interactions between the terbium ions brings essential overestimation of the calculated magnetization. Accounting for dipole-dipole and exchange interactions, we are able to describe satisfactorily the magnetization behavior in fields up to 4 T. In high magnetic fields directed along the C_4 or C_3 axes, the calculated magnetization is overestimated. Simulations were carried out for samples of cylindrical shape with the demagnetizing factor N=0.5. Note that the exchange interaction plays dominant role in formation of local magnetic fields while the dipole-dipole interaction brings relatively small contributions. In agreement with the experimental data [8], simulations reveal a small anisotropy of the magnetization of single crystals with "easy" direction being [100]. However, contrary to the experimental data [8, 23], the difference between calculated magnetizations $M(B,T)_{B\|[111]} - M(B,T)_{B\|[110]}$ is positive. The sign of this difference is stable against possible variations of the parameters of the exchange interaction between the nearest neighbor Tb^{3+} ions consistent with the temperature dependences of the isotropic bulk and anisotropic site susceptibilities in $Tb_2Ti_2O_7$ [17].

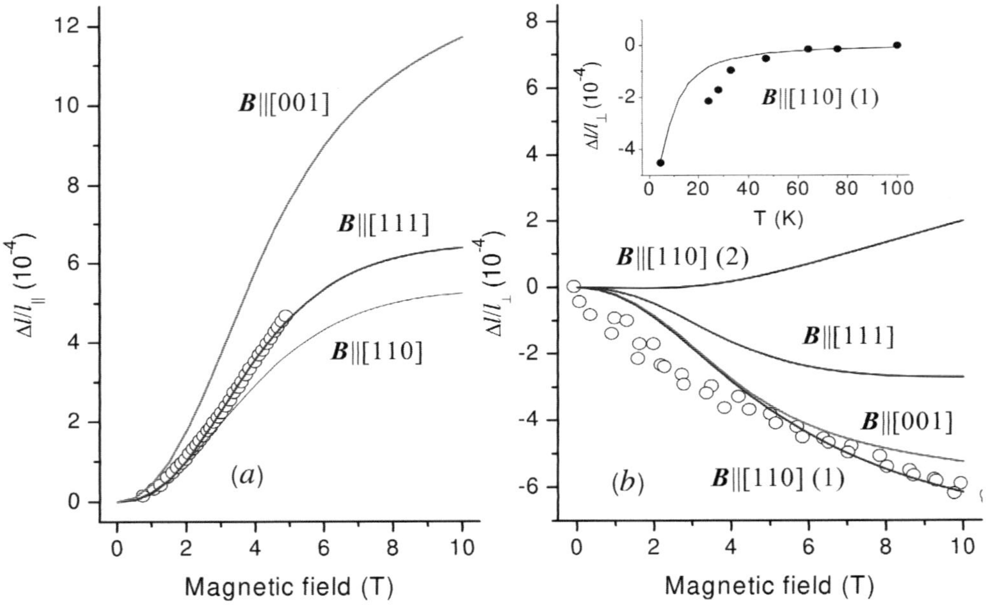

Figure 3. Calculated field dependences of the longitudinal (*a*) and transversal (*b*) parastriction of single crystals at temperature 4.2 K for $\boldsymbol{B}\|[111]$ and 4.5 K for $\boldsymbol{B}\|[001]$ and $\boldsymbol{B}\|[110]$. Experimental data from [8] (*a*) and [13] (*b*) are represented by symbols. Inset: Calculated and measured [13] temperature dependences of the transversal parastriction $\Delta l / l_{[001]}$ in the field $\boldsymbol{B}\|[110]$, $B = 6$ T.

When varying independently the parameters of the effective field (3), we have obtained $\lambda_{\|} = -0.092$ T/μ_B, $\lambda_{\perp 1} = -0.058$ T/μ_B, $\lambda_{\perp 2} = -0.01$ T/μ_B; these values have been used in all simulations which we discuss here. Accounting for the allowed by symmetry contributions into the effective magnetic fields due to Dzyaloshinskii-Moriya interaction [24] does not help to solve the problem with the sign of

$M(B,T)_{B\|[111]} - M(B,T)_{B\|[110]}$. The absolute values of these differences are comparable to the positive contributions into the magnetization due to the magnetoelastic interaction (the inverse parastriction effect).

Figure 1(*b*) shows the field dependences of the magnetization (averaged over a sphere) of the polycrystalline sample at different temperatures. The calculated magnetization curves nearly matche experimental data [25] down to temperature of 1.75 K.

Relative values of magnetic moments of the Tb^{3+} ions at nonequivalent sites are very sensitive to the exchange coupling constants. Calculated field dependences of magnetic moments for magnetic field along the rhombic [110] axis at temperature of 1.6 K are shown in figure 2 (*a*). Figures 2 (*b*) and 2 (*c*) show the angles θ and φ between the moments and the [001] and [100] axes of the cubic unit cell, respectively. Moments at α-sites (*n*=1, 2; the magnetic field is declined from the local trigonal axis by the angle of ~35.3 degrees) relatively quickly saturate. As it is clear from the $\theta(B)$ dependence, as the magnetic field increases, α-moments slowly rotate from their local easy axis toward the field direction [110] lying in the (001) plane, and the component of the α-site moment along [001] becomes smaller. In agreement with experimental data [26], in the magnetic field $\boldsymbol{B}\|[110]$, the moments at β-sites (*n*=3, 4; the field is perpendicular to the local trigonal axis) undergo a "spin-melting" at $B = 1.1$ T where their magnitude vanishes. Moreover, the β-moments flip from a direction opposite to magnetic field to the direction parallel to field (see figure 2 (*c*)) at the same critical field. Note that simulations without taking into account the exchange interactions predict monotonous increasing of the β-moments with increasing magnetic field.

Figure 3 shows calculated field and temperature dependences of the forced magnetostriction of single crystals. It is seen that parastriction of single crystals depends strongly on a direction of the external magnetic field. Similarly to the magnetization, the longitudinal parastriction has the largest values in magnetic fields directed along the tetragonal axes. At liquid helium temperatures the transversal parastriction is negative and about two times less in magnitude than the longitudinal parastriction in the fields along the tetragonal and trigonal axes in agreement with the measurements in powder samples [8]. We have found very large anisotropy of the transversal parastriction for the magnetic field directed along the C_2 axis (in particular, [110], see figure 3 (*b*)). In this case, the transversal parastriction has extreme values along the C_2 axis ([1-10]) normal to the field (maximum, positive value) and along the C_4 axis ([001], minimum, negative value). Calculated temperature dependences of the parastriction (not shown) are also different for different directions of the applied field. With increasing temperature and correspondingly increasing population of the first excited doublet state of the Tb^{3+} ions, the longitudinal parastriction in magnetic field applied along the [111] axis quickly decreases and changes its sign at $T \approx 16$ K. With further increasing temperature, it remains negative, but the absolute value slowly deceases. However, the longitudinal parastriction in magnetic fields parallel to C_4 and C_2 axes remains positive at all temperatures. The calculated temperature dependence of the transversal parastriction in the field $\boldsymbol{B}\|[110]$ agrees with experimental data [13] (see Inset in figure 3 (*b*)), but we obtained significantly overestimated absolute values of the transversal parastriction for $\boldsymbol{B}\|[001]$ which, according to [13], should be about two times less than for $\boldsymbol{B}\|[110]$.

Volume parastriction $\Delta V / V = e(A_{1g})$ is determined, similarly to the contributions into the lattice thermal expansion and the elastic bulk modulus $C(A_{1g})$, by parameters $2B_p^k(\Gamma_1^1) + B_p^k(\Gamma_1^2)$. Using values of these parameters presented in table 2, we obtained positive volume parastriction at all temperatures in agreement with the results of measurements in powder samples [8]. The calculated crystal lattice expansion $\delta a(24\text{ K}, 4.2\text{ K})/a \approx 0.45 \cdot 10^{-4}$ caused by the interaction between the lattice and the terbium subsystem does not agree with the anomalous negative thermal expansion below 20 K which has been reported for a crystal sample [27] but is consistent with the just published results of x-ray powder diffraction experiments [28]. Despite the correctly estimated thermal expansion, we obtained rather small contributions into the bulk modulus (absolute values are less than one GPa as compared with difference of ~22 GPa between the bulk moduli at 50 and 5 K measured in [15]). Note that the absolute value of the negative contribution into the bulk modulus due to linear electron-deformation interaction should have a maximum at $T \approx 12$ K (the full symmetrical local deformations $e'(\Gamma_1^j)$ mix two lower doublets in the energy spectrum of the Tb^{3+} ions but do not split them), however, the corresponding minimum in the temperature dependence of $C(A_{1g})$ has not been revealed in [14] and [15].

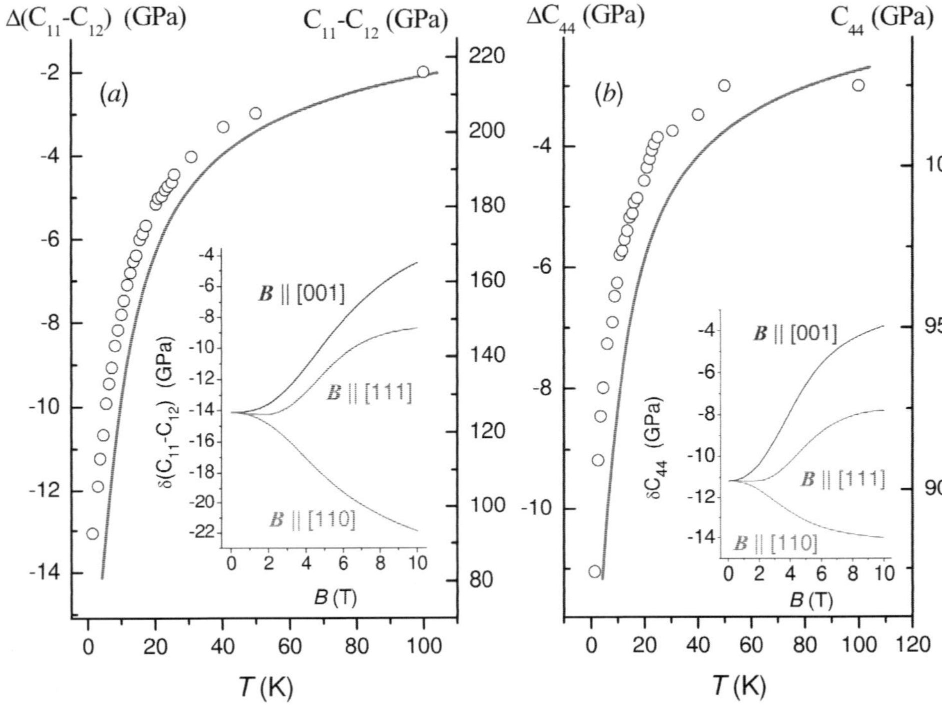

Figure 4. Temperature dependences of elastic constants measured in [15] (symbols) and calculated contributions into the elastic constants caused by linear electron-deformation interaction. Insets: calculated changes of elastic constants at temperature 4.2 K in magnetic fields directed along the C_4, C_3 and C_2 symmetry axes.

Temperature dependences of the elastic moduli $C(E_g)$ and $C(F_{2g})$ measured in [15] as well as the calculated contributions to these moduli induced by the electron-deformation interaction (10) are shown in figure 4. The calculated change of $C(E_g)$ in the temperature range from 100 down to 4.2 K is smaller than the measured one by an order of magnitude, but calculated and measured changes of the $C(F_{2g})$ modulus in the same temperature range have comparable values. Thus, it is possible that renormalization of the elastic multipolar susceptibilities of the Tb^{3+} ions due to intersite interactions through the dynamic lattice deformations and second order terms in the electron-deformation interaction play essential role in responses of the magnetic subsystem on the lattice strains. Simulations of the changes of the elastic moduli in external magnetic fields, $\delta C_{ij}(\boldsymbol{B}) = C_{ij}(T,\boldsymbol{B}) - C_{ij}(T,\boldsymbol{B}=0)$, at low temperatures have allowed us to reveal rather large anisotropy of magnetoelastic effects (see Insets in figure 4): in agreement with the data presented in [14], the magnetic field directed along the C_4 axis suppresses the lattice softening caused by the electron-deformation interaction, however, the field directed along the C_2 axis induces additional remarkable decrease of the elastic moduli $C(E_g)$ and $C(F_{2g})$ and may give rise to structural phase transition.

4. Conclusion

In the present work, a model has been derived to describe spectral, magnetic and magnetoelastic properties of terbium titanate pyrochlore single crystals (we have not discussed in details the forced magnetostriction in powder samples because of possible porosity). The model operates with three phenomenological parameters of the anisotropic exchange interaction between the nearest neighbor terbium ions, six crystal field parameters and thirty parameters which determine modulation of the crystal field by lattice strains. The parameters of the electron-lattice interaction have been estimated in the framework of the semi-phenomenological exchange charge model of the crystal field in ionic crystals, but six parameters of the quadrupolar component of the modulated crystal field have been varied to fit the experimental data on single crystal deformations in the magnetic fields available from literature. The fitting was performed by making use of the elastic constants of $Tb_2Ti_2O_7$ determined recently from ultrasonic measurements [15].

We reproduced satisfactorily main features of the temperature and field dependences of the magnetization and parastriction in $Tb_2Ti_2O_7$ which had been published earlier. Thus we have good reasons to believe that the giant anisotropy of the transversal parastriction and softening of the crystal lattice in magnetic fields parallel to the rhombic symmetry axes ([110], in particular) revealed from our simulations are real properties of $Tb_2Ti_2O_7$. However, some problems concerning more exact quantitative description of the magnetization, parastriction and changes of elastic constants with temperature remain unsolved. Additional studies of effects due to mixing of sublattice displacements corresponding to Raman active A_{1g}, E_g and four F_{2g} lattice vibrations [28] with macroscopic strains, intersite multipolar interactions and two-particle terms in the electron-deformation interaction are necessary, but these tasks are out of scopes of the present work.

The authors acknowledge support by the RFBR grant №09-02-00930 and by the Kazan Federal University in the frameworks of the project F11-22.

References

[1] Gardner J S, Gingras M J P and Greedan J E 2010 *Rev. Mod. Phys.* **82** 53
[2] Greedan J E 2006 *J. Alloys Comp.* **408-412** 444
[3] Gardner J S, Dunsiger S R, Gaulin B D, Gingras M J P, Grredan J E, Kiefl R F, Lumsden M D, MacFarlane W A, Raju N P, Sonier J E, Swainson I and Tun Z 1999 *Phys. Rev. Lett.* **82** 1012
[4] Gardner J S, Keren A, Ehlers G, Stock C, Segal E, Roper J M, Fak B, Stone M B, Hamman P R, Reich D H and Gaulin B D 2003 *Phys. Rev.* B **68** 180401
[5] Kao Y-J, Enjalran M, Del Maestro A, Molavian H R and Gingras M J P 2003 *Phys. Rev.* B **68** 172407
[6] Kaplan M D and Vekhter B G 1995 *Cooperative phenomena in Jahn-Teller crystals* (New-York: Plenum Press)
[7] Aleksandrov I V, Mamsurova L G, Pukhov K K, Trusevich N G and Scherbakova L G 1981 *JETP Lett* **34** 68
[8] Aleksandrov I V, Lidskii V L, Mamsurova L G, Neigauz M G, Pigal'skii K S, Pukhov K K, Trusevich N G and Scherbakova L G 1985 *ZhETF* **89** 2230 (*Sov. Phys. JETP* **62** 1287)
[9] Mumsurova L G, Pigal'skii K S and Pukhov K K 1986 *Pis'ma ZhETF* **43** 584 (*Sov. Phys. JETP Lett.* **43** 755)
[10] Mumsurova L G, Pigal'skii K S, Pukhov K K, Trusevich N G and Scherbakova L G 1988 *ZhETF* **94** 209 (*Sov. Phys. JETP* **67** 550)
[11] Bumagina L A, Kritov V I, Khasanov A Kh and Malkin B Z 1981 *Sov. Phys. JETP* **53** 792
[12] Gaulin B D, Gardner J S, McClarty P A and Gingras M J P 2011 *Phys. Rev.* B **84** 140402(R)
[13] Ruff J P C, Islam Z, Clancy J P, Ross K A, Nojiri H, Matsuda Y H, Dabkowska H A,

Dabkowski A D and Gaulin B D 2010 *Phys. Rev. Lett.* **105** 077203
[14] Nakanishi Y, Kumagai T, Yoshizawa M, Matsuhira K, Takagi S and Hiroi Z 2011 *Phys Rev* B **83** 184434
[15] Luan Y 2011 *Elastic properties of complex transition metal oxides studied by Resonance Ultrasound Spectroscopy*, University of Tennessee PhD thesis, http//trace.tennessee.edu/utk-graddis/993
[16] Lau G C, Muegge B D, McQueen T M, Duncan E L and Cava R J 2006 *J. Solid State Chem.* **179** 3126
[17] Malkin B Z, Lummen T T A, van Loosdrecht P H M, Dhalenne G and Zakirov A R 2010 *J. Phys.: Condens. Matter* **22** 20603
[18] Stevens K W H 1952 *Proc. Phys. Soc.* A **65** 209
[19] Malkin B Z 1987 Crystal Field and Electron-Phonon Interaction in Rare-Earth Ionic Paramagnets in *Spectroscopy of Solids Contaning Rare Earth Ions* eds Kaplyanskii A A and Macfarlane R M (North-Holland, Amsterdam) chapter 2 pp 13–50
[20] Malkin B Z, Zakirov A R, Popova M N, Klimin S A, Chukalina E P, Antic-Fidancev E, Goldner Ph., Aschehoug P and Dhalenne G 2004 *Phys. Rev.* B **70** 075112
[21] Maczka M, Sanjuán M L, Fuentes A F, Hermanowicz K and Hanuza J 2008 *Phys. Rev.* B **78** 134420
[22] Apetrei A, Mirebeau I, Goncharenko I and Crichton W A 2007 *J. Phys.: Condens. Matter* **19** 376208
[23] Yasui Y, Kanada M, Ito M, Harashina H, Sato M, Okumura H, Kakurai K and Kadowaki H 2002 *J. Phys. Soc. (Japan)* **71** 599
[24] Thompson J D, McClarty P A, Ronnow H M, Regnault L P, Sorge A and Gingras M J P 2011 *Phys. Rev. Lett.* **106** 187202
[25] Mirebeau I, Bonville P and Hennion M 2007 *Phys. Rev.* B **76** 184436
[26] Sazonov A P, Gukasov A, Mirebeau I, Cao H, Bonville P, Grenier B and Dhalenne G 2010 *Phys. Rev.* B **82** 174406
[27] Ruff J P C, Gaulin B D, Castellan J P, Rule K C, Clancy J P, Rodriguez J and Dabkowska H A 2007 *Phys. Rev. Lett.* **99** 237202
[28] Goto K, Takatsu H, Taniguchi T and Kadowaki H 2011 *J Phys. Soc. (Japan)* in press, arXiv:1109.1980 [cond-mat]
[29] Lummen T T A, Handayani I P, Donker M C, Fausti D, Dhalenne G, Berthet P, Revcolevschi A and van Loosdrecht P H M 2008 *Phys. Rev.* B **77** 214310

Journal of Physics: Conference Series **324** (2011) 012037 doi:10.1088/1742-6596/324/1/012037

On the Interpretation of the level structure of the Ground $3d^5$ Manifold of Mn III, Fe IV, Co V and Ni VI

A M Leushin
Kazan (Volga Region) Federal University, Kremlevskaya ul. 18,
Kazan, 420008, Russia

E-mail: amleushin@gmail.com

Abstract. The level structure of the ground $3d^5$ configuration of Mn^{2+}, Fe^{3+}, Co^{4+} and Ni^{5+} ions was theoretically interpreted by means of a least-squares fit of the energy parameters to the observed values within the framework of the single-configuration approximation. In the Hamiltonian in addition to real electrostatic, spin-orbit, and spin-spin interactions, electrostatic and spin-orbit interactions correlated by configuration mixing were included. It was shown that the correct positions of almost all the energy levels are determined when the Hamiltonian includes the terms of the lineal (two-body operators) and nonlinear (three-body operators) theory of the configuration interaction. The most correct theoretical description of the experimental spectra was obtained by taking into account relativistic interactions and correlation effects of spin-orbit interactions. Adjustable parameters of the interactions included into the Hamiltonian were found.

1. Introduction

The theory of the energy levels of iron-group and rare-earth paramagnetic ions with open d and f shells is usually developed within the framework of the single-configuration approximation. Furthermore, in most cases, among all the interactions occurring in a free many-electron atom, only the strongest interactions are taken into account, such as the electrostatic repulsion of electrons and the spin-orbit interaction. In this case, consistent quantum-mechanical calculations yield positions of the energy levels that differ from the experimental positions by several tens or, sometimes, hundreds of reciprocal centimeters. Because of this, the energy spectra are very often analyzed using semi empirical methods, with which the best possible agreement between the theory and experiment is achieved by varying the radial integrals considered as free parameters.

Such semi-empirical variants of the theory allow one to leave the framework of the single-configuration approximation and take into account the correlation effects, as well as all relativistic interactions. This is done by replacing the real Hamiltonian matrix with the matrix of some effective energy operator, which, in addition to the known interactions binding the states of the main configuration, includes the contributions of configuration interactions, which leads to the appearance of new two- [1-6] and three-body [7, 8] operators. As a result, the total number of necessary radial parameters in the effective Hamiltonian describing the d^n configuration of a free ion becomes equal to 11 and increases to 19 for the f^n configuration.

Nevertheless, the effective Hamiltonian with 19 operators and parameters of a free atom with an open 4f shell is frequently used in practical calculations for interpretation of the observed spectra of rare-earth ions in crystals and allows one to find the positions of theoretical energy levels, which often differ from the experimentally determined levels only by several reciprocal centimeters [9-11]. As for the elements of the iron group, which have an open 3d shell, their optical spectra have never been described using the total effective Hamiltonian with 11 parameters and operators. At best, one uses a free-ion Hamiltonian containing four parameters [12, 13], namely, B and C, characterizing the electrostatic interaction of electrons, ζ_{3d}, representing the spin-orbit interaction, and α, associated with the configuration interaction. The reason for this is that, in addition to narrow lines, the spectra of transition metals in crystals always contain broad bands of electron-vibrational character [14] and, to explain their origin, one has to include into the Hamiltonian terms describing the interactions of an ion with crystal lattice vibrations, which are stronger than the neglected interactions in the free ion. At the same time, the effects of weak (relativistic and configuration) interactions are often observed against the background of broad bands and, moreover, the positions of the bands themselves sometimes cannot be explained without taking into account these interactions. In such cases, it is necessary to use a more precise free-ion Hamiltonian.

In many papers [6, 15-18], the behavior of free ions of the iron group was described with the use of Hamiltonians with more than four parameters. The highest accuracy in interpretation of the optical spectra was reached using a Hamiltonian obtained on the basis of the group-theoretical considerations and containing 17 effective orthogonal operators [19, 20]. However, due to the great number of orthogonal operators, the complexity of their physical interpretation, and the lack of experimental data, the technique of orthogonal operators will hardly find wide use in description of the behavior of impurity ions in crystals in the near future.

In the present study, we propose an interpretation of the level structure of the $3d^5$ ground manifold of Mn III, Fe IV, Co V and Ni VI, some of which can serve as dopants to impart important technological properties to various crystals. We will describe theoretically the experimental spectra of such ions [21], using an 11-parameter Hamiltonian in order to determine the role of various interactions and to use the data obtained for the analysis of the optical spectra of these ions in crystals.

2. Hamiltonian of a many-electron atom

To describe the states of the $3d^n$ configurations, we will write the effective free-ion Hamiltonian in a form similar to that used for rare-earth ions with $4f^n$ configurations:

$$H = \sum_{k=0,2,4} F^k f_k + \zeta_{3d} A_{so} + \alpha L(L+1) + \beta G(R_5) + \sum_{k=2,3} T^k t_k + \sum_{k=2,4} M^k m_k + \sum_{k=2,4} P^k p_k \quad (1)$$

The first term in (1) represents the electrostatic repulsion of electrons (H_{EL}), characterized by the radial Slater integrals $F^k(3d, 3d)$. The next term is the ordinary spin-orbit interaction of electrons (H_{SO}), determined by the parameter ζ_{3d}. Then, we have the terms pertaining to the theory of the linear configuration interaction (CI) (H_{LCI}), characterized by the parameters α and β [4]. The operator $\alpha L(L+1)$, where L is the angular momentum of the ion, adequately represents the relativistic orbit-orbit interactions as well [22]. The factor $G(R_5)$ in the term $\beta G(R_5)$ is the Casimir operator of the R_5 group. The fifth term allows us to take into account the nonlinear CI effects (H_{NLCI}) (the effective three-body operators), determined by configurations with single-electron excitations. These interactions are characterized by the parameters T^2 and T^3 [8]. The next term describes the spin-spin (H_{SS}) and spin-other-orbit (H_{SOO}) relativistic interactions, determined by the radial Marvin integrals M^k(3d, 3d) [23]. The last term in (1) corresponds to the electrostatically correlated spin-orbit interaction of electrons (H_{ELSO}) [5, 6], characterized by the parameters P^2 and P^4, which were defined in [19].

The matrix elements of the operators A_{SO} and m_k, dependent on the angular variables, were calculated by the standard technique of irreducible tensor operators, *nj* symbols, and fractional parentage coefficients. A detailed description of the necessary mathematical apparatus can be found in monographs [24-28]. The electrostatic interaction matrices f_k and the reduced matrix elements of the double tensor operator $V^{(11)}$ necessary for A_{SO} were taken from the tables given in [29]. For the eigenvalues of the Casimir operator of the R_5 group, we used the expression [24]

$$\langle (w_1 w_2)|G(R_5)|(w_1 w_2)\rangle = \frac{1}{6}[w_1(w_1+3)+w_2(w_2+1)],$$

where w_1 and w_2 are the parameters characterizing the irreducible representations of the R_5 group, which are also given in the tables of [29] for all the states of the configurations under study. The matrix elements of the three-body operators t_k of the nonlinear CI theory were taken from [8]. The reduced matrix elements of the p_k operators in H_{ELSO} were expressed in terms of the matrix elements of the orthogonal operators z_i introduced in [19]. The total matrix of Hamiltonian (1) was composed using all the states of the $3d^5$ configuration present in the scheme of the total angular momentum $SLJM_J$.

3. Interpretation of the optical spectra

To interpretation theoretically the spectra of ions with l^n configurations, we have used a program (developed by E. N. Irinyakov) for self-consistent comparison of the theoretical (E^T) and experimental (E^E) data for energy levels by minimizing the residual $F(x_1,...x_n)=\sum_i W_i(E_i^T(x_1,...x_n)-E_i^E)^2$, where x_i are the variable parameters and E_i and W_i are the energy and the weight factor of the *i*th level. The program allows one to describe the experimental energy levels most correctly and to find the best set of parameters under the condition of an unambiguous correlation between the theoretical and the experimental energy levels. It is this situation that we consider since, for all the ions the positions of all the 37 energy levels of the ground $3d^5$ configuration are found from the analysis of the optical spectra and the energy levels are unambiguously attributed to the suggested Russell-Saunders states.

As an example, table 1 shows the interaction parameters calculated for the ion Ni^{5+} by four variants (I, II, III, and IV) differing by the number of terms taken into account in Hamiltonian (1). The first column of table 1 lists the parameters used and some characteristics of the calculation accuracy. Here, the Racah parameters B and C are related to the Slater integrals F^k by the formulas $B=(9F^2-5F^4)/441$ and $C=5F^4/63$; n is the number of experimentally measured energy levels (coinciding with the number of levels in the spectrum of the $3d^5$ configuration of the Ni^{5+} ion) and m is the number of variable parameters necessary for the calculation of the rms error σ for each calculation variant $\sigma=\sqrt{\sum_i^n (E_i^T-E_i^E)^2/(n\text{-}m)}$; and $\Delta_m=\max|E^{\grave{O}}-E^E|$ is the absolute value of the maximum deviation of the calculated energy of the level E^T from the experimental energy E^E.

It is seen from table 1 that both the rms error σ and the deviation Δ_m decrease considerably as the terms H_{LCI}, H_{NLCI}, H_{SS}, H_{SOO}, and H_{ELSO} with the parameters β, T^2 , T^3, M^0, M^2, P^2, and P^4 are successively taken into account. From variant I to variant IV, the error σ decreases by a factor of about 3. The first calculation version, which takes into account only the H_{EL} and H_{SO} interactions, yields incorrect order for several multiplets. When only one of the H_{LCI} terms with the parameter α are taken into account (variant I), the maximum deviation Δ_m is observed for the level $^2S_{1/2}$; the maximum deviation is observed for the lowest levels $^2F_{7/2}$ and Δ_m becomes considerably smaller as we add the term with the parameter β (variant II). At the same time the error σ decreases approximately 2-fold.

Table 1. Numerical values of different sets of adjustable parameters (in cm^{-1}) for the $3d^5$ configuration of the Ni^{5+} ion

Parameter	I	II	III	IV
B	1419	1419	1420	1420
C	5176	5215	5208	5210
ξ	894	901	906	932
α	100	108	109	109
β		-245	- 222	- 232
T^2			-7.3	-7.4
T^3			24	22
M^0				3.41
M^2				3.21
P^2				65
P^4				47
m	4	5	7	11
σ	138.9	65.6	61.4	58.1
Δ_m	381 ($^2S_{1/2}$)	157 ($^2F_{7/2}$)	122 ($^2F_{7/2}$)	136 ($^2F_{7/2}$)

Note: n (I-IV) = 37; n and m are dimensionless integers.

In the calculations by variants I and II, the theory predicts a wrong fine structure of the ^4G (the $^4G_{11/2}$, $^4G_{5/2}$, $^4G_{9/2}$ and $^4G_{7/2}$ instead of the $^4G_{5/2}$, $^4G_{7/2}$, $^4G_{11/2}$ and $^4G_{9/2}$) and ^4D (the $^4D_{3/2}$ and $^4D_{5/2}$ levels are interchanged) terms. The introduction of the H_{NLCI} operators (variant III) no changes the situation, leading to the incorrect position of the some energy levels but to a small decrease in σ. The maximum deviation Δ_m in this case is observed also for the $^2F_{7/2}$ level but it becomes considerable less. The improvement of the theoretical description of the experimental spectra is achieved by taking into account the relativistic interactions (H_{SS} and H_{SOO} operators) and the correlation effects of the spin-orbit interaction (H_{ELSO} operators). The introduction of these operators leads to correct position of all the energy levels. Table 2 presents the theoretical E_i^T and experimental E_i^E energies of all the levels together with the difference $\Delta_i = E_i^T - E_i^E$, enabling one to see how well the theory agrees with the experiment in the case of variant IV. Table 2 also gives the compositions of the state terms in the wave functions of the given levels. The parameters corresponding to the best theoretical description of the experimental energy levels and the values of σ obtained in the calculations by variant IV for all the ions considered are given in table 3. This table also repeats for comparison the parameters for the Ni^{5+} ion. The agreement between the theory and experiment for the other ions of the isoelectronic sequence of the $3d^5$ configuration is somewhat worse to that obtained for the Ni^{5+} ion (table 2). For all the other ions theory predicts a wrong fine structure of the second and third ^2D (numbering of the same name terms in order of increasing of energy) and ^2P terms. Moreover for the ions Fe^{3+} and Co^{4+} a wrong fine structure is found for ^4F and second ^2G terms and for ions Mn^{2+} and Fe^{3+} the $^2I_{11/2}$ and $^2I_{13/2}$ levels and $^2F_{5/2}$ and $^2F_{7/2}$ levels (second ^2F term) are interchanged. The largest deviation Δ_m is obtained for the energy level $^2D_{5/2}$ (second term) of Mn^{2+} ion for the energy level $^2D_{3/2}$ (third term) of Fe^{3+} ion and for the energy level $^2F_{7/2}$ (second term) of Co^{4+} ion. The reason of these differences between the theory and the experiments is apparently that the this theory is strictly speaking only applicable to the cases of well isolated ground configurations of equivalent electrons. From the ions considered this situation

Journal of Physics: Conference Series **324** (2011) 012037 doi:10.1088/1742-6596/324/1/012037

Table 2. Energy levels (in cm^{-1}) and state compositions (in %) for the $3d^5$ configuration of the Ni^{5+} ion obtained using the parameters of variant IV

J	E^T	E^E	Δ	State compositions
5/2	0.0	0.0	0.0	$99.81\,^6S + 0.19\,^4P$
11/2	41967.4	41920.9	46.5	$99.74\,^4G + 0.26\,^2H$
5/2	42029.5	42003.6	25.9	$99.16\,^4G + 0.73\,^2F1 + 0.07\,^4F + 0.02\,^4D$
9/2	42048.8	42023.2	25.6	$99.82\,^4G + 0.11\,^4F + 0.07\,^2H + 0.01\,^2G1$
7/2	42057.3	42035.1	22.2	$99.55\,^4G + 0.30\,^2F1 + 0.11\,^4F + 0.02\,^4D$
5/2	45876.6	45884.2	-7.6	$89.01\,^4P + 9.49\,^4P + 0.86\,^2D3 + 0.35\,^2D1$
3/2	46126.9	46104.4	22.5	$91.86\,^4P + 7.88\,^4D + 0.17\,^2D3 + 0.07\,^2D1$
1/2	46340.9	46324.8	16.1	$99.08\,^4P + 2.80\,^4D + 0.12\,^2S$
7/2	50321.4	50331.0	-9.6	$99.21\,^4D + 0.41\,^4F + 0.34\,^2F1 + 0.04\,^4G$
1/2	50662.2	50643.5	18.7	$97.19\,^4D + 2.80\,^4P + 0.01\,^2P$
5/2	50793.8	50777.6	16.2	$89.89\,^4D + 9.41\,^4P + 0.36\,^4F + 0.11\,^2D3$
3/2	50801.6	50780.5	21.1	$91.83\,^4D + 7.88\,^4P + 0.19\,^4F + 0.07\,^2D2$
11/2	61182.4	61196.0	-13.6	$98.24\,^2I + 1.74\,^2H + 0.01\,^4G$
13/2	61269.0	61279.5	-10.5	$100.00\,^2I$
5/2	64083.2	64152.4	-69.2	$50.60\,^2D3 + 30.71\,^2F1 + 15.81\,^2D2 + 1.31\,^4F$
3/2	65166.1	65173.5	-7.4	$70.33\,^2D3 + 21.79\,^2D2 + 7.62\,^4F + 0.24\,^4P$
7/2	67173.4	67085.1	88.3	$95.42\,^2F1 + 2.47\,^4F + 1.33\,^2G2 + 0.31\,^2F2$
5/2	68540.6	68444.9	95.7	$51.31\,^4F + 38.29\,^2F1 + 7.62\,^2D2 + 2.40\,^2D1$
9/2	68600.7	68551.2	49.5	$95.08\,^4F + 4.61\,^2G2 + 0.20\,^2H + 0.11\,^4G$
7/2	68824.9	68801.7	23.2	$94.58\,^4F + 2.79\,^2F1 + 1.20\,^2F2 + 0.72\,^2G2$
3/2	69246.9	69173.5	73.4	$92.15\,^4F + 6.21\,^2D3 + 1.44\,^2D1 + 0.19\,^4D$
5/2	69392.4	69447.0	-54.6	$45.82\,^4F + 29.82\,^2F1 + 17.47\,^2D3 + 4.72\,^2D1$
9/2	72850.4	72908.8	-58.4	$74.24\,^2H + 23.57\,^2G2 + 2.16\,^4F + 0.03\,^4G$
11/2	73702.9	73756.6	-53.7	$98.00\,^2H + 1.76\,^2I + 0.25\,^4G$
7/2	74636.2	74627.7	8.5	$97.77\,^2G2 + 0.97\,^2F1 + 0.96\,^4F + 0.22\,^2F2$
9/9	75455.7	75441.7	14.0.	$71.77\,^2G2 + 25.50\,^2H + 2.66\,^4F + 0.05\,^2G1$
5/2	79438.7	79391.4	47.3	$97.87\,^2F2 + 1.09\,^4F + 0.66\,^2D2 + 0.36\,^2F1$
7/2	79472.3	79608.3	-136.	$98.09\,^2F2 + 1.46\,^4F + 0.18\,^2G1 + 0.17\,^2F1$
1/2	86469.7	86532.1	-62.4	$99.75\,^2S + 0.12\,^4P + 0.12\,^2P$
3/2	96530.6	96461.2	69.4	$99.82\,^2D2 + 0.08\,^4D + 0.07\,^2D1 + 0.03\,^2D3$
5/2	96566.2	96566.5	-0.3	$99.18\,^2D2 + 0.65\,^2F2 + 0.12\,^4D + 0.03\,^2D1$
9/2	107893.6	107848.3	45.3	$99.94\,^2G1 + 0.06\,^2G2 + 0.01\,^4G$
7/2	107905.0	107887.0	18.0	$99.73\,^2G1 + 0.18\,^2F2 + 0.09\,^2G2 + 0.01\,^4G$
3/2	129969.2	129951.6	17.6	$99.15\,^2P + 0.62\,^2D1 + 0.23\,^2D3$
1/2	130008.1	130025.9	-17.8	$99.87\,^2P + 0.12\,^2S + 0.01\,^4D$
5/2	140880.4	140922.2	-41.8	$76.64\,^2D1 + 23.30\,^2D3 + 0.04\,^2D2 + 0.01\,^4F$
3/2	140907.7	141006.8	-99.1	$76.01\,^2D1 + 23.03\,^2D2 + 0.84\,^2P + 0.10\,^2D2$

take place only for Ni^{5+} because the energies of levels of its ground configuration stretch for 141006 cm^{-1}, but the energies of the levels closest of the excited $3d^44s$ configuration begin with 307843 cm^{-1}.

For the Co^{4+} ion the interval between energies of the levels of the ground and fist excited $3d^4 4s$ configurations is equal to 80410 cm^{-1} and for the Fe^{3+} ion it already is equal only 19508 cm^{-1}. However, the levels of the $3d^5$ ground configuration of Mn^{2+} ion, beginning with approximately 62000 cm^{-1}, are strongly overlapped with the levels of the $3d^4 4s$ configuration, while the upper $^2D_{3/2}$ level of ground configuration locate at 89543 cm^{-1}. In this context, the $3d^5$ configuration of the Mn^{2+}, Fe^{3+} and Co^{4+} ions, apparently, can not be considered as well isolated.

Table 3. Numerical values of parameters (in cm^{-1}) for the $3d^5$ configuration of the ions Mn^{2+}, Fe^{3+}, Co^{4+} and Ni^{5+} obtained in variant IV

Ion	B	C	ξ	α	β	T^2	T^3	M^0	M^2	P^2	P^4	σ
Mn^{2+}	917	3224	345	65	207	21	17	2.72	1.22	117	99	77.7
Fe^{3+}	1101	3946	486	81	-14	17	20	3.01	1.33	96	81	65.6
Co^{4+}	1264	4598	685	96	-145	0.1	21	3.12	2.94	70	59	55.0
Ni^{5+}	1420	5210	932	109	-232	-7	22	3.41	3.21	65	47	58.1

During the calculations within version IV, we found that the iterative procedure with variations in all parameters very rapidly converged to different minima of the merit function values; however, the sets of parameters of magnetic interactions obtained in this case were physically unacceptable. In particular, the Marvin integral M^2 was sometimes larger in magnitude than the integral M^0, or some of these integrals turned out to be negative. At the same time, it is known that all integrals M^k are positive by definition and, in addition, they should decrease in magnitude with increasing k. Sometimes, on the contrary, the integrals M^k satisfied the above-mentioned conditions but the parameters P^k, characterizing the H_{ELSO} interaction, demonstrated anomalous behavior; exactly parameter P^4 was larger in magnitude than the parameter P^2. Physically, since the matrix elements of H_{ELSO} are products of the matrix elements of H_{EL} and H_{SO}, the parameters P^k should behave like analogous Slater integrals F^k, i.e., should be positive and decrease with an increase in k. The existence of many sets of parameters at which approximately identical minimum values of the merit function are obtained apparently indicates that the problem of anomalous behavior parameters M^k or P^k is a purely mathematical problem related to the difficulty of searching for a physically acceptable minimum of a function of many variables. In this case, one could try to solve this problem through reasonable choice of the initial values of the parameters and use such an algorithm in iterations that would not allow the program to leave at any price the local minimum of the merit function. We imposed the following condition on the initial parameters M^k and P^k: their ratios M^2/M^0 and P^4/P^2 should correspond to those obtained in [30] in the interpretation of optical spectra of iron-group ions with $3d^3$ and $3d^7$ ground configurations. As an algorithm, we used the Fletcher–Powell gradient method [31], which provides the slow but reliable achievement of the nearest minimum. In the initial stage of the calculations, we varied all parameters except for M^2 and P^4 and, after attainment of the intermediate minimum, all the parameters including M^2 and P^4 were varied. The procedure rapidly converged to a new local minimum at the values of the parameters that are listed in table 3.

4. Discussion of the results

For most interactions, the parameters given in table 3 in general agree with the data obtained by similar calculations in other studies. It makes no sense to compare our results with the data of these papers in more detail because different authors used different, not our, sets of parameters. However, we may conclude that the use of Hamiltonian (1) allowed us to obtain the most correct semi empirical description of the experimental spectra. Further progress in such studies can be achieved, as was shown in

[20] on the example of the Cr^{3+} and Ni^{3+} ions, only by additionally taking into account the correlation effects of magnetic interactions using the Hamiltonian containing 17 effective orthogonal operators. As follows from Table 3, most of the parameters are positive and monotonically increase with increasing atomic number Z of the element in isoelectronic sequence. The β parameter of the linear CI theory and the electrostatically correlated spin-orbit interaction parameters P^2 and P^4, in contrast, decreases with increasing Z. The β parameter is positive at the beginning of sequence and then changes its sign.

It should be noted that, in addition to the four calculation variants considered above, we performed calculations in which, along with the parameters of variant III, either the relativistic H_{SS} and H_{SOO} interactions or the correlation H_{ELSO} effects were additionally taken into account. In another variant, we varied all the parameters of variant IV but excluded the H_{SS} interaction from the matrix of Hamiltonian (1). As a result, it was found that the theoretical interpretation of the spectra by the two variants (taking into account either only the parameters M^0 and M^2 or only P^2 and P^4) was considerably worse as compared to variant IV. In contrast, the calculation excluding the H_{SS} interaction gives even a slightly better coincidence between the theory and experiment. These calculations show that, for the description of the fine structure of the spectra of iron-group elements to be successful, it is necessary to take into account all the spin-dependent interactions simultaneously. The same conclusion was drawn by the authors of [6] from the analysis of the spectra of doubly ionized atoms of the iron-group elements Me III (Me = Ti, V, Cr, Mn, Fe, Co, and Ni).

5. Conclusions

The energy levels of the ground configuration of free ions Mn^{2+}, Fe^{3+}, Co^{4+} and Ni^{5+} are satisfactorily interpreted theoretically only if the Hamiltonian contains the two- and three-body operators representing the effects of correlation of the electrostatic interaction of electrons by configuration mixing. The best agreement between the theory and experiment is achieved when all the spin-dependent relativistic interactions and the correlation effects of the spin-orbit interaction are additionally and simultaneously taken into account. The found adjustable parameters of all the interactions monotonically change with the atomic numbers of the elements in the isoelectronic sequence. The current practice of analysis of the behavior of iron-group ions in crystals using a free-ion Hamiltonian that involves terms with only four parameters (B, C, ξ, and α) seems to be erroneous since it is incapable of predicting correctly the levels of even a free ion. Such calculations may lead to wrong conclusions concerning the crystal-field effects and the electron-phonon interaction. Most appropriate is, evidently, the method based on the use for a free ion of an 11-parameter Hamiltonian, with the parameters of the weak interactions being fixed while the parameters of the crystal field and strong interactions are varied.

Acknowledgments

This work was supported by the Kazan Federal University (project F11-22). The author is grateful to E.N. Irinyakov for his help in computer programming of calculations.

References

[1] Trees R E 1951 *Phys. Rev.* **83** 756
[2] Trees R E 1952 *Phys. Rev.* **85** 382
[3] Trees R E 1964 *J. Opt. Soc. Am.* **54** 651
[4] Rajnak K and Wybourne B G 1963 *Phys. Rev.* **132** 280
[5] Judd B R, Crosswhite H M and Crosswhite H 1968 *Phys. Rev.* **169** 130
[6] Pasternak A and Goldshmidt Z B 1972 *Phys. Rev.* **A6** 55
[7] Judd B R 1966 *Phys. Rev.* **141** 4
[8] Feneuille S 1966 *C. R. Acad. Sci. (Paris)* **262** 23

[9] Carnall W T, Goodman G I, Rajnak K and Rana R S 1989 *J. Chem. Phys.* **90** 3443
[10] Tanner P A, Kumar V V R K, Jayasakar C K and Reid M F 1994 *J. Alloys Comp.* **215** 349
[11] Wells J P R, Dean T and Reeves R J 2002 *J. Luminesc.* **96** 239
[12] Sviridov D T and Smirnov Yu F 1977 *Theory of the Optical Spectra of Transition-Metal Ions* (Moscow: Nauka) [in Russian]
[13] Chang Y M, Rudowicz C and Yeung Y Y 1994 *Comp. Phys.* **8** 583
[14] Rebane K K 1968 *Impurity Spectra of Solids: Elementary Theory of Vibrational Structure* (Moscow: Nauka) [in Russian] 1970 (New York: Plenum Press)
[15] Pasternak A and Goldshmidt Z B 1974 *Phys. Rev.* **A9** 1022
[16] Judd B R, Hansen J E and Raassen A J J 1982 *J. Phys.* **B15** 1457
[17] Uylings P H M 1984 *J. Phys.* **B17** 2375
[18] Uylings P H M, Raassen A J J and Wyart J F 1984 *J. Phys.* **B17** 4103).
[19] Judd B R 1985 *J. Phys.* **B18** 2327
[20] Hansen J E, Judd B R, Raassen A J J and Uylings P H M 1997 *Phys. Rev. Lett.* **78** 3078
[21] Ralchenko Yu, Kramida A E, Reader J and NIST ASD Team (2010) *NIST Atomic Spectra Database* (ver.4.0.1) [Online]
[22] Wybourne B G 1964 *J. Chem. Phys.* **40** 1457
[23] Marvin H H 1947 *Phys. Rev.* **71** 102
[24] Judd B R 1963 *Operator Techniques in Atomic Spectroscopy* (New York: McGraw-Hill)
[25] Wybourne B G 1965 *Spectroscopic Propertie of rare Earths* (London, Sydney: John Wiley and Sons, Inc, N.Y)
[26] Yutsis A P and Savukinas A Yu 1973 *Mathematical Foundations of the Atomic Theory* (Vilnius: Mintis) [in Russian]
[27] Sobel'man I I 1977 *Atomic Spectra and Radiative Transitions* (Moscow: Nauka [in Russian], 1979 Berlin: Springer)
[28] Nikitin A A and Rudzikas Z B 1983 *Principles of the Theory of the Spectra of Atoms and Ions* (Moscow: Nauka) [in Russian]
[29] Nielsen C W and Koster G F 1963 *Spectrocopic Coefficients for the* p^n, d^n, *and* f^n *Configurations* (Cambridge: MIT Press)
[30] Leushin A M and Irinyakov E N 2006 *Optics and Spectroscopy* **100** 326
[31] Fletcher R and Powell M J D 1963 *Comp. J.* **6** 163

Spectral dependence of nonlinear optical absorption of silica glass with copper nanoparticles

A N Golubev[1], S I Nikitin[1], M A Smirnov[1] and A L Stepanov[1,2]

[1]Kazan Federal University, Kremlevskaya 18, Kazan, Russia
[2]Zavoisky Physical-Technical Inst. of RAS, Sibirsky Trakt 10/7, Kazan, Russia

E-mail: golubevandrei@yandex.ru

Abstract. The nonlinear optical properties of silica glass with copper nanoparticles synthesized by ion implantation were investigated by z-scan method in nanosecond time scale. The reverse saturation absorption (RSA) at the wavelength range of 450-540 nm and saturation absorption (SA) at 550-585 nm were observed. It was supposed that the two-photon electron absorption from bound of *d*-states determined the RSA effect and the SA is due to saturation of plasmon excitation.

1. Introduction

Looking for new photonic materials and the design of ultrafast optical devices with these materials are determining the progress of modern optoelectronics. Composite materials based on optical transparent dielectrics with metal nanoparticles (MNPs) are prospective for this purpose. Such composites can be used for light signal controlling in dielectric waveguides as ultrafast nonlinear optical switchers, light modulators and optical limiters [1]. Additionally to application aims there are still some fundamental questions in nature of nonlinear optical effects in spite of numerous studies of the composites with MNPs. The development of optimal methods for MNP synthesis to produce composites with controlled optical properties is also the actual technological task.

Two decades ago ion implantation as a method for MNP synthesis in dielectrics was suggested for fabrication of novel nonlinear optical materials. The first paper devoted to investigation of nonlinear optical properties of dielectric materials with ion-synthesized MNPs was published in 1991 [2]. In this work the Au nanoparticles in SiO_2 glass were studied. At 532 nm wavelength, which is near of plasmon absorption maximum of Au nanoparticles (λ_{max}=525 nm), the giant value of $\chi^{(3)} \sim 10^{-7}$ esu was observed. The same order of the real part of $\chi^{(3)}$ in soda-lime silicate glass with implanted Cu nanoparticles - $Re(\chi^{(3)})=1.3 \cdot 10^{-7}$ esu was detected at 532 nm [3]. These values measured with nanosecond laser pulses are five orders higher than one for the standard nonlinear material CS_2 with $Re(\chi^{(3)}) \sim 10^{-12}$ esu. Researches of nonlinear optical properties of materials with MNPs were continued on Cu:SiO_2, Ag:SiO_2 and Au:SiO_2 for different spectral areas in picosecond time scale [4-8].

In general the nonlinear optical properties of dielectrics with MNPs dependent on many parameters: nanoparticles size, size distribution and shape, concentration, dielectric environment etc. For different synthesis conditions of MNPs composites based on same SiO_2 matrix can demonstrate none equivalent nonlinear optical properties. For example, in the work [9] it was shown that in SiO_2 glass containing Cu nanoparticles fabricated by ion-exchange process after thermal annealing at 500-650^{0}C reveals different kind of nonlinear absorptions (nonlinear absorption was measured with z-scan technique for the laser pulse duration of 7 ns at 532 nm). Thermal annealing at 500^{0}C initiates the SA

Published under licence by IOP Publishing Ltd

of the composite whereas after treatment at 600 and 650 ^{0}C the RSA was detected. The SA in the sample was changed to RSA at higher laser intensities also. Such changes of nonlinear absorption were also observed for SiO_2 with implanted Cu nanoparticles for picosecond laser pulses (55 ps) at the same wavelength [8]. Thus, nonlinear optical properties of dielectrics with MNPs are determined not only by chemical compositions but also by a selected synthesis method and by an arrangement of nanoparticle structure in the samples. This happened because technological processes of MNP synthesis are not still known well.

The aim of present article is to study of nonlinear absorption of composite such as SiO_2 glass with ion-implanted Cu nanoparticles in wide spectral area around plasmon resonance absorption in nanosecond pulse laser regime.

2. Experimental method

Composite material on basis of silica glass with copper nanoparticles was synthesized by ion implantation using Wikham accelerator (Sussex University, UK). Implantation was performed by one-charged Cu^+ ions with energy of 160 keV, dose of $8\cdot10^{16}$ ion/cm^2 and current density of 10 μA/cm^2. Thickness of layer with MNPs was about 100 nm and average particle size was ~ 5 nm [10]. Transmission spectra of Cu:SiO_2 sample and virgin glass (figure 1) were measured using dual-beam spectrophotometer SPECORD M40 in 200-900 nm wavelengths range. Registered absorption band at 560 nm of the implanted glass (figure 1) corresponds to surface plasmon resonance in copper nanoparticles [11].

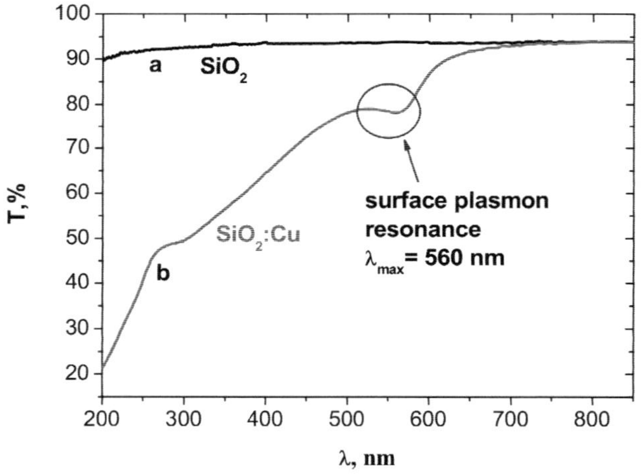

Figure 1. Transmission spectrum of silica glass (a) and silica glass with copper nanoparticles (b).

For measurement of nonlinear optical absorption the *z*-scan method was applied [12]. The *z*-scan setup is presented in figure 2. In experiments Solar LS LX325 Ti:Sapphire and dye lasers were used. These lasers were pumped with second harmonic of Nd:YAG Solar LS LQ829 laser and second or third harmonics of Nd:YAG Solar LS LQ129, respectively. After some modifications of the lasers the laser beams have the almost Gaussian profile. Laser pulse widths on half-height were measured using the high-speed photodetector and they were equals to 3-6 ns for dye laser, 12 ns for Nd:YAG and 10-30 ns for Ti:Sapphire laser.

Laser beam was focused by lens with a focal length of 18 cm. Spots radii in beam waist were 10-15 μm. Sample was mounted on motorized translation stage. Signals from photodiodes were digitized using 14-bit ADC. Value of normalized transmission $T(z)$ was defined as ratio of the amplitude of the signals from photodiode 2 to photodiode 1 and was averaged by 300-500 pulses at the each sample position z. Obtained dependences of $T(z)$ are plotted as functions of sample position z respect to the focus of lens. For elimination thermal effects pulse repetition rate was not exceed 10 Hz.

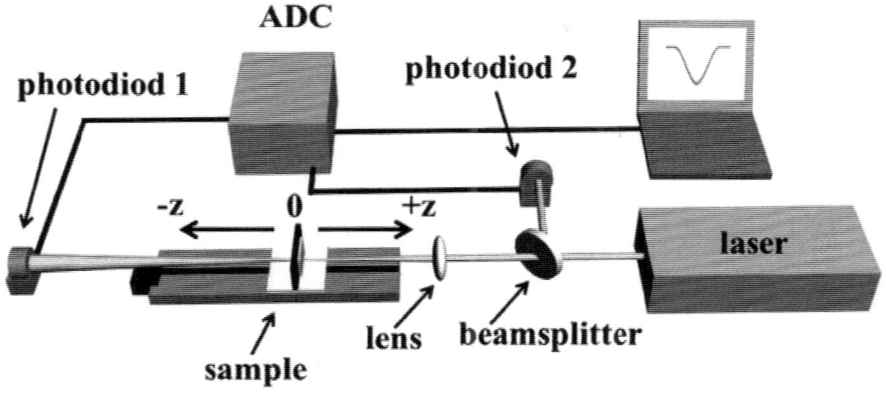

Figure 2. Experimental z-scan setup.

3. Results and discussion

The study of the nonlinear optical properties of the silica glass with copper nanoparticles was done in the broad spectral range from 450 to 585 nm. In dependence on a laser wavelength the $Cu:SiO_2$ sample demonstrates the RSA or the SA. In figure 3 results of the $T(z)$ measurements of the sample for two wavelengths – 510 and 560 nm are presented. At these wavelengths the values of the nonlinear RSA (figure 3a) and SA (figure 3b) are maximal. It is important to point out, that the shape of $T(z)$ like presented on figure 3a can be caused by optical damage of the sample, which appears as a result of the overheating with following destruction of MNPs under laser irradiation. In this case a decreasing $T(z)$ near the lens focus will be induced by increasing of scattered laser emission on the optically destructed region. When the laser beam diameter becomes comparable with this destructed region contribution to $T(z)$ is most significant. However, it was measured that laser damage threshold of nanoparticles in the our setup was around 4 μJ. Therefore all measurements were done at smaller laser energy.

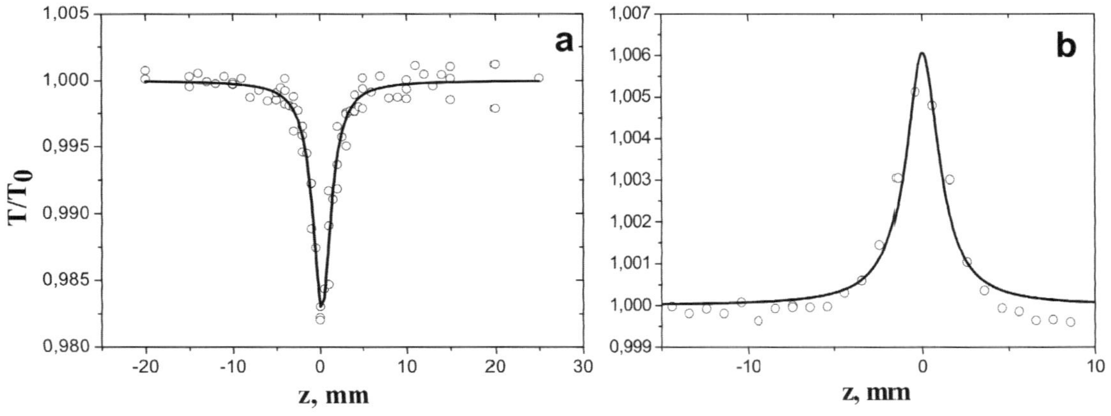

Figure 3. The measured normalized transmission - open circle of $Cu:SiO_2$ as function of sample position at z for different wavelengths λ_{ex}: a.- reverse saturable absorption at 510 nm and b.- saturable absorption at 560 nm. Fitting was done using equation (5) – solid line.

Absorption coefficient in the case of only the RSA is written by

$$\alpha_{RSA}(I) = \alpha_1 + \beta_1 I. \tag{1}$$

The SA can be presented as

$$\alpha_{SA}(I) = \frac{\alpha_2}{1 + I/I_s}, \tag{2}$$

where α_i – the linear absorption coefficients, I – the laser intensity, I_s – the saturation intensity and the nonlinear absorption coefficient $\beta_1 > 0$. In an approximation of the small nonlinearity ($I << I_s$), which was realized in present experiments, the coefficient of the SA (2) will be

$$\alpha_{SA}(I) = \alpha_2 - \alpha_2 I/I_s = \alpha_2 + \beta_2 I, \text{ where } \beta_2 = -\alpha_2/I_s < 0 \tag{3}$$

Thus in the case of small nonlinearity the expressions of nonlinear absorption coefficients for RSA and SA are similar but they differ with sign of the β_i.

Taking into account equations (1) and (3), in the general case the $T(z)$ in z-scan method can be obtained by solving equation [12]:

$$dI / dz = -(\alpha + \beta I)I, \tag{4}$$

where z' - propagation depth in the sample, α and β – are linear and nonlinear absorption coefficients taking into account the all possible transitions, respectively. For Gaussian beam the dependence $T(z)$ is given by expression:

$$T(z) = 1 - \frac{I_0 L_{eff} \beta}{1 + z^2/z_0^2}, \tag{5}$$

where I_0' - maximum intensity of the laser beam on its axis in the waist, L_{eff} – effective thickness of layer with MNPs, $L_{eff} = (1 - e^{-\alpha L})/\alpha$ ($L \cong 100$ nm for the present sample); z_0 – diffraction length. Also in such approximation it is possible to write:

$$I_0 L_{eff} = 1 - \frac{(1 - e^{-\alpha L})E}{\alpha \pi^{3/2} \sqrt{2} \omega_0^2 \tau}, \tag{6}$$

where E – laser pulse energy, τ – laser pulse duration, ω_0 – size of the laser beam waist.

It was mentioned above that for the Cu:SiO_2 sample the SA and the RSA were observed at different wavelengths at same laser energy. It means that the absorption coefficient consists of the two parts corresponding to the SA and the RSA and the contributions from these parts depend on wavelength. Using linear coefficient α obtained from transmission measurements (figure 1) the nonlinear part of the absorption coefficient (β) was estimated by fitting with the expression (5). Collected data for different wavelength are presented in figure 5.

As seen from figure 4 the kind of the nonlinear absorption in Cu:SiO_2 sample depends on wavelength: the RSA prevails in spectral range of 450-540 nm, inverse situation was observed at 550 – 585 nm where the main contribution is from the SA. The SA dominates for spectral area where the linear absorption is determined mainly by the surface plasmon resonance in copper nanoparticles (figure 4). The RSA was observed when the linear absorption is described by electron excitations from the *d*-zone of copper particles.

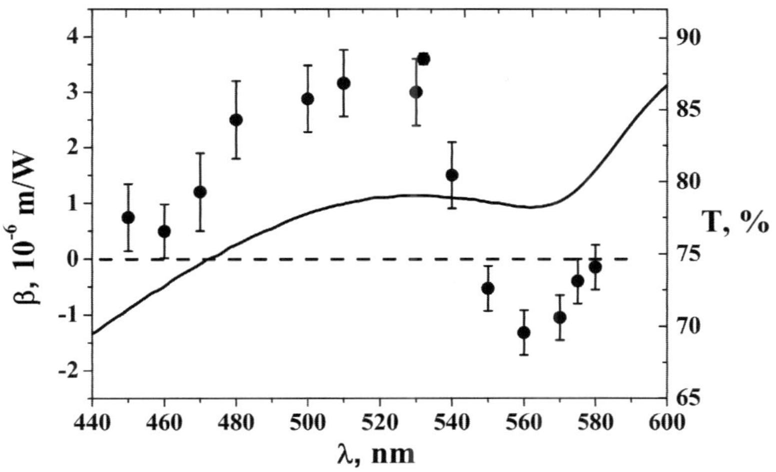

Figure 4. Transmission (T) of the $Cu:SiO_2$ (solid line) and dependence of nonlinear optical absorption coefficient β on laser wavelengths (black circles).

In literature the several mechanisms of the optical nonlinearity of the MNPs were assumed [13, 14], which are determined due to intraband, interband electron transition or free carrier absorption. Obtained in present experiment results evidence that different kind absorptions are responsible for the nonlinear effects of $Cu:SiO_2$ sample. The spectral dependence of nonlinear coefficient β allows to suppose that the mechanism responsible for the RAS at 450-480 nm wavelengths is the two-photon absorption which is determined by electron transitions from the bound *d*-states to conduction band of copper nanoparticles. On the other hand, the SA effect is based on saturation of the plasmon excitation in MNPs. For deeper understanding of the proposed mechanisms the time-resolved measurements of the nonlinear absorption are planning to be done.

4. Conclusion

In nanosecond time scale of laser irradiation in a spectral area near surface plasmon resonance absorption band of copper nanoparticles in silica glass the RSA in range of 450-540 nm and the SA in range of 550-585 nm were observed for the first time. It was supposed that the RSA is determined by two-photon absorption due to electron transitions from bound of *d*-states to conduction band of copper nanoparticles and the SA effect is due to saturation of plasmon excitation.

Acknowledgments

This work was supported in part by the Ministry of Education and Science of the Russian Federation (No. 02.740.11.0797) and the Russian Foundation for Basic Research (No. 11-02-90420-Ukr).

5. References

[1] Haglund Jr 1974 Quantum-dot composites for nonlinear roptical applications *Optics of small particles, interfaces and surfaces* ed. R Hummel and P Wissmann (London: CRC Press) chapter 2 pp 192-224
[2] Fukumi K, Chayahara A and Kadono K 1991 *Japan. J. Appl. Phys.* **30** L742-L744
[3] Tsuji H, Kido S, Gotoh Y and Ishikawa J 2000 *Rev. Sci. Instr.* 2000 **71** 804-806
[4] Haglund Jr, Magruder R, Morgan S, Henderson D, Weller R, Yang L and Zuhr R 1992 *Nucl. Instr. Meth. Phys. Res.* **65** 405-411
[5] Haglund Jr, Yang L, Magruder H, Wittig J, Becker K, and Zuhr R 1993 *Opt. Lett.* **18** 373-375

[6] Magruder R, Yang Li, Haglund Jr, White C, Yang L, Dorsinville R and Alfano R 1993 *Appl. Phys. Lett.* **62** 1730-1733
[7] Ganeev R, Ryasnyansky A, Stepanov A and Usmanov T 2004 *Phys. Stat. Sol. B.* **241** 935-944
[8] Ganeev R, Ryasnyansky A, Stepanov A and Usmanov T 2004 *Phys. Stat. Sol. B.* **241** R1-R4-944
[9] Karthikeyan B, Anija M, Sandeep S, Nadeer M and Philip R 2008 *Opt. Comm.* **281** 2933-37
[10] Stepanov A, Valeev V, Nuzhdin V and Faizrakhmanov I 2009 *Phys. Solid. State* 2009 **51** 1912-18
[11] Kreibig U and Volmer M 1995 *Optical properties of metal clusters* (Berlin: Springer)
[12] Sheik-Bahae M, Said A, Wei T, Hagan D and Stryland E 1990 *EEE J. Quant. El.* **26** 760-769
[13] Hache F, Ricard D and Girard C 1988 *Phys. Rev. B* **38** 7990–96
[14] Hache F, Ricard D, Flytzanis C and Kreibig U 1988 *Appl. Phys. A* **47** 347–357

Crystal electric field parameters for Yb^{3+} ion in $YbRh_2Si_2$

A S Kutuzov and A M Skvortsova

Kazan Federal University, Kremlevskaya, 18, Kazan 420008, Russian Federation

E-mail: Alexander.Kutuzov@gmail.com

Abstract. The tetragonal crystal electric field parameters for Yb^{3+} ion in $YbRh_2Si_2$ are determined from the analysis of the literature data on angle-resolved photoemission, inelastic neutron scattering and electron paramagnetic resonance.

1. Introduction

$YbRh_2Si_2$ and $YbIr_2Si_2$ belong to the class of heavy-fermions Kondo-lattice compounds which attracted the interest of scientists due to the very peculiar magnetic, thermal and transport properties. The discovery of electron paramagnetic resonance (EPR) in the Kondo lattices $YbRh_2Si_2$ and $YbIr_2Si_2$ below the thermodynamically measured Kondo temperatures was very unexpected [1-3] and stimulated different approaches for its explanation [4-7]. In particular, in [6, 7] is shown that the idea of a collective spin motion of quasi-localized Yb *f*-electrons and wide-band electrons successfully explains the EPR signal existence. In [8] the static magnetic susceptibility of $YbRh_2Si_2$ and $YbIr_2Si_2$ were also successfully studied, based on entirely local properties of Yb^{3+} ions in a tetragonal crystal electric field (CEF). Until recently the available experimental data were not sufficient to unambiguously determine the CEF parameters and therefore eigenfunctions of Yb^{3+} ion [9] (but attempts have been made [10, 11]). Though the energy splittings between CEF states of Yb^{3+} ion in $YbRh_2Si_2$ are known from inelastic neutron scattering (INS) experiments [12], but the symmetry identification of CEF levels was only recently carried out by performing angle-resolved photoemission spectroscopy (ARPES) experiments [13]. This new information together with INS and EPR data give us an opportunity to find one set of tetragonal CEF parameters for Yb^{3+} ion in $YbRh_2Si_2$, as reported in details in this paper.

Besides, Yb^{3+} ion was used as a probe in EPR experiments in high-temperature superconductors, in particular $YBa_2Cu_3O_x$ ($6 < x < 7$) compounds, in which yttrium was partially substituted by ytterbium [14-18]. The samples with oxygen content $x < 6.4$ have a tetragonal crystal symmetry. So, the problem of Yb^{3+} ion in tetragonal CEF seems to be actual and important.

2. Yb^{3+} ion in tetragonal CEF

A free Yb^{3+} ion has a $4f^{13}$ configuration with one term ^2F. The spin-orbit interaction splits the ^2F term into two multiplets: $^2F_{7/2}$ with $J = 7/2$ and $^2F_{5/2}$ with $J = 5/2$, where J is the value of the total momentum $\mathbf{J} = (J_x, J_y, J_z)$. Multiplets are separated by about 1 eV [19]. As the spin-orbit coupling is much stronger than the CEF in the case of rare earth, we will consider only the ground multiplet $^2F_{7/2}$ with states $|J = 7/2, M_J\rangle \equiv |M_J\rangle$, where M_J is the eigenvalue of J_z. If z axis coincides with tetragonal axis of CEF and x and y axes are directed along the twofold axes then the Hamiltonian of the Yb^{3+} ion interaction with the tetragonal CEF could be written via equivalent operators $O_k^q(\mathbf{J})$ [19] as follow:

Published under licence by IOP Publishing Ltd

$$H = \alpha B_2^0 O_2^0 + \beta(B_4^0 O_4^0 + B_4^4 O_4^4) + \gamma(B_6^0 O_6^0 + B_6^4 O_6^4)\,, \tag{1}$$

where B_k^q are the CEF parameters, $\alpha = 2/63$, $\beta = -2/1155$, $\gamma = 4/27027$ [19].

From the group theory it is known that the two-valued irreducible representation $D^{7/2}$ of rotation group contains two two-dimensional irreducible representations Γ_7^t and Γ_6^t of the double tetragonal group: $D^{7/2} = 2\Gamma_7^t + 2\Gamma_6^t$ [19]. Therefore the states of Yb^{3+} in the tetragonal CEF are four Kramers doublets. As the decomposition of $D^{7/2}$ includes twice each of representations Γ_7^t and Γ_6^t, the matrix of operator (1) could be expressed via two two-dimensional matrices

$$\begin{pmatrix} 2C_1 & C_3 \\ C_3 & 2C_2 \end{pmatrix} \text{ and } \begin{pmatrix} 2A_1 & A_3 \\ A_3 & 2A_2 \end{pmatrix}, \tag{2}$$

the former corresponding to bases $|5/2\rangle$, $|-3/2\rangle$ and $|-5/2\rangle$, $|3/2\rangle$, the latter corresponding to bases $|7/2\rangle$, $|-1/2\rangle$ and $|-7/2\rangle$, $|1/2\rangle$. It is convenient to introduce parameters C, A and D:

$$C = C_1 - C_2 = 4B_2^0/21 + 40B_4^0/77 - 560B_6^0/429\,, \quad A = A_1 - A_2 = 4B_2^0/7 + 8B_4^0/77 + 80B_6^0/143\,, \tag{3}$$

$$D = -C_1 - C_2 = A_1 + A_2 = 2B_2^0/21 - 64B_4^0/77 - 160B_6^0/429\,,$$

where $C_1 + C_2 + A_1 + A_2 = 0$ as traces of O_k^q are equal to zero. C_3 and A_3 are

$$C_3 = -8\sqrt{3}B_4^4/77 - 80\sqrt{3}B_6^4/1287\,, \qquad A_3 = -8\sqrt{35}B_4^4/385 + 80\sqrt{35}B_6^4/3003\,. \tag{4}$$

Let us define eigenvectors of matrices (2) $(c_{1,2}\,, \pm c_{2,1})$ and $(a_{1,2}\,, \pm a_{2,1})$ via angular parameters φ_7 and φ_6 which correspond to Γ_7^t and Γ_6^t symmetries:

$$c_1 = \cos(\varphi_7/2), \quad c_2 = \sin(\varphi_7/2) \quad \text{and} \quad a_1 = \cos(\varphi_6/2), \quad a_2 = \sin(\varphi_6/2). \tag{5}$$

Since matrices (2) are diagonal in the bases of their eigenvectors we can find the relations between our angular parameters and CEF parameters:

$$\tan\varphi_7 = \frac{C_3}{C} = -\frac{2}{\sqrt{3}}\left(\frac{117B_4^4 + 70B_6^4}{143B_2^0 + 390B_4^0 - 980B_6^0}\right), \quad \tan\varphi_6 = \frac{A_3}{A} = -\frac{2}{3}\sqrt{\frac{7}{5}}\left(\frac{39B_4^4 - 50B_6^4}{143B_2^0 + 26B_4^0 + 140B_6^0}\right), \tag{6}$$

it is enough to take $-\pi/2 \le \varphi_{6,7} \le \pi/2$. The eigenenergies E_k, wave functions and g-factors of Kramers doublets are given in table 1. In this table ${}^k\Gamma_7^t$ and ${}^k\Gamma_6^t$ are symmetry symbols, where $k = 1..4$ is the number of Kramers doublet. The arrow ↑ or ↓ in wave functions corresponds to the upper or lower sign and denotes up and down effective spin projection. They have been chosen such that $\langle\uparrow| J_+ |\downarrow\rangle \ne 0$, where $J_+ = J_x + iJ_y$. Moreover, the phases of the wave function have been chosen as $\theta|\uparrow\rangle = |\downarrow\rangle$, where θ is a time reversing operator [19]. In g-factors left and right indexes correspond to the upper and lower signs; $g_J = 8/7$ is the Lande g-factor.

Table 1. Energies, wave functions and g-factors of Yb^{3+} ion in tetragonal crystal electric field.

$E_{1,2} = -D \pm C/\cos\varphi_7$	$E_{3,4} = D \pm A/\cos\varphi_6$
$\|{}^1\Gamma_7^t\ \uparrow,\downarrow\rangle = \pm c_1\,\|\pm 5/2\rangle \pm c_2\,\|\mp 3/2\rangle$	$\|{}^3\Gamma_6^t\ \uparrow,\downarrow\rangle = \pm a_1\,\|\mp 7/2\rangle \pm a_2\,\|\pm 1/2\rangle$
$\|{}^2\Gamma_7^t\ \uparrow,\downarrow\rangle = \mp c_2\,\|\pm 5/2\rangle \pm c_1\,\|\mp 3/2\rangle$	$\|{}^4\Gamma_6^t\ \uparrow,\downarrow\rangle = \mp a_2\,\|\mp 7/2\rangle \pm a_1\,\|\pm 1/2\rangle$
$g_{\parallel}({}^{1,2}\Gamma_7^t) = g_J(5c_{1,2}^2 - 3c_{2,1}^2) = g_J(1 \pm 4\cos\varphi_7)$	$g_{\parallel}({}^{3,4}\Gamma_6^t) = g_J(a_{2,1}^2 - 7a_{1,2}^2) = -g_J(3 \pm 4\cos\varphi_6)$
$g_{\perp}({}^{1,2}\Gamma_7^t) = \mp 4\sqrt{3}g_J c_1 c_2 = \mp 2\sqrt{3}g_J \sin\varphi_7$	$g_{\perp}({}^{3,4}\Gamma_6^t) = -4g_J a_{2,1}^2 = -2g_J(1 \mp \cos\varphi_6)$

The Zeeman energy $g_J\mu_B\mathbf{HJ}$ in the basis $|\uparrow\rangle$, $|\downarrow\rangle$ of each doublet could be represented by matrix

$$H_{\text{Zeeman}} = g_{\parallel}\mu_B H_z S_z + g_{\perp}\mu_B(H_x S_x + H_y S_y)\,, \tag{7}$$

where

$$g_{\parallel} = 2g_J \langle \uparrow | J_z | \uparrow \rangle , \qquad g_{\perp} = g_J \langle \uparrow | J_+ | \downarrow \rangle \tag{8}$$

and **H** is the magnetic field, **S** is the effective spin operator with $S = 1/2$, μ_B is the Bohr magneton, $g_{\parallel}$ and $g_{\perp}$ are g-factors when the field is applied parallel and perpendicular to the tetragonal z-axis, respectively (table 1).

In the case of cubic symmetry $B_2^0 = 0$, $B_4^4 = 5B_4^0$ and $B_6^4 = -21B_6^0$, so that $\tan\varphi_7 = -\sqrt{3}$, $\tan\varphi_6 = -\sqrt{35}$, $c_1 = \sqrt{3}/2$, $c_2 = -1/2$, $a_1 = \sqrt{7/12}$, $a_2 = -\sqrt{5/12}$. In accordance with expansion $\Gamma_8 = \Gamma_7^t + \Gamma_6^t$ [19] the doublets ${}^2\Gamma_7^t$ and ${}^3\Gamma_6^t$ merge into a cubic quartet Γ_8 with energy $E(\Gamma_8) = -16B_4^0/77 + 1280B_6^0/429$. The doublets ${}^1\Gamma_7^t$ and ${}^4\Gamma_6^t$ turn into cubic doublets Γ_7 and Γ_6 with energies $E(\Gamma_7) = 144B_4^0/77 - 320B_6^0/143$ and $E(\Gamma_6) = -16B_4^0/11 - 1600B_6^0/429$ and with isotropic g-factors $g(\Gamma_7) = 3g_J = 3.429$ and $g(\Gamma_6) = -7/3g_J = -2.667$, respectively (see figure 1). Here $\Gamma_{6,7,8}$ are irreducible representations of double cubic group [19].

As g-factors of each doublet depend only on one parameter φ_6 or φ_7 (table 1) we can find the equation relating $g_{\parallel}$ and $g_{\perp}$. Figure 1 represents the diagram of g-factors. The solid and dashed parts of the line $g_{\parallel} + 2g_{\perp} + 7g_J = 0$ correspond to the doublets ${}^4\Gamma_6^t$ and ${}^3\Gamma_6^t$, the solid and dashed parts of the ellipse $(g_{\parallel} - g_J)^2/4 + g_{\perp}^2/3 = 4g_J^2$ correspond to the doublets ${}^2\Gamma_7^t$ and ${}^1\Gamma_7^t$. The line and the ellipse meet in the point $(-g_J, -3g_J)$ marked by a star.

On the g-diagram (figure 1) we marked the experimental g-points which correspond to the absolute values of $YbRh_2Si_2$ g-factors measured in the EPR experiments at 5 K [1, 2] (four points with different signs of $g_{\parallel}$ and $g_{\perp}$). A slight difference between experimental and theoretical g-factors can be explained mainly by taking into account the Kondo interaction, i.e. an exchange coupling between the $4f$-electrons of the Yb^{3+} ion and conduction electrons [8]. This g-diagram do not allow us to choose between ground doublets ${}^2\Gamma_7^t$ and ${}^4\Gamma_6^t$ as the value of g-shift is unknown. But in recent ARPES experiments on $YbRh_2Si_2$ it was determined that the ground doublet is ${}^2\Gamma_7^t$ [13]. Points on the ellipse (figure 1) nearest to the experimental g-points ($g_{\parallel} = -0.17$, $g_{\perp} = \pm 3.561$) correspond to $\varphi_7 = \pm 1.2660$ ($g_{\parallel} = -0.229$, $g_{\perp} = \pm 3.777$). These values of φ_7 we will use in our further calculations.

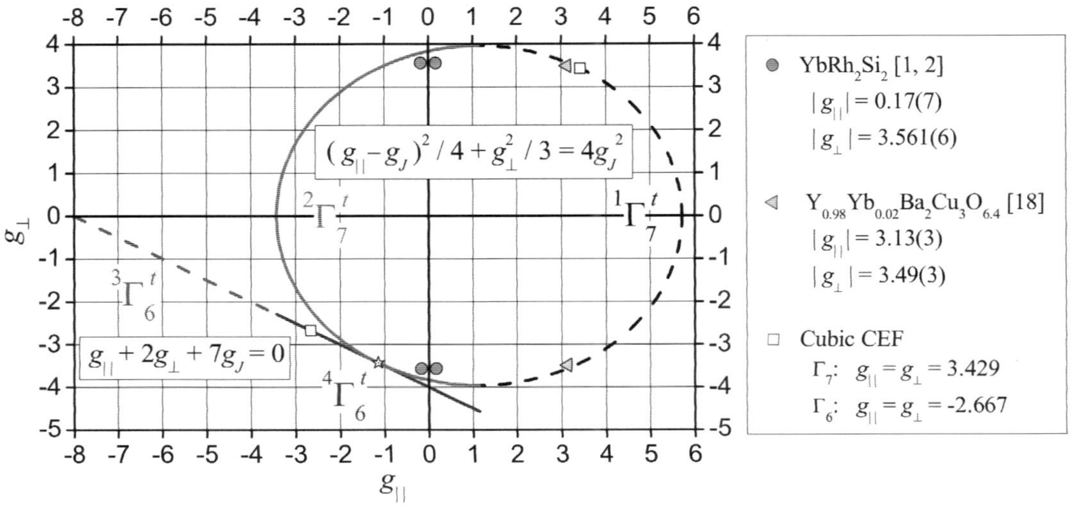

Figure 1. Diagram of g-factors of Yb^{3+} ion in tetragonal crystal electric field.

3. Determination of CEF parameters

Let us find all possible sets of tetragonal CEF parameters for the given exited doublets energies $\Delta_1 < \Delta_2 < \Delta_3$. It follows from (3) that

$$B_2^0 = 3A/2 + C/2 + D/2\,, \quad B_4^0 = A/16 + 5C/16 - D\,, \quad B_6^0 = 39A/160 - 91C/160 - 13D/40 \quad (9)$$

and from (4) that

$$B_4^4 = -7\sqrt{35}A_3/16 - 35\sqrt{3}C_3/16\,, \qquad B_6^4 = 117\sqrt{35}A_3/160 - 273\sqrt{3}C_3/160\,. \quad (10)$$

Taking one of the doublets with energy E_k (table 1) as the ground, defining the differences of doublets energies as $E_{mk} = E_m - E_k$ and solving this system of linear equations we can express C, A and D through E_{mk}. Substituting relations $A_3 = A \tan \varphi_6$ and $C_3 = C \tan \varphi_7$ into (10) and then C, A and D into (9) and (10) we find:

$$B_2^0 = \frac{1}{8}b + \frac{3}{4}b_6 \cos\varphi_6 + \frac{1}{4}b_7 \cos\varphi_7\,,$$

$$B_4^0 = -\frac{1}{4}b + \frac{1}{32}b_6 \cos\varphi_6 + \frac{5}{32}b_7 \cos\varphi_7\,, \qquad B_4^4 = -\frac{7\sqrt{35}}{32}b_6 \sin\varphi_6 - \frac{35\sqrt{3}}{32}b_7 \sin\varphi_7\,, \quad (11)$$

$$B_6^0 = -\frac{13}{160}b + \frac{39}{320}b_6 \cos\varphi_6 - \frac{91}{320}b_7 \cos\varphi_7\,, \qquad B_6^4 = \frac{117\sqrt{35}}{320}b_6 \sin\varphi_6 - \frac{273\sqrt{3}}{320}b_7 \sin\varphi_7\,,$$

where b, b_6 and b_7 are determined in table 2. To use (11) we have to choose the ground doublet and the exited doublets sequence to express energy differences E_{mk} in table 2 through experimental values $\Delta_1 < \Delta_2 < \Delta_3$. Angular parameters φ_6 and φ_7 can take the values $-\pi/2 \le \varphi_{6,7} \le \pi/2$ independently, the energy scheme does not depend on them. Formulas (11) are suitable for Yb^{3+} ion in any tetragonal CEF. Now we apply (11) for Yb^{3+} ion in $YbRh_2Si_2$.

Table 2. b, b_6 and b_7 in (11).

Ground doublet	b	b_6	b_7
${}^1\Gamma_7'$	$E_{31} - E_{21} + E_{41}$	$E_{31} - E_{41}$	$-E_{21}$
${}^2\Gamma_7'$	$E_{32} - E_{12} + E_{42}$	$E_{32} - E_{42}$	E_{12}
${}^3\Gamma_6'$	$E_{43} - E_{13} - E_{23}$	$-E_{43}$	$E_{13} - E_{23}$
${}^4\Gamma_6'$	$E_{34} - E_{14} - E_{24}$	E_{34}	$E_{14} - E_{24}$

According to INS experiments on $YbRh_2Si_2$ [12] the excited doublets energies approximately equal to $\Delta_1 = 17$, $\Delta_2 = 25$, $\Delta_3 = 43$ meV relative to the ground doublet. The Kramers doublets sequence was recently determined from ARPES experiments [13]: ${}^2\Gamma_7'$, ${}^4\Gamma_6'$, ${}^1\Gamma_7'$, ${}^3\Gamma_6'$. Above we found φ_7 from experimental g-factors of ${}^2\Gamma_7'$ ground doublet. So, we need only the value of φ_6 to find CEF parameters using formulas (11).

Let us consider INS experiments [12] in more detail. Magnetic part of INS spectrum of $YbRh_2Si_2$ at $T = 1.5$K [12] is presented on figure 2 (we have digitized data from Fig. 2 of [12]). Three peaks correspond to the transitions between ground doublet and excited doublets. In [12] magnetic response has been fitted quite well by a sum of three Lorentzians. We have also fitted the data from [12] by formula used in [20-22]:

$$S_{\mathrm{mag}}(\varepsilon) \propto \frac{\varepsilon}{1 - \exp(-\varepsilon / kT)} \sum_{i=1,2,3} I_i \frac{1 - \exp(-\Delta_i / kT)}{\Delta_i} F(\varepsilon - \Delta_i, L_i)\,, \quad (12)$$

where ε is the energy transferred from the neutron to the sample, I_i are the intensities of the CEF transitions, L_i are the half-width of Lorentzians

$$F(\varepsilon, L) = \frac{1}{\pi}\frac{L}{L^2 + \varepsilon^2}. \tag{13}$$

The solid line in figure 2 indicates the fit, the fitted values of Δ_i, L_i and I_i are given in the second row of table 3.

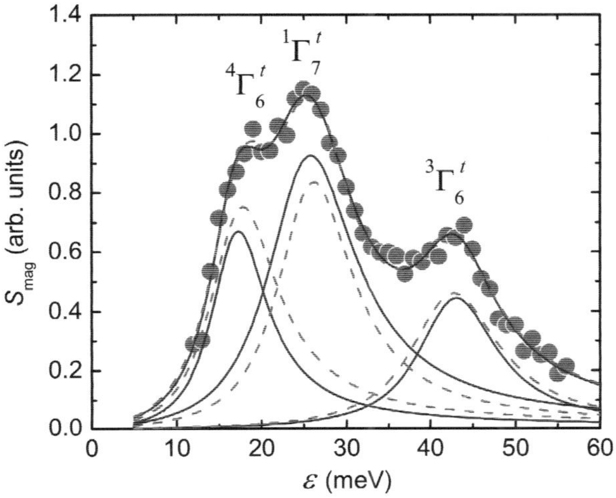

Figure 2. Magnetic part of the inelastic neutron scattering spectrum of $YbRh_2Si_2$ at $T = 1.5$ K (digitized data from Fig. 2 of [12]). The solid line indicates a fit by expression (12) with Δ_i, L_i and I_i as independent parameters (see the second row in table 3). The dashed line indicates a fit by (12) with Δ_i, L_i and φ_6 as independent parameters and with I_i calculated by using (15) (see the third row in table 3).

Table 3. Kramers doublets energies Δ_i (meV), half-width of Lorentzians L_i (meV), CEF transition intensities I_i, angular parameter φ_6 for $YbRh_2Si_2$ from the data fits according to expression (12) as shown in figure 2; angular parameter φ_7 from the *g*-factors. Second and third rows correspond to solid and dashed lines in figure 2.

Δ_1	Δ_2	Δ_3	L_1	L_2	L_3	I_1	I_2	I_3	φ_6	φ_7
16.8	25	42.7	4.1	6.3	5.7	0.47	1	0.44		
17.1	25.6	42.3	4.9	5.4	6.4	0.63	0.78	0.51	0.5361	−1.2660

In accordance with [20-22] for polycrystalline sample or powder the intensity of transition from ground doublet Γ to excited doublet Γ' is given by

$$I_{\Gamma'} \propto \sum_{\alpha\sigma\sigma'} \left|\langle\Gamma\sigma| J_\alpha |\Gamma'\sigma'\rangle\right|^2, \tag{14}$$

where $\alpha = x, y, z$ and $\sigma, \sigma' = \uparrow, \downarrow$. For transitions from $^2\Gamma_7^t$ ground doublet to excited doublets $^4\Gamma_6^t$, $^1\Gamma_7^t$ and $^3\Gamma_6^t$ we obtain

$$I_{^4\Gamma_6^t} = (\sqrt{7}c_2a_2 + \sqrt{15}c_1a_1)^2 = I(\varphi_7)\cos^2\left(\frac{\varphi_6 - \gamma(\varphi_7)}{2}\right) \equiv I_1,$$
$$I_{^1\Gamma_7^t} = 32c_1^2c_2^2 + 12(c_2^2 - c_1^2)^2 = 4(3 - \sin^2\varphi_7) \equiv I_2, \quad (15)$$
$$I_{^3\Gamma_6^t} = (\sqrt{7}c_2a_1 - \sqrt{15}c_1a_2)^2 = I(\varphi_7)\sin^2\left(\frac{\varphi_6 - \gamma(\varphi_7)}{2}\right) \equiv I_3,$$

where

$$I(\varphi_7) = 11 + 4\cos\varphi_7, \qquad \gamma(\varphi_7) = 2\arctan\left[\sqrt{7/15}\tan(\varphi_7/2)\right]. \quad (16)$$

These intensities as the functions of still undefined parameter φ_6 are plotted in figure 3. Experimentally observed relation $I_3 < I_1 < I_2$ takes place for φ_6 approximately lying within the interval $0.25 < \varphi_6 < 0.65$. The fitted values of intensities are in the ratio $I_1 : I_2 : I_3 = 0.47 : 1 : 0.44$ (see the second row in table 3). The calculation of intensities using (15) gives the nearest (but not near enough) result with $\varphi_6 = 0.6155$: $I_1 : I_2 : I_3 = 0.75 : 1 : 0.71$.

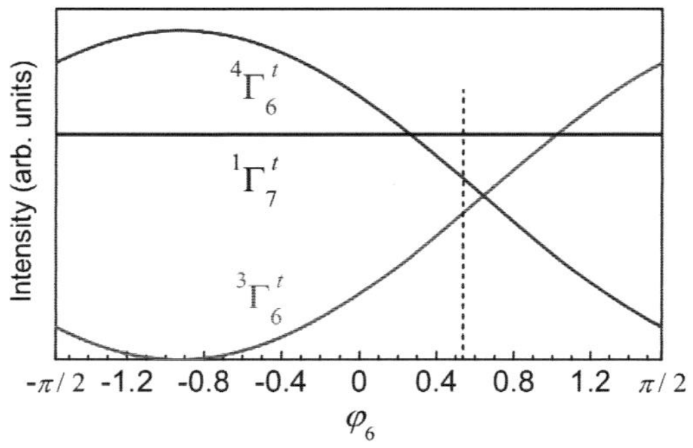

Figure 3. Intensities of the CEF transitions in the INS spectrum calculated by using (15) with $\varphi_7 = -1.2660$. The vertical line correspond to $\varphi_6 = 0.5361$.

We made also another fit using formula (12) with fitting parameters Δ_i, L_i and φ_6 (see the third row in table 3). In (12) the intensities $I_i(\varphi_6)$ were calculated by (15) assuming $\varphi_7 = -1.2660$. The experimental data was also very well fitted (see dashed line in the figure 2). The fitted value of $\varphi_6 = 0.5361$ is marked on figure 3 by vertical line.

We took obtained in this fitting procedure values of Δ_i and φ_6, determined earlier value of φ_7 and then we calculated CEF parameters with the help of formulas (11). Corresponding CEF parameters are presented in table 4.

Table 4. Crystal electric field parameters B_k^q (meV) for Yb^{3+} ion in $YbRh_2Si_2$.

B_2^0	B_4^0	B_4^4	B_6^0	B_6^4
22.4	−6.6	±29.6	−2.3	±63.9

If we simultaneously change the signs of φ_6 and φ_7 the intensities (15) stay the same but B_4^4 and B_6^4 (11) change the signs. In table 4 positive signs of B_4^4 and B_6^4 correspond to the φ_6 and φ_7 given in table 3 whereas negative signs of B_4^4 and B_6^4 correspond to φ_6 and φ_7 with the opposite signs. So, used INS experimental data do not allow to determine the signs of B_4^4 and B_6^4. Unfortunately, the relation $g_\perp \sim \sin\varphi_7$ (see table 1) also do not help us to choose the signs of B_4^4 and B_6^4. The reason is that in usual EPR experiments only the absolute values of g-factors are defined, therefore we have to consider two points on g-diagram (figure 1) with opposite sings of $g_\perp$ and φ_7.

In the papers [10, 11] the sets of CEF parameters for Yb^{3+} ion in $YbRh_2Si_2$ were found by the least squares fit to the experimental energy levels and g-factors (these sets of CEF parameter could be obtained from (11), see details in [9]). But these sets of CEF parameters do not agree with other experimental data. Though one set of CEF parameters from [10] agrees with doublets sequence established in ARPES experiments but gives almost zero intensity of INS transitions between ground doublet ${}^2\Gamma_7^t$ and excited doublet ${}^3\Gamma_6^t$, i.e. this set reproduces only two peaks instead of three ones experimentally observed in INS experiment (figure 2). Another set of CEF parameters from [10] and CEF parameters from [11] correspond to the ground doublet ${}^4\Gamma_6^t$ instead of ${}^2\Gamma_7^t$ in a contradiction with ARPES findings.

4. Summary

We analytically calculated the tetragonal CEF splitting of the ground ${}^2F_{7/2}$ multiplet of Yb^{3+} ion. Four Kramers doublets energies and doublets wave functions are expressed via five CEF parameters. We also analytically solved the inverse problem, i.e. all possible sets of tetragonal CEF parameters that satisfy the given experimental energy scheme of Yb^{3+} ground multiplet ${}^2F_{7/2}$ are found (see (11)). The diagram of possible g-factors of Yb^{3+} ion in tetragonal CEF is plotted (figure 1).

We determined the values of CEF parameters for Yb^{3+} ion in $YbRh_2Si_2$ (table 4) using energy intervals and intensities of CEF transitions between Kramers doublets taking from fitting of INS data [12], the Kramers doublets sequence obtained from ARPES experiments [13] and the g-factors found from EPR experiments [1, 2].

Acknowledgments

This work was supported by the Volkswagen Foundation (I/84689), and partially by the Swiss National Science Foundation IZ73Z0_128242/1 and by the Ministry of Education and Science of the Russian Federation. We wish to acknowledge professor B.I. Kochelaev for bringing our interest to this problem and continuous interest to the results.

References

[1] Sichelschmidt J, Ivanshin V A, Ferstl J, Geibel C and Steglich F 2003 *Phys. Rev. Lett.* **91** 156401
[2] Sichelschmidt J, Wykhoff J, Krug von Nidda H-A, Ferstl J, Geibel C and Steglich F 2007 *J. Phys.: Condens. Matter* **19** 116204
[3] Sichelschmidt J, Wykhoff J, Krug von Nidda H-A, Fazlishanov I I, Hossain Z, Krellner C, Geibel C and Steglich F 2007 *J. Phys.: Condens. Matter* **19** 016211
[4] Abrahams E and Wölfle P 2008 *Phys. Rev.* B **78** 104423
[5] Schlottmann P 2009 *Phys. Rev.* B **79** 045104
[6] Kochelaev B I, Belov S I, Skvortsova A M, Kutuzov A S, Sichelschmidt J, Wykhoff J, Geibel C and Steglich F 2009 *Eur. Phys. J.* B **72** 485–489
[7] Belov S I, Kutuzov A S and Kochelaev B I, these proceedings
[8] Kutuzov A S, Skvortsova A M, Belov S I, Sichelschmidt J, Wykhoff J, Eremin I, Krellner C, Geibel C and Kochelaev B I 2008 *J. Phys.: Condens. Matter* **20** 455208
[9] Kutuzov A S and Skvortsova A M 2009 *Magn. Reson. Solids* **11** 7–13
[10] Leushin A M, Ivanshin V A and Kurkin I N 2007 *Phys. Solid State* **49** 1417–21
[11] Leushin A M and Ivanshin V A 2008 *Physica* B **403** 1265–67

Journal of Physics: Conference Series **324** (2011) 012039 doi:10.1088/1742-6596/324/1/012039

[12] Stockert O, Koza M M, Ferstl J, Murani A P, Geibel C and Steglich F 2006 *Physica* B **378–380** 157–158
[13] Vyalikh D V, Danzenbächer S, Kucherenko Yu, Kummer K, Krellner C, Geibel C, Holder M G, Kim T K, Laubschat C, Shi M, Patthey L, Follath R and Molodtsov S L 2010 *Phys. Rev. Lett.* **105** 237601
[14] Likodimos V, Guskos N, Wabia M and Typek J 1998 *Phys. Rev.* B **58** 8244–47
[15] Aminov L K, Ivanshin V A, Kurkin I N, Gafurov M R, Salikhov I Kh, Keller H and Gutmann M 2001 *Physica* C **349** 30–34
[16] Gafurov M R, Aminov L K, Kurkin I N and Izotov V V 2005 *Supercond. Sci. Technol.* **18** 352–355
[17] Gafurov M R, Kurkin I N and Kurzin S P 2005 *Supercond. Sci. Technol.* **18** 1183–89
[18] Maisuradze A, Shengelaya A, Kochelaev B I, Pomjakushina E, Conder K, Keller H and Müller K A 2009 *Phys. Rev.* B **79** 054519
[19] Abragam A and Bleaney B 1970 *Electron Paramagnetic Resonance of Transition Ions* (Oxford: Clarendon Press)
[20] Holland-Moritz E, Wohlleben D and Loewenhaupt M 1982 *Phys. Rev.* B **25** 7482–7502
[21] Muzychka A Yu 1998 *JETP* **87** 162–174
[22] Goremychkin E A, Muzychka A Yu and Osborn R 1996 *JETP* **83** 738–746

Journal of Physics: Conference Series **324** (2011) 012040 doi:10.1088/1742-6596/324/1/012040

Ab initio simulation of helium inside carbon nanotubes

D B Timerkaeva[1,2], V Stutzmann [1], B Minisini [1], D A Tayurskii [2]

[1] Institut Supérieur des Matériaux et Mécaniques Avancés du Mans, 44 Av. Bartholdi, Le Mans, 72000, France

[2] Institute of Physics, Kazan Federal University, 18 Kremlevskaya st., Kazan, 420018, Russia

E-mail: dtimerkaeva@gmail.com

Abstract. In present work we consider the complex behaviour of quantum liquids like liquid He-4 inside carbon nanotubes. Interactions between helium atoms and carbon atoms of the short-length atomistic model and model with periodical boundary conditions of carbon nanotube were studied via *ab initio* quantum simulations. Effects of geometrical confinement of the tube on the He behaviour inside CNT (13,0) have been explored. Nanotubes with typical average diameter of 10 angstroms are under consideration.

1. Introduction

Nowadays various nanostructures like carbon nanotubes (CNT's), fullerenes, quantum dots and others continue to be one of the hottest research areas in the world of science [1]. CNT's were first observed by soviet scientists Radushkevich and Lukyanovich in 1950 and the first transmission electron microscopy pictures of the hollow structures of the carbon filaments were published in 1952 [2]. The work remained basically unknown and in 1991 carbon nanotubes have been rediscovered by Iijima [3]. The chain of investigations in this sphere lasts until our days. The interest is driven by the possibility of variety commercial applications [1], [4]-[5]. Among them, the storage of hydrogen inside nanotubes and nanotube bundles was extensively studied [6]-[9] and opened an important branch of investigations in the field of gas-nanotube interacting systems. The previous works have revealed that the gases in confined geometry are ruled by new physics and new effects [10]-[20].

The natural quasi one-dimensional system is realized via atoms encapsulation inside small radius nanotubes, inside the interstitial channels of nanotube bundles and via their adsorption on the grooves of the bundle surface. Two-dimensional systems can be formed on the inner and outer surface of freestanding nanotubes and on the outer surface of nanotube bundles whereas three-dimensional anisotropic condensed phase was considered as a result of interaction of helium atoms between neighboring interstitial channels [10].

In the confined geometry conditions, helium atoms ^4He and ^3He form bound states, so-called dimers, trimers and so on [14]-[16]. It is known, that such dimers should be bound extraordinarily weak in unconfined space. The binding energy value is up to four times smaller than the interatomic potential well [17]-[18]. This leads to the uncommon high length of the bond of the order of several van der Waals radiuses and can open a new branch of quantum chemistry, where giant molecules with practically destroyed bonds act as a building material [19]. In addition, lots of fundamental questions dealing with the general problem of existing He_n clusters appear. Finally, there is a category of interesting problems, which deal with the existence possibility of n-mers in helium liquids including

Journal of Physics: Conference Series **324** (2011) 012040 doi:10.1088/1742-6596/324/1/012040

thin films. This is a huge field remains largely unexplored even if several examples, discussed by Ghassib [20], have illustrated that such clusterization can lead to interesting results.

In this context the problem of a correct description of helium in the confined geometry at nanoscale length has been emerged. So far the experimental studies face significant difficulties, the majority of studies dealing with helium inside nanotubes [21]-[22] use theoretically calculated Stan and Cole adsorption potential [23] based on the van der Waals interaction between helium atom and nanotube. Stan and Cole [23] in their study proposed a smooth walls nanotube model with van der Waals parameters extracted from the experiments on helium adsorption on planar graphite [24]. They can be used to evaluate for example the binding energies [25]-[26] or the adsorption coverage [27] but most of all these methods can provide a powerful tool to have an opportunity to "look inside" the nanotube and analyze gas and liquid behavior whereas experiments are difficult to carry on.

In present work we consider the complex behavior of quantum liquids like liquid He-4 and liquid He-3 inside carbon nanotubes. Interaction between helium atoms and carbon atoms of the short-length atomistic model of CNT and the model with periodical boundary conditions are studied via *ab initio* quantum simulations. Effects of geometrical confinement of the tube on the He behavior as well as effects of helium presence on the various electronic properties of the tube are explored. Nanotubes with typical average diameter of 10 angstroms are under consideration.

2. Method

The average diameter of experimentally produced CNTs has a value around 1 nm [28]. Several possible chiralities with theoretically calculated diameter close to 1nm can be considered. Among them we have chosen a single-walled (13,0) zigzag nanotube. This choice was motivated by the small periodicity of the zigzag nanotubes compare to chiral ones. We have performed the calculations on the models with and without periodic boundary conditions presented on the Fig. 1, 2 and Fig. 3 respectively. Dangling bonds on the edges of short models were closed by the hydrogen atoms. Short atomistic nanotube models were employed to perform Hartree-Fock, post-Hartree-Fock and ONIOM calculations, while the periodic structure was used in DFT calculations.

2.1. Hartree-Fock, post-Hartree-Fock and ONIOM methods.

Gaussian03W [29] software was used for the Hartree-Fock and Post-Hartree-Fock method calculations. Compare to the DFT, Hartree-Fock method deals with the wave functions of electrons. The later allows to compute the exchange coulomb interaction precisely, and post-Hartree-Fock method allows to take into account the electron correlations accurately. The DFT method due to handling the density distribution function in the calculations leads to approximate calculations of both exchange and correlation interactions and doesn't take into account van der Waals interactions properly. However, post-HF method are time consuming and consequently gives reliable values but only for small size systems. In our case this disadvantage forces us to reduce the length of nanotube model.

One helium atom was situated inside the tube. As a trial locations we have chosen a central helium location ($r = 0$Å) and a location at a distance $r = 3{,}33$Å away from CNT center. The influence of the method and the basis set on the results of calculations was studied. The two-layers ONIOM technique [30] allows to perform the calculations of the subsystem with a cheaper cost. Here helium atom is related with "higher layer" and is treated with HF or post-HF methods, whereas the CNT is treated with DFT-LDA method with various basis sets. Performing the full DFT method in this software is associated with difficulties: some parameters for helium atom are absent. The obtained results are presented in Table 1.

The difference in energies between two helium-CNT configurations gives qualitative information about the adsorption potentials calculated with various methods.

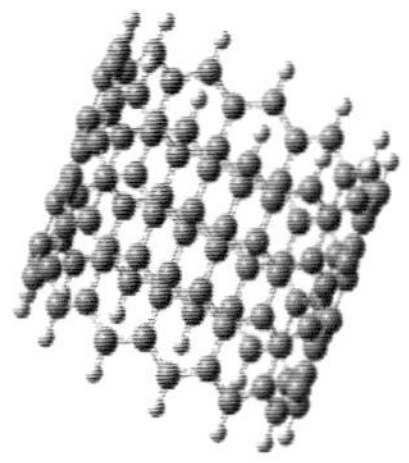

Figure 1. Short tube model CNT(13,0) $l = 9{,}22$Å.

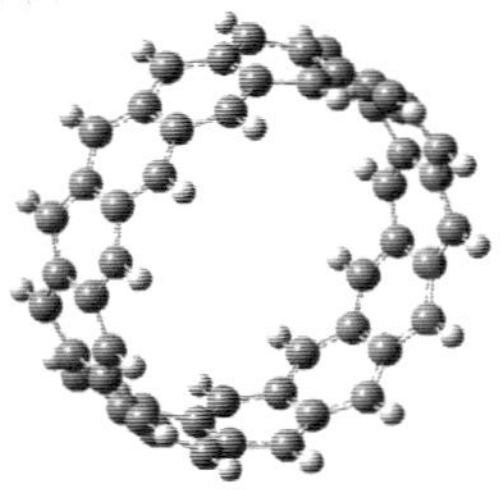

Figure 2. Short tube model CNT(13,0) $l = 4{,}96$Å.

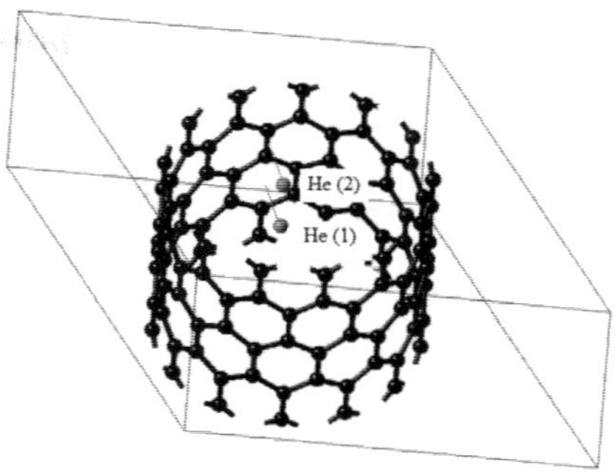

Figure 3. Periodic cell corresponding to infinite CNT model.

According to obtained results increasing the basis set for the helium atom hasn't an influence on the energy difference in ONIOM calculations, however, it is clear that the basis set 6-31+G(d) applied to nanotube walls provide a higher energy difference value ΔE than 3-21G. The Moller-Plesset second order perturbation theory (MP2) [31] in case of low basis set for "lower level" doesn't improve the values. An almost zero difference (0,00008Ha) is obtained in the case of 6-31+G(d) basis set for "lower layer". The comparison between the ONIOM and pure HF results confirms the importance of taking into account the correlation interaction between carbon atoms in such calculations. ONIOM calculations lead to a greater energy difference in comparison with the results obtained with only HF method.

Table 1. Comparison the total energy of the system for two trial helium atom dispositions: E_{border} corresponds to helium atom location close to the border and E_{center} corresponds to helium atom location close to nanotube axis.

Methods (CNT b.s./He b.s.)	E_{border} (Ha)	E_{center} (Ha)	ΔE (Ha)
DFT-LDA(3-21G)/HF(3-21G)	-1977,502014	-1977,520270	-0,018256
DFT-LDA (6-31+G(d))/HF(3-21G)	-1988,478964	-1988,501593	-0,022629
DFT-LDA (3-21G)/HF(6-31+G(d))	-1977,521495	-1977,539751	-0,018256
DFT-LDA (3-21G)/MP2(3-21G)	-1977,513511	-1977,531767	-0,018256
DFT-LDA (3-21G)/HF(6-31+G(d))	-1977,521495	-1977,539751	-0,018256
DFT-LDA (3-21G)/MP2(6-31+G(d))	-1977,532695	-1977,550951	-0,018256
DFT-LDA (6-31+G(d))/HF(6-31+G(d))	-1988,498445	-1988,521074	-0,022629
DFT-LDA (6-31+G(d))/MP2(6-31+G(d))	-1988,509645	-1988,532354	-0,022709
DFT-LDA (3-21G)/HF(3-21G)	-1977,502014	-1977,520270	-0,018256
HF 3-21 G	-1975,557694	-1975,594138	-0,036444
HF 6-31 + G (d)	-1986,770631	-1986,811064	-0,040433

Of course, the perfect outcome can be obtained with the best parameters set. In our case it should be the post-HF method with basis set 6-31+G(d) applied to the whole system. But the limitation in resources forces us to use less powerful techniques. During the testing of the methods, serial calculations were performed using HP Z600 workstations only for two configurations, and some of them took important time period (up to two weeks). It is not acceptable for the important quantity of calculations we have to perform in order to have some information concerning the interactions

Journal of Physics: Conference Series **324** (2011) 012040 doi:10.1088/1742-6596/324/1/012040

between Helium atom and nanotube. That is the reason why we perform the following calculations on the short model tube with the simple 3-21G basis set.

2.2. *DFT method.*

A full DFT method was realized using VASP (Vienna Ab-initio Simulation Package) software in MedeA [32] interface, which allows to use the periodic boundary conditions, and consequently we can use an atomistic infinite model. The nanotube structures were created with Materials Studio 4.4 software [33]. The energy cutoff of plane wave basis set is chosen up to 500eV. For the exchange and the correlation contributions the Generalized Gradient Approximation of Perdew-Burke-Ernzerhof (PBE-GGA) [34] was used. Number of k-points is a critical parameter in that kind of simulations, but increasing leads to higher price of the simulations. On the other hand, the high symmetry of the system allows to decrease the value of this parameter. Depending on the periodic cell parameter along the tube axis k-mesh 1×1×3 or k-mesh 1×1×2 were used. As a convergence criterion for the SPE calculations 10^{-5} eV on the total energy was used.

The cell parameters were taken to correspond to an intertube distance of 5,45 Å. We were interested in the freestanding tube, but higher intertube distance is incompatible with our computing resources. After the first 3 optimization steps (containing two steps of cell volume, cell shape and atomic position optimization and as a third step only atomic position optimization), we doubled the cell size in the axis direction and perform optimization again. Nanotube size and cell parameters are presented in the Table 2. Increasing of the cell was performed to exclude the helium-helium interaction while adsorption potential curves calculations. The periodic boundary conditions being used, it is necessary that the cell size be much larger than parameter $\sigma = 2{,}57\text{Å}$ [20]. Larger size of the cell also permits to reduce number of k-points in axis directions from three to two. Other parameters were kept without changes.

Table 2. Parameters of the CNT models after optimization.

			Cell parameters					
Model	D (Å)	N_{at}	*a* (Å)	*b* (Å)	*c*(Å)	α(°)	β(°)	γ(°)
1 CNT(13,0) initial	10.180	52	15.640	15.640	4,260	90.000	90.000	120.000
1 CNT(13,0) final	10.247	52	15.732	15.737	4.285	89.998	89.995	119.981
2 CNT(13,0) initial	10.247	104	15,732	15.737	8.570	89.998	89.995	119.981
2 CNT(13,0) final	10.250	104	15.735	15.740	8.561	89.998	89.996	119.982

3. Results and discussion

On the Fig. 4 the potential curves for helium atom inside carbon nanotube (13,0) (R=5,13 Å) as a function of the distance from nanotube center are presented. Set of energy calculations were performed for the helium-CNT system with different methods. The results were compared with the Stan and Cole potential [23]. Values for the DFT methods were rescaled to obtain the same tube radius as in the MP2 calculations. The obtained depth of potential well (~3K) in the case of MP2 method isn't as deep as in theoretical calculations (~100K) and the minimum position is shifted to the axis direction (1,5 Å compare to 2,3 Å [23]). The same tendency can be noticed in the DFT results. Moving helium atom along the radius from the center to the wall directions, three adsorption sites were studied: adsorption to the center of hexagonal ring, named "hole"-site, adsorption to the node center, "node"-site, and finally to the carbon atom, "edge"-site. It should be mentioned that only short-range (chemical) interactions were taken into account, and the dispersion interactions are badly represented due to the limitation of the functional.

The minimum of the potential wells for the "hole"-site and "node"-site adsorptions have practically the same values. And the minimum position is found at a distance $r \approx 1{,}7\text{Å}$, while the minimum position of the potential curve for "edge"-site adsorption shifted towards the axis direction and equal to $r \approx 1,6$Å . The depth of the potential well is of 20K smaller than those coming from Stan and Cole calculations [23].

Using various methods potential curves of different depth values and different minimum positions can be obtained. Whatever the method of calculation used, the influence on the results is important and few reasons can be mentioned to explain these discrepancies. For MP2 calculations very short nanotube segment was used. In addition the presence of hydrogen atoms at the end and the edge effects can lead to discrepancy in energy values. It was already mentioned above that basis set strongly influences the energy values, but higher is the basis set and higher are the resources requirements. On the other hand DFT with the lower computational cost leads to reasonable results. The carbon nanotube model with periodical boundary conditions is alike the infinite tube model considered by Stan and Cole. Stan and Cole's semi-empirical potential is a van der Waals potential summarized over infinite smooth tube walls. The input parameters for [23] were taken from the helium adsorption experiment on graphite plane [24]. Therefore some effects caused by structure rearrangement can be lost.

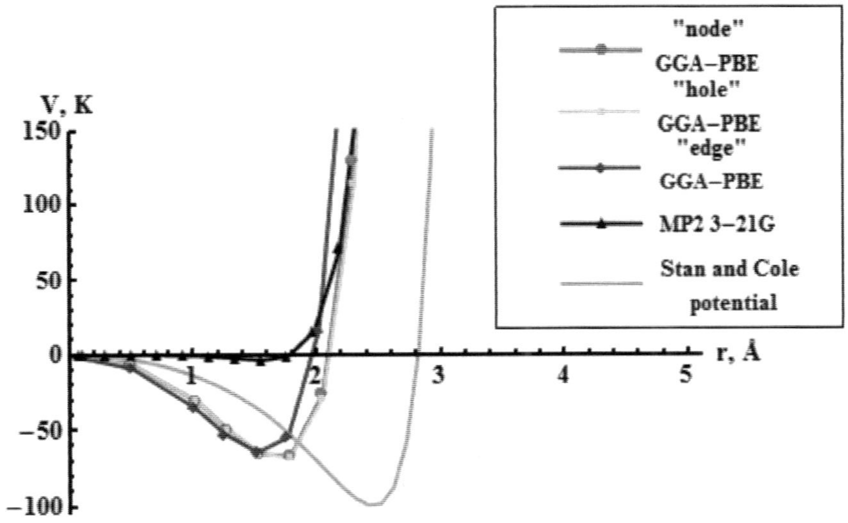

Figure 4. Adsorption potentials for helium atom inside CNT (13,0) with radius 5,13 Å (black) 3-21G (MP2), (blue, red, green) DFT GGA-PBE, (orange) Stan and Cole potential [23].

Presented on the Fig. 5a and Fig 5b slices of electron density distribution confirm that the helium atoms presence cannot affect the electronic properties of the tube. Here the electronic density distribution is studied for two configurations, the central position and the position close to the border, corresponding to the highest probability. It can be noticed that in both cases we had the same spherical shape for helium and the same shape for the electronic density distribution for the tube. If the helium atom goes closer to the border, then the electronic shells overlap, and this fact tells us about the very strong repulsion. That is the reason why the probability of the helium atom presence closer to the border is very low. The presence of some helium atoms cannot really affect the electronic properties of the tube except in the cases of very high pressure; in particular the HOMO-LUMO gap value of the tube remains the same when helium is situated in the central part up to 2,25Å far away from the center.

Transport properties of the CNT (13,0) were explored using the DFT GGA-PBE scheme. Optimization calculations for the helium atom position inside optimized CNT were performed for two initial states, presented on the Fig. 3. From the geometrical point of view it is clear, that the helium position He (2) is not at a local minimum, but at an unstable equilibrium position.

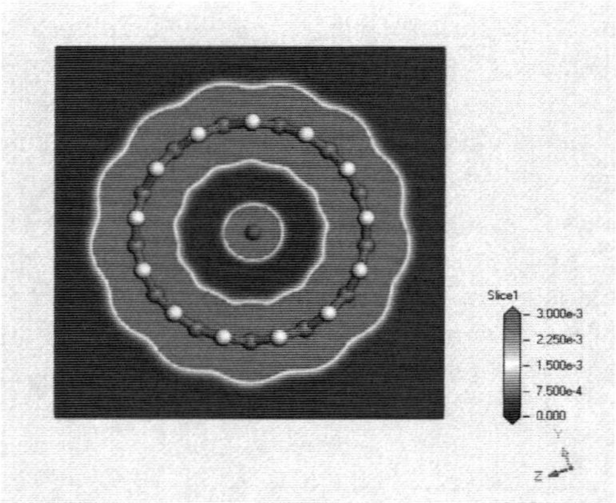

Figure 5a. Electron density distribution: helium atom situated in the centre of CNT (13,0).

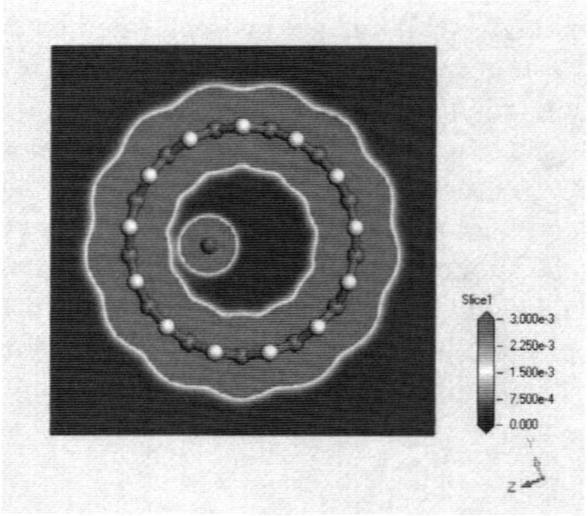

Figure 5b. Electron density distribution: helium atom situated in the most preferable position inside CNT (13,0).

To have an idea about the transportation and the energy barriers of helium atoms inside CNTs, helium atom is allowed to flow from He (1) position to He (2) position (see Fig. 3). Set of 10 single point energy calculations were performed with 0,21 Å step from minimum position He (1) to He (2) (see Fig. 6). An energy barrier of 12K which can be treated as activation energy was obtained. This is the energy value required to a linear flow of helium atom. Another flow possibility is a zigzag-motion along the CNT walls. From energetic consideration, this case is more preferable with an energy barrier of about 1K (difference between energy minimums of the adsorption potential curves presented on fig.4). In this case helium moves as a snake that bypasses the carbon atoms.

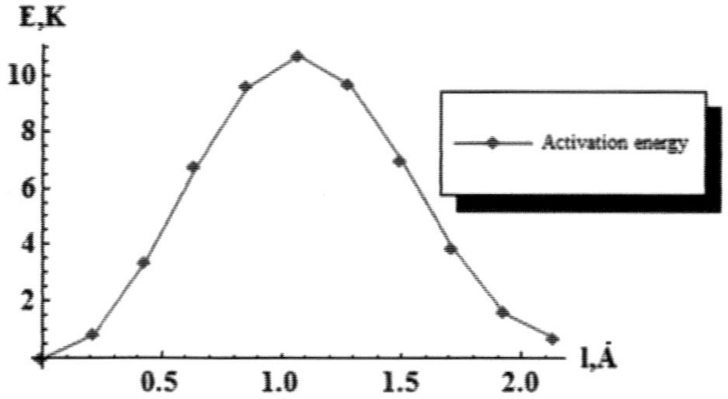

Figure 6. Activation energy.

Table 3. Binding energies of the helium atom situated inside the CNT (13,0)

Helium atom position	E_{CNT-He} (eV)	E_{He} (eV)	E_{CNT} (eV)	E_{bind} (eV)	E_{bind} (K)
He (1)	-950,631	0,0133	-950,625	-0,0195	-225,96
He (2)	-950,630	0,0133	-950,625	-0,0179	-207,65

An important property of interest is the cohesive energy (or binding energy) of helium atom with CNT. This value can be found from the expression: $E_{bind} = E_{He+CNT} - E_{CNT} - E_{He}$, where E_{He+CNT} – total energy value of carbon nanotube-helium molecule system with helium situated on its energetically most preferable position, E_{CNT} – total energy of the tube without helium and E_{He} – the energy of helium atom at the same periodic cell. The binding energy values of this two optimized positions of helium atom inside the CNT are presented in the Table 3. He (1) position corresponds to the minimum of the "hole"-site potential curve, while He (2) corresponds to the minimum of the "node"-site potential curve. Negative values of binding energy indicate that the adsorbed state is more profitable than separated state as well as the "hole"-site adsorption is more preferable than the "node"-site adsorption.

In summary, we have performed first-principle calculations on the adsorption and transport properties of nanotube with helium molecule inside. Adsorption potentials were obtained via post HF MP2 and DFT GGA-PBE calculations. Comparison with the theoretical Stan and Cole potential shows a shift of the minimum position to the center. The straightforward flow requires to overcome an energy barrier around 12K, while the zigzag-type motion requires an energy corresponding to 1K. Obtained binding energy values shows that the adsorbed state is profitable compare to the separated state.

Acknowledgement
The work was supported by the Ministry of Science and Education of Russia. D.B.T. thanks Region Pays de la Loire for partial individual financial support (MOA0021).

References
[1] Bhushan B et al *Springer Handbook of Nanotechnology* **2010**
[2] Radushkevich L and Lukyanovich V *Zh. Phys. Khim.* 1952 **26** 88
[3] Iijima S *Lett. to Nature* 1991 **354** 56
[4] Maiti A *Microelectronics Journal* 2008 **39**-2 208
[5] Jung H Y, Jung S M et al. Applied Physics Letters 2007 **90** 153114
[6] Nikitin A, Li X., Zhang Z et al. *Nano Lett* 2008 **8-1** 162
[7] Xia Y et al. *Phys Rev B* 2005 **71** 075412
[8] Han S S, Lee H M *Carbon* 2004 **42** 2169
[9] Du A J and Smith S C *Nanotechnology* 2005 **16** 2118
[10] Carraro C *Phys. Rev. B* 2000 **61** R16351
[11] Aichinger M, Kilić S, Krotscheck E and Vranješ L *Phys Rev B* 2008 **70** 155412
[12] Stan G *J Low Temp Phys* 2009 **157** 374
[13] Mayol R, Ancilotto F, Barranco M, Hernández E S and Pi M *J Low Temp Phys* 2008 **148** 851
[14] Kilpatrick J E and Kilpatrick M F *J. Chem. Phys.* 1951 **19** 930
[15] Pais A and Uhlenbeck G E *Phys. Rev.* 1959 **116** 250
[16] Ghassib H B *Z. Phys. B. Condensed Matter* 1984 **56** 91
[17] Feltgen1 R, Kirst H, Köhler K A et al. *J. Chem. Phys.* 1982 **76** 2360
[18] Uang Y-H and Stwalley W C *J. Chem. Phys.* 1982 **76** 5069
[19] Stwalleya W C *Cont. Phys.* 1978 **19** 65
[20] Ghassib H B *Z Phys B- Condensed Matter* 1984 **56** 91
[21] M.C. Gordillo, J. Boronat, J Low Temp Phys (2009) 157: 296–323
[22] D. Hirashima ´ K. Yamashita, JOURNAL OF LOW TEMPERATURE PHYSICS, Volume 158, Numbers 1-2, 112-117
[23] Stan G and Cole M W *Surf. Sci.* 1998 **395** 280
[24] Cole M W and Klein J R 1983 **124** 547
[25] Zhao J, Buldum A, Han J and Lu J P *Nanotechnology* 2002 **13** 195

[26] Chang H, Lee J D, Lee S M and Lee Y H *Appl Phys Lett* 2001 **79-23** 3863
[27] Kuznetsova A, Yates J T, Simonyan V V, Johnson J K, Huffman C B and Smalley R E *J Chem Phys* 2001 **115** 6691
[28] Journet C and Bernier P *Applied physics A: Materials science & processing* 1998 **67**-1 1
[29] http://www.gaussian.com/
[30] Frisch M O et al. *Gaussian 03 revision C.02 Gaussian Inc.: Wallingford CT* 2004
[31] Møller C. and Plesset M S *Phys. Rev.* 1934 **46** 618
[32] http://www.materialsdesign.com/medea
[33] http://accelrys.com/events/webinars/materials-studio-44/
[34] Perdew J P, Burke S and Ernzerhof M *Phys Rev Lett* 1996 **77** 3865

AUTHOR INDEX